Microalgae Biotechnology for Food, Health and High Value Products

Md. Asraful Alam • Jing-Liang Xu
Zhongming Wang
Editors

Microalgae Biotechnology for Food, Health and High Value Products

Springer

Editors
Md. Asraful Alam
School of Chemical Engineering
Zhengzhou University
Zhengzhou, Henan, China

Jing-Liang Xu
School of Chemical Engineering
Zhengzhou University
Zhengzhou, Henan, China

Zhongming Wang
Biomass Energy and R&D
Guangzhou Institute of Energy
Conversation-Chinese Academy of Sciences
Guangzhou, Guangdong, China

ISBN 978-981-15-0168-5 ISBN 978-981-15-0169-2 (eBook)
https://doi.org/10.1007/978-981-15-0169-2

This Springer imprint is published by the registered company Springer Nature Singapore Pte Ltd.
The registered company address is: 152 Beach Road, #21-01/04 Gateway East, Singapore 189721, Singapore

This book is dedicated to my beloved parents, M.A. Sobhan and Rofia Sobhan*.* *Actually, my parents live in a village in Bangladesh, so science books may not change their lives or be within their reach. But, my love towards them will remain noted for as long as this book lasts.*

— Md. Asraful Alam

Preface

Microalgae are predominantly fascinating sources of products ranging from human nutrition, food/feed additives, cosmetics, medicines, and health supplements to emerging sources including bioplastics, bio-based polymers, and vaccination agents for aquaculture organisms. The global market for microalgae products was valued at US$ 32.60 Bn in 2017 and estimated to reach US$ 53.43 Bn by 2026 (https://www.credenceresearch.com/report/algae-products-market). The major products currently being commercialized or under consideration for commercial extraction include human nutrition, animal and aquatic feed, phycobilins, β-carotene, polysaccharides, polyunsaturated fatty acids, vitamins, sterols, stable isotope biochemicals, biologically active molecules of antimicrobial, antiviral, antibacterial, and anticancer drugs for use in human and animal health; however, more new items are likely to be produced in the following decade. Moreover, commercial algae production is being lauded recently as a new agricultural phenomenon that can provide sustainable feedstocks of proteins and oils for hundreds of new products, absorb millions of tons of carbon dioxide, treat wastewater, and become a driver of economic growth around the world. Our previous book *Microalgae Biotechnology for Development of Biofuel and Wastewater Treatment* addressed microalgae, which represent a very promising biomass resource for wastewater treatment and producing biofuels; in that edition, we have vastly presented the culture and harvesting technology of microalgae in fresh or in wastewater either in open or close system. This book, *Microalgae Biotechnology for Food, Health, and High Value Products* has focused only on the various applications of microalgae ranging from human nutrition, food/feed additives, cosmetics, medicine and health, soil improvement, etc. It provides coverage of relevant, up-to-date research assembled by a group of contributors who are dedicated to the advancement of microalgae applications for human kind. We believe that this book will be greatly helpful for commercial algae producers, algae product developers, scientific researchers, students, or community people who are dedicated to the advancement of microalgae biotechnology for health, diet, nutrition, cosmetics, biomaterials, etc.

Henan, China — Md. Asraful Alam
Henan, China — Jing-Liang Xu
Guangdong, China — Zhongming Wang

Contents

Part I
Microalgae in Food Product Development

Chapter 1
Food and High Value Products from Microalgae: Market Opportunities and Challenges

Khondokar M. Rahman

Abstract Microalgae are a potential source of molecules for a wide range of food and novel high-value products and have good market opportunities. They can be used in biofuels, health complements, feed, medicine and cosmetics. The development of innovative and sustainable technologies with minimum energy inputs is required for large-scale cultivation and downstream processing of lipids and hydrocarbons in order for the production to be economically viable. In addition, the viability of bioenergy production from microalgae biomass is contingent on the net energy gain of the overall process, with exhaustive utilization of algal biomass for biofuel and other co-products for feed, food, and chemicals. The energy output from the biomass as fuel has to be greater than the energy required to produce and process the algae. Microalgae produce a comprehensive variety of bioproducts such as enzymes, pigments, lipids, sugars, vitamins and sterols. Moreover, its capability to alter atmospheric CO_2 into beneficial products such as lipids, carbohydrates, metabolites and proteins cannot be overstated. The key challenges appear to be high cost of operation, infrastructure and maintenance, selection of algal strains with high protein contents, dewatering and commercial scale harvesting. Optimizing the manufacture and commercialization of microalgae value products depend also on numerous factors (such as market and financial affairs). There is limitation of authentic and reliable data and statistics of microalgae market opportunities which make it difficult to assess their actual potential. Long-term research is needed to develop systems for the production of sustainable algal-based products, as sustainability is a key concern especially for food, feed and fuel.

Keywords Microalgae · High value product · Market challenges · Opportunities

---

K. M. Rahman (✉)
School of Engineering and the Built Environment, Birmingham City University, Birmingham, UK
e-mail: Khondokar.Rahman@bcu.ac.uk

M. A. Alam et al. (eds.), *Microalgae Biotechnology for Food, Health and High Value Products*, https://doi.org/10.1007/978-981-15-0169-2_1

1 Introduction

Algae which grow through photosynthesis are a diverse group of simple plants that can range from microscopic (microalgae) to large seaweed (macroalgae). They play a vital role in many ecosystems. Algae are disseminated globally specially in the sea, in both fresh and wastewater.

Microalgae are unicellular, but some are larger, multicellular organisms (Ozkurt 2009), for example macroalgae (or seaweed) (Fig. 1.1). It is almost recognized in the academic community that microalgae are not viable for fuels commercially or sustainably due to the cost and nutrient inputs.

A number of microalgae have been examined for their prospect of value-added products with extraordinary pharmacological and biological potentials, animal and fish feed, cosmetics, chemicals and polymer, pollution control, etc. (Khan et al. 2018). Biotechnology for the cultivation of microalgae has developed since the middle of the last century for the production of biofuel when oil price increased dramatically but now numerous commercial applications have been recorded. The most widespread application of microalgal culture has been through aquaculture for farming of marine animals, including finfish, crustaceans and molluscs. There is potential to develop multi-trophic systems and utilize microalgae for the purpose of bioremediation of, for example wastewater or aquaculture wastes, which reuses the wastewaters for the growth of micro- and macroalgae. Wastewater provides nutrients, for example ammonia, nitrite, nitrate, dissolved organic nitrogen and phosphate (Abe et al. 2002), which can be used for the production of microalgae. Microalgae are photosynthetic with the potential to produce hydrocarbons and lipids whilst harnessing the energy of the sun and sequestering carbon dioxide. This chapter presents the outcomes of a review of current scientific, economic and market expansions in the various field of products derived from microalgae for food, feed and high-value products (HVPs). It delivers the existing and future projections on topics related to the field of microalgae. The areas covered in this study are as follows:

1. Microalgae Background and a Brief History of Algal Research.
2. Algae to Algae Product: A Value Chain.
3. High-Value Product from Microalgae.
4. Microalgae Product and Market.
5. Future Prospects of Microalgae Industry.

1.1 Microalgae Characteristics and Composition: Background

Microalgae is a rich carbon source, and this can be utilized in biofuels, health supplements, pharmaceuticals and cosmetics (Das et al. 2011). They also have potential to bioremediate wastewater and CO_2 sequestration (Alam and Wang 2019). A wide range of bioproducts, as well as different materials such as polysaccharides, lipids,

Fig. 1.1 Microalgae (DOE: U.S. Department of Energy and Wageningen University 2016)

pigments, proteins, vitamins, and antioxidants, are produced from microalgae (Brenna and Owende 2010). Advances in cultivation technologies and techniques and genetic engineering offer potential to expand their application and processing for high-value products.

Industrial-scale cultivation of microalgae to produce bioproducts and bioenergy has received increased attention in recent decades and its applications have been widespread. (Plaza et al. 2009). Algae is produced and sold directly as food and nutrient supplement, while it's treated goods are used in both the biopharmaceuticals and cosmetics industries (Luiten et al. 2003; Borowitzka 2013; Pulz and Gross 2004).

Microalgae and cyanobacteria are recognized as marketable sources of HVPs, for example β-carotene, astaxanthin, pigments and extracts from algae, which are used in cosmetics. Algal products are categorized into three types based on their market values, and they are: high-value, medium-value and low-value products (Fig. 1.2).

1.2 A Brief History of Algal Research

Algae have been used as a food source and for treatment of various illnesses for over two thousand years in China, Japan, Tawian and Australia (Gao 1998). Regarding microalgae use as an energy fuel, Meier (1955) and Oswald and Golueke (1960) recommended the use of microalgal cell from biological compound fraction for the production of methane via anaerobic digestion (AD). The detection of a number of microalgae species that can produce relatively high concentrations of lipids similar to the cellular oil under certain evolution circumstances times back to the 1940s (USDOE 2010).

Microalgae contain higher concentrations of lipids compared to any terrestrial plant. An average lipid content differs from 1% to 70% while under specific functioning circumstances few of them can reach up to 90% of oil weight by dry biomass weight (Mata et al. 2010; Georgianna and Mayfield 2012). There have been a number of serious global crises in oil production. For example, the two worst crises in the oil and energy sector were in 1973 and 1979 during the Yom Kippur War and

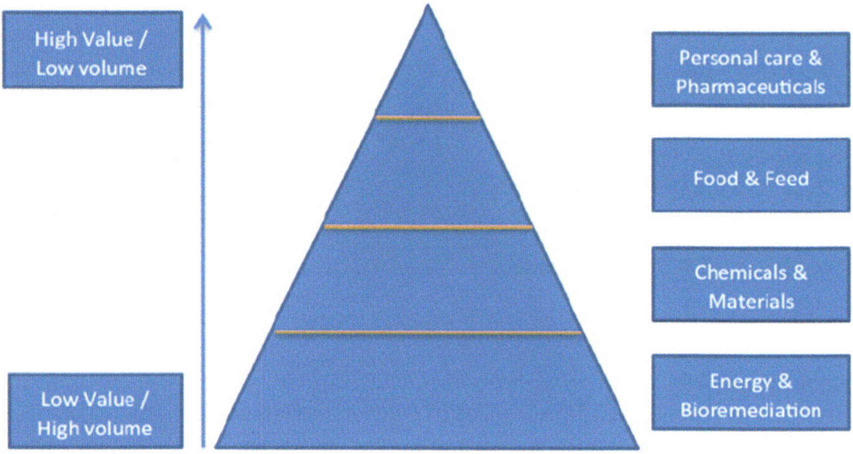

Fig. 1.2 Value pyramid: market value of microalage products (Voort et al. 2015)

the Iranian Revolution. This initiated disruptions in Middle Eastern oil dissemination (Oil Squeeze 2008; Duncan 2001). Research into microalgae as an alternative biofuel emerged in 1978 after an oil crisis struck. The use of microalgae for biofuel production has gained enormous research interests in recent years, mainly due to the ability to photosynthetically convert CO_2 into potential biofuel biomass, and food, feedstocks, and high value biochemical (Zeng et al. 2012).

By evaluating the physiology and biochemistry of microalgae, researchers (Hounslow 2016) narrowed their focus to potential lipid activates. For an example, under nutrient stress, lipid accumulation is favoured, and triacylglycerol (TAG) is formed as the main element. A number of studies have stated that most microalgal species can improve lipid accumulation and undergo transformation under nutrient stress.

There is a wide diversity of microalgae species with respect to the size and shape of the organisms that perform photosynthesis in eukaryotic cells. These eukaryotic microorganisms are crucial for life on earth. Planktonic algae, living in the oceans, perform nearly half of global photosynthesis (Behrenfeld and Falkowski 1997). Algal protein is a prospective source of fish and animal feed and this types of protein has been projected to get a related profile of amino acid (Gross 2013). For example, cyanobacterium *Arthrospira platensis* and many other marketable species of the single-celled green alga have 42–70% of dry weight protein (Milovanovic et al. 2012; Plaza et al. 2009) and these microalgae also have an amino acid that relates well with egg, albumin and soya (Williams and Laurens 2010) particularly comprising all of the vital amino acids which humans are unable create in their body but need to obtain from food (Gantar and Svircev 2008). Chlorophyll is a useful bioactive compound, and the process of its extraction from marine microalgae begins with dewatering and desalting the highly diluted microalgal culture (Hosikian et al. 2010).

2 Algae to Algae Product: A Value Chain

A 'value chain' has been defined by Bush (2019) as 'an interrelated functioning action businesses perform throughout the process of transforming raw materials into finished products'. The value chain and the development of high-value products from microalgae depends on composition, applications, formulation, production scale and comparative reference. An impression of the whole value chain required to derive energy from both macroalgae and microalgae. The stages of the value chain include cultivation, harvesting and biomass pre-treatment, and this includes washing, purification, dewatering and drying.

2.1 Nature and Characterization of Algae and Algae Products

Microalgae are versatile and can produce various composites like proteins, carbohydrates, lipids, carotenoids and different vitamins and minerals. The relative composition of these are dependent upon species and growth conditions (Koller et al. 2014; Hamed 2016). Various studies have been conducted on the characteristics of microalgae and to elucidate the key compounds of commercial interest. These include, for example, *Chlorella vulgaris*, *Arthrospira platensis*, *Nannochloropsis* sp. and *Phaeodactylum tricornutum* (Tibbetts et al. 2015; Matos et al. 2016); other species are *Dunaliella salina* (Muhaemin and Kaswadji 2009), *Haematococcus pluvialis* red stages (Shah et al. 2016) and green stages (Kim et al. 2015) and *Scenedesmus almeriensis* (Sánchez et al. 2008).

At present *Chlorella vulgaris*, *Arthrospira platensis*, *Haematococcus pluvialis* and *Dunaliella salina* are supported for the food, feed and nutraceutical sectors. Similarly, carotenoids and lipids, especially omega-6 and omega-3, find superior uses than existing in many markets, for example in fish and animal food, aquaculture, in protein and in cosmetics industry (Molino et al. 2018).

2.2 Principles of Value Chains: A Bio-Based Economies

The microalgae value will mostly depend on the following variables: (1) composition, (2) purity level, (3) applications and (4) formulation (Vieira 2016). The value of a product increases with its level of refinement or processing.

2.2.1 Value Depends on Composition

The chemical composition of microalgae includes: lipid at 20–30%, protein is around 50%, carbohydrate 20–30% and other compounds account for approximately 5%. The market value depends on the concentration and amount of the essential amino acids, polysaccharides, polyunsaturated fatty acid (PUFAs) and amount of essential vitamins (Vieira 2016).

2.2.2 Value Depends on Applications

Likewise, the value depends upon the application, for example what is the purpose of the use of algae and compared to similar alternative sources of this product. This includes the current application as food, feed, medicine and cosmetics and the additional application as fuels, fertilizers, wastewater treatment and chemicals (Vieira 2016).

2.2.3 Value Depends on Formulation

The value of the microalgae-based product depends upon its formulation, for example is it granular, small or bigger particle size, a paste, dry or wet. A number of dry algal products are used for food and feed, aquaculture and pharmaceuticals (Vieira 2016). This product value chain also depends on comparative reference—for example, microalgae high-value product in comparison source of algae link with others: soy and fish (oil and meal).

3 High-Value Products from Microalgae

Microalgae are progressively playing an important role in the development of cosmeceuticals, alternative of pharmaceuticals and high-value foods. High-value extracts from microalgae include: proteins, lipids, carbohydrates, different pigments, vitamins and anti-oxidants used in different sectors, such as cosmetics, nutrition and pharmaceuticals.

3.1 Algae: Human Food

Japan has a well-established market for edible products (algae soup) (Fig. 1.3a) from algae (Holdt and Kraan 2011) and according to Wellinger (2009) China and Japan have the highest production and consumption of dry algae globally. There is utilization as whole cell (e.g. nori or others) and extracts, which can be used as an ingredient.

Fig. 1.3 (**a**) Cooking healthy soup from algae (Fresh Designpedia 2019). (**b**) Algae as medicine in a tablet format

3.1.1 Agar from Algae

Agar is a polysaccharide derived from seaweeds (macroalgae) of the rhodophyceae class, and the most commonly used solidifying agent (Mesbah and Wiegel 2006). Agar formulates double helices that combined to form a gel or balm, holding water within the opening (Tiwari and Troy 2015). This water-holding capacity makes this gelatinous substance a good thickener. Throughout Japan, Korea and China it was used as a desert and in other food ingredients, however, presently it is also important as a gelatine for growing organisms in medical and scientific studies.

3.1.2 Alginates from Algae

Alginates are a polysaccharide derived from Phaeophyceae, which are brown algae containing alginic acid salts and are the raw ingredients in alginates manufacture. It is a polymer made up of arrangements of α- β-D-mannuronic acid and L-guluronic acid. They are widely used in various food, beverage, printing, textile and pharmaceutical industries as a thickening agent, stabilizer, emulsifier and chelating agent (Hay et al. 2013).

3.2 Pharmaceuticals and Health-Related Products

The application of algae-based products (e.g. from cyanobacteria) in the different sectors (e.g. pharmaceuticals) has increased globally (Sathasivam et al. 2019). Interest in microalgae-based antibiotics and pharmacologically active compounds (Fig. 1.3b) is also growing. There are a variety of products in the pharmaceutical sector derived from algae and their application such as: antivirals, antimicrobials and antifungals, and drugs and therapeutic proteins. The pharmaceuticals products derived from microalgae include: omega-3 fatty acids, EPA, DHA, beta carotene and astaxanthin (Sathasivam et al. 2019).

3.2.1 Omega-3 Fatty Acids/PUFA

Acids, for example EPA and DHA, from algae have received considerable attention because of their involvement in the inhibition and action of numerous diseases, including, thrombosis, atherosclerosis and arthritis. Although marine fish oil remains the traditional source of both EPA and DHA, there are studies that seem to indicate that greater quantity of EPA and some DHA can be gained from algae. Omega-3 and omega-6 are the two most important fatty acids, PUFAs have two important groups with a long chain of carbon atoms: carboxyl and methyl groups (Harris 2010). Microalgae is capable of supplying omega-3 fatty acids at high concentrations. Different species, for example *Crypthecodinium*, *Thraustochytrium* and *Schizochytrium*, contain the omega-3 fatty acid DHA, and the species of *Phaeodactylum, Chlorella* and *Monodus* contain Eicosapentaenoic acid (EPA). The human body can only form carbon–carbon double bonds after the ninth carbon from the methyl end of a fatty acid (Jones and Rideout 2014). Alpha-linolenic acid (ALA) is an essential fatty acids, and they must be obtained from the diet and that cannot be synthesized in the body (Jones and Papamandjaris 2012).

3.2.2 Astaxanthin

Astaxanthin is a xanthophyll carotenoid present in a variety of microalgae and this is the most abundant natural pigment existing in nature (Ambati et al. 2014). The main source of astaxanthin for human consumption is obtained via digestion of seafood or extracted from the microalga *Haematococcus pluvialis*. It is estimated that the global astaxanthin market is approximately US$257 million (Khattar et al. 2009). The majority of astaxanthin produced from microalgae is utilized for aquaculture, most of them in fish coloration. The astaxanthin market size was assessed at US$555.4 million in 2016 (Market Research Report 2017). Astaxanthin is mainly utilized by the salmon feed industry due to their potent antioxidant activity; it would also be beneficial for patients with cardiovascular, immune, inflammatory and neuro-related diseases (Wu et al. 2015).

Astaxanthin is produced commercially, but naturally occurs from the microalgae species *Hematococcus pluvialis*. Astaxanthin, which produced commercially, is most commonly used is in fish farming and it conveys colouring to farmed salmonids and crustaceans (Koller et al. 2014) (Fig. 1.4). It has been used widely in the nutraceutical and pharmaceutical industry due to its antioxidant activity and fortification (Cardozo et al. 2007). The marketable astaxanthin is led by the synthetic one since the market value for the natural one is higher than the synthetic version (Pérez-López et al. 2014; Li et al. 2011). The demand for the production of astaxanthin from the natural sources has increased because the demand for health-related food to the consumer has increased.

3.2.3 Beta-Carotene

Beta-carotene is a member of a family of molecules found in microalgae and plays a vital role in health because of its antioxidant exchanges to vitamin A. Beta-carotene is a good source of vitamin A, and the provitamin A function of beta-carotene contributes to vitamin A intake (Grune et al. 2010). The concentration of carotenoids (Table 1.1) in most algae is 0.1–2%, but *Dunaliella*, if grown in the right circumstances of high salinity and light strength, could yield up to 14% of beta-carotene. Global market revenue of beta-carotene is a

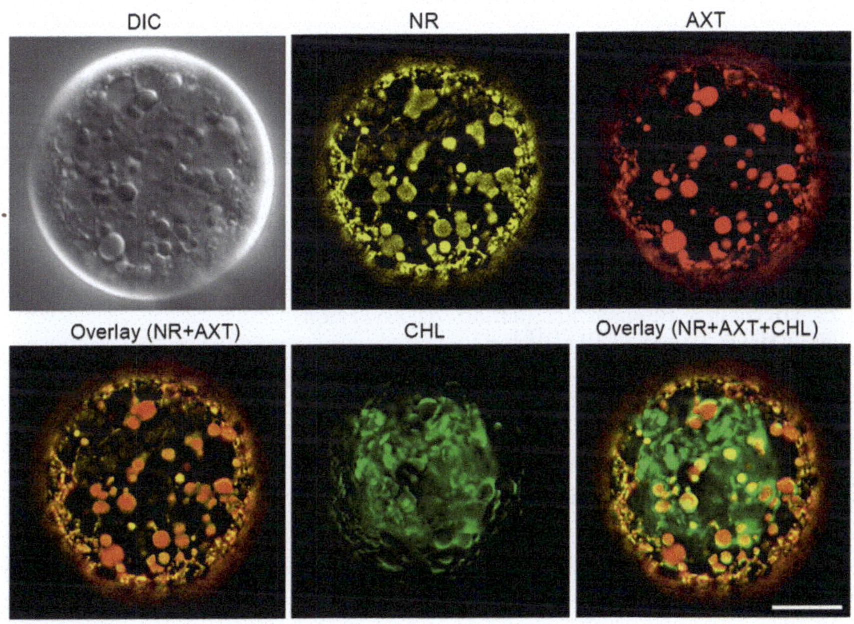

Fig. 1.4 Lipids, astaxanthin and chlorophyll in an astaxanthin-rich *H. pluvialis* cell. *NR* Nile Red, *AXT* astaxanthin, and *CHL* chlorophyll (Wayama et al. 2013)

Table 1.1 Ideal conditions of beta-carotene production by *D. salina* (Hermawan et al. 2018)

Processing condition	Reactor configuration	Productivity
Temperature: 25 °C; pH: 7.5 ± 0.5	Semi-continuous outdoor, closed tubular (55 L)	Biomass: 2 g m^{-2} d^{-1} Total carotenoids: 102.5 ± 33.1 mg m^{-2} d^{-1} (β-carotene: 10% of biomass)
Temperature: 29 °C pH: 7.5–8.6	Semi-continuous outdoor	The optimum Dissolved Oxygen (DO) 6.3–6.9 mg/L
Temperature: 30 °C; pH: 7.5	Continuous turbidostat, flat-panel (2.5 L)	β-Carotene: 13.5 mg L^{-1} d^{-1} (15.0 pg $cell^{-1}$)
Temperature 30 °C; pH: 7.5	Continuous turbidostat, flat-panel (1.9 L) with in situ extraction	β-Carotene: 0.7 mg L^{-1} d^{-1} β-Carotene: 8.3 mg L^{-1} d^{-1} (8.9 pg $cell^{-1}$)

CAGR (Compound Annual Growth Rate) of 3.5% and is US$224 million in 2018 (Transparency Market Research 2018).

Naturally, beta-carotene, which is a carotenoid compound, is derived from algae *Dunaliella salina*. The bioactivity of beta carotene is comparatively higher and this is why it is broadly used in medicine and used as provitamin A supply, precise growth and sight. Due to its antioxidants characteristics, beta-carotene is considered an inhibitor of some genes and it exhibits anticancer properties (Berman et al. 2014; Harasym and Oledzki 2014; Zhang et al. 2016).

3.3 Algae in Nutraceuticals

Algae and their extracts play an important role in the nutraceutical industry. The key species currently utilized include *Spirulina* and Chlorella, and products related to algae nutraceuticals include poly unsaturated fatty acids (PUFAs, such as DHA, ARA, GAL and EPA), beta-carotene and Astaxanthin. Dried *Spirulina* used in food supplements contains about 60% protein, with a composition which is rich in all vital amino acids (Nicoletti 2016).

Spirulina and *Chlorella* species are leading the worldwide microalgae market as they are getting acceptance in food- and health-related supermarkets and stores (Koyande et al. 2019). The two popular usable species of *spirulina* are *Arthrospira platensis* and *Arthrospira maxima* (Tomaselli 1997). Their cell wall is similar to the Gram-positive bacteria, subsequently they comprise of glucosamines and muramic acid linked to peptides (Falquet 1997). The wall of *Spirulina* is not digestible, is fragile and makes the cell content readily available to digestive enzymes.

3.4 Animal and Fish Feed

In aquaculture, microalgae are used as vital sources of live feeds and supplements for larval and juvenile animals, which include oyster spat, juvenile abalone and rotifer. The product related to food from algae is aquaculture feed, shrimp feed, shellfish diet and livestock feed. Microalgae are a significant direct and indirect feed source for the primary growing phases of many farmed finfish, shellfish and invertebrate species (Shields and Lupatsch 2012).

3.5 Algae in Cosmetics

Many algae species are used to produce high-value cosmetics. They are used as thickening agents, antioxidants and water-binding agents in cosmetics. Irish moss, AlgaVia proteins, different types of vitamins, sugar, starch, different micro-and macro-nutrients are other forms of algae. These are useful for skin, either as ointments or antioxidants and the associated invention are algae in cosmetics, carrageenan and alginates. The most common applications are in skin care, protection from sun and hair care, toothpaste, shaving cream, lotions and antibacterial creams (Pimentel et al. 2018).

3.6 Algae Chemicals

Next to the extraction of oil from algae—they commonly referred as algae cake and these cake can be used as organic manures instead of chemical or inorganic fertilizers. Microalgae as a chemical and the related products are chlorophyll, phycocyanin, fucoxanthin and the potential applications are: defoamers, inks, algae based resins, stable isotopically labelled compounds, dyes and colourants (Kruus 2017).

3.6.1 Phycocyanin

It is one of the major pigment molecules (chemical formula $C_{165}H_{185}N_{20}O_{30}$) found within *Spirulina*. The dietary and aesthetic values of phycocyanin have been well known (Romay et al. 2003). It is a dye of blue colour which fits to Phycobilli proteins found in blue green algae.

3.6.2 Phycoerythrin

This is a red protein pigment complex produced by blue-green microalgae *Arthrospira platensis*. This is generally used as natural colouring agent in the food industry (Taufiqurrahmi et al. 2017). The price of any chemicals (e.g. phycocyanin) is an important issue as it states its market capability. The cost of cultivation is one of the main factors that determine the price of phycocyanin. The carotenoids produce by certain forms of microalgae, particularly red ones, or rhodophyta, comprise phycocyanin and phycoerythrin (Becker 1994).

3.6.3 Fucoxanthin

Fucoxanthin (with formula $C_{42}H_{58}O_6$) is an explicit carotenoid found in brown seaweeds with extraordinary natural properties. Ishimozuku (*Sphaerotrichia divaricata*), an edible brown alga, has morphology that is almost indistinguishable to that of Okinawa-mozuku. Ishimozuku contains various anti-oxidant ingredients and is more likely to improve human health (Maeda et al. 2018).

4 Microalgae Product and Market

Economic feasibility depends upon the market value and production costs (e.g., biomass production costs and biorefinery costs). The algal product types are: hydrocolloids, carotenoids, and omega-3 PUFA, and their applications are in food and feed, nutraceutical, cosmetics and chemicals.

4.1 Market Opportunities: Algae-Derived Products

Commercial-scale cultivation of microalgae has already improved, even though the total production cost of some microalgae biomass is more than the industrial expectation. To develop the competitiveness of microalgae-based products, many aspects of their value chain need to be improved. The technical and the economic aspects, including maintenance, market awareness and following regulations, are important factors for wider algal market opportunities and scope. It is important to understand the regulatory framework, environmental and ecological law, and risk management for the development of microalgae based market product.

4.2 Important Determining Market Opportunities and Challenges

Algae biomass contains not only proteins and lipids. In fact, even if these products currently represent most of the algae market, further high-value bioproducts can be obtained by processing algae in efficient ways, such as pigments for food coloration; pharmaceutics and phenolics as cosmetics, skin care emollients and sun protection products. The European Biomass Industry Association (EUBIA) is currently working on this front, preparing projects and strategies to make algae biorefinery a real competitive solution. At present the manufacture of food and feed from microalgae in European countries assures 5% of global market (Enzing et al. 2014). Currently the United States, Asia and Oceania dominate the market but Europe could also be another leading country in microalgae-based bioproducts in the next decade but needs to focus on specific policies to launch specific targets.

4.3 Microalgae-Based Products: Market Opportunities and Key Challenges

4.3.1 Market Scope of Spirulina

This is a microscopic and filamentous cyanobacterium whose name originates from the spiral or helical nature of its fibres and it has been stated that it was used during the Aztec civilization (Dillon et al. 1995). *Spirulina* or *Arthrospira* is a blue-green alga which achieved fame after it was successfully used by NASA as a dietary supplement for spacemen on planetary tasks. A number of studies exploring the efficacy and the potential clinical applications of *Spirulina* in treating several diseases have suggested anticancer, antiviral and anti-allergic effects (Karkos et al. 2011).

The market for food companies was not well organized and control in 1970–1980 when *Spirulina* was introduced on the market. The current main providers of *spirulina* are located in the USA (Earthrise), Hawaii and Thailand (Bosschaert 2002). Regarding the market scope of *Spirulina*, the logistics are simple, and Merredin's location is ideal for logistics purposes.

4.3.2 Market Scope of Omega-3 Fatty Acids/PUFA

Omega-3 PUFA market is predicted to grow at a CAGR of 13.5% (Market Watch 2018). Omega-3 poly unsaturated fatty acid (PUFA) is a type of necessary fatty acid which cannot be synthesized by the human body and which needs intake through omega-3-rich food. Omega 3 PUFA is used to enhance the cardiovascular and cognitive functioning of the human body. The ingredients of Omega-3 PUFA are sourced from various fish oil, krill oil, chia seed, flaxseed and other plant sources (Market Research Future 2019). The Global Omega-3 PUFA Market is segmented into Europe, North America and Asia Pacific along with rest of the world (RoW). Among these, North America holds the major market share in the global Omega-3 PUFA market both in terms of value and volume. This is attributed by the rising number of health conscious people in the USA and also presence of companies in the North American region such as Unilever, Abbott, and Nestle SA. Also, vital companies are introducing new products in the North American region in order to retain their existing customers and also to acquire new customers. The Asia Pacific region is expected to grow immensely during the forecast period (Market Research Future 2019). According to FAO statistics (2010) the market size of EPA and DHA are 300 million and 1.5 billion USD, respectively, and the price is 0.2–0.5 USD/gm and 18–22 USD/gm, respectively.

Over 75% of the manufacture volume of microalgae was used in the health food marketplace as nutritional enhancements (Chacon-Lee and Gonzalez-Marino 2010). The algae-based valued food additives and ingredients, e.g. DHA, represent a rising market. Martek's (now DSM) algae-derived DHA is found in 99% of all baby foods in the USA (Eckelberry 2011).

4.3.3 Market Scope of Astaxanthin

The demand for natural astaxanthin is now increasing and it has been increased into a billion dollar market potential of nutraceuticals (Martín et al. 2008). The Global Astaxanthin Market is segmented into five regions: (1) Asia Pacific, (2) North America, (3) Europe, (4) Latin America and (5) Middle East and Africa. North America holds a major share of the market due to the growing demand for carotenoid pigments in feed, supplements, food, health care products, and others. The feed segment is predicted to witness a rapid growth in the market due to growing consumption of *Haematococcus pluvialis* to produce high standard astaxanthin in countries such as the USA, Canada and Mexico. It is estimated the health care products section is set to witness a higher growth in market share during the forecast period 2017–2023 (Market Watch 2019). It is projected that the astaxanthin market is set to spectate a higher growth owing to growing use of carotenoid pigments during the forecast period (Market Watch 2018). According to Shah et al. (2016) the total value of astaxanthin synthesis is estimated at more than $200 million, which corresponds to 130 metric tons of product annually. The average market price is above USD 2000 per kilogram, and the cost is estimated at USD 1000 per kilogram (Shah et al. 2016; Milledge 2011). Recent investigations showed that microalgae-derived astaxanthin was only responsible for less than 1% of the total commercialized market, ascribed to its higher product prices than those of synthetic products and technological challenges of large-scale cultivation and harvesting of microalgae (Koller et al. 2014). The chemical components and structures of natural astaxanthin improve its bioavailability and biological activities (especially antioxidant activity), which highlight the commercial value of the carotenoid. Due to the growing market demand on natural astaxanthin, the market value is estimated at US$1.1 billion and projected to reach 670 metric tons by 2020 (Shah et al. 2016). In the case of *Haematococcus* astaxanthin, the market value is estimated in the range US$2500/kg to US$15,000/kg, depending on products' purity (Koller et al. 2014; Pérez-López et al. 2014). The market size and country wise market sector and future market potential of Astaxanthin are presented in Tables 1.2 and 1.3, respectively.

4.3.4 Market Scope of Beta-Carotene

Europe held the largest market share of beta-carotene in 2017 (Market Watch 2019). The growth in the food and beverages (Fig. 1.5), pharmaceutical and personal care industry has caused a rise in the beta-carotene market. Food and medicine industries in developing countries, for example China and India, is contributing to the increase in demand for beta carotene market in Asia Pacific (Table 1.4). The market share of

Table 1.2 Market size and potential market of microalgae products (Molino et al. 2018)

Market sectors/area	Market size (year 2009) (Million US dollar)	Potential market (year 2020) (Million US dollar)
Colouring agents	300	800
Antioxidant nutraceuticals	30	300
Pharmaceuticals/ chemicals	Developing	500
Cosmetics/makeups	Emerging/rising	30

Table 1.3 Prominent countries having market sectors and future market for Astaxanthin

Company	Location	Production capacity of pure Astaxanthin
Alga Technologies	Israel	N/A
Cyanotech	Hawaii	N/A
Stone Forest Astaxanthin Biotech Co Ltd	China	1200 kg/year
Yunnan Baoshan Zeyuan Microalgae Health Technology Co., Ltd	China	500 kg/year
Jingzhou Natural Astaxanthin Inc	China	N/A
Beijing Gingko Group	China	900 kg/year
Yunnan Alphy Biotech Co Ltd	China	600 kg/year
Yunnan SGYJ Biotech Co Ltd	China	400 kg/year
Algaetech International	Malaysia, Indonesia	N/A
Parry Nutraceuticals	India	N/A
Mera Pharmaceuticals Inc.	Hawaii	N/A
Fuji Chemicals	Japan, Sweden	N/A
Valensa International	Florida	N/A

N/A: data not available

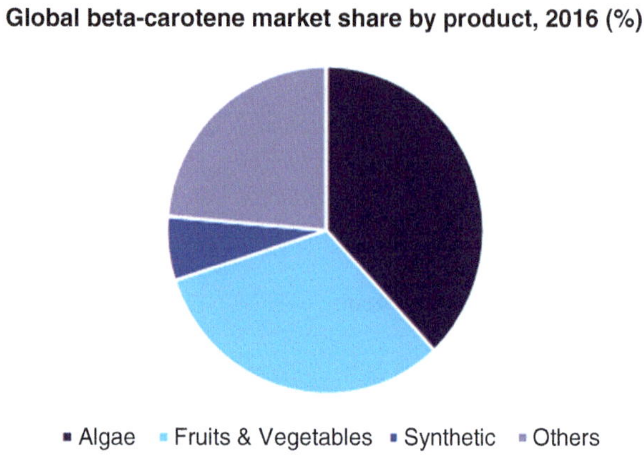

Fig. 1.5 Market share of beta-carotene from algae (Market Research Report 2016)

Table 1.4 Prominent companies in the beta-carotene market (Market Watch 2019)

Company	Location/area
Aqua Carotene	USA, New Zealand
Cognis Nutrition and Health	Australia
Fuqing King Dnarmsa Spirulina Co., Ltd.	China
Cyanotech	Hawaii, USA, UK
Nikken Sohonsha Corporation	Japan
Tianjin Lantai Biotechnology	China
Parry Nutraceuticals	Asia/India
Seambiotic	Israel
Muradel	Australia

beta-carotene in 2017 is estimated as US$247 million, but this amount was increased to US$285 million by 2015 which is a CAGR of 1.8% (Rastogi et al. 2017).

4.3.5 Market Scope of Phycoerythrin

Phycobiliproteins is a protein produced commercially from *Spirulina* and the red microalgae *Porphyridium* and *Rhodella* (Becker 1994; Singh et al. 2005; Borowitzka 2013; Spolaore et al. 2006). Ultrafiltration was used to isolate phycoerythrin protein from *Grateloupia turuturu* following cell homogenization, which was testified to retain 100% of the protein without denaturation (Denis et al. 2009). Phycoerythrin can establish a significant amount of the complete protein content in red algae, with levels of 1.2% reported for *P. palmata* (Wang et al. 2001).

The key companies in phycoerythrin market are: Europa Bioproducts, Sigma-Aldrich, Jackson Immuno Research, Thermo Fisher Scientific, SETA BioMedicals, Binmei Biotechnology, Algapharma Biotech, Phyco-Biotech,

Norland Biotech, Columbia Bioscience and Dainippon Ink and Chemicals. The main regions which play a vital role in phycoerythrin market are: North America, Europe, China, Japan, Middle East and Africa, India and South America. The most important types of Phycoerythrin products: PE545, R-phycoerythrin and B-phycoerythrin (Market Watch 2019). The prominent commercial market sources of Phycoerythrin are: Europa-bioproducts, Cambridge, UK, and Invitrogen, NY, America.

4.4 European Guidelines on Marketing of Microalgae Products

The key challenges of marketing microalgae products in the European Union are: biomass production cost, technical breakthroughs, access to venture capital and regulatory, academic and industrial training (Fig. 1.6). It is very important to understand, monitor, observe and practice the regulation for food and feed prior to use and marketing.

The market size based on the amount of nutrient obtained from microalgae are still less compared to the ones derived from cereals and other crops. But still the sector has seen an impressive and exceptional growth. Even with the challenges due to the climatic situations together with the inadequate domestic demand and the difficulty of the EU Novel Food regulation, a survey discovered that the EU can improve its market position in the next decade (Vigani et al. 2015). The general

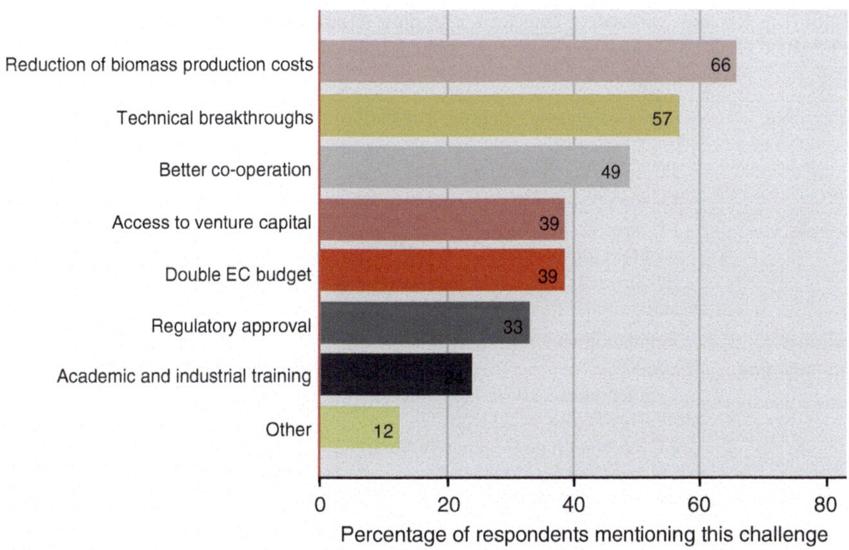

Fig. 1.6 Key challenges in the development of microalgae market in the EU (Salimbeni 2014)

principles and requirements of EU food law are controlled under Regulation (EC) 178/2002. Other relevant regulations include (Parker et al. 2014):

- Regulation (EC) 258/97 on novel foods and ingredients.
- Regulation (EC) 767/200 on the marketing of feed materials and compound feed.
- Regulation (EC) No 41/2009.
- Regulation (EU) No 1169/2011.
- Commission Implementing Regulation (EU) No 828/2014 (European Union 2014).
- Directive 89/107/EEC on food additives.
- Regulation (EC) 1831/2003 on the authorization, and labelling of feed additives.
- Regulation (EC) No 852/2004 on food hygiene.
- Regulation 183/2005 on feed hygiene.
- Regulation 1829/2003 on GMOs for food and feed.
- Regulation (EU) No 828/2014.

A significant number of other European Union regulations and directives cover the production and distribution of animal feedstuffs, while Regulation 183/2005 is the key measure for algal production. Market introduction of food products using the whole microalgae (e.g. *Spirulina* or *Chlorella*) or microalgal ingredients which are incorporated into pasta with the green algae colour are subject to food safety regulations that apply to all food products. Production through engineering technology and marketing of food and feed from microalgae are mainly regulated into the European Community by the Food Safety Regulation (EC 178/2002) and the Novel Food Regulation (EC 258/97) (Enzing et al. 2014) (Table 1.5).

Table 1.5 EU and USA safety guidelines regarding investigation, manufacture and market of microalgal products for food and feed applications (Modified from Enzing et al. 2014)

	Europe	United States
Research	– Commission Implementing Regulation (EU) 2017/2470. – EC directives 2009/41/EC (GM algae). – EC directive 2001/18/EC (GM algae).	– NIH rDNA Guidelines. – EPA Standards, under the Toxic Substances Control Act (TSCA).
Production	– EC directives 2009/41/EC (contained use of GM algae). – EC directive 2001/18/EC (deliberate release of GM algae).	– TSCA Environmental Release Application (TERA). – Microbial Activity TSCA. – USDA Plant Protection Act.
Market introduction	– Regulation on Food Safety (EC 178/2002). – Regulation EC 2017/2470 Directive 2002/46/CE (Borowitzka 2013). – EC Regulation on Genetically Modified Food and Feed (EC 1829/2003). – Regulation EC2017/2470 and 2002/46/CE. – EC Regulation on Nutrition and Health claims made on foods (EC 1924/2006).	– Food, Drug and Cosmetic Act. – Dietary Supplement Health and Education Act.

5 Future Prospects of the Microalgae Industry

To ensure the sustainability of the microalgae biorefinery process, an advanced microalgae biorefinery needs to be applied through the production of manifold products in the form of HVPs and biofuel. Algae can play an important role in the biobased economy and they are efficiently cultivated in places that are unsuitable for agriculture and where nature is not harmed (Wolkers et al. 2011). Recently, microalgae cultivation has received increased attention because this could synthesize larger amounts of HVPs, such as pigments, vitamins, PUFAs, anti-oxidants and many more.

Algae are considered the most diverse group of organisms on earth and therefore have potential for advanced use in the future, however based on its demand and potential supply and technology. Algae and algae-derived products have future prospects in the industrial sector based on the demand in this sector and having the capability of algae for industrial use. The use of algae-based products in pharmaceuticals (bioactive and new drugs), nutraceuticals (probiotic, antioxidants) and also use in industry as biofuels, chemical/enzyme, new cosmetics/cosmeceuticals which is currently limited, has a wider scope to expand. Future prospects of product areas of algae are: nutrient-rich food, bioenergy, bioactive medicine, novel enzymes, specialist chemicals, biofertilizer, clean insecticides and bioremediation. Despite potential demand, supply and technical expertise, technological issues hinder the future prospects of algal applications. The major constrains include: engineering difficulty in production, formulation of invention, strain stability and productivity. Refining of algal fuel and bioproducts technology from an experimental scale to a profitable level is possible by overcoming the challenges and limitations associated with this technology.

6 Conclusion

The nature, characteristics and composition of algae are diverse and the reason product types and formulations are different. Based on an extensive overview on microalgal research, it was found that significant research on microalgae, microalgal product and their market potential have been conducted at the lab, pilot and in large scales. The value chain of microalgae and algal products is a biobased economy and the value depends on its composition, application and formulations. The potential high-value products (HVPs) from microalgae are: food, agar, alginates, astaxanthin, beta-carotene, omega-3 fatty acids, phycocyanin, phycoerythrin and fucoxanthin; these are used mainly in pharmaceuticals, nutraceuticals, cosmeceuticals and industrial sectors. High value product that extracted from microalgae improve the economics in a biorefinery approach and have its market scope and opportunities. However it need to understand whether it is market driven or technology driven. Production economics such as the cost effectiveness of the system is also important for further investments. The technology also needs to be robust and reliable for its market flow.

Microalgae is not currently considered economically viable for the production of fuel but has proven potential for the production of food and high-value products. This chapter provides insights on the broader algae-based value chain, high value products derived from microalgae, their market scope and opportunities and biorefinery of microalgae for the production of high-value products. The proper and modern use of technology and tools gives higher yield, comfort of operation and helps economic processing. The potential for production of multiple products from microalgae has led to more effective and efficient invention pathways in the use of materials and energy. An understanding of the environmental impacts and legislations is important in the evaluation of technology and economic performance of a biorefinery system and its marketing. From an algal value chain analysis, it was found that value depends on composition, application, formulation, production scale and comparative reference. To increase the market opportunities of algae-derived products, it is important to determine market opportunities and challenges and find a sustainable solution. It is very important to practice and understand the rules, regulations and legislation related to algae in high-value products and their supply chain stages. There is no formal regulation related to algae and algae high-value products in many developing countries, but there are a few regulations in the EU, USA and other parts of the developed world. These regulations are on novel food and ingredients, feed material and compound food, food additives, food hygiene, and GMOs for food and feed. It is important to emphasize on research, production and market introduction for better prospects in the microalgae industry.

References

Abe, K., Imamaki, A., & Hirano, M. (2002). Removal of nitrate, nitrite, ammonium and phosphate ions from water by the aerial microalga *Trentepholia aurea*. *Journal of Applied Phycology, 14*, 129–134.

Alam, M. A., & Wang, Z. (Eds.). (2019). *Microalgae biotechnology for development of biofuel and wastewater treatment* (pp. 1–655). New York, NY: Springer.

Ambati, R. R., Phang, S. M., Ravi, S., & Aswathanarayana, R. G. (2014). Astaxanthin: Sources, extraction, stability, biological activities and its commercial applications—A review. *Marine Drugs, 12*(1), 128–152. https://doi.org/10.3390/md12010128.

Behrenfeld, M. J., & Falkowski, P. G. (1997). Photosynthetic rates derived from satellite-based chlorophyll concentration. *Limnology and Oceanography, 42*(1), 1–20.

Borowitzka, M. A. (2013). High-value products from microalgae their development and commercialization. *Journal of Applied Phycology, 25*, 743–756.

Bosschaert, T. (2002). *Spirulina*. Plant Research, University of Western Australia, School of Architecture, Student of Delft University of Technology, Faculty of Industrial Design Engineering.

Becker, E. W. (1994). *Microalgae biotechnology and microbiology*. Cambridge: Cambridge University Press. isbn:978-0-521-06113.

Berman, J., et al. (2014). Nutritionally important carotenoids as consumer products. *Phytochemistry Reviews, 14*, 727–743.

Brenna, L., & Owende, P. (2010). Biofuels from microalgae—A review of technologies for production, processing, and extractions of biofuels and co-products. *Renewable and Sustainable Energy Review, 14*, 557–577.

Bush, B. (2019). *How lead gen marketers can use value chains to make better decisions. Linked in report.* https://www.linkedin.com/pulse/how-lead-gen-marketers-can-use-value-chains-make-better-becky-bush/

Cardozo, K. H. M., et al. (2007). Metabolites from algae with economic impact. *Comparative Biochemistry and Physiology Part C: Toxicology and Pharmacology, 146*(1–2), 60–78.

Chacon-Lee, T. L., & Gonzalez-Marino, G. E. (2010). Microalgae for healthy foods—Possibilities and challenges. *Comprehensive Reviews in Food Science and Food Safety, 6*, 655–675.

Das, P., Aziz, S. S., & Obbard, J. P. (2011). Two phase microalgae growth in the open system for enhanced lipid productivity. *Renewable Energy, 36*(9), 2524–2528.

Dillon, J. C., Phuc, A. P., & Dubacq, J. P. (1995). Nutritional value of the alga *Spirulina. World Review of Nutrition and Dietetics, 77*, 32–46.

Denis, C., Massé, A., Fleurence, J., & Jaouen, P. (2009). Concentration and pre-purification with ultrafiltration of a r-phycoerythrin solution extracted from macro-algae grateloupia turuturu: Process definition and up-scaling. *Separation and Purification Technology, 69*, 37–42.

DOE (U.S. Department of Energy). (2016). *National algal biofuels technology review*. Washington, DC: U.S. Department of Energy, Office of Energy Efficiency and Renewable Energy, Bioenergy Technologies Office.

Duncan, R. C. (2001). The peak of world oil production and the road to the Olduvai Gorge. *Population and Environment, 22*(5), 503–522.

European Union. (2014). Commission implementing regulation (EU) no: 828/2014. *The Official Journal of the European Union.*

Eckelberry, R. (2011). Algae—Food or chemical grade. *Algae Industry Magazine.* http://www.algaeindustrymagazine.com/algae-business-algae-food-or-chemical-grade/

Enzing, C., Ploeg, M., Barbosa, M., & Sijtsma, L. (2014). *Microalgae-based products for the food and feed sector: An outlook for Europe. Scientific and policy reports* (Vol. 82, pp. 19–37). Luxembourg: EU Publications.

FAO (2010). FAO fisheries and aquaculture report no. 978. Report of the join FAO/WHO expert consultation on the risk and benefits of fish consumption. Rome, 25–29 January 2010.

Falquet, J. (1997). *The nutritional aspects of Spirulina.* Genève: Antenna Technology.

Fresh Designpedia. (2019). *Algae eating and healthy stay—What you should know about the algae.* Retrieved August 28, 2019, from https://www.freshdesignpedia.com/trends/algae-eating-and-healthy-stay-what-you-should-know-about-the-algae.html

Gantar, M., & Svircev, Z. (2008). Microalgae and cyanobacteria: Food for thought. *Journal of Phycology, 44*, 260–268.

Gao, K. (1998). Chinese studies on the edible blue-green alga, Nostoc flagelliforme: A review. *Journal of Applied Phycology, 10*, 37–49.

Georgianna, D. R., & Mayfield, S. P. (2012). Exploiting diversity and synthetic biology for the production of algal biofuels. *Nature, 488*(7411), 329–335.

Gross, M. (2013). *Development and optimization of algal cultivation systems.* Graduate theses and dissertations, Iowa State University. 13138p. 4.

Grune, T., Lietz, G., Palou, A., et al. (2010). Beta-carotene is an important vitamin A source for humans. *The Journal of Nutrition, 140*, 2268S–2285S.

Harasym, J., & Oledzki, R. (2014). Effect of fruit and vegetable antioxidants on total antioxidant capacity of blood plasma. *Nutrition, 30*, 511–517.

Harris, W. S. (2010). Omega-3 fatty acids. In P. M. Coates, J. M. Betz, M. R. Blackman, et al. (Eds.), *Encyclopedia of dietary supplements* (2nd ed., pp. 577–586). London: Informa Healthcare.

Hamed, I. (2016). The evolution and versatility of microalgal biotechnology: A review. *Comprehensive Reviews in Food Science and Food Safety, 2016*(15), 1104–1123.

Hay, I. D., Rehman, Z. U., Moradali, M. F., Wang, Y., & Rehm, B. H. A. (2013). Microbial alginate production, modification and its applications. *Microbial Biotechnology, 6*, 637–650.

Hermawan, J., Masithah, E. D., Tjahjaningsih, W., Abdillah, A. A. (2018). Increasing β-carotene content of phytoplankton *Dunaliella salina* using different salinity media. IOP Conference Series: Earth and Environmental Science.

Holdt, S. L., & Kraan, S. (2011). Bioactive compounds in seaweed: Functional food applications and legislation. *Journal of Applied Phycology, 23*, 543–597.

Hosikian, A., Lim, S., Halim, R., & Danquah, M. K. (2010). Chlorophyll extraction from microalgae: A review on the process engineering aspects. *International Journal of Chemical Engineering, 2010*, 1–11.

Hounslow, E. (2016). *Salt stress in two Chlamydomonas species: Novel insights into biofuel production from microalgae*. PhD thesis, University of Sheffield, UK.

Jones, P. J. H., & Rideout, T. (2014). Lipids, sterols, and their metabolites. In A. C. Ross, B. Caballero, R. J. Cousins, K. L. Tucker, & T. R. Ziegler (Eds.), *Modern nutrition in health and disease* (11th ed.). Baltimore, MD: Lippincott Williams & Wilkins.

Jones, P. J. H., & Papamandjaris, A. A. (2012). Lipids: Cellular metabolism. In J. W. Erdman, I. A. Macdonald, & S. H. Zeisel (Eds.), *Present knowledge in nutrition* (10th ed., pp. 132–148). Washington, DC: Wiley-Blackwell.

Karkos, P. D., Leong, S. C., Karkos, C. D., Sivaji, N., & Assimakopoulos, D. A. (2011). *Spirulina* in clinical practice: Evidence-based human applications. *Evidence-based Complementary and Alternative Medicine, 2011*, 19. https://doi.org/10.1093/ecam/nen058.

Khan, M. I., Shin, J. H., & Kim, J. D. (2018). The promising future of microalgae: Current status, challenges, and optimization of a sustainable and renewable industry for biofuels, feed, and other products. *Microbial Cell Factories, 17*(1), 36. https://doi.org/10.1186/s12934-018-0879x.

Khattar, J. I. S., Singh, D. P., & Kaur, G. (2009). *Algal biology and biotechnology*. New Delhi: I.K. International.

Koller, M., Muhr, A., & Braunegg, G. (2014). Microalgae as versatile cellular factories for valued products. *Algal Research, 6*, 52–63.

Koyande, A. K., Chew, K. W., Rambabu, K., Tao, Y., Chu, D.-T., & Show, P.-L. (2019). Microalgae: A potential alternative to health supplementation for humans. *Food Science and Human Wellness, 8*(1), 16–24.

Kim, J. H., Affan, M. A., Jang, J., Kang, M. H., Ko, A. R., Jeon, S. M., Oh, C., Heo, S. J., Lee, Y. H., Ju, S. J., et al. (2015). Morphological, molecular, and biochemical characterization of astaxanthin-producing green microalga *Haematococcus* sp. KORDI03 (haematococcaceae, chlorophyta) isolated from Korea. *Journal of Microbiology and Biotechnology, 25*, 238–246.

Kruus, M. (2017). *Purification, biomass production and cryopreservation of aero-terrestrial microalgae and cyanobacteria*. Helsinki Metropolia University of Applied Sciences (thesis).

Li, J., Zhu, D., Niu, J., Shen, S., & Wang, G. (2011). An economic assessment of astaxanthin production by large scale cultivation of *Haematococcus pluvialis*. *Biotechnology Advances, 29*(6), 568–574.

Luiten, E. E., Akkerman, I., Koulman, A., Kamermans, P., Reith, H., Barbosa, M. J., Sipkema, D., & Wijfels, R. H. (2003). Realizing the promises of marine biotechnology. *Biomolecular Engineering, 20*, 429–439.

Milovanovic, I., Misan, A., Saric, B., Kos, J., Mandic, A., Simeunovic, J., et al. (2012). Evaluation of protein and lipid content and determination of fatty acid profile in selected species of cyanobacteria. In *Proceedings of the 6th Central European Congress on Food, CE Food, Novi Sad, Serbia, 23–26 May 2012*.

Maeda, H., Fukuda, S., Izumi, H., & Saga, N. (2018). Anti-oxidant and fucoxanthin contents of brown alga *Ishimozuku* (Sphaerotrichia divaricata) from the West Coast of Aomori, Japan. *Marine Drugs, 16*, 255. https://doi.org/10.3390/md16080255.

Market Research Report. (2017). Astaxanthin market analysis by source (Natural [Yeast, Krill/Shrimp, Microalgae] and Synthetic), by product (dried biomass/powder, oil, soft gels, liquid), by application, and segment forecasts, 2018–2025.

Market Watch. (2019). Beta carotene market size, share 2019. Global Beta Carotene Market Report, 2019.

Market Research Report. (2016). *Beta-carotene market analysis by source (algae, fruits & vegetables, & synthetic), by application (food & beverages, dietary supplements, cosmetics, & animal feed) and segment forecasts to 2024*. https://www.grandviewresearch.com/industry-analysis/beta-carotene-market.

Mata, T. M., Martins, A. A., & Caetano, N. (2010). Microalgae for biodiesel production and other applications: A review. *Renewable and Sustainable Energy Reviews, 14*(1), 217–232.

Matos, Â. P., Feller, R., Moecke, E. H. S., de Oliveira, J. V., Junior, A. F., Derner, R. B., & Sant'Anna, E. S. (2016). Chemical characterization of six microalgae with potential utility for food application. *Journal of the American Oil Chemists' Society, 93*, 963–972.

Market Research Future. (2019). *Omega-3 PUFA market scope 2019, global industry analysis, key players, size, share, growth, trends and forecast to 2023*. https://www.marketwatch.com/press-release/omega-3-pufa-market-scope-2019-global-industry-analysis-key-players-size-share-growth-trends-and-forecast-to-2023-2019-02-22

Market Watch. (2018). *Astaxanthin market challenges, key players, industry segments, development, opportunities, forecast report 2021*. https://www.marketwatch.com/press-release/astaxanthin-market-challenges-key-players-industry-segments-development-opportunities-forecast-report-2021-2018-05-23

Martín, J. F., Gudiña, E., & Barredo, J. L. (2008). Conversion of β-carotene into astaxanthin: Two separate enzymes or a bifunctional hydroxylase-ketolase protein? Microbial Cell Factories 7:3.

Mesbah, N. M., & Wiegel, J. (2006). Isolation, cultivation and characterization of alkalithermophiles. *Methods in Microbiology, 35*, 451–468.

Milledge, J. J. (2011). Commercial application of microalgae other than as biofuels: A brief review. *Reviews in Environmental Science and Bio/Technology, 10*(1), 31–41.

Meier, R. L. (1955). Biological cycles in the transformation of solar energy into useful fuels. In F. Daniels & A. Duffie (Eds.), *Solar energy research* (pp. 179–183). Madison, WI, University of Wisconsin Press.

Molino, A., Iovine, A., Casella, P., Mehariya, S., Chianese, S., Cerbone, A., Rimauro, J., & Musmarra, D. (2018). Microalgae characterization for consolidated and new application in human food, animal feed and nutraceuticals. *International Journal of Environmental Research and Public Health, 15*(11), E2436. https://doi.org/10.3390/ijerph15112436.

Muhaemin, M., & Kaswadji, R. F. (2009). Biomass nutrient profiles of marine microalgae *Dunaliella salina*. *Journal Penelitian Sains, 13*, 64–67.

Nicoletti, M. (2016). Review microalgae nutraceuticals. *Food, 5*, 54. https://doi.org/10.3390/foods5030054.

Ozkurt, I. (2009). Qualifying of safflower and algae for energy. *Energy Education Science and Technology Part A, 23*, 145–151.

Oswald, W. J., & Golueke, C. G. (1960). Biological transformation of solar energy. *Advances in Applied Microbiology, 2*, 223–262. https://doi.org/10.1016/S0065-2164(08)70127-8.

Oil Squeeze. (2008). Business: Oil squeeze. http://content.time.com/time/magazine/article/0,9171,946222,00.html

Parker, B., Malin, G., Benson, D., & Schlarb-Ridley, B. (2014). *Regulatory factsheet 17—algae as a feedstock for energy. EnAlgae project output WP2A10.18, 1 pp*. www.enalgae.eu/public-deliverables.htm

Pérez-López, P., et al. (2014). Life cycle assessment of the production of the red antioxidant carotenoid astaxanthin by microalgae: From lab to pilot scale. *Journal of Cleaner Production, 64*, 332–344.

Pimentel, F., Alves, R., Rodrigues, F., & Oliveira, M. P. P. (2018). Macroalgae-derived ingredients for cosmetic industry—an update. *Cosmetics, 5*(1), 2.

Plaza, M., Herrero, M., Cifuentes, A., & Ibanez, E. (2009). Innovative natural functional ingredients from microalgae. *Journal of Agricultural and Food Chemistry, 57*, 7159–7170.

Pulz, O., & Gross, W. (2004). Valuable products from biotechnology of microalgae. *Applied Microbiology and Biotechnology, 65*, 635–648.

Rastogi, R. P., Datta, M., & Pandey, A. (2017). *Book: Algal green chemistry—Recent progress in biotechnology*. Amsterdam: Elsevier.
Romay, C., González, R., Ledón, N., Remirez, D., & Rimbau, V. (2003). C-phycocyanin: Abiliprotein with antioxidant, anti-inflammatory and neuroprotective effects. *Current Protein & Peptide Science, 4*, 207–216.
Salimbeni, A. (2014). European biomass industry association, open workshop on microalgae market, 12 Nov 2014, Brussels.
Sánchez, J. F., Fernández, J. M., Acién, F. G., Rueda, A., Pérez-Parra, J., & Molina, E. (2008). Influence of culture conditions on the productivity and lutein content of the new strain *Scenedesmus almeriensis*. *Process Biochemistry, 43*, 398–405.
Sathasivam, R., Radhakrishnan, R., Hashem, A., Elsayed, F., & Allah, E. F. (2019). Review microalgae metabolites: A rich source for food and medicine. *Saudi Journal of Biological Sciences, 26*, 709–722.
Singh, S., Kate, B. N., & Banerjee, U. C. (2005). Bioactive compounds from cyanobacteria and microalgae: an overview. *Critical Reviews in Biotechnology, 25*(3), 73–95.
Shah, M. M. R., Liang, Y., Cheng, J. J., & Daroch, M. (2016). Astaxanthin-producing green microalga *Haematococcus pluvialis*: From single cell to high value commercial products. *Frontiers in Plant Science, 7*, 531.
Shields, R. J., & Lupatsch, I. (2012). Algae for aquaculture and animal feeds. SCHWERPUNKT, Technikfolgenabschätzung – Theorie und Praxis 21. Jg., Heft 1.
Spolaore, P., Joannis-Cassan, C., Duran, E., & Isambert, A. (2006). Commercial applications of microalgae. *Journal of Bioscience and Bioengineering, 101*, 87–96.
Tibbetts, S. M., Milley, J. E., & Lall, S. P. (2015). Chemical composition and nutritional properties of freshwater and marine microalgal biomass cultured in photobioreactors. *Journal of Applied Phycology, 27*, 1109–1119.
Tiwari, B. K., & Troy, D. (2015). *Seaweed sustainability: Food and non-food applications*. Waltham, MA: Elsevier.
Taufiqurrahmi, N., Religia, P., Mulyani, G., Suryana, D., Ichsan, Tanjung, F. A., et al. (2017). Phycocyanin extraction in *Spirulina* produced using agricultural waste. *IOP Conference Series: Materials Science and Engineering 206*, 012097.
Tomaselli, L. (1997). Morphology, ultrastructure and taxonomy of Arthrospira (*Spirulina*) maxima and Arthospira (*Spirulina*) platensis. In A. Vonshak (Ed.), *Spirulina platensis (Arthrospira): Physiology, cell biology and biotechnology* (pp. 1–16). London: Taylor and Francis.
Transparency Market Research. (2018). Beta-carotene market (source—fruits & vegetables, algae & fungi, synthetic; end use—food, aquaculture feed, poultry & pet feed, dietary supplements, pharmaceuticals, cosmetics)—global industry analysis, size, share, growth, trends, and forecast 2019–2027.
USDOE. (2010). *National algal biofuels technology roadmap*. Washington, DC: U.S. Department of Energy, Office of Energy Efficiency and Renewable Energy, Biomass Program.
Vieira, V. V. (2016). *The role of the value-chain for the development of high-value products from microalgae*. Algal Biomass Summit 2017, European Algal Biomass Association, UABA.
Vigani, M., Parisi, C., Rodríguez-Cerezo, E., Barbosa, M. J., Sijtsma, L., Ploeg, M., & Enzing, C. (2015). Food and feed products from micro-algae: Market opportunities and challenges for the EU. *Trends in Food Science & Technology, 42*, 81–92. https://doi.org/10.1016/j.tifs.2014.12.004.
Voort, M. P. J., van der Vulsteke, E., de Visser, C. L. M. (2015). *Marco-economics of algae products. Output report WP2A7.02 of the EnAlgae Project, Swansea*. Retrieved August 17, 2019, from http://www.enalgae.eu/publicdeliverables.htm.
Williams, P. J. L. B., & Laurens, L. M. (2010). Microalgae as biodiesel & biomass feedstocks: Review & analysis of the biochemistry, energetics and economics. *Energy and Environmental Science, 3*, 554–590.

Wang, G., Sun, H., Fan, X., & Tseng, C. (2001). Large-scale isolation and purification of R-phycoerythrin from red alga *Palmaria palmata* using the expanded bed adsorption method. *Acta Botanica Sinica, 44*, 541–546.

Wayama, M., Ota, S., Matsuura, H., Nango, N., Hirata, A., Kawano, S., et al. (2013). Three-dimensional ultrastructural study of oil and Astaxanthin accumulation during encystment in the green alga *haematococcus pluvialis*. *PLoS One, 8*(1), e53618.

Wellinger, A. (2009). IEA bioenergy: Algal biomass, does it save the world? Short reflections. Task. 37.

Wolkers, H., Barbosa, M., Kleinegris, D. M. M., Bosma, R., & Wijffels, R. H. (2011). Microalgae: The green gold of the future? In P. Harmsen (Ed.), *Large-scale sustainable cultivation of microalgae for the production of bulk commodities* (Vol. 1, p. 34). Wageningen: ProPress.

Wu, H., Niu, H., Shao, A., Wu, C., Dixon, B. J., Zhang, J., & Wang, Y. (2015). Astaxanthin as a potential neuroprotective agent for neurological diseases. *Marine Drugs, 13*(9), 5750–5766. https://doi.org/10.3390/md13095750.

Zhang, Z.-Q., Cao, W.-T., Liu, J., Cao, Y., Su, Y.-X., & Chen, Y.-M. (2016). Greater serum carotenoid concentration associated with higher bone mineral density in Chinese adults. *Osteoporosis International, 27*, 1593–1601.

Zeng, X., Danquah, M. K., Chen, X. D., & Yinghau, L. (2012). Microalgae bioengineering: From CO_2 fixation to biofuel production. *Renewable and Sustainable Energy Review, 15*, 3252–3260.

Chapter 2
Microalgae as a Mainstream Food Ingredient: Demand and Supply Perspective

Alex Wang, Kosmo Yan, Derek Chu, Mohamed Nazer, Nga Ting Lin, Eshan Samaranayake, and James Chang

Abstract The value of future market of microalgae for food can be in billions of dollars, currently with microalgae-based nutritional supplement products such as *Spirulina*, *Chlorella Euglena* power or tablet and *astaxanthin* from *Haematococcus pluvialis*, already available on the shelves of selected stores in various parts of the world. Although microalgae have potential to further become mainstream food ingredients, they are facing many challenges before they are accepted as part of regular meals. In this chapter, various nutritional values of microalgae, such as protein, lipids, and vitamins, are examined in detail from the perspective of food application. The topic is a diverse one. While we focus on the inherent nutritional values from algae to demonstrate their suitability for the daily consumption, we also concurrently look through the lens of cultivation technology and biotechnology, as well as through analysis of bioavailability, in attempt to make clear path towards the goal of algae production. Demand analysis is made subsequently from functional aspect of microalgae and the derivatives as sources of food colorants and consumer segments where vegetarian and elder population are likely the earlier beneficiaries. Throughout the chapter, examples of microalgae-containing products are given at selected markets, highlighting the regional characteristics with a proper emphasis on nutritional values. Only with the suitable positioning of the values of algae in the target consumers and enabling of various technologies can microalgae become a significant part of the mainstream food ingredient market.

Keywords Microalgae food · Consumer segmentation · Demand and supply

---

A. Wang (✉) · K. Yan · D. Chu · M. Nazer · N. T. Lin · E. Samaranayake · J. Chang
Geb Impact Technology Company Limited, Hong Kong, China
e-mail: awang@gebimpact.com

M. A. Alam et al. (eds.), *Microalgae Biotechnology for Food, Health and High Value Products*, https://doi.org/10.1007/978-981-15-0169-2_2

1 Introduction

The global population on Earth is expected to surge to nearly ten billion by 2050. In 2016, approximately 55% of the inhabitants live in urban settlement. According to the United Nations, this percentage is expected to rise to 60% by 2030. The demand for food will inherently experience explosive growth. The global nutraceutical market, for example, which includes functional food and beverage ingredients, dietary supplements, personal care, and pharmaceuticals, valued at nearly US$200 billion in 2015, is expected to increase to nearly US$300 billion by 2021 at a compound annual growth rate of over 7% (Transparent 2018).

From a more granular perspective, meat (protein) consumption is driving growth in emerging markets at a compounded annual growth rate of 3–4%. Increasing affluence in the rising middle class is one of the primary contributors to this phenomenon. Ultimately, the inevitable question needs to be addressed as how our society would sustainably feed the global population with what we are going to have as food sources. Scientists and companies involved in plant-based proteins around the world are searching for and trying to come up with sustainable resources, to fill the gap and meet rising demand. Beyond meat (protein), the demand for other nutrition will increase drastically due to ever-increasing human activities.

The microalgae (i.e., the prokaryotic cyanobacteria and the eukaryotic microalgae) are an extremely diverse collection of organisms. Microalgae are ubiquitous throughout the world. It has persisted and thrived in a diverse natural environment. The ability to propagate rapidly, use light energy efficiently, fix atmospheric CO_2, and produce more biomass per hectare than vascular plants ensures its ability to overcome extreme habitation conditions. The single-cell organism, a rich source of biologically active compounds, has existed since prehistoric times. This highly nutritious biomass content has only been recognized as such and widely cultured at large industrial scale, for health benefits in recent years. Microalgae have demonstrated the potential to meet the population's needs for a more sustainable food supply, specifically with respect to protein, for years to come (Caporgno and Mathys 2018; Alam and Wang 2019). The beauty of this plentiful and yet humble resource is that it does not infringe upon our existing food supply; instead, it adds to the choices and variety of health benefits, a perfectly augmented sustainable food supply that will provide adequate nutritional balance for generations.

1.1 Historical Perspective of Algae as Food

The origin of microalgae as a food ingredient is not a recent phenomenon. It has served a vital role in the upstream and downstream of our food chain. We have been consuming our aquamarine food source, which feeds upon algae as a primary source of nutrients. Omega-3 (fish oil) from the cell organism enters our food chain through

Fig. 2.1 Aztecs harvesting *Spirulina* from Lake Texcoco

fish (Ji et al. 2015). Microalgae, a primary source of the aquamarine community, have been an integral part of our food chain.

Research showed that humans have used microalgae as a food source or nutritional supplements for hundreds of years. Aztecs (1300–1521 AD) and other Mesoamericans before Conquista consumed *Spirulina* from Lake Texcoco, as a dry cake known as tecuitlatl (Digs 2013). Figure 2.1 depicts Aztecs harvesting tecuitlatl, also known as green mud. Since the *Spirulina* species requires highly salty water, dams were built to prevent spillover of freshwater during the flooding. Eventually, the dikes were destroyed during the Spaniard conquest, and the environment of Lake Texcoco was no longer suitable for the growth of *Spirulina*, resulting in the loss of diet for the native Mexicans.

For centuries, the population in Chad has been harvesting *Spirulina* from Lake Kossorom for food "dihe" on the daily basis (Abduqader et al. 2000). Lake Chad has similar alkalinity to Lake Texcoco. In approximately 900 AD, the Kanembu discovered the nutritional value of *Spirulina* which they called dihe. As shown in Fig. 2.2, a woman in Chad is harvesting *Spirulina* in the shallow water. Although malnutrition is found throughout Central Africa, in Chad, where "dihe" is part of the diet, malnutrition is negligible.

More recently, to alleviate the suffering of children with acute malnutrition in Bangui, Central African Republic, the nuns who helped serve women and children with pre- and postnatal care in St. Joseph Health Centre literally took the matter into their own hands. They utilized the technical skills and formula which a French pharmacist provided to cultivate vitamin-rich green *Spirulina* in their own backyard, as shown in Fig. 2.3. The results were beyond anyone's expectations: "none of our

Fig. 2.2 A woman in Chad is harvesting *Spirulina*

Fig. 2.3 Nuns grow *Spirulina* in Central African Republic to fight malnutrition with ingenuity (Central Africa 2015)

babies die anymore, we have huge success with this." Sister Margherita's eyes beamed with satisfaction when she relays.

1.2 Recent Development of Production of Algae as a Mainstream Food Ingredient

Spirulina, one of the most popular microalgae (cyanobacteria), contains all the essential amino acids (EAAs) plus minerals like iron. It is a good source of plant-based protein. With over 60% of protein in biomass content, it is comparable to seven times (7×) the amount from soybeans, on the same area of land. In one study in which *Spirulina* were administered to malnourished kids with ages between 6 months to 5 years old in the Democratic Republic of Congo, results have shown significant improvement in their health conditions (Matondo et al. 2016). Other extensive exploratory work is done on plant-based protein (Soeder and Pabst 1970) in the 1970s that helped countries accelerate the commercialization path of *Chlorella* and *Spirulina*, as health food supplements in Japan, Taiwan, and Mexico.

Although rapid advancement in food science and technology has been made in the twentieth century, such as genetically modified (GM) food crops, chemical preservatives, artificial flavorings, and colorants, which all make our everyday foodstuff more attractive and enjoyable, the consequential ailments and arisen health-related issues also affect our quality of life. The need for replacement of these unnatural ingredients coupled with meeting of the demand of additional food sources for increasing population provides the ample opportunity for microalgae to become a viable contender. Microalgae have been a recognized source of naturally produced high-value chemicals including carotenoids (Borowitzka 2013), long-chain polyunsaturated fatty acids (PUFAs) (Martins et al. 2013), and food colorant from microalgae (Begum et al. 2015). High-quality nutritional biomass incorporates a comprehensive range of bioactive extracts from microalgae including antioxidant (Sansone and Brunet 2019), antibiotic (Falaise et al. 2016), antiviral, anticancer, anti-inflammatory, antihypotensive, and other activities.

Table 2.1 shows the US Food Labeling Report from an independent food testing lab on the dry biomass from axenic *Euglena* culture produced at Geb Impact. The protein content exceeds 55% of the dry biomass and nearly 10% is lipids. In this

Table 2.1 US Food Labeling Report from a third-party food testing lab on the dry biomass from axenic *Euglena* culture grown at Geb Impact facilities in Hong Kong

Test	Results	Method
Moisture (g/100 g)	2.46	GB 5009.3-2010
Ash (g/100 g)	10.3	In-house method by gravimetric technique
Total dietary fiber (g/100 g)	9.0	AOAC 985.29 (2005)
Protein (g/100 g)	56.3	In-house method by Kjeldahl method
Total fatty acid as triglycerides (g/100 g)	9.2	AOAC 996.06 (2005)
Sat. fat (g/100 g)	4.72	AOAC 996.06 (2005)
Trans. fat (g/100 g)	0.309	AOAC 996.06 (2005)
Sugar (glucose, galactose, fructose, sucrose, maltose, lactose) (g/100 g)	2.6	In-house method by high-performance liquid chromatography-RI technique
Added sugar 3 (g/100 g)	0	Information provided by customer
Sodium (mg/100 g)	233	In-house method by inductively coupled plasma-AES technique
Iron (mg/100 g)	40.3	
Calcium (mg/100 g)	76.4	
Potassium (mg/100 g)	768.0	
Vitamin D (mcg/100 g)	19	
Cholesterol (mg/100 g)	<1	AOAC 994.10, 2005

chapter, various key nutrition ingredients from microalgae are to be analyzed in more detail, comparison being drawn with those of modern food; the focus is given on the microalgae and their derivatives applied for mainstream food application.

2 Content and Quality of Microalgae Protein for Food

Protein is one of the macronutrients required for the human body. They are chains of amino acids linked together by peptide bonds. There are two types of amino acid, namely, the EAAs and non-essential amino acids (NEAAs). According to WHO/FAO/UNU recommendations regarding human's requirements of EAAs (WHO/FAO/UNU 2002), humans must include EAA in their diet since the EAAs cannot be synthesized in the human body.

Historically, microalgae have not been cultivated as a source of food stock. The long-term health impact on human development is less known. Although a plethora of proteins and amino acids are present in various microalgae biomass, their contribution to human health warrants further investigation, in particular, due to the fact the dry biomass of certain strains contains over 50% of their weight as protein, one of the major nutrition sources for humans (Bleakley and Hayes 2017).

In this section, microalgae will be compared with other conventional food for being a source of dietary protein. The content of protein in microalgae will be discussed, and the quality of microalgae protein will be assessed in terms of amino acid profile and bioavailability. At last, the application of microalgae protein other than nutrient source will also be discussed.

Table 2.2 Comparison of protein content from various food origins (Koyande et al. 2019)

Food origin	Protein content (% dry matter)
Beef	17.4
Fish	19.2–20.6
Chicken	19–24
Peanut	26
Wheat germ	27
Parmesan cheese	36
Skimmed milk powder	36
Soybean flour	36
Beer yeast	45
Whole egg	47
Chlorella sp.	50–60
Spirulina sp.	60–70

2.1 Protein Content in Microalgae

Among various food sources, microalgae contain a significant protein content. As shown in Table 2.2, meat such as beef contains about 17% of protein, and plant such as soybean flour contains around 36% of protein in their dry biomass. In comparison, the dry biomass of *Chlorella* sp. and *Spirulina* sp. consists of 70% of protein (Koyande et al. 2019). *Chlorella* and *Arthrospira* accumulate high-quality proteins, both species having well-balanced amino acid profiles. The amino acids of both species are similar to other conventional protein sources such as egg and soybean.

Some of the advantages of microalgae proteins over other currently used protein sources include (1) low requirement for area of land, compared to animal-based proteins (<2.5 m per kg of protein compared to 47–64 m^2 for pork, 42–52 m^2 for chicken, and 144–258 m^2 for beef production); (2) low requirement for area of land, for other plant-based proteins used for food and feed, such as soybean meal, pea protein meal, and others; (3) usage of nonarable land for cultivation; (4) minimal freshwater consumption; and (5) potential replacement of non-sustainable soy-based protein sources. Microalgae can produce high protein yield. Their protein yield is 4–15 tons/Ha/year. This is higher than the protein yield from wheat, pulse legumes, and soybean, which are 1.1 tons/Ha/year, 1–2 tons/Ha/year, and 0.6–1.2 tons/Ha/year, respectively (van Krimpen et al. 2013).

Apart from *Chlorella* and *Spirulina* mentioned above, there are other microalgae species widely used in food production including *Dunaliella terticola*, *Dunaliella* salina, and *Aphanizomenon flos-aquae* due to their high protein content and nutritional value (Soletto et al. 2005). However, the current global market in microalgae food product is being dominated by *Chlorella* and *Spirulina* species, due to their high protein content and nutrient value and, more importantly, ease to grow (Chronakis and Madsen 2011). The composition of the major microalgae species is listed in Table 2.3, where *Spirulina* has a higher protein content than other microalgae species.

2.2 Quality of Microalgae Protein

The protein quality of a certain food can be graded by their content of EAAs and their bioavailability. For humans, there are nine EAAs including phenylalanine, valine, threonine, tryptophan, methionine, leucine, isoleucine, lysine, and histidine. As shown in Table 2.4, microalgae have a comparable profile of EAA content when compared with a typical dietary protein source such as egg and soybean (Becker 2007).

Apart from the amino acid profile, the quality of the protein in microalgae was being studied by their bioavailability. For a nutrient component to be utilized by the human body, it needs to travel from the food matrix into the human body. When the food protein is released into the GI tract, they will first be denatured by the acidic environment in the stomach. Proteolytic enzymes including pepsin from the

Table 2.3 Comparison of composition of various microalgae species (Koyande et al. 2019)

	Composition (% dry matter)		
Microalgae species	Protein	Lipids	Carbohydrates
Anabaena cylindrical	43–56	4–7	25–30
Aphanizomenon flos-aquae	62	3	23
Chaetoceros calcitrans	36	15	27
Chlamydomonas reinhardtii	48	21	17
Chlorella vulgaris	51–58	14–22	12–17
Chlorella pyrenoidosa	57	2	26
Diacronema vlkianum	57	6	32
Dunaliella salina	57	6	32
Dunaliella bioculata	49	8	4
Euglena gracilis	39–61	22–38	14–18
Haematococcus pluvialis	48	15	27
Isochrysis galbana	50–56	12–14	10–17
Porphyridium cruentum	28–39	9–14	40–57
Prymnesium parvum	28–45	22–38	25–33
Scenedesmus obliquus	50–56	12–14	10–17
Scenedesmus dimorphus	8–18	16–40	21–52
Scenedesmus quadricauda	47	1.9	21–52
Spirogyra sp.	6–20	11–21	33–64
Spirulina maxima	60–71	6–7	13–16
Spirulina platensis	46–63	4–9	8–14
Synechococcus sp.	63	11	15
Tetraselmis maculata	52	3	15

stomach, trypsin, and chymotrypsin from pancreas will cleave the food proteins into polypeptides. The polypeptides will then be broken down by peptidases into small peptides and amino acids. Amino acids are transported into enterocyte in the intestine, across the basolateral membrane, and eventually enter the circulation and can be metabolized (van der Wielen et al. 2017).

Bioavailability assesses the digestion in the GI tract, food absorption, metabolism, distribution within the human body via circulation, and bioactivity of the nutrients on the host (Carbonell-Capella et al. 2014). Numerous methods are developed to measure the bioavailability of food protein, including protein efficiency ratio (PER) and biological value (BV) (Becker 2007; Hoffman and Falvo 2004). PER is calculated by measuring the weight gain in rats after a feeding trial of the candidate food, as compared to casein protein. This is to measure how effective the candidate protein can stimulate animal growth (Hoffman and Falvo 2004).

BV is used to measure the incorporation of food protein nitrogen into the nitrogen within the test subject body. The measurement is done by measuring the amount of nitrogen loss via excretion of the animal including urine and feces, in order to predict the amount of nitrogen retention in the subject (Fixsen and Jackson 1932).

Table 2.4 Amino acid profile of microalgae compared with conventional food sources (weight% in protein) (Becker 2007)

Source	Ile	Leu	Val	Lys	Phe	Tyr	Met	Cys	Try	Thr	Ala	Arg	Asp	Glu	Gly	His	Pro	Ser
WHO/FAO	4.0	7.0	5.0	5.5	6.0				1.0									
Egg	6.6	8.8	7.2	5.3	5.8	4.2	3.2	2.3	1.7	5.0	–	6.2	11.0	12.6	4.2	2.4	4.2	6.9
Soybean	5.3	7.7	5.3	6.4	5.0	3.7	1.3	1.9	1.4	4.0	5.0	7.4	1.3	19.0	4.5	2.6	5.3	5.8
Chlorella vulgaris	3.8	8.8	5.5	8.4	5.0	3.4	2.2	1.4	2.1	4.8	7.9	6.4	9.0	11.6	5.8	2.0	4.8	4.1
Dunaliella bardawil	4.2	11.0	5.8	7.0	5.8	3.7	2.3	1.2	0.7	5.4	7.3	7.3	10.4	12.7	5.5	1.8	3.3	4.6
Scenedesmus obliquus	3.6	7.3	6.0	5.6	4.8	3.2	1.5	0.6	0.3	5.1	9.0	7.1	8.4	10.7	7.1	2.1	3.9	3.8
Arthrospira maxima	6.0	8.0	6.5	4.6	4.9	3.9	1.4	0.4	1.4	4.6	6.8	6.5	8.6	12.6	4.8	1.8	3.9	4.2
Spirulina platensis	6.7	9.8	7.1	4.8	5.3	5.3	2.5	0.9	0.3	6.2	9.5	7.3	11.8	10.3	5.7	2.2	4.2	5.1
Aphanizomenon sp.	2.9	5.2	3.2	3.5	2.5	–	0.7	0.2	0.7	3.3	4.7	3.8	4.7	7.8	2.9	0.9	2.9	2.9

Table 2.5 Comparative data on biological value (BV), digestibility coefficient (DC), net protein utilization (NPU), and protein efficiency ratio (PER) of different processed algae (Becker 2007)

Algae	Processing	BV	DC	NPU	PER
Casein		87.8	95.1	83.4	2.50
Egg		94.7	94.2	89.1	–
Scenedesmus obliquus	DD	75.0	88.0	67.3	1.99
Scenedesmus obliquus	DS	72.1	72.5	52.0	1.14
Scenedesmus obliquus	Cooked–SD	71.9	77.1	55.5	1.20
Chlorella sp.	AD	52.9	59.4	31.4	0.84
Chlorella sp.	DD	76.0	88.0	68.0	2.00
Spirulina sp.	SD	77.6	83.9	65.0	1.78
Spirulina sp.	DD	68.0	75.5	52.7	2.10

AD air-dried, *DD* drum dried, *SD* sun dried

As shown in Table 2.5 (Becker 2007), the microalgae being tested have lower bioavailability than conventional protein source. Some microalgae possess a cellulosic cell wall, which is indigestible for humans. The protein from the untreated sample of these strains such as *Chlorella* thus has lower bioavailability than the conventional protein source.

The bioavailability of the microalgae protein can be enhanced by disrupting the cell wall. Physical methods including crushing, grinding, and heating have been reported to enhance the digestibility of the algae protein (Becker 2007; Marrion et al. 2003). Mechanical disruption of the cell wall is also applied in a commercial setting to produce microalgae food supplement with enhanced bioavailability (Nakayama 1991).

On the other hand, microalgae such as *Spirulina* do not possess a cellulosic cell wall and thus have a higher digestibility (Gutiérrez-Salmeán et al. 2015).

2.3 Applications of Microalgae Protein Other than the Nutrient Source

Phycobiliproteins (PBPs) are produced in cyanobacteria including *Spirulina*. Due to its brilliant color, it is used as a natural food dye (Babu et al. 2006). Phycocyanin, a blue PBP isolated from *Spirulina*, is widely used in food including chewing gum and jellies (Santiago-Santos et al. 2004).

Apart from food colorant, the microalgae protein has been proposed with other health benefits, including anticancer effect. Glycoprotein produced by *Chlorella vulgaris* has shown anticancer ability in cell-based assay (Hasegawa et al. 2002). With the notion that intact protein absorption is observed in some condition (Gardner 1988), the potential of dietary microalgae protein being anticancer exists, and more studies are required to confirm.

Besides the intact microalgae protein, the peptides produced by enzymatic hydrolysis of the microalgae protein are reported with anticancer activity (Abd

Table 2.6 Possible bioactivities of peptides from selected algae, adopted with modification (Samarakoon and Jeon 2012)

Marine algae	Possible bioactivity	Proteolytic enzymes, fermenting microorganisms, or others	Bioactivc amino acids or peptide sequences	IC_{50} values[a]
Navicula incerta			Acidic amino acids; Glu-, Asp-, Lys-, Arg-	
	Antioxidative:			
	DPPH	Pepsin		196 μg/mL
	Hydroxyl	α-chymotrypsin		102 μg/mL
	Superoxide	Neutrase		196 μg/mL
Navicula incerta	Hepatic fibrosis inhibitory effect	Papain	Pro-Gly-Trp-Asn-Gln-Trp-Phe-Leu Val-Glu-Val-Leu-Pro-Pro-Ala-Glu-Leu	
Chlorella vulgaris	Antioxidative: superoxide radical	Pepsin	Val-Glu-Cys-Iyr-Gly-Pro-Asn-Arg-Pro-Glu-Phe	7.5 μM
Chlorella vulgaris	ACE inhibitory	Pepsin	Val-Glu-Cys-Iyr-Gly-Pro-Asn-Arg-Pro-Glu-Phe	29.6 μM
Chlorella vulgaris	Anti-proliferation	Pepsin	Val-Glu-Cys-Iyr-Gly-Pro-Asn-Arg-Pro-Glu-Phe	70.7 μM
Chlorella vulgaris	ACE-I inhibitory	Pepsin	Ile-Val-Val-Glu	315.3 μM
			Ala-Phe-Leu	63.8 μM
			Phe-Ala-Leu	26.3 μM
			Ala-Glu-Leu	57.1 μM
			Val-Val-Pro-Pro-Ala	79.5 μM
Spirulina platensis	ACE-I inhibitory	Pepsin	Ile-Ala-Glu	34.7 μM
			Phe-Ala-Leu	26.2 μM
			Ala-Glu-Leu	57.1 μM
			Ile-Ala-Pro-Gly	11.4 μM
			Val-Ala-Phe	35.8 μM

IC_{50} values: the concentration of peptide required to inhibit 50% of the activity

El-Hack et al. 2019). Dolastins are peptides isolated from *Lyngbya* sp. and *Symploca* sp. They showed anticancer activity on a cell-based assay for ovarian cancer and colon cancer (Aherne et al. 1996). Other peptides derived from microalgae are also reported with different bioactivities such as antihypertensive and antioxidative, as shown in Table 2.6 (Samarakoon and Jeon 2012).

Together with delivery tools to transport the protein or the peptide to their target site (Liu et al. 2019), it is possible to develop microalgae-derived peptides for pharmaceutical purposes. More investigation on this topic is merited.

3 Lipids from Microalgae as Part of Human Diet

Lipids, a group of nonpolar organic molecules, are essential macronutrient for all known living organisms including humans. A balanced intake of lipids is important for human body development, energy source, and homeostasis maintenance. Among the numerous types of lipids, fatty acids, as the major components determining the functional properties of lipids, are the basic molecules that can be absorbed via the intestine and assimilated into human biological activity. Some essential fatty acids, especially long-chain PUFAs like omega-3 and omega-6, are not synthesized in human metabolism and are needed exogenously. Microalgae as a primary producer in the aqua environment are the major source of omega-3 and omega-6 oils to all kinds of higher trophic levels including fishes that produce fish oils. PUFAs are critical to maintaining human health in different aspects such as heart disease prevention, brain development, and aging deceleration. Compared to fish oils and plant-based oils for the source of edible oils like PUFAs, oils from microalgae sources have valuable and competitive potentials. To fully exploit this potential of microalgae as an alternative source of PUFAs to meet the demand of the market, microalgae processing technology and relevant equipment need to be further developed.

3.1 Chemical Structures of Lipids

Lipids, including fats and oils in nature, are products of esterification forming a -COOC- (ester) functional group between two organic molecules (Li-Beisson et al. 2016). A typical example such as a triglyceride consists of a glycerol molecule forming three ester bonds with three fatty acids between each OH (hydroxyl) group from glycerol and the single COOH (carboxylic acid) group of fatty acids via a condensation, in which the reaction has three H_2O molecules eliminated in dehydration reaction (H of hydroxyl group and OH of COOH are removed to combine as H_2O) (Fig. 2.4). Free fatty acids are organic compounds with general formula $H_2OCH_3(CH_2)nCOOHCH_3(CH_2)nCOOH$, where "n" can be from 2 to 28 and is always an even number. The carbon chain can be either saturated or unsaturated; the latter one has at least one C=C double bond which causes kink structure in the chain (Cohen et al. 2011).

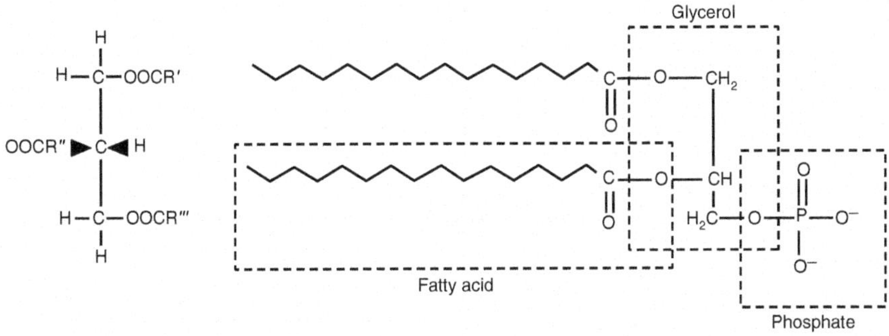

Fig. 2.4 Lipid molecules. Triacylglycerol (NL) on the left. Phospholipid (polar lipid) on the right. R′, R″, and R‴ in the triacylglycerol molecule represent fatty acid chains. Phospholipid molecule is negatively charged (Chen et al. 2018)

3.2 The Functions of Lipids in a Human Body

Lipids are one of the main ingredients in the human diet and contribute to supply energy and essential nutrients such as essential fatty acids (FAs), cholesterol, and lipid-soluble vitamins. Thus, they are critical to maintaining overall health. Depending on individual differences in diet, personal lipid consumption ranges from 50 g up to 150 g/day (Meynier and Genot 2017). As a source of energy, lipid provides 9 kcal per 1 g consumed, which is two times higher than the energy content of protein or carbohydrate (4 kcal/g). Excess fats are stored in the fat tissue, and fatty acids are released for respiration when alternative energy source other than carbohydrate is required.

Equally important as for energy purpose, phospholipids, triglycerides, and cholesterol of lipid members also function as structural building blocks of the membrane of cells that are fundamental units of the body. Besides quantitative lipid intake, the qualitative aspect of lipids that are differentiated by the property of fatty acids is also critical for a healthy state. For example, the carbon chain length and the saturation state of the fatty acids as the main components of phospholipids and triglycerides determine the arrangement of the cell membrane and its fluidity. A shorter chain and higher unsaturated fatty acids promote flexibility of membranes and thereby affect fundamental biological functions such as dynamic of a particle in and out across the cell membrane barrier. Also, many important micronutrients like vitamins A, D, E, and K are insoluble in water and rely on fat as a carrier for its absorption in the intestine, delivery in blood, and involvement in metabolism in a body including growth and development, part of which is illustrated in Fig. 2.5.

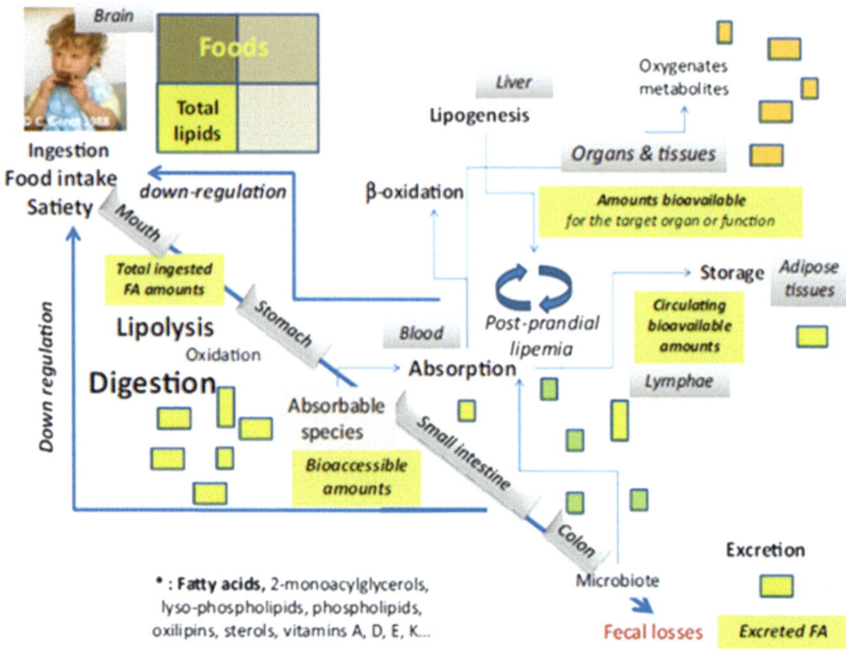

Fig. 2.5 Schematic overview of lipid digestion and metabolism highlighting crucial steps and some regulation pathways (Meynier and Genot 2017)

3.3 Essential Fatty Acids for Humans: PUFAs

Fatty acids can be grouped into saturated, monounsaturated, and PUFAs according to the hydrogen saturation degree of their carbon chain. Two groups of PUFAs, known as omega-3 and omega-6 that differ in the first C=C double bond at carbon atom numbers 3 and 6, are exceptionally important bioactive substances for benefits on human health and growth promotion as our bodies cannot synthesize the PUFAs to meet our own needs.

Types of omega-3 fatty acids include α-linolenic acid (18:3, $n-3$; ALA), eicosapentaenoic acid (20:5, $n-3$; EPA), and docosahexaenoic acid (22:6, $n-3$; DHA) that exceptionally originate from marine oils (Doughman et al. 2007); types of omega-6 fatty acids include linoleic acid (18:2, $n-6$; LA), ɣ-linolenic (18:3, $n-6$; GLA), and arachidonic acid (20:4 $n-6$; ARA) which are involved in human physiology. Evidence has shown that PUFAs take the most important roles in cardiovascular disease prevention (increased "good" HDL cholesterol) (Miller et al. 2011), cancer and type 2 diabetes, blood clotting, hypertension, macular degeneration, rheumatoid arthritis, osteoporosis, anti-inflammation, hormone-like homeostasis (prostaglandins), brain (60% fat and DHA/EPA-rich) development, and eye vision protection (retina is DHA-rich).

Nowadays, an intake of 250 mg to 2 g of EPA and DHA daily is recommended to prevent coronary heart or PUFA-deficient-related aging diseases. On the supply side, seafood such as oily fish including salmon and tuna is used for commercial production of both DHA and EPA. However, fishes do not produce PUFAs themselves; instead, they obtain PUFAs via bioaccumulation from the food chain by preying on small fishes that rely on consuming zooplankton as food rather than directly consuming phytoplankton (*microalgae*) that primarily produce PUFAs (Arbex et al. 2015).

3.4 Advantages of Algae as a Source of Lipid Production

Microalgae are photosynthetic microorganisms in freshwater or marine and contribute to more than 32% of global photosynthesis and nearly half of regeneration of the atmospheric oxygen. Lists of major factors favor using microalgae for oil production, for example, they have higher photosynthetic efficiency; around 1.8 kg carbon dioxide is needed to produce 1 kg biomass (Adamczyk et al. 2016). Limited land is needed thus not to compete with arable land for crop cultivation; free CO_2 from the atmosphere and formulated nitrogen and phosphate are easily and repeatedly utilized anywhere with sunlight and appropriate temperature, thus no reliance on the soil condition as do plants. Microalgae cultivations do not aggravate the greenhouse effect and water pollution (no pesticides needed); saltwater or brackish water can be used depending on species. The quality of microalgal oil is comparable to plant-based oil; high proliferation rate (a couple of days) can bioaccumulate 20–50% lipid content of total biomass depending on the strain selected and cultivation method (Roy and Pal 2015) (Ma et al. 2016). Theoretically, the annual oil yield by selecting an oil-rich microalgal species can be as much as 19,000–57,000 L per acre, which is almost 60 times higher than the plant source oil production of best-performing species; other nutrient-rich by-products can be harvested (Farag and Price 2013).

As a good source of primary producers for oil production, microalgae are credible sources for PUFA production including as a source of omega-3 and omega-6 in all fish, which chemical structure is shown in Fig. 2.6. However, as an intermediate PUFA source, market supply of fish oils will be limited due to the bottleneck caused by overfishing problem in fishery industry. Additionally, chemical or plastic contamination polluting aqua ecosystem's environment also affects the fish oil quality, for example, unacceptable odor of oil. Many evidences support that plant-based omega-3 are as nutritional and therapeutic as the omega-3 in fish oil. Clinical trials of EPA or DHA from specific microalgae indicate similar efficacies from fish oil on promoting human health such as lowering cardiovascular risk by reducing the level of plasma triglycerides and oxidative stress. In addition, studies demonstrate that microalgae oil can have a higher concentration of omega-3 such as DHA than fish oil does (Skulas-Ray et al. 2011). Important minerals from microalgae oil like iodine can also be a good source for human biological need. Consideration of vegan's preference for nonanimal food products as their diet and plant-based food from diverse types of microalgae biomass provides additional options for the sources of

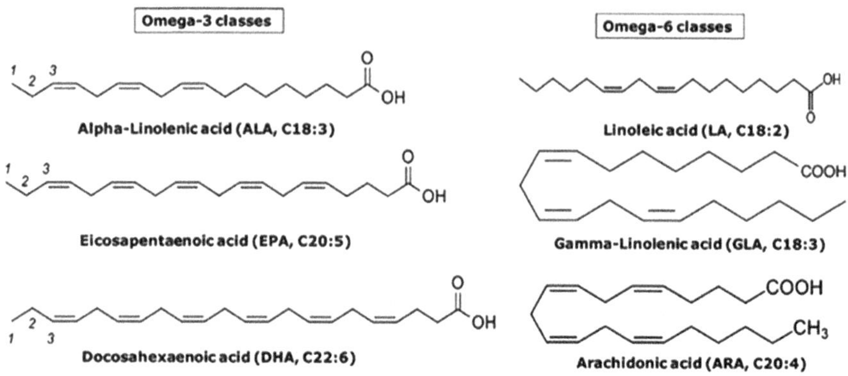

Fig. 2.6 Chemical structures of the most common polyunsaturated fatty acids found in microalgae (Matos 2017)

essential lipids.

Table 2.7 lists lipid contents from various microalgae species together with protein contents and carbohydrate contents. As an important primary photosynthetic source of essential lipids, both more environmentally friendly and sustainable microalgae can be a good alternative or even better source of PUFAs than fish oils and plant-based oil for meeting a growing PUFA market demand.

3.5 *Lipid Contents, Quality of PUFAs, and Commercial Value in Microalgae*

Like their functions in most other cells, lipids are major constituents of microalgae cells for their basic biological functions such as energy storage, signaling, and building blocks of cell membranes. However, due to the high diversity of microalgae, lipid contents of microalgae vary in a wide range as abovementioned. Under specific conditions, the oil content of some species may be as high as 80% of the dry biomass (Roy and Pal 2015) (Table 2.7). For example, the most well-studied lipid producer *Botryococcus braunii* can produce lipids of 75% (w/w) of total cell mass. Not surprisingly, their lipid products are varied in quality with different carbon chain length (C12–C22) of saturated, monounsaturated, polyunsaturated, or even branched fatty acids (Sun et al. 2018). Besides saturated fatty acids (SFAs) such as myristic (C14:0) and palmitic (C16:0) acids or monounsaturated fatty acid (MUFA) like oleic acid (C18:1) for the industrial application, algal oils also contain highly valuable PUFAs that can be potentially used as ingredients in nutraceutical, pharmaceutical, and therapeutic industries. It is worth mentioning that the neutral lipid content of the total lipid in the microalgae biomass also varies among different species (Table 2.8), which can serve different purposes of metabolic processes.

Table 2.7 Protein, carbohydrate, and lipid contents of few microalgae (Roy and Pal 2015)

Algae	Protein (%)	Carbohydrates (%)	Lipid (%)
Spirulina platensis	50–65	8–14	4–9
Chlorella sp.	51–58	12–17	14–22
Scenedesmus sp.	50–56	10–52	12–14
Dunaliella sp.	49–57	4–32	6–8
Synechococcus sp.	63	15	11
Euglena sp.	39–61	14–18	14–20
Prymnesium sp.	28–45	25–33	22–38
Anabaena sp.	48	25–30	4–7
Chlamydomonas sp.	43–56	2.9–17	14–22
Porphyridium sp.	28–39	50–57	
Spirulina maxima	60–71	13–16	6–7
Spirogyra	6–20	33–64	11–21
Tetraselmis	52	15	16–45
Pavlova	24–29	6–9	9–14
Enteromorpha intestinalis	6.15	30.58	7.13
Rhizoclonium riparium	21.09	15.34	3.37
Lola capillaris	40.87	22.32	4.05
Ulva lactuca	8.44	35.27	4.36
Catenella repens	8.42	28.96	5.29
Polysiphonia mollis	16.59	25.81	5.79

Table 2.8 Lipid content of selected microalgae strains (Ma et al. 2016)

Species	Total lipid content (% of DW)	Neutral lipid content (% of total lipid)
Nannochloropsis	37–60	23–58
Isochrysis	25–33	80
Dunaliella salina	23	30
Haematococcus pluvialis	16–35	50–59
Neochloris oleoabundans	2–47	23–73
Phaeodactylum tricornutum	20–30	–
Crypthecodinium cohnii	20	–
Spirulina platensis	7.6–8.2	–
Tetraselmis maculate	8	–
Scenedesmus obliquus	12–14	–

Microalgae are reported to produce edible oil known as single-cell oil (SCO) mostly containing different subtypes of LC-PUFAs (Matos 2017). Omega-3 oils include *C. vulgaris* (ALA producer); *Isochrysis galbana*, *Nannochloropsis oculata*, and *Phaeodactylum tricornutum* (EPA producers); and *Crypthecodinium cohnii* and *Schizochytrium limacinum* (DHA producers). Omega-6 oils include *Arthrospira platensis* (GLA producer) and *Porphyridium cruentum* (ARA producer). However,

Table 2.9 Microalgal long-chain fatty acids useful in food application (Matos 2017)

Fatty acid fraction	Microalgae source	Application of the fatty acid	Daily intake recommendation for humans (mg)
Omega-3			
Alpha-linolenic acid (ALA)	*Chlorella vulgaris*	Nutritional supplement (single-cell oil)	1000–2000
Eicosapentaenoic acid (EPA)	*Nannochloropsis oculata*, *Phaeodactylum tricornutum*, *Monodus subterraneus*, *Isochrysis galbana*	Nutritional supplement, Psychotherapeutic medication, brain development for children, cardiovascular health	250–500
Docosahexaenoic acid (DHA)	*Schizochytrium limacinum*, *Crypthecodinium cohnii*, *Pavlova lutheri*	Food supplement, important for brain and eye development of fetus and children, significant for cardiovascular health, adult dietary supplement in food	250–500
Omega-6			
Gamma-linolenic acid (GLA)	*Arthrospira platensis*	Nutritional supplements, anti-inflammatory, autoimmune diseases	500–750
Arachidonic acid (ARA)	*Porphyridium cruentum*, *Mortierella alpina*, *Parietochloris incisa*	Nutritional supplements, anti-inflammatory, muscle anabolic formulations (bodybuilder)	50–250

the ratio of specific PUFA production differs depending on the microalgae species of interest. For example, *C. cohnii* exclusively produces DHA, *Nannochloropsis oculata* and *Phaeodactylum tricornutum* contain mainly EPA, and *Porphyridium cruentum* contains predominantly ARA (Doughman et al. 2007).

The Food and Agriculture Organization of the United Nations suggests a daily intake of 0.25–0.5 g omega-3 PUFAs for human nutrition. Table 2.9 shows microalgal long-chain fatty acids useful in food application with daily recommended dosage for humans of omega-3 and omega-6. In 2014, the market consumption of omega-3 PUFAs worldwide was around 0.13 million metric tons and valued at 2.5 billion USD (Matos 2017). The demand in 2020 is believed to double with a market value of around 5 billion USD. In the PUFA market, EPA is not only used in fishery industry such as salmon farming and needed for juvenile development, but the EPA-rich oils are also combined with DHA for infant formulation or nutritional supplements that are critical for structural maintenance of the human brain and eye. The inclusion of DHA into an infant formula is becoming popular and recommended by nutrition organizations. DHA functions similarly as EPA in body development, and its anti-inflammatory effects are also required for the development of human fetus and production of healthy breast milk. For example, one of the key leading microalgae companies in this field, Martek Biosciences Corporation, has developed a patent

Table 2.10 Long-chain polyunsaturated fatty acids of some nontoxic algae versus fish oil supplements (Doughman et al. 2007)

Kingdom	Phyla	Species name (common)	DHA	EPA	AA
Plantae	Chlorophyceae	(Green algae)	0%	2.5%	0.1%
Plantae	Rhodophyta	(Red algae)	0%	20%	4%
Chromista	Heterokonteae	(Yellow green)	0.5%	27%	6%
Chromista	Heterokonteae	*Schizochytrium*	37.4%	2.8%	1.0%

to produce DHA-rich oil from *C. cohnii* and *Schizochytrium* sp. for food and pharmaceutical purposes.

For the omega-6 lipid market, Exsymol is a major producer of cosmetic product that suppresses antiaging of skin based on GLA lipids extracted from *Spirulina*. GLA lipid is known to be a precursor for C20 eicosanoids involved in synthesizing prostaglandin. Its functions in a human body are also associated with positive health effects including anti-inflammatory effects, boost of immune system, and diabetes control. ARA, another omega-6 lipid, is an important component of cell membrane especially needed in the development of brain, muscles, and liver. In addition, it is also involved in vasodilation and anti-inflammatory processes. It is recommended to take 50–250 mg daily as a supplement for maintaining a healthy state of adults. Not just as a food element, ARA can be used as active product extracted from red microalgae such as *Porphyridium cruentum* for commercial cosmetic product (Exsymol), which is believed to benefit the skin such as resistance to extreme conditions (Matos 2017).

Long-chain polyunsaturated fatty acids from microalgae have been compared with that from fish oil as the fishing industry has been strained for various environmental concerns (Table 2.10). It is suggested that PUFA from microalgae can be potential replacement of that from fish oil (Doughman et al. 2007).

3.6 Extraction and Limitations of Microalgae Oil

To meet the demand for microalgae-related products, cultivation of microalgae and scaled-up production require a series of scientific-/engineer-based knowledge since microalgae are highly diverse group with more than hundreds of thousands of species. Selection of appropriate microalgae to produce economically feasible production of oil is a must. The high varieties in species or even strains imply cultivation technique needs to be adjusted anytime to fit the production of the microalgae of specific interest. For microalgae oil as part of human diet, the issue of phycotoxin contaminations is a major concern to be addressed. For example, *Spirulina* biomass can be polluted with microcystins that are produced by another cyanobacteria *Microcystis* growing in the same environment. Microalgae dinoflagellates including *Alexandrium* and *Karenia* are neurotoxin producer of saxitoxin and brevetoxin (Matos 2017). Also, once the biomass is harvested, the extraction of microalgae oil

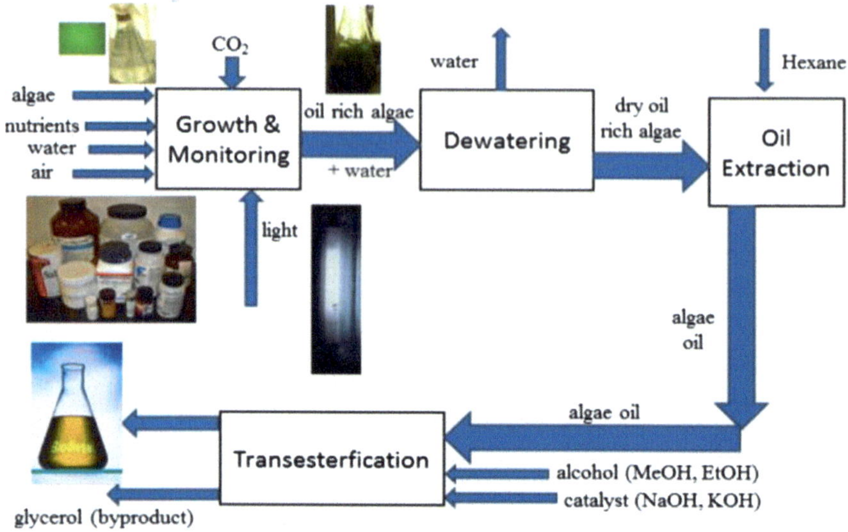

Fig. 2.7 Process diagram of producing biodiesel from algae (Farag and Price 2013)

is another challenge to be overcome to fully take advantage of its benefits. To extract microalgae oil, different strategies are developed to fit various types of microalgae, to mechanically disrupt the physical cell wall barriers with methods including bead beating, enzymatic disruption, chemical method, ultrasonication, microwave, etc. (Chen et al. 2018). Subsequently, extraction can be done with a nonpolar solvent like hexane or chloroform combining with ethyl ester, water, and methanol (Farag and Price 2013). Recently, non-solvent extractions like using supercritical fluid extraction (carbon dioxide) have been developed but costs are high. Additional preservation step is needed to prevent isolated PUFAs from being oxidized in a chain reaction, which could produce undesirable, devalued, highly odorous oil. The extracted oil needs to be further processed to guarantee no solvent remained or any heavy metal contamination. In general, the steps involved for oil extraction are time-consuming and include high operation cost such as higher power consumption (Fig. 2.7).

4 Vitamins

Vitamins are precursors of enzyme cofactors that drive essential biosynthesis in the human body. Humans essentially need 13 different vitamins to stay healthy. Yet, the human body can only endogenously synthesize two of them (vitamin D and niacin). This suggests that humans must obtain the rest of the 11 vitamins from the diet. The 11 vitamins include vitamin A, vitamin B (thiamine, riboflavin, pantothenic acid, pyridoxine, biotin, folic acid, and cobalamin), vitamin C, vitamin E, and vitamin

Table 2.11 Vitamin content of four marine microalgae (Fabregas and Herrero 1990)

	T. suecica	*I. galbana*	*D. tertiolecta*	*C. stigmatophora*
Vitamin A	493,750	127,500	137,500	82.30
Tocopherol (E)	421.8	58.2	116.3	669.0
Thiamine (B1)	32.3	14.0	29.0	14.6
Riboflavin (B2)	19.1	30	31.2	19.6
Pyridoxine (B6)	2.8	1.8	2.2	1.9
Cobalamin (B12)	0.5	0.6	0.7	0.6
Folic acid	3.0	3.0	4.8	3.1
Pantothenic acid	37.7	9.1	13.2	21.4

K. Since microalgae contain all these 11 vitamins, it can be considered as a source of vitamin that can help fulfill our daily needs and combat vitamin deficiency in regions or diets where access to vitamins is limited. Table 2.11 shows 8 of 11 essential vitamins from microalgae. It is to be noted that the fat-soluble vitamins extracted from microalgae need to be ingested together with food rich in lipid to facilitate their absorption into the body.

This section aims at addressing the following questions: (1) Which species of microalgae is rich in a specific type of vitamin? (2) How is the vitamin synthesized/ obtained by the microalgae? (3) Is the vitamin present bioavailable to humans? (4) How cultivation conditions and harvesting stage affect the level of the vitamin in microalgae?

4.1 Vitamin A

Microalgae cannot synthesize vitamin A. However, they can synthesize provitamin A carotene, the precursor of vitamin A. For *Spirulina* sp., 1 g of it contains 0.9 mg of all-trans β-carotene. Studies have shown that *Spirulina* β-carotene is bioavailable, with a conversion factor to vitamin A of 4.5 to 1 in adults (Tang and Suter 2011). It has also been proven that regular intake of *Spirulina* sp. as a supplement can increase serum retinol (vitamin A1) level (Tang and Suter 2011). Retinol is needed for the prevention of night blindness and can possibly suppress tumor growth. For *Dunaliella* sp., 13.8% of its dry weight is β-carotene which is much higher than that of *Spirulina* sp. (Tang and Suter 2011). Yet, the current use of *Dunaliella* sp. focuses mainly on food coloring. Studies revealed that the cultivation condition and harvesting stage affect the provitamin A content of algae. At least for *Nannochloropsis* sp., algae grown under 12:12 L:D (light/dark cycle) contain double the amount of β-carotene when compared to that grown under 24:0 h L:D, and algae harvested at log phase contain only half of the β-carotene content of that harvested at stationary phase (Table 2.12) (Brown et al. 1999). On the other hand, urea serves as the best nitrogen source in the culture medium if higher algal β-carotene content is desired (Abalde et al. 1991).

Table 2.12 Vitamin content of three microalgae grown under 12:12 L:D and harvested at the log phase as well as that of *Nannochloropsis* sp. grown under different lighting conditions and harvested at different stages (Brown et al. 1999)

Vitamin	Units	*Tetraselmis* sp. *CS-362*	*Pavlova pinguis*	*Stichococcus* sp. *CS-92*	*Nannochloropsis* sp. *CS-246*			LSD
					12:12, log	12:12, stationary	24:0, log	
β-Carotene	mg/g	1.05 ± 0.03	0.6 ± 0.03	0.37 ± 0.03	0.50 ± 0.01	1.1 ± 0.20	0.29 ± 0.04	0.07
α-Tocopherol (E)	mg/g	0.07 ± 0.005	0.14	0.16 ± 0.01	0.29	0.18	0.35 ± 0.02	0.02
Thiamine (B1)	μg/g	109 ± 18	36 ± 5	29 ± 2	70 ± 8	51 ± 6	70 ± 10	14
Riboflavin (B2)	μg/g	26 ± 4	50 ± 10	25	25 ± 3	48 ± 3	62 ± 2	7
Folates	μg/g	20 ± 2	23	24 ± 3	17	26	18 ± 8	n.s.
Pyridoxine (B6)	μg/g	5.8 ± 0.4	8.4	17	3.6	6.0	9.5 ± 0. 5	0.5
Cobalamin (B12)	μg/g	1.95 ± 0.05	1.7	1.95 ± 0.05	1.7	1.0	0.85 ± 0.35	n.s.
Biotin	μg/g	1.3 ± 0.7	1.9	1.3	1.1	1.0	0.95 ± 0.35	n.s.

4.2 Vitamin B

4.2.1 Vitamin B1 (Thiamine)

Tetraselmis sp. is rich in vitamin B1 (Table 2.12). The amount of vitamin B1 is much more than that found in conventional food considered to be rich in thiamine such as carrot and cow liver (Table 2.13). This makes microalgae an attractive source of thiamine supplement. Vitamin B1 is involved in glucose metabolism and the maintenance of proper nerve function.

Not all microalgae can endogenously synthesize thiamine. Only microalgae like *C. reinhardtii* which contains all the genes involved in thiamine metabolic pathway can synthesize thiamine endogenously (Croft et al. 2006). Synthesis of thiamine in algae begins with the synthesis of thiazole ring and pyrimidine ring separately, followed by the connection of the two parts by a methylene bridge (Fig. 2.8). For thiamine auxotrophs, thiamine cannot be synthesized since some of the genes involved in the abovementioned biosynthetic pathway are lost. In this case, thiamine needs to be obtained from external sources like symbiosis or scavenging (Tandon et al. 2017). In the case of symbiosis, the algae coexist with bacteria (e.g., *E. coli*) with the bacteria providing the algae with thiamine and the algae providing photosynthate to the bacteria (Tandon et al. 2017). In other words, when culturing microalgae, one needs to determine if the species under cultivation is thiamine-auxotrophic or not. If the species is thiamine-auxotrophic, extra component needs to be added into the culture medium. It is interesting to note this "extra component" may not necessarily be thiamine due to differences in specificity on thiamine requirement by different algal species (Tandon et al. 2017). For instance, some species can make use of thiazole ring or pyrimidine ring of thiamine to fulfill its thiamine requirement, while some like *Emiliania huxleyi* can make use of 4-amino-5-hydroxymethyl-2-methylpyrimidine (HMP), a thiamine analog to do so (Tandon et al. 2017).

Table 2.13 Vitamin content of conventional food which presents maximum value for some vitamins (Fabregas and Herrero 1990)

	Orange	Carrot	Soy flour	Cow liver
Vitamin A[a]	14,728	175,438–1,052,631	1538	659,793
Tocopherol[b]	17.82	39.47		34.36
Thiamine[b]	6.20	11.40	8.46	9.27
Riboflavin[b]	2.32	5.26	3.07	96.21
Pyridoxine[b]	9.30	16.66	6.12	34.36
Cobalamin[b]	0	0	0	1.03
Folic acid[b]	2.66	1.48		1.71
Pantothenic acid[b]	15.50	70.17	19.35	171.82
Biotin[b]	0.07	0.06		6.87

[a]I.U./kg dry weight
[b]mg/kg dry weight

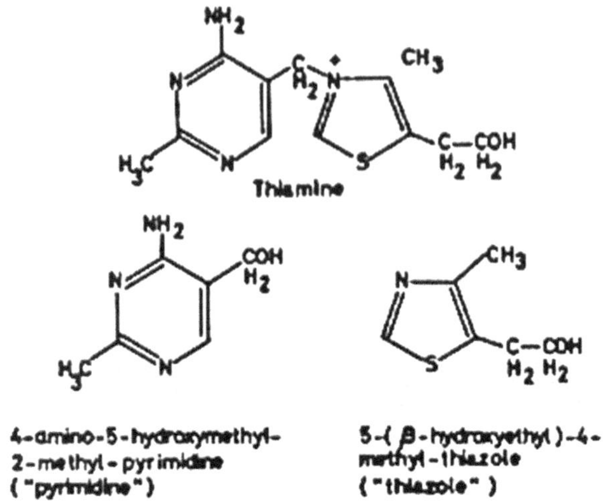

Fig. 2.8 Structure of thiamine (Feenstra n.d.)

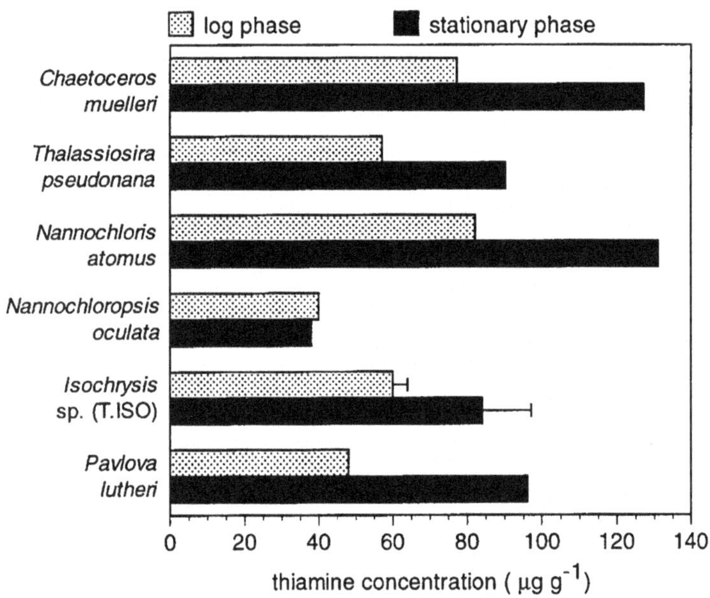

Fig. 2.9 Content of thiamine in the log and stationary phase of six microalgae (Brown et al. 1999)

Studies revealed that for most microalgae, the thiamine content of microalgae harvested at stationary phase is much more than that harvested at log phase (Fig. 2.9). However, lighting condition during cultivation (i.e., 12:12 L:D vs 24:0 h L:D) does not affect the thiamine content of microalgae (Brown et al. 1999) (Table 2.14).

Table 2.14 Suggested daily value of vitamins (SDVV) according to FDA (The Food and Drug Administration 2019)

Vitamin	What it does	Where is it found	Daily value[a]
Biotin	– Energy storage. – Protein, carbohydrate, and fat metabolism.	– Avocados. – Cauliflower. – Eggs. – Fruits (e.g., raspberries). – Liver. – Pork. – Salmon. – Whole grains .	300 mcg
Folate/folic acid *(important for pregnant women and women capable of becoming pregnant)*	– Prevention of birth defects. – Protein metabolism. – Red blood cell formation.	– Asparagus. – Avocado. – Beans and peas. – Enriched grain products (e.g., bread, cereal, pasta, rice). – Green leafy vegetables (e.g., spinach). – Orange juice.	400 mcg
Pantothenic acid	– Conversion of food into energy. – Fat metabolism. – Hormone production. – Nervous system function. – Red blood cell formation.	– Avocados. – Beans and peas. – Broccoli. – Eggs. – Milk. – Mushrooms. – Poultry. – Seafood. – Sweet potatoes. – Whole grains. – Yogurt .	10 mg
Riboflavin	– Conversion of food into energy. – Growth and development. – Red blood cell formation.	– Eggs. – Enriched grain products (e.g., bread, cereal, pasta, rice). – Meats. – Milk. – Mushrooms. – Poultry. – Seafood (e.g., oysters). – Spinach	1.7 mg
Thiamine	– Conversion of food into energy. – Nervous system function.	– Beans and peas – Enriched grain products (e.g., bread, cereal, pasta, rice). – Nuts. – Pork. – Sunflower seeds. – Whole grains	1.5 mg

(continued)

Table 2.14 (continued)

Vitamin	What it does	Where is it found	Daily value[a]
Vitamin A	– Growth and development. – Immune function. – Reproduction. – Red blood cell formation. – Skin and bone formation. – Vision.	– Cantaloupe. – Carrots. – Dairy products. – Eggs. – Fortified cereals. – Green leafy vegetables (e.g., spinach and broccoli). – Pumpkin. – Red peppers. – Sweet potatoes.	5000 IU
Vitamin B6	– Immune function. – Nervous system function. – Protein, carbohydrate, and fat metabolism. – Red blood cell formation.	– Chickpeas. – Fruits (other than citrus). – Potatoes. – Salmon. – Tuna.	2 mg
Vitamin B12	– Conversion of food into energy. – Nervous system function. – Red blood cell formation .	– Dairy products. – Eggs. – Fortified cereals. – Meats. – Poultry. – Seafood (e.g., clams, trout, salmon, haddock, tuna).	6 mcg
Vitamin D nutrient of concern for most Americans	– Blood pressure regulation. – Bone growth. – Calcium balance. – Hormone production. – Immune function. – Nervous system function.	– Eggs. – Fish. – Fish liver oil. – Fortified cereals. – Fortified dairy products. – Fortified margarine. – Fortified orange juice. – Fortified soy beverages (soymilk) .	400 IU
Vitamin E	– Antioxidant. – Formation of blood vessels. – Immune function.	– Fortified cereals and juices. – Green vegetables (e.g., spinach and broccoli). – Nuts and seeds. – Peanuts and peanut butter. – Vegetable oils.	30 IU
Vitamin K	– Blood clotting. – Strong bones.	– Green vegetables (e.g., broccoli, kale, spinach, turnip greens, collards, Swiss chard, mustard greens).	80 mcg

[a]The Daily Values are the amounts of nutrients recommended per day for Americans 4 years of age or older

4.2.2 Vitamin B2 (Riboflavin)

Riboflavin is involved in the catabolism of carbohydrates, proteins, and fats to supply energy to our body. When compared with other microalgae, *Pavlova pinguis* is significantly rich in riboflavin, containing 50 µg of vitamin B2 per gram (Table 2.12). Studies revealed that different cultivation conditions affect the amount of riboflavin present in microalgae. At least for *Nannochloropsis* sp., the algae cultured under 24:0 h L:D contain 2.5-fold more riboflavin than that cultured under 12:12 L:D. Harvest stage affects the riboflavin content of microalgae as well. In summary for *Nannochloropsis* sp., the algae harvested at log phase contain only half of the riboflavin content of that harvested at stationary phase (Brown et al. 1999).

4.2.3 Vitamin B5 (Pantothenic Acid)

T. suecica is rich in pantothenic acid when compared with other microalgae, containing 37.7 mg per kg dry weight of vitamin B5 (Table 2.11). This is much higher than that of some conventional food considered to be rich in vitamin B5 such as oats, cheese, salmon, and soybeans. Vitamin B5 is involved in the production of blood cells and, together with other vitamin B, produces energy for the body from food (Table 2.15).

Studies revealed that pantothenic acid content of microalgae decreases continuously throughout cultivation. Therefore, algae should be harvested at an earlier stage if a higher content of algal pantothenic acid is desired (Pratt and Johnson 1966).

4.2.4 Vitamin B6 (Pyridoxine)

Stichococcus sp., when compared with other microalgae, is rich in pyridoxine. It contains 17 µg of vitamin B6 per gram (Table 2.12). Vitamin B6 is needed for the synthesis of neurotransmitters and myelin and is, therefore, important for the nervous system. Studies revealed that *Nannochloropsis* sp. grown at 24:0 L:D has a higher (more than twofold) pyridoxine content than that grown at 12:12 L:D (Brown et al. 1999).

Table 2.15 Pantothenic acid content of *Chlorella* sp. (Pratt and Johnson 1966)

Item	Pantothenic acid
C. vulgaris, dried 5-day harvest	2779
C. vulgaris, dried 7-day harvest	1318
C. vulgaris, dried 21-day harvest	309
C. pyrenoidosa, dried 5-day harvest	1609
C. pyrenoidosa, dried 7-day harvest	920
C. pyrenoidosa, dried 21-day harvest	316

4.2.5 Vitamin B7 (Biotin)

Biotin is needed to produce energy from food and is important for the healthy growth of hair, nails, and skin. In the algal kingdom, only a small proportion of microalgae are biotin auxotrophs (e.g., *Dictyostelium discoideum* and *Entamoeba histolytica*). All of them have a complex plasmid and are also either auxotrophic to cobalamin/thiamine or both. Biotin auxotrophy is caused by the loss of a single gene in the biotin biosynthetic pathway; however, the lost gene is different from auxotrophs to auxotrophs. For biotin auxotrophs, biotin needs to be obtained from external sources, mostly through the engulfment of bacteria by phagocytosis (Croft et al. 2006). On the other hand, studies revealed that different lighting conditions (24:0 L:D and 12:12 h L:D) have no effect on the biotin content of *Nannochloropsis* sp. (Brown et al. 1999).

4.2.6 Vitamin B9 (Folate)

In terms of mg per kg dry weight, microalgae generally contain more folate than oranges, a fruit conventionally considered to be vitamin B9 rich (Tables 2.11, 2.12, and 2.13). Folate is needed to produce new cells and maintain DNA stability. Study revealed that unlike plants, different algal species make use of the same isoform of a gene in regulating the same step in the biosynthesis of folate. Moreover, in algae, the cellular localization of the enzymes involved in the biosynthesis of folate is fluctuating among different species and is, therefore, less predictable than that in plants (Gorelova et al. 2019). On the other hand, different lighting conditions (24:0 L:D and 12:12 h L:D) do not have an effect on the folate content of *Nannochloropsis* sp. (Brown et al. 1999).

4.2.7 Vitamin B12 (Cobalamin)

It is known that vitamin B12 is needed for DNA repairing and can reduce the chance of having breast cancer. Microalgae serves as an attractive B12 supplement to vegetarian as plants do not synthesize B12. Certain algae species (e.g., *Spirulina* sp., *Chlorella* sp., and *Pleurochrysis carterae*) are rich in cobalamin. However, the cobalamin present is not necessarily produced by the algae itself. The ability to endogenously synthesize cobalamin mainly depends on whether the methionine synthase of the algae is cobalamin dependent or not (Croft et al. 2005). Interestingly, there are some algae (e.g., *C. reinhardtii*) that possess both cobalamin-dependent and cobalamin-independent methionine synthase (Croft et al. 2005). In this case, it is found that the algae will be a cobalamin auxotroph; it will use cobalamin-dependent methionine synthase in the presence of exogenous vitamin B12 and use cobalamin-independent methionine synthase in the absence of exogenous vitamin B12. Of course, the presence of other cobalamin-dependent enzymes (e.g., vitamin B12-dependent ribonucleotide reductase) further determines whether the algae are a cobalamin auxotroph or not (Croft et al. 2005). In nature, the external cobalamin source of cobalamin-auxotrophic algae is cobalamin-prototrophic bacteria. Studies reveal that in such a case, the algae and bacteria coexist in a mutualistic relationship; bacteria supply cobalamin to algae, while algae supply organic carbon source (generated from photosynthesis) to bacteria to support their growth (Croft et al. 2005). However, the cobalamin from bacteria is in a form called pseudo-cobalamin which is not bioavailable to algae. The differences between B12 and pseudo-B12 is that B12 has 5,6-dimethylbenzimid azole (DMB) as the lower axial ligand, while pseudo-B12 has an adenine group as the ligand; in the presence of pseudo-B12 and DMB, auxotrophic algae can convert pseudo-B12 to B12 via a process called microalgal remodeling, making the cobalamin bioavailable to themselves (Fig. 2.10). In other words, when producing axenic cultures of algae, one needs to consider whether the algae are a cobalamin auxotroph or not. If the algae are a cobalamin auxotroph, vitamin B12 has to be added into the medium to support the growth of the algae.

Previous studies revealed that the bioavailability of algal B12 to humans varies (Watanabe et al. 2002). Cobalamin of *Spirulina* sp. is not bioavailable because it exists in the form of pseudo-cobalamin. Intrinsic factor (the enzyme responsible for the absorption of cobalamin in the ileum) in human strictly recognizes the structure of cobalamin (Watanabe et al. 2002). Its affinity to pseudo-cobalamin is 79–87% lower than that to authentic cobalamin; therefore, most ingested pseudo-cobalamin will eventually be excreted in the urine (Watanabe et al. 2002). Currently, there is still controversy over whether absorbed pseudo-cobalamin is an antagonist of authentic cobalamin and blocks B12 metabolism or not. Yet, for safety sake, when culturing *Spirulina* sp. for the use as food ingredients, it is advised to culture it in the absence of cobalt. This is because cobalamin is a cobalt-containing tetrapyrrole, and studies have revealed that the absence of cobalt in culture medium can significantly reduce the pseudo-cobalamin content in *Spirulina* sp. (Watanabe et al. 2002). On the other hand, cobalamin of *Chlorella* sp. and *Pleurochrysis carterae* is bioavailable because it exists in the form of authentic B12 (Watanabe et al. 2002).

4.3 Vitamin E (Tocopherol)

Vitamin E is an antioxidant and is also involved in maintaining the integrity of cell membranes. *Euglena gracilis* Z is microalgae rich in vitamin E, with 97% of the tocopherol it contains in the form of α-tocopherol, the form of vitamin E with the highest physiological activity. Studies have been carried out to find out how the vitamin E content of *Euglena gracilis* Z can be maximized. Results of the studies concluded that vitamin E content of *Euglena gracilis* Z can be maximized by cultivating the algae in α-tocopherol production medium (which contains 2% glucose, 1.2% peptone, inorganic salts, thiamine, and cyanocobalamin) at pH 5 with the addition of L-tyrosine, homogentisate, ethanol, and peptone at the fifth day of cultivation (Table 2.16). L-Tyrosine increases the tocopherol content because it shares the same biosynthetic pathway with tocopherol. Therefore, its direct addition into

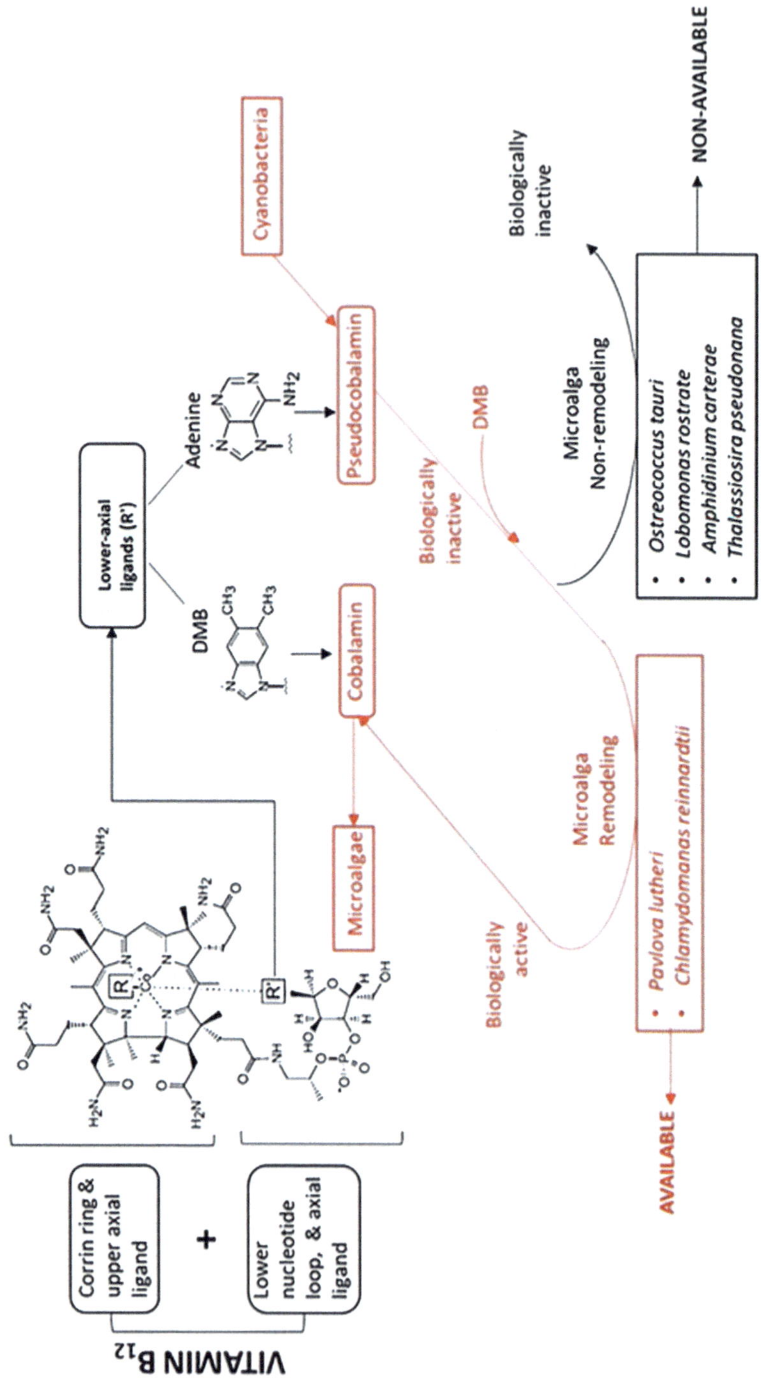

Fig. 2.10 Microalgal remodeling (Tandon et al. 2017)

Table 2.16 The amount of vitamin E produced by *Euglena gracilis Z* under different conditions (Tani 1989)

Medium	Cell yield (g/L)	α-Tocopherol produced (mg/L)	(mg/g dry cells)
KH medium	7.0	2.0	0.3
α-Tocopherol production medium	9.0	9.9	1.1
Added L-tyrosine	10.7	14.8	1.3
Added homogentisate	11.9	17.0	1.4
Added L-tyrosine, homogentisate, and ethanol	11.4	49.8	4.3
Fed L-tyrosine, homogentisate, ethanol, and peptone	28.4	143.6	5.1

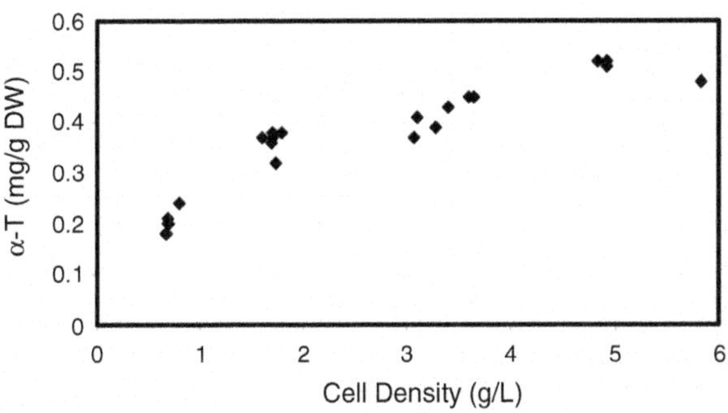

Fig. 2.11 The effect of decreasing light availability per cell on α-tocopherol content of *D. tertiolecta* (Carballo-Cárdenas et al. 2003)

the medium will, in turn, reduce its own biosynthesis and promote the biosynthesis of vitamin E. As for homogentisate, it increases the production of tocopherol because it is a precursor of tocopherol (Tani 1989). The tocopherol content can further be enhanced by increasing the light intensity. Such an act can induce the production of vitamin E to rescue oxidative damages of the thylakoid membrane (Carballo-Cárdenas et al. 2003). Oxygen starvation and low temperature also play a role in increasing the vitamin E content of *Euglena gracilis Z* (Abalde et al. 1991).

The vitamin E content of *D. tertiolecta* and *T. suecica* is comparable to that of *Euglena gracilis Z.* Therefore, studies have also been carried out to enhance the vitamin E content in these two species of algae. It is found that for *D. tertiolecta*, a decrease in light availability per cell increases the tocopherol content of the algae (Fig. 2.11). This is because a decrease in light availability occurs as the culture grows. Therefore, higher production of vitamin E is needed to deal with oxidative processes such as cell senescence which occurs as the culture grows (Carballo-Cárdenas et al. 2003). Moreover, nitrate serves as the best source of nitrogen if a higher algal tocopherol content is desired (Abalde et al. 1991). As for *T. suecica*, in case nutrient supply is sufficient, changes in light availability per cell have no effect on the vitamin E content of the algae (Fig. 2.12). However, vitamin E content can be enhanced by the addition of nitrate and phosphate (Carballo-Cárdenas et al. 2003).

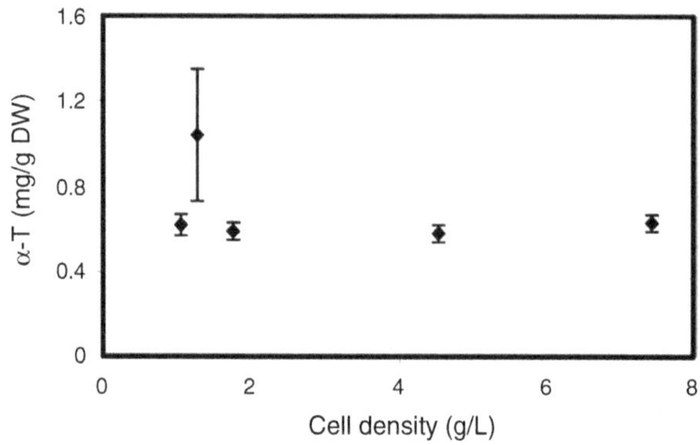

Fig. 2.12 The effect of increasing cell density (decreasing light availability per cell) on α-tocopherol content of *T. suecica* when nutrient supply is sufficient (Carballo-Cárdenas et al. 2003)

4.4 Vitamin K1, Phylloquinone

Anabaena cylindrica is rich in vitamin K1, containing 200 μg of vitamin K1 per g of dry weight, three times SDVV of vitamin K (Table 2.14). This is also much higher than the vitamin K1 content of conventional food considered to be rich in vitamin K1 such as spinach and parsley. Vitamin K1 can help prevent osteoporosis and cardiovascular disease. There are several advantages of using microalgae instead of chemical methods to produce vitamin K1. First, microalgae only produce the bioavailable *E*-isomer of phylloquinone, whereas chemical methods produce a mixture of *E*-isomer and non-bioavailable *Z*-isomer. Moreover, the production of vitamin K1 using microalgae does not require extreme temperature and pressure and is more sustainable. There are also advantages of using microalgae instead of plants to produce phylloquinone—higher growth rate, simpler genome, and easier screening (Tarento et al. 2018).

Studies revealed that the vitamin K1 productivity of *Anabaena cylindrica* can be increased by fourfold by increasing light intensity and nitrate concentration in the culture medium (Fig. 2.13). Nitrate works by increasing the concentration of PSI which uses phylloquinone for electron transfer (Tarento et al. 2018).

5 Algae and Its Extract as Food Colorant

Generally, colorants are added to food and beverages to make them look more appealing and attractive. Colorants can be natural or synthetic. Synthetic colorants are mainly derivatives of coal tar which are obtained during coal distillation. Due to safety fears, many of the synthetic colorants are banned in many countries. Some of these synthetic dyes contain lead acetate which is toxic to the nervous system, while

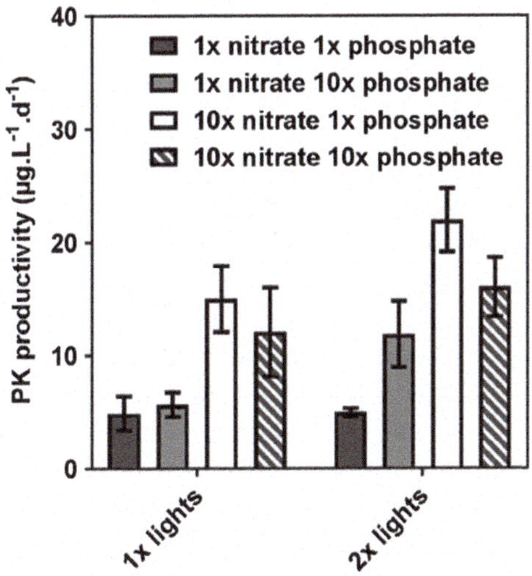

Fig. 2.13 Phylloquinone productivity of *Anabaena cylindrica* under different conditions (Tarento et al. 2018)

(a) (b)

Fig. 2.14 (**a**) Chinese rice cakes with algae pigments. (**b**) Cocktail drink with microalgae as colorant. (Photo Courtesy of Ceco International)

some other synthetic colors are allergens and irritants, and some others are known carcinogens. Therefore, there is an increasing demand for natural colors for use in food products. Figure 2.14 shows Chinese rice cakes and cocktail drink with microalgae and extracts for the color enhancement. Although there are wide selections of plant extracts serving as natural food colorants, pigments such as carotenoids and PBPs from microalgae are good alternatives for natural colors, providing unique nutritional properties such as antioxidant and protein. In addition, using microalgae for the production of colors has several advantages, including controllable production, easier extraction, higher yields, no lack of raw materials, and no seasonal variations.

5.1 Carotenoids

Carotenoids: In 2010, the market for carotenoids is valued at US$1.2 billion (data from industry and trade sources). A significant portion of the output is chemically synthesized. Only two compounds, β-carotene and astaxanthin, represent a major part of the natural production of carotenoids. In the 1980s, the commercial world started production of β-carotene from microalgae in open ponds. *Dunaliella bardawil* was selected for its high β-carotene content, and growth process naturally precludes natural contamination. The market price for the natural version at the time supports the commercialization of the compound.

Carotenoids function as accessory, light-harvesting pigments and protect the photosynthetic apparatus against photodamage (Ben-Amotz et al. 1987). It is the largest group of green algae. Carotenoids are richly colored molecules consisting of a class of more than 600 naturally occurring organic pigments synthesized by plants. Most of the natural carotenoids have more cis-β-carotene (synthetic carotenoids are dominated by trans-β-carotene). Xanthophyll, astaxanthin, has many applications in nutraceuticals and food and feed industries. Species like *Chlorella*, *Chlamydomonas*, *Dunaliella*, and *Haematococcus* produce carotenoids as part of their biomass and are rich sources of carotenoids, while *H. pluvialis* represents the richest biological source of carotenoids (BGG 2016).

Dunaliella is another important source for the production of β-carotene. There are several advantages of using *Dunaliella* to obtain β-carotene. This includes increased absorption by the human body, high efficiency, and isomeric composition, and it can be produced up to 14% of dry weight of the biomass rapidly (Metting 1996). Factors which increase the production of carotenoids include high salinity, photosynthetic photon flux density (PPFD; >500 μmol/s) (Brown et al. 1997), low growth temperature, far-red light, and nutrient limitation.

Apart from high PPFD, all other factors lead to an increase of carotenoids with a reduction in the growth rate. It can be said that there is an inverse relationship that exists between β-carotene and the specific growth rate. Starvation of nitrate and sulfate resulted in the accumulation of β-carotene in the wild-type *Dunaliella* sp. (Becker 2004); carotene/Chl ratio of *D. salina* increased by 33 times under nitrogen starvation conditions in aqueous two-phase systems.

Besides *Dunaliella*, *Chlorella zofingiensis* and *Muriellopsis* are also suitable candidates for carotenoid production. At present, carotenoid production from green microalgae refers only to astaxanthin and β-carotene from *H. pluvialis* and *D. salina*. Besides these, cyanobacteria *Synechocystis* sp. PCC6803 and *Synechococcus* sp. PCC702 have also been considered as suitable organisms for genetic modification with the aim of enhancing β-carotene accumulation (Vernass 2004). Despite all the advantages, there are still several disadvantages of using microalgae as a source of natural colorant. These disadvantages include inadequate process control, low efficiency, high CO_2 consumption, contamination issues, and the requirement of high amounts of salt, water, and solar radiation required for optimal production.

To overcome these inadequacies, alternative strategies of operating systems or culturing parameters are needed. Such strategies involve using extensive open ponds, natural ponds, wheel-driven ponds, photobioreactors, and different culture medium to increase the beta-carotene production. Innovative cultivation approaches can also enhance the productivities of the microalgae. During one study by Raja et al., De Walne's medium was used in the cultivation of *Dunaliella*; with the addition of *Sargassum wightii* and *Ulva lactuca* which are extracted from seaweed, a significant increase in the growth rate and β-carotene yield of were achieved (Raja et al. 2004).

5.2 *Phycobiliproteins*

Phycobilin: Estimated total available market for PBP products is greater than US$60 million. This number represents early 2010 numbers. Current market value would far exceed this due to rapid commercialization and an expansion in capacity. *Spirulina (Arthrospira)*, an excellent source of phycocyanin, has long held the major share of global production. Estimate annual global production exceeds 5000 tons. New capacity from producers in Asia is rapidly coming onstream to meet market demand.

PBPs are a group of colored, water-soluble proteins present in cyanobacteria and certain algae. Isolated as pigment-protein complex, PBPs can be safely used in food coloring and in cosmetics as they are nontoxic and noncarcinogenic. Figure 2.15 shows drinks and break spread that use "the blue protein" extracted from *Spirulina*. PBPs belong to two families (α and β) with each polypeptide covalently attached with chromophores.

Based on the presence of different chromophores, the chromoproteins can be categorized into four groups: (1) phycoerythrin (PE) λ_{max} 480–570 nm, (2) phycocyanin (PC) λ_{max} 590–630 nm, (3) phycoerythrocyanin (PEC) λ_{max} 630–665 nm, and (4) allophycocyanin (APC) λ_{max} 620–665 nm. Figure 2.16 shows the physical structure of phycobilisomes, with the presence of PE, PC, and APC.

Fig. 2.15 Drinks and bread spread with phycocyanin. (Photo Courtesy of Ceco International)

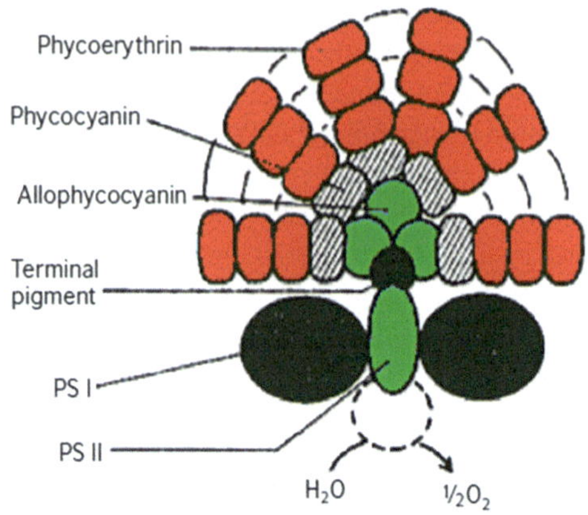

Fig. 2.16 Structure of phycobilisome (Gurpreet et al. 2009)

Phycocyanin derived from *S. platensis* is used as a natural pigment in food items such as chewing gums, dairy products, and jellies (Santiago-Santos et al. 2004). The blue, purple, and pink pigments developed by them have applications in foods that do not require heating, such as desserts, ice creams, sweets, cakes, decorations, milkshakes, etc. The blue color produced from the red microalga *Porphyridium aerugineum* resisted change with pH and was stable in light but was sensitive to heat. These properties are important for their use in food item since many food items are acidic, particularly drinks and confectionery items. This blue color was added to beverages such as Pepsi without heat application and did not lose their color for one month at room temperature (Sekar and Chandramohan 2008). The color was very stable in dry preparations.

PE gives yellow fluorescence which can be applied for a range of food items that fluoresce such as transparent lollipops made from sugar solution, dry sugar drop candies for cake decoration, soft drinks, and alcoholic beverages.

5.3 Microalgae as Food Colorant: In Butter Cookies

A study (Gouveia et al. 2007) used *Chlorella vulgaris* biomass on butter cookies as a coloring ingredient. The effects of *C. vulgaris* integrated onto the cookie surface were monitored for 3 months. The color of the cookie remained stable throughout the storage period. These *C. vulgaris*-enriched cookies appeared to have more appealing green tones which is more evident with the increasing microalgal concentration (Fig. 2.17).

Furthermore, a correlation between the increase in firmness and an increase of microalgal biomass was evident, thus making the texture more favorable (Fig. 2.18). Increased firmness will lead to higher biscuit structure protection. The biomass incorporation was not detected or negatively associated with the taste of the biscuits; this was confirmed by a sensory evaluation performed by 15 untrained individuals.

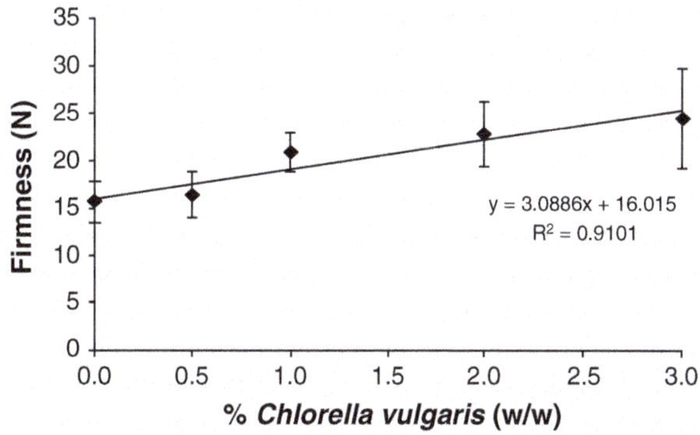

Fig. 2.17 Cookies with *Chlorella vulgaris* biomass (0.0–3.0% w/w) (Gouveia et al. 2007)

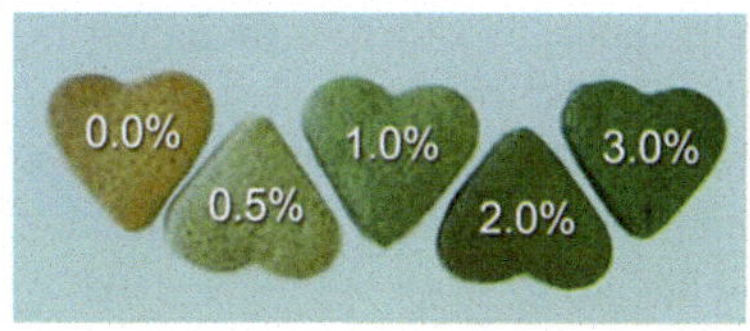

Fig. 2.18 Firmness values of colored cookies, with different concentrations of *Chlorella vulgaris* biomass (Gouveia et al. 2007)

The study investigates how microalgae can be blended to nonperishable food product and present itself in natural color. *Chlorella* sp. has been used as part of the ingredient to utilize its green color. The study has found that nonperishable tonalities in the product were welcomed by the consumers and hence concludes that additional products with microalgae as ingredients are possible (Gouveia et al. 2007).

In summary, as the world needs safer and sustainable alternatives for almost all aspects of life, it should come as no surprise that there is an increasing demand for the use of microalgae as a food colorant. Microalgae contain minimal harmful substances (allergens, irritants, and carcinogens) that are usually found in many synthetic colorants. In addition, microalgae are easier to extract, produce higher yields, do not lack raw materials, and have no seasonal variations—making them sustainable. Carotenoids and PBPs are the prominent pigments from microalgae that are used as food colorants. Species such as *Chlorella*, *Chlamydomonas*, *Dunaliella*, and *Haematococcus* are rich in carotenoids, and the production of carotenoids can be improved by factors such as high salinity, PPFD, low growth temperature, far-red light, and nutrient limitation. Phycocyanin produces a wide range of colorful pigments such as blue, purple, and pink which are used as colorants in dairy products, jellies, beverages, desserts, and alcoholic drinks. Apart from improving the appeal of food providing rich colors to them, using microalgae as a food colorant has the added nutritional benefit of acting as an antioxidant and improves the texture of the food item without altering its taste.

6 Demand from Vegetarian, Ethnic, Religious, and Special Interest Groups

6.1 Vegetarianism

There is a growing trend for vegetarian and vegan diets nowadays, especially in Western countries due to increasing evidence of epidemiological evidence that suggests improvement and maintenance of health from such diets. Vegetarian, by definition, does not include in their diet animal products such as red meat, poultry, fish, even eggs, and/or dairy products, although the exact diet practice is dependent on the type of vegetarianism (Allès et al. 2017). The growing vegetarian movement does not necessarily imply that more people are converting to a strict vegetarian diet only, but rather there is an increasing number of people opting for more nonanimal plant-based products or "switching" between animal and nonanimal products in their diet. They have been coined the term "flexitarian" (Grebow 2019). The increasing trend of vegetarian/vegans encompasses both the scrupulous vegetarians and the niche group of people who have opted for increasing proportion of their diet to consist of plants. These discriminant groups of people, though they bear the fruits of a vegetarian diet on their health and well-being, also face nutritional concerns which include vitamin B12 deficiency, insufficient omega-3 and omega-6 fatty acids, and protein deficiencies. A possible godsend for this distinct group is microalgae. Microalgae as aforementioned are packed with many of the nutrients that are needed to complete a vegetarian/vegan diet while at the same are a more environmentally friendly and sustainable food source option, especially with respect to protein demand, that can cope with the increasing environment and nutrition-conscious public.

Population following a vegetarian and vegan diet have a lack of sources of food with EAAs as most of the plant-derived proteins do not have complete EAA profiles which is a major concern for a plant-based diet as this can often lead to issues of protein deficiencies. According to the American Dietetic Association and Dietitians of Canada, individuals who exercise with moderate to intense activity require 1.3–1.7 g protein per kg of body weight per day to repair and add muscle tissue to the body (Koyande et al. 2019). As a result, microalgae are an ideal food source for vegetarians as it is an excellent source of EAAs. Some examples of microalgae species with very high protein content, even higher than most animal-based proteins per gram, are *Chlorella vulgaris* and *Spirulina platensis*. *Chlorella* and *Spirulina* species have been reported to constitute up to 70% of its biomass to contain well-balanced EAA content required for human consumption (Koyande et al. 2019). The table below can be used to compare the amino acid profile of microalgae proteins to conventional meat sources. The amino acid content in some microalgae is comparable to high protein content sources consumed widely across the world. Isoleucine, valine, lysine, tryptophan, methionine, threonine, and histidine are some of the amino acids that are present in microalgae in comparable quantities to conventional protein-rich sources.

Apart from macronutrients, micronutrients are also essential for survival. An important group of micronutrients is vitamins which plant-based vegetarian diets are often rich in. However, some nutrients of concern in the diet of vegetarians include vitamin B12, vitamin D, omega-3 fatty acids, calcium, iron, and zinc (Craig 2010). Table 2.17 shows comparison of average content of linoleic in various oil products. Vegetables and fruits lack essential vitamin B12 (cobalamin) as plants do not synthesize or require it. Vitamin B12 deficiency is a possibility when relying on vegetables for daily nutritional needs. Microalgae have been noted to contain high amounts of vitamins. In 1990, according to a study conducted by Fabregas and Herrero, it was found that microalgae *Dunaliella tertiolecta* can synthesize vitamin B12 (Fabregas and Herrero 1990). It has also been reported 9–18% of *Chlorella* strains are rich sources of vitamin B12 (Islam et al. 2017; Alsenanai et al. 2015). Watanabe et al. reported that although *Spirulina* species are capable of synthesizing vitamin B12, *Chlorella* species have better bioavailability. They suggest that microalgae-derived foods could possibly provide the necessary nutrients to counteract any deficiencies that may result from a restrictive vegetarian diet (Watanabe et al. 2002).

Table 2.17 Comparison of average content of linoleic (LA 18:2 $n-6$) and α-linolenic (ALA 18:3 $n-3$) acids in selected oil product (g/100 g fat) (Narinder et al. 2014)

Sr. No.	Oil	LA 18:2 ($n-6$)	ALA 18:3($n-3$)	Total unsaturated fatty acids
1	Soybean	50.8	6.8	80.7
2	Cotton seed	50.3	0.4	69.6
3	Corn	57.3	0.8	82.8
4	Safflower	73.0	0.5	86.3
5	Sunflower	66.4	0.3	88.5
6	Sesame	40.0	0.5	80.5
7	Olive	8.2	0.7	81.4
8	Peanut	31.0	1.2	77.8
9	Rapeseed (zero erucic acid)	22.2	11.0	88.0
10	Rapeseed (high erucic acid)	12.8	8.6	88.4
11	Cocoa butter	2.8	0.2	36.0
12	Coconut	1.8	–	7.9
13	Palm kernel	1.5	–	12.9
14	Palm	9.0	0.3	47.7
15	Almond	18.2	0.5	87.7
16	Cashew	17.0	0.4	73.8
17	Chestnut	35.0	4.0	75.5
18	Walnut	61.0	6.7	86.5
19	Butter	2.3	1.4	32.6

Table 2.18 shows the comparison of fatty acids from animals and vegetables. Compared to nonvegetarians, vegetarians (particularly vegans) tend to have lower blood levels of omega-3 fatty acids, EPA, and DHA (Craig 2010). The vegetarian population omega-3 indices up to 60% lower than those who consume marine products (Craddock et al. 2017). Omnivorous populations often obtain DHA and EPA from consuming fish and animal products; microalga is a possible alternative source of PUFAs like omega-3 and omega-6 for this niche population. Several studies have investigated the effects of DHA (dose ranges from 172 mg/day to 2.14 g/day) via algal supplementation on reasonable sample size ranges of 20–108 participants. These studies monitored the DHA levels by measurements that included serum total phospholipid DHA (PL DHA), platelet total PL DHA, red blood cell DHA (RBC DHA), and omega-3 indices. In all studies, there was a positive correlation in serum, plasma, platelet, and RBC DHA fractions and omega-3 indices and algal DHA supplementation. The omega-3 indices were reported to have increased by 55–82%, while the DHA serum total phospholipids and platelet phospholipids increased by 238–246% and 209–225%, respectively, within groups supplemented with algal DHA (Craddock et al. 2017). As a result, it illustrates the fact that consumption of algal sources of DHA in vegetarian populations greatly increased the levels of circulating DHA. DHA supplementation via algae is a viable option in the vegetarian population to address the long-chain PUFA deficiencies.

Furthermore, microalgae have an added advantage to meet the needs of the growing vegetarian population in an environmental aspect. Using microalgae as a source of bulk protein production is a viable option as it has much lower land requirements compared to animal-based proteins, lower than 2.5 m^2 per kg of protein compared to the 42–52 m^2 per kg of chicken and 144–258 m^2 per kg of beef production (Caporgno and Mathys 2018). The land usage is also lower than when compared with other plant-based protein source cultivation such as soybean meal, pea protein meal, and others (Smetana et al. 2017). Apart from that usage of nonarable land for cultivation, minimal freshwater for consumption and possibility of growing in seawater are added environmental benefits that come with microalgae-based bulk protein production. With several advantages over other currently used protein sources, microalgae provide a much more sustainable means to meet the growing vegetarian population and its dietary needs (Sathasiran et al. 2019; Barba 2018).

6.2 Special Interest Groups: Athletes, Elderly

Microalgae are used in combating chronic diseases, neurodegenerative and cardiovascular diseases, cancer, diabetes, and obesity due to chronic inflammation and oxidative stress. Algal extracts have exhibited anti-inflammatory activities by inhibiting the production of pro-inflammatory cytokines and eicosanoids. These algal metabolites work in a diverse manner such as modulation of enzyme activities, regulation of cellular activities, and interference of two major signaling pathways. There is also an added benefit of algal extracts having antioxidant properties. Ease

Table 2.18 Fatty acids produced from animals and vegetables (Craig 2010)

Fatty acid		Coconut oil (%)	Palm oil (%)	Soybean oil (%)	Sunflower oil (%)	Olive oil (%)	Butter fat (%)	Beef tallow (%)	Pork lard (%)	Chicken fat (%)	Lard oil (%)	Fish oil (%)
C-4:0	Butyric						3.5					
C-6:0	Capronic	0.5					2					
C-8:0	Caprylic	7					1.5					
C-10:0	Capric	6					3					
C-12:0	Lauric	46					3.5					
C-14:0	Myristic	18.5	1		1		11	3.5	1.5	1	1.5	6
C-14:1	Myristoleic						1	0.5				0.5
C-15:0	Pentadecanoic						1	1				0.5
C-15:1	Pentadecenoic											0.5
C-16:0	Palmitic	9.5	43	10.5	6	11.5	28	27	25.5	23	20.5	13.5
C-16:1	Palmitoleic					1	2.5	2.5	2.5	5	3.5	7.5
C-16:2	Hexadecadienic											0.5
C-16:3	Hexadecatrienic											0.5
C-16:4	Hexadecatetraenic											1
C-17:0	Margaric						1	2.5	0.5			0.5
C-17:1	Heptadecenoic						0.5	0.5			0.5	
C-18:0	Stearic	3	5	4	4	2.5	10	22	17.0	5.5	6	2.5
C-18:1	Oleic	7.5	38.5	22	20.5	75	25	36.5	40	40	51.5	14
C-18:2	Linoleic	2	11	54.5	68	9	3	2.5	12	21	13.5	1.5
C-18:3	Linolenic			7.5		1	0.5	0.5	1.0	2.5	1	1
C-18:4	Octadecatetraenoic											3
C-19:0	Nonadecanoic											0.5
C-20:0	Arachidic		0.5	0.5		0.5	0.5					0.5

(continued)

Table 2.18 (continued)

Fatty acid		Coconut oil (%)	Palm oil (%)	Soybean oil (%)	Sunflower oil (%)	Olive oil (%)	Butter fat (%)	Beef tallow (%)	Pork lard (%)	Chicken fat (%)	Lard oil (%)	Fish oil (%)
C-20:1	Gadoleic						0.5	0.5	1	0.5	1	11.5
C-20:2	Eicosadienoic								0.5		0.5	
C-20:5	Eicosapentaenoic											8.5
C-22:0	Behenic			0.5								
C-22:1	Erucic											14
C-22:5	Docosapentaenoic											1
C-22:6	Docosahexaenoic											8.5
Saturated fatty acids		90.5	49.5	15.5	11	14.5	65	56	44.5	30.5	28	24
Monounsaturated fatty acids		7.5	38.5	22	20.5	76	29.5	40.5	43.5	45.5	56.5	48
Polyunsaturated fatty acids		2	11	62	68	9.5	3.5	3	13.5	23.5	15	25.5

of digestion and bioavailability of algal products among the elderly are other key reasons that support the use of algae as a supplement for combating diseases related to aging.

Diet plays an important role in well-being because it not only plays a role in providing nutrients to meet the metabolic requirement but also a role in modulating various bodily functions. In that sense, it has a very significant impact on deterring or benefiting some diseases. The effect of diet on some chronic diseases has been reported in scientific papers, showing the extraordinary possibility of diet on improving and supporting our health (Levin and Fleurence 2018). These types of food that have demonstrated improving the state of health or well-being or reducing risk of diseases can be considered as functional foods. Having said that, older people may have more difficulties in getting the necessary nutrients they need from their diet due to aging. Elderly are also more susceptible to chronic diseases with age, making dietary changes using functional foods to prevent these chronic conditions even more paramount. Foregoing the information mentioned, it is apparent microalgae are a possible functional food that could meet the various needs of the elderly due to their ease of digestion, bioavailability, and richness in micro- and macronutrients with reduced caloric intake.

Microalgae have many components that favor supplementation to the elderly that includes natural pigments and several bioactive compounds which have antitumor, anti-inflammatory, anti-obesity, and neuroprotective properties. Table 2.19 shows some of the microalgae that have been incorporated into the food products, most of which are suitable for vegetarians and elderly populations.

7 Marketing of Microalgae: Supply Meets Demand

7.1 Algae Product Forms

Microalgae, as a food ingredient, are a versatile element capable of crossing cultural, age, and social divide. It clearly shows the potential to meet the population's needs in its nutritional values for more sustainable food solutions. It is not suggested that readers abandon or totally replace current dietary practices, knowing the single-cell organism provides additional choices in meeting the demand for food stock and providing more options for healthy lifestyle practices.

The development of microalgae-based food industry is based upon producing and utilizing microalgae for innovative functional food products. Other than the high protein content, balanced amino acid profile, and plethora of vitamins, incorporation of microalgae into foods adds an additional level of potential benefits for human health due to the presence of bioactive compounds.

Consumers who are looking for natural health and food supplements are fully aware of the benefits microalgae bring for their overall health and well-being, with increasing new research results on the benefits that microalgae can provide that are

Table 2.19 Microalgae incorporated into various food products

Product	Microalgae incorporation	Addition	Benefit
Oil/water emulsion	*C. vulgaris* green and *C. vulgaris* orange (after carotenogenesis)	2% w/w	Techno-functional properties
Oil/water emulsions	*C. vulgaris* green, *C. vulgaris* orange (after carotenogenesis) and *H. pluvialis* (red, after carotenogenesis)	*C. vulgaris*: 0.25–2.00% w/w *H. pluvialis*: 0.05–2.00% w/w	Coloring and nutritional properties (antioxidative activity)
Vegetarian food gels	*C. vulgaris*, *H. pluvialis*, *A. maxima, and D. vlkianum*	0.75% w/w	Techno-functional and nutritional properties (antioxidative activity, ω-3 PUFAs)
Vegetarian food gels	*A. maxima and D. vlkianum*	0.1% w/w	Techno-functional and nutritional properties (ω-3 PUFAs)
Vegetarian food gels	*H. pluvialis* and *A. maxima*	0.75% w/w	Techno-functional properties
Frozen yogurt	*Arthrospira* sp.	2–8% w/w	Nutritional properties
Dairy products (fermented milk)	*A. platensis*	3 g/L	Nutritional properties
Natural and probiotic yogurt	*A. platensis*	0.1–0.8% w/w	Techno-functional properties and nutritional properties
Yogurt	*Chlorella* sp.	0.25% w/w extract powder and 2.5–10.0% extract liquid	Techno-functional properties and nutritional properties
Processed cheese	*Chlorella* sp.	0.5 and 1.0% w/w	Techno-functional properties and nutritional properties
Cookies	*C. vulgaris*	0.5, 1.0, 2.0, and 3.0% w/w	Coloring agent
Biscuits	*I. galbana*	1 and 3% w/w	Techno-functional properties and nutritional properties (ω-3 PUFAs)
Biscuits	*A. platensis*	*A. platensis*: 0.3, 0.6, and 0.9% Phycocyanin extract: 0.3% w/w to wheat flour	Nutritional properties
Biscuits	*A. platensis*	1.63, 3, 5, 7, 8.36% w/w	Techno-functional and nutritional properties (protein, fiber content, and antioxidative activity)

(continued)

Table 2.19 (continued)

Product	Microalgae incorporation	Addition	Benefit
Biscuits	*A. platensis*, *C. vulgaris*, *T. suecica*, and *P. tricornutum*	2 and 6% w/w	Techno-functional properties and nutritional properties (antioxidative activity)
Cookies	*H. Pluvialis*	Astaxanthin powder 5, 10, and 15% w/w	Techno-functional properties and nutritional properties (antioxidative activity)

Fig. 2.19 Chinese mooncakes with pigments from algae and/or algae extract. (Ceco International)

coming out of the labs. However, the algae market has not taken off as one has hoped; there seems to be a hard glass ceiling for the marketers and corporations to break through, one of which is the format which these products are marketed.

There is no wonder the Aztecs call *tecuitlatl* mud cake. The description is hardly appealing nor appetizing. The root of market pull really comes from the desire "not have to tolerate the nasty taste of existing food supplements," but in a format of desirable food products, we can consume and enjoy on the daily basis.

One of the viable value propositions is to integrate high-quality nutritional biomass into familiar food products in the existing diet, working closely with Food & Beverage (F&B) to complement and cleverly mask the undesirable fishy taste with a natural flavoring that appeals to the regional taste palates.

Figure 2.19 shows seasonal mooncakes prepared for the Chinese Mid-Autumn Festival. A traditional Chinese dessert delicacy is enjoyed while the family gathers around for a meal. Exclusive vegan sources can be employed, with low plant-based natural sweeteners to meet the health-conscious consumers. Current reception from our test market was remarkable and overwhelming. The plan is to produce noodles, bread, pate, and sausages with microalgae and natural vegan ingredients commercially for the vegan market and healthy lifestyle market segments.

7.2 Production Challenges

At the current state of market development, the overall cost related to utilizing microalgae-derived products as food substitutes has yet to reach the desired level. From the production perspective, the lack of economies of scale for cultivation and processing is a major hurdle. With the development of advanced technology, automation, and artificial intelligence, these obstacles can be significantly lowered. The benefits to humanity (health) and the environment and the sustainability of food supply, with the growing population, outstrip the challenges and hurdles presented.

There have been many reviews on the production challenges for microalgae. To food application, safety and cost are of primary concern. Figure 2.20 shows some of the current algae production experience, which includes (1) contamination control with the need for axenic culture for food application, (2) photobioreactors that are better designed and monitored for repeatability and reliability, (3) maintaining low-cost operation with intelligent energy management, (4) cost-effective harvesting and dewatering technologies, and (5) extraction technology for various functional high-value ingredients (not shown).

Fig. 2.20 (**a**) Micrograph of axenic *Euglena* culture produced at Geb facility (×200 Mag). (**b**) Indoor cultivation of *Chlorella* in a Phyco-Flow (Varicon Aqua). (**c**) Encystment of *Haematococcus pluvialis*. (**d**) Ground dry powder of *Euglena*. All produced at Geb Impact facilities in Hong Kong

8 Conclusion

Microalgae have come a long way from ad hoc food stock for small indigenous population to systematic studies and commercialization as potential mainstream food supply. In the chapter, attempts have been made to highlight the most prominent features of microalgae as food ingredients and the applications. Algae cultivation techniques are also described in detail together with biotechnological understanding. The values of nutrition from microalgae to human health have been clearly studied and demonstrated. Like application of any new technology, from earlier adopters to earlier mainstream users, there is a chasm to cross; algae-based food as mainstream ingredients is without exception. The demand from interested customers highly depends on cohesive and collective marketing of the potential values to the right consumers with the right products with the right pricing points at the right time. The dawn with commercialization of algae as mainstream food would likely come sooner rather than later as the projected population growth and environmental mandates dwell on us.

Acknowledgments We would like to thank the Hong Kong Innovation and Technology Commission (ITC) for the support of three of the authors through Post-Doctoral Hub Program PsH/045/19 and Research Program IsP/210/18 & IsP/153/19.

References

Abalde, J., Fabregas, J., & Herrer, C. (1991). β-Carotene, vitamin C and vitamin E content of the marine microalga *Dunaliella tertiolecta* cultured with different nitrogen sources. *Bioresource Technology, 38*(2–3), 121–125.

Abd El-Hack, M. E., Abdelnour, S., Alagawany, M., Abdo, M., Sakr, M. A., Khafaga, A. F., Mahgoub, S. A., Elnesr, S. S., & Gebriel, M. G. (2019). Microalgae in modern cancer therapy: Current knowledge. *Biomedicine & Pharmacotherapy, 111*, 42–50.

Abduqader, G., Barsanti, L., & Tredici, M. R. (2000). Harvesting of *Arthrospira platensis* from Lake Kossorom (Chad) and its household usage among the Kanembu. *Journal of Applied Phycology, 12*, 493–498.

Adamczyk, M., Lasek, J., & Skawinska, A. (2016). CO2 biofixation and growth kinetics of *Chlorella vulgaris* and *Nannochloropsis gaditana*. *Applied Biochemistry and Biotechnology, 179*(7), 1248–1261.

Aherne, G. W., Hardcastle, A., Valenti, M., Bryant, A., Rogers, P., Pettit, G. R., Srirangam, J. K., & Kelland, L. R. (1996). Anti-tumour evaluation of dolastatins 10 and 15 and their measurement in plasma by radioimmunoassay. *Cancer Chemotherapy and Pharmacology, 38*(3), 225–232.

Alam, M. A., & Wang, Z. (Eds.). (2019). *Microalgae biotechnology for development of biofuel and wastewater treatment* (pp. 1–655). Singapore: Springer.

Allès, B., Baudry, J., Méjean, C., Touvier, M., Péneau, S., Hercberg, S., & Kesse-Guyot, E. (2017). Comparison of sociodemographic and nutritional characteristics between self-reported vegetarians, vegans, and meat-eaters from the NutriNet-Santé study. *Nutrients, 9*(9), 1023.

Alsenanai, F., et al. (2015). *Nutraceuticals from microalgae: Nutraceuticals and functional foods in human health and disease prevention*. St Lucia, QLD: University of Queensland.

Arbex, A. K., et al. (2015). The impact of fatty acid to human health. *Journal of Endocrine and Metabolic Diseases, 2015*(5), 98–104.
Babu, B. R., Rastogi, N. K., & Raghavarao, M. S. (2006). Mass transfer in osmotic membrane distillation of phycocyanin colourant and sweet lime juice. *Journal of Membrane Science, 212*, 58–69.
Barba, F. J. (2018). Trends in microalgae incorporation into innovative food products with potential health benefits. *Frontiers in Nutrition, 5*, 58.
Becker, W. (2004). The nutritional value of microalgae for aquaculture. In A. Richmond (Ed.), *Microalgae for aquaculture. Handbook of microalgal culture* (pp. 380–391). Oxford: Blackwell.
Becker, E. W. (2007). Micro-algae as a source of protein. *Biotechnology Advances, 25*(2), 207–210.
Begum, H., Yusoff, F. M. D., Banerjee, S., Khatoon, H., & Shariff, M. (2015). Availability and utilization of pigments from microalgae. *Critical Review in Food Science and Nutrition, 56*(13), 2209–2222.
Ben-Amotz, A., Gressel, J., & Avron, M. (1987). Massive accumulation of phytoene induced by norflurazon in *Dunaliella bardawil* (chlorophyceae) prevents recovery from photoinhibition. *Journal of Phycology, 23*, 176–181.
BGG. (2016). Health benefits and production methods of natural astaxanthin. bggworld.com
Bleakley, S., & Hayes, M. (2017). Algal proteins: Extraction, application, and challenges concerning production. *Food, 6*, 33.
Borowitzka, M. (2013). High-value products from microalgae—Their development and commercialization. *Journal of Applied Phycology, 25*(3).
Brown, M., Mular, M., Miller, I., et al. (1999). The vitamin content of microalgae used in aquaculture. *Journal of Applied Phycology, 11*(3), 247–255.
Brown, M. R., Jeffrey, S. W., Volkman, J. K., et al. (1997). Nutritional properties of microalgae for mariculture. *Aquaculture, 151*, 315–331.
Caporgno, M. P., & Mathys, A. (2018). Trends in microalgae incorporation into innovative food products with potential health benefits. *Frontier in Nutrition, 5*, 58.
Carballo-Cárdenas, E. C., Tuan, P. M., Janssen, M., & Wijffels, R. H. (2003). Vitamin E (alpha-tocopherol) production by the marine microalgae *Dunaliella tertiolecta* and *Tetraselmis suecica* in batch cultivation. *Biomolecular Engineering, 20*(4–6), 139–147.
Carbonell-Capella, J. M., Buniowska, M., Barba, F. J., Esteve, M. J., & Frígola, A. (2014). Analytical methods for determining bioavailability and bioaccessibility of bioactive compounds from fruits and vegetables: A review. *Comprehensive Reviews in Food Science and Food Safety, 13*(2), 155–171.
Central Africa. (2015). http://news.algaeworld.org/2015/04/nuns-growing-*spirulina*-in-central-african-republic-to-fight-malnutrition-with-ingenuity/
Chen, Z., Wang, L., Qiu, S., & Ge, S. (2018). Determination of microalgal lipid content and fatty acid for biofuel production. *Bio-Medical Research International, 2018*, 17.
Chronakis, I. S., & Madsen, M. (2011). Algal proteins, handbook of food proteins. In G. O. Phillips & P. A. Williams (Eds.), *Woodhead publishing series in food sciences, technology and nutrition* (pp. 353–394).
Cohen, G., Riahi, Y., & Sasson, S. (2011). Lipid peroxidation of poly-unsaturated fatty acids in normal and obese adipose tissues. *Arch Physiology Bio-chemistry, 117*(3), 131–139.
Craddock, J. C., Neale, E. P., Probst, Y. C., & Peoples, G. E. (2017). Algal supplementation of vegetarian eating patterns improves plasma and serum docosahexaenoic acid concentration and omega-3 indices: A systematic literature review. *Journal of Human Nutrition and Dietetics., 30*, 693–699.
Craig, W. J. (2010). Nutrition concerns and health effects of vegetarian diets. *Nutrition in Clinical Practice., 25*(6), 613–620.
Croft, M. T., Lawrence, A. D., Raux-Deery, E., Warren, M. J., & Smith, A. G. (2005). Algae acquire vitamin B12 through a symbiotic relationship with bacteria. *Nature, 438*(7046), 90–93.
Croft, M. T., Warren, M. J., & Smith, A. G. (2006). Algae need their vitamins. *Eukaryote Cell, 5*(8), 1175–1183.

Digs, C. (2013). http://charliesdigs.blogspot.com/2013/07/spirulina-aztec-food-supplement.html
Doughman, S. D., Krupanidhi, S., & Sanjeev, C. B. (2007). Omega-3 fatty acids for nutrition and medicine: Considering microalgae oil as a vegetarian source of EPA and DHA. *Current Diabetes Review, 3*(3), 198–203.
Farag, I., & Price, K. (2013). Resources conservation in microalgae biodiesel production. *International Journal of Engineering and Technical Research, 1*, 49–56.
Fabregas, J., & Herrero, C. (1990). Vitamin content of four marine microalgae. Potential use as a source of vitamins in nutrition. *Journal of Industrial Microbiology, 5*(4), 259–263.
Falaise, C., et al. (2016). Antimicrobial compounds from eukaryotic microalgae against human pathogens and diseases in aquaculture. *Marine Drugs, 14*(9), 159.
Feenstra, W. J. (n.d.). *Biochemical genetics of thiamine biosynthesis*. https://www.arabidopsis.org/ais/1965/fees-1965-aaglt.html
Fixsen, M. A., & Jackson, H. M. (1932). The biological values of proteins: A further note on the method used to measure the nitrogenous exchange of rats. *The Biochemical Journal, 26*(6), 1919–1922.
Gardner, M. L. (1988). Gastrointestinal absorption of intact proteins. *Annual Review of Nutrition, 8*, 329–350.
Gorelova, V., Bastien, O., De Clerck, O., Lespinats, S., Rébeillé, F., & Van der Straeten, D. (2019). Evolution of folate biosynthesis and metabolism across algae and land plant lineages. *Nature, 9*, 5731.
Gouveia, L., et al. (2007). *Chlorella vulgaris* biomass used as coloring source in traditional butter cookies. *Innovative Food Science & Emerging Technologies., 8*, 433–436.
Grebow, J. (2019). Plant products need to target flexitarians. Nutritional Outlook.
Gurpreet, K., Khattar, J. I. S., Singh, D. P., Yadvinder, S., & Jeevesh, N. (2009). *Microalgae: A source of natural colours* (pp. 129–150). New Delhi: I. K. International Publishing House.
Gutiérrez-Salmeán, G., Fabila-Castillo, L., & Chamorro-Cevallos, G. (2015). Nutritional and toxicological aspects of *Spirulina (arthrospira). Nutricion Hospitalaria, 32*(1), 34–40.
Hasegawa, T., Matsuguchi, T., Noda, K., Tanaka, K., Kumamoto, S., Shoyama, Y., & Yoshikai, Y. (2002). Toll-like receptor 2 is at least partly involved in the antitumor activity of glycoprotein from *chlorella vulgaris*. *International Immuno-pharmacology, 2*(4), 579–589.
Hoffman, J. R., & Falvo, M. J. (2004). Protein—Which is best? *Journal of Sports Science and Medicine, 3*(3), 118–130.
Islam, M. N., Alsenanai, F., & Schenk, P. M. (2017). *Microalgae as a sustainable source of nutraceuticals, chapter 1*. Hoboken, NJ: John Wiley & Sons.
Ji, X. J., et al. (2015). Omega-3 biotechnology: A green and sustainable process for omega-3 fatty acids production. *Frontier of Bioengineering and Biotechnology, 3*, 158.
Koyande, A. K., Chew, K. W., Rambabu, K., Tao, Y., Chu, D. T., & Show, P. L. (2019). Microalgae: A potential alternative to health supplementation for humans. *Food Science and Human Wellness, 8*(1), 16–24.
Li-Beisson, Y., Nakamura, Y., & Harwood, J. (2016). Lipids: From chemical structures, biosynthesis, and analyses to industrial applications. *Subcell Biochemistry, 86*, 1–18.
Liu, X., Wu, F., Ji, Y., & Yin, L. (2019). Recent advances in anti-cancer protein/peptide delivery. *Bioconjugate Chemistry, 30*(2), 305–324.
Ma, X. N., et al. (2016). Lipid production from *Nannochloropsis*. *Marine Drugs, 14*(4), pii: E61.
Marrion, O., Schwertz, A., Fleurence, J., Guéant, J. L., & Villaume, C. (2003). Improvement of the digestibility of the proteins of the red alga *palmaria palmata* by physical processes and fermentation. *Nahrung, 47*(5), 339–344.
Martins, D. A., et al. (2013). Alternative sources of *n*-3 long-chain polyunsaturated fatty acids in marine microalgae. *Marine Drugs, 11*(7), 2259–2281.
Matondo, F. K., et al. (2016). Spirulina supplements improved the nutritional status of undernourished children quickly and significantly: Experience from Kisantu, the Democratic Republic of the Congo. *International Journal Of Pediatrics, 2016*, 1296414.

Matos, Â. P. (2017). The impact of microalgae in food science and technology. *Journal of the American Oil Chemists' Society, 94*(11), 1333–1350.
Metting, F. B. (1996). Biodiversity and application of microalgae. *Journal of Industrial Microbiology, 17*, 471–489.
Meynier, A., & Genot, C. (2017). Molecular and structural organization of lipids in foods: Their fate during digestion and impact in nutrition. *OCL, 24*(2), D202.
Miller, M., et al. (2011). Triglycerides and cardiovascular disease: A scientific statement from the American Heart Association. *Circulation, 123*(20), 2292–2333.
Nakayama, H. (1991). *Method of disrupting the chlorella cell wall by cell rupture*. U.S. patent no. 5330913.
Pratt, R., & Johnson, E. (1966). Production of pantothenic acid and inositol by *Chlorella vulgaris* and *C. pyrenoidosa*. *Journal of Pharmaceutical Sciences, 55*(8), 799–802.
Narinder, K., Vishal, C., & Anil K., G. (2014). Essential fatty acids as functional components of foods—A review. *Journal of Food Science and Technology, 51*(10), 2289–2303.
Raja, R., Anbazhagan, C., Ganesan, V., et al. (2004). Efficacy of *Dunaliella salina* (volvocales chlorophyta) in salt refinery effluent treatment. *Asian Journal of Chemistry, 16*, 1081–1088.
Roy, S. S., & Pal, R. (2015). Microalgae in aquaculture: A review with special references to nutritional value and fish dietetics. *Proceedings of the Zoological Society, 68*(1), 1–8.
Samarakoon, K., & Jeon, Y. (2012). Bio-functionalities of proteins derived from marine algae—A review. *Food Research International, 48*(2), 948–960.
Santiago-Santos, M. C., Ponce-Noyola, T., Olivera-Ramirez, R., Ortega-Lopez, J., & Canizares-Villanueva, R. O. (2004). Extraction and purification of phycocyanin from *Calothrix* sp. *Process Biochemistry, 39*(12), 2047–2052.
Sansone, C., & Brunet, C. (2019). Promises and challenges of microalgal antioxidant production. *Antioxidants, 8*, 199.
Sathasiran, R., Radhakrishnan, R., Hashem, A., & Abd_Allah, E. A. (2019). Microalgae metabolites: A rich source for food and medicine. *Saudi Journal of Biological Science, 26*(4), 709–722.
Sekar, S., & Chandramohan, M. (2008). Phycobiliproteins as a commodity: Trends in applied research, patents and commercialization. *Journal of Applied Phycology, 20*, 113–136.
Skulas-Ray, A. C., et al. (2011). Dose-response effects of omega-3 fatty acids on triglycerides, inflammation, and endothelial function in healthy persons with moderate hypertriglyceridemia. *The American Journal of Clinical Nutrition, 93*(2), 243–252.
Smetana, S., Sandmann, M., Rohn, S., Pleissner, D., & Heinz, V. (2017). Autotrophic and heterotrophic microalgae and cyanobacteria cultivation for food and feed: Life cycle assessment. *Bioresources Technology, 245*(Pt A), 162–170.
Soeder, C. J., & Pabst, W. (1970). *Massive cultivation of microalgae: Results and prospects*. Chem: Springer.
Soletto, D., Binaghi, L., Lodi, A., Carvalho, J. C. M., & Converti, A. (2005). Batch and fed-batch cultivations of *Spirulina platensis* using ammonium sulphate and urea as nitrogen sources. *Aquaculture, 243*(1), 217–224.
Sun, X. M., et al. (2018). Microalgae for the production of lipid and carotenoids: A review with focus on stress regulation and adaptation. *Biotechnology for Biofuels, 2018*(11), 272.
Tandon, P., Jin, Q., & Huang, L. (2017). A promising approach to enhance microalgae productivity by exogenous supply of vitamins. *Microbial Cell Factories, 16*(1), 219.
Tang, G., & Suter, P. M. (2011). Vitamin A, nutrition, and health values of algae: *Spirulina*, *Chlorella*, and *Dunaliella*. *Journal of Pharmacy and Nutrition Sciences, 1*, 111–118.
Tani, Y. (1989). Algal and microbial production of vitamin E. *Biotechnology of Vitamins, Pigments and Growth Factors, 1989*, 95–104.
Tarento, T. D. C., McClure, D. D., Vasiljevski, E., Schindeler, A., Dehghani, F., & Kavanagh, J. M. (2018). Microalgae as a source of vitamin K1. *Algal Research, 36*, 77–87.
The Food and Drug Administration. (2019). *The FDA recommendation of suggested daily vitamin value*. https://www.accessdata.fda.gov/scripts/InteractiveNutritionFactsLabel/factsheets/Vitamin_and_Mineral_Chart.pdf

Transparent. (2018). www.transparencymarketresearch.com/pressrelease/global-nutraceuticals-product-market.htm

van der Wielen, N., Moughan, P. J., & Mensink, M. (2017). Amino acid absorption in the large intestine of humans and porcine models. *The Journal of Nutrition, 147*(8), 1493–1498.

van Krimpen, M. M., Bikker, P., van der Meer, I. M., van der Peet-Schwering, C. M. C., Vereijken, J. M. (2013). Cultivation, processing and nutritional aspects for pigs and poultry of European protein sources as alternatives for imported soybean products, Wageningen UR Livestock Research Report 662.

Vernass, W. F. J. (2004). Targeted genetic modification of cyanobacteria: New biotechnological applications. In A. Richmond (Ed.), *Handbook of microalgal culture: Biotechnology and applied phycology* (pp. 312–351). Bodwin, Cornwall: Blackwell Science MPG Books.

Watanabe, F., Takenaka, S., Kittaka-Katsura, H., Ebara, S., & Miyamoto, E. (2002). Characterization and bioavailability of vitamin B12-compounds from edible algae. *Journal of Nutritional Science and Vitaminology, 48*, 325–331.

WHO/FAO/UNU Expert Consultation. (2002). *A joint report, protein and amino acid requirements in human nutrition, WHO technical report series 935*. Geneva: United Nations University, World Health Organization.

Chapter 3
Microalgal Pigments: A Source of Natural Food Colors

Emeka G. Nwoba, Christiana N. Ogbonna, Tasneema Ishika, and Ashiwin Vadiveloo

Abstract Naturally sourced colorants and dyes are currently gaining demand over synthetic alternatives due to an increase in consumer awareness brought forward by health and environmental issues. Microalgae are unicellular organisms which are microscopic in size and represent major photosynthesizers with the ability to efficiently convert available solar energy to chemical energy. Due to their distinct advantages over terrestrial plants such as faster growth rates, ability to grow on non-arable land, and diversity in the production of various natural bioactive compounds (e.g., lipids, proteins, carbohydrate, and pigments), microalgae are currently gaining promise as a sustainable source for the production of natural food-grade colorants. The versatility of microalgae to produce various pigments (e.g., chlorophylls, carotenoids, xanthophylls, and phycobiliproteins) that can be commercially exploited as a source of natural colorant is there to be explored. Various growth factors such as temperature, pH, salinity, and light in terms of both quality and quantity have been shown to significantly impact pigment production. In this chapter, we comprehensively review the characteristics of microalgal pigments and factors that affect pigment production in microalgae while evaluating the overall feasibility of exploiting them as a natural source of food colorants.

Keywords Food colorants · Microalgae · Pigments · Sustainability

E. G. Nwoba · A. Vadiveloo (✉)
Algae Research and Development Centre, Murdoch University, Murdoch, WA, Australia
e-mail: Ashiwin.Vadiveloo@murdoch.edu.au

C. N. Ogbonna
Department of Plant Science and Biotechnology, University of Nigeria, Nsukka, Nigeria

T. Ishika
Algae Research and Development Centre, Murdoch University, Murdoch, WA, Australia

Department of Microbiology, Jashore University of Science and Technology, Jashore, Bangladesh

M. A. Alam et al. (eds.), *Microalgae Biotechnology for Food, Health and High Value Products*, https://doi.org/10.1007/978-981-15-0169-2_3

1 Introduction

Colorants are often required to improve the visible appearance of food. These colorants used can either be sourced naturally or be artificially produced. However, in recent times, the demand for naturally sourced colorants has been significantly increasing due to various health concerns that have been associated with the use of artificial colorants such as attention deficit hyperactivity disorder and other cases of hyperactivity (Schab and Trinh 2004; Jacobson and Kobylewski 2010). As such, Farré et al. (2010) established a significant shift in the demand for naturally sourced pigments over those that are chemically produced.

Due to the inherent disadvantages of synthetic colorants and the health benefits associated with natural colorants sourced from pigments, there has been great interest in exploiting the production of these natural pigments from different sources at a commercial scale. Initially, pigments were conventionally sourced from only plants and animals (Timberlake and Henry 1986). However, an increase in demand has necessitated the use of fast-growing microorganisms which are not only robust but also serve as a reliable source for the production of natural pigments (Ogbonna 2016b). Many species of microorganisms such as bacteria (Heer and Sharma 2017), fungi (Ogbonna 2016a, b; Ogbonna et al. 2017), cyanobacteria, and microalgae (Soares et al. 2016) have been well documented for the production of natural pigments.

Among the other candidates, microalgae and cyanobacteria have attracted great interest as a natural source for the production of pigments due to the ease of their cultivation, fast growth rates, and their diversity in producing a wide range of pigments naturally. They are able to produce a variety of pigments of various color shades and biological activities. These include chlorophylls, carotenoids, xanthophylls, and phycobiliproteins. Despite containing multiple pigments, only chlorophylls are produced at a reasonable concentration (1–2% g g^{-1} dry weight) by microalgae cells under natural conditions (Wang and Chen 2008; Li et al. 2008). Most other desired pigments found in algae cells under normal growth conditions are typically at low concentrations (e.g., below 0.5% g g^{-1} dry weight), making them not economically competitive enough against chemically synthesized alternatives (Wang and Chen 2008; Mulders et al. 2014).

However, some of these pigments have high commercial value (β-carotene, astaxanthin, and c-phycocyanin), and their feasibility at large scale from various microalgae sources are currently being explored (Fig. 3.1).

2 Microalgae and Pigments

Microalgae represent a wide variety of microscopic photosynthetic organisms. Their pigment systems have evolved from primitive "plastid" which they acquired from oxygen-evolving ancestral photosynthetic cyanobacterium (Jeffrey et al.

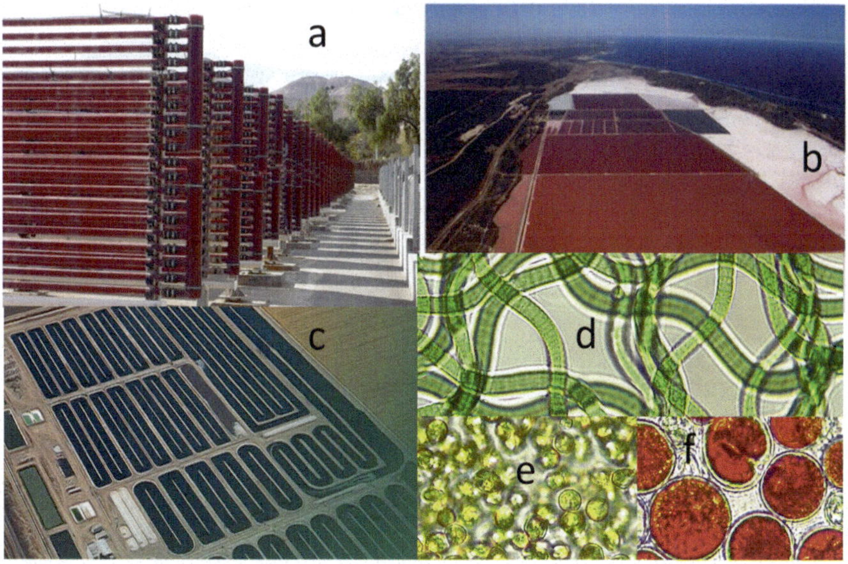

Fig. 3.1 Commercial-scale production of astaxanthin (**a**), β-carotene (**b**), and c-phycocyanin (**c**) from *Haematococcus pluvialis* (**f**), *Dunaliella salina* (**e**), and *Arthrospira platensis* (**d**), respectively

2011). A series of further evolution resulted in the present photosynthetic microalgae derivatives consisting of a variety of photosynthetic pigments that assist in photosynthesis and photoprotection (Jeffrey et al. 2011). The availability of light is seen to significantly vary in different aquatic environments and according to changes in weather or climate. The diversity in their inherent pigment composition allows microalgae to survive and successfully adapt to the constantly changing light conditions. In general, photosynthetic pigments in microalgae are typically classified into three groups which are chlorophylls, carotenoids, and phycobiliproteins (Gantt and Cunningham Jr 2001).

2.1 Microalgae Classification and Their Pigment Composition

The concentration of pigments can significantly vary among different species of microalgae. Pigments approximately make up 1–14% of the dry weight of microalgae which again is species-specific. Most species of microalgae have a chlorophyll content of 0.5–1% and a carotenoid content of 0.1–0.2% (Spolaore et al. 2006). Chlorophyll *a* is the major pigment found in all microalgae which is vital for photosynthesis. Most green microalgae (chlorophytes) contain chlorophyll *a* and *b* and carotenoids, while microalgae with brownish appearance (Bacillariophyceae) generally contain chlorophyll *a* and *c* and carotenoids. Phycobiliproteins are only present in cyanobacteria, red algae, and cryptophytes, while dinoflagellates are typically composed of chlorophyll *a* and *c* and the carotenoid known as peridinin, as major photosynthetic pigments. Most pigments are usually distributed as protein-pigment complexes within the thylakoid membranes of microalgal chloroplast (Gantt and Cunningham Jr 2001). Pigment composition is among the most important criteria in differentiating the various algal classes. Table 3.1 highlights the distribution of pigments in various classes of microalgae. It is important to note that chlorophylls and carotenoids are generally fat-soluble molecules, whereas phycobilins are water-soluble pigments (Gantt and Cunningham Jr 2001).

3 Application of Microalgal Pigments as Natural Food Colorants

As highlighted earlier, the shift in demand brought forward by consumers for the production of naturally derived products for application in the food industry has necessitated significant innovation (Nwoba et al. 2019b). Food color has always remained one of the pivotal organoleptic properties that precisely impact consumer decision on the selection and acceptance of desired food. Food color offers sensorial attractiveness and visual perception that affects the consumers food selection decision and preference (Martins et al. 2016; Shim et al. 2011). Despite all food products having their own natural color, processing techniques as well as storage conditions including factors such as light, temperature, air, and moisture have been shown to significantly affect the color of the final product (Martins et al. 2016).

Therefore, food colorants remain an integral component of the production process aimed toward the masking of unpleasant food attributes. Food colorants can be typically categorized as dyes, pigments, or specific substances which can directly or indirectly impart color to foods, drugs, or cosmetics or directly to the human body (FDA 2019). Based on this definition, food colorants are broadly divided into either synthetic (artificial) or natural colorants. Synthetic food colorants (made from petroleum, organic acids, and inorganic chemicals) have been widely used over a long period; however, they are currently restricted due to evidential cases of toxicity, serious side effects and allergic and neurocognitive reactions at short-, medium-, and long-term bases (Dias et al. 2015; Laokuldilok et al. 2016). As such, naturally derived food colorants are progressively replacing synthetic colorants in the food manufacturing industries due to consumer demand (Carocho et al. 2015; Rodriguez-Amaya 2019). Besides their efficacy in providing color and organoleptic properties to foods, food colorants of natural origins are considered safer than those derived from chemical synthesis (Carocho et al. 2014). Also, naturally derived food colorants perform both antioxidative and preservative roles, as well as provide functional properties being the driver of what is currently sought after as functional foods (Carocho et al. 2014; Rodriguez-Amaya 2019; Bagchi 2006). The bioactive properties of natural food colorants, which include anticancer, anticholesterolemic,

Table 3.1 Pigment composition of microalgae (Thrane et al. 2015; Gantt and Cunningham Jr 2001; Humphrey 1980; Nwoba et al. 2019a)

Pigments	Microalgae												Spectrum absorbed (nm)
					Chromophytes								
	Cyanophytes	Prochlorophytes	Chlorophytes	Rhodophytes	B	C	R	X	H	E	Cr	Dinophytes	
Chlorophylls													
Chlorophyll *a*	√	√	√	√	√	√	√	√	√	√	√	√	438,670
Chlorophyll *b*	/	√	√	/	/	/	/	/	/	/	/	/	457, 645
Chlorophyll *c*	/	/	/	/	√	√	√	√	√		√	√	446, 628
Chlorophyll *d*	√	/	/	√	/	/	/	/	/	/	/	/	445, 693
Carotenoids													
Carotene													
α-carotene	√	√	/	/	/	/	/	/	/	/	/	/	
β-carotene	√	/	/	/	/	/	/	/	/	/	/	/	453
Xanthophyll													
Alloxanthin	/	/	/	/	/	/	/	/	/	/	√	/	464
Antheraxanthin	/	/	√	/	/	√	/	/	/	/	/	/	446
Diadinoxanthin	/	/	√	/	√		√	√	/	√	/	√	448
Diatoxanthin	/	/	√	/	/	/	/	/	/	/	/	√	453
Fucoxanthin	/	/	/	/	√	√	√	/	√	/	/	/	443
Lutein	/	/	√	/	/	/	/	/	/	/	/	/	447
Neoxanthin	/	/	√	/	/	/	/	/	/	/	/	/	437
Vaucheriaxanthin	/	/	/	/	/	/	/	√	/	√	/	/	
Violaxanthin	/	/	√	/	/	/	/	/	/	/	/	/	437
Zeaxanthin	√	√	√	√	/	√	/	/	/	/	/	/	450–453
Peridinin	/	/	/	/	/	/	/	/	/	/	/	√	475
Phycobiliproteins													
Allophycocyanin	√	/	/	√	/	/	/	/	/	/	/	/	650–655

(continued)

Table 3.1 (continued)

	Microalgae													
					Chromophytes									
Pigments	Cyanophytes	Prochlorophytes	Chlorophytes	Rhodophytes	B	C	R	X	H	E	Cr	Dinophytes	Spectrum absorbed (nm)
Phycocyanin	√	/	/	√	/	/	/	/	/	/	√	/	620–630
Phycoerythrin	√	√	/	√	/	/	/	/	/	/	√	/	575–580

B Bacillariophyceae, *C* Chrysophyceae, *R* Raphidophyceae, *X* Xanthophyceae, *H* Haptophyceae, *E* Eustigmatophyceae, *Cr* Cryptophyceae; "/", not present antidiabetic, and anti-inflammatory actions, have been well documented (Wang et al. 2015; Nwoba et al. 2019a). Sources of naturally occurring food colorants include vegetables, flowers, insects (cochineals, aphids), fruits, leaves, and microorganisms such as fungi and bacteria (Martins et al. 2016).

Although the demand for the utilization of natural color ants in food (and other applications such as pharmaceuticals, cosmetics, textiles, printing) is on the rise due to the toxicity of many synthetic dyes, they can be limited by some shortcomings. Besides their poor tinctorial strength and persistence, low stability, and most importantly their higher production cost, the production of natural dyes is also restricted in terms of sustainability. A potential solution to some of these major challenges can be achieved through the exploitation of novel or unconventional sources of colorants such as microalgae. Microalgae boast a wide array of naturally colored pigments and could be considered as a sustainable option for the production of food colorants. Considering that microalgal cultivation is both environmentally friendly and sustainable, they do not only offer great potential as a source of natural food dyes but also represent a cost-favorable option.

As of recent, secondary metabolites sourced from microalgae are gaining tremendous economic importance and are of high biotechnological interest to researchers. In stark contrast to primary metabolites, these compounds do not directly mediate the growth and the reproduction of these photosynthetic organisms. Among the most sought-after products are metabolites (pigments) such as phycocyanin, phycoerythrin, allophycocyanin, astaxanthin, α-, β-, and ϒ-carotene, lutein, lycopene, violaxanthin, chlorophyll *a*, canthaxanthin, and a host of other substances produced by algae cells. These pigments are responsible for the visual appearance (color) of different algae (Table 3.1) (Nwoba et al. 2019a). The diversity of pigments (color) in microalgae primarily supports its applications in food as well as cosmetics and pharmaceuticals. Interestingly, many of these pigments in microalgae can be produced in concentrations higher than that of vascular plants (Fig. 3.2) or can only be found in microalgae (e.g., phycobiliproteins, fucoxanthin).

In comparison to terrestrial plants, microalgae have higher areal productivities due to their faster growth rates and are cost-effective (does not need arable land) while they can be also sustainably grown using seawater or wastewater (freshwater is limited), and finally the biomass produced is suitable for biorefinery application (co-produces several other high-value macromolecules) (Nwoba et al. 2019a). It is interesting to note that the pigment composition can significantly vary according to individual sub-strains of a particular species in which each strain is characterized by its own unique appearance (color) (Fig. 3.3). Despite the various appealing attributes of microalgae, only two different carotenoids (β-carotene from *Dunaliella salina* and astaxanthin from *Haematococcus pluvialis*) and one phycobiliprotein (phycocyanin from *Arthrospira platensis*) are currently produced at a commercial scale (Borowitzka 2013). These commercially exploited colorants have high yields and market values to warrant economically feasible production from microalgae. Figure 3.4 shows the images of commercial products that are currently produced from algal pigments.

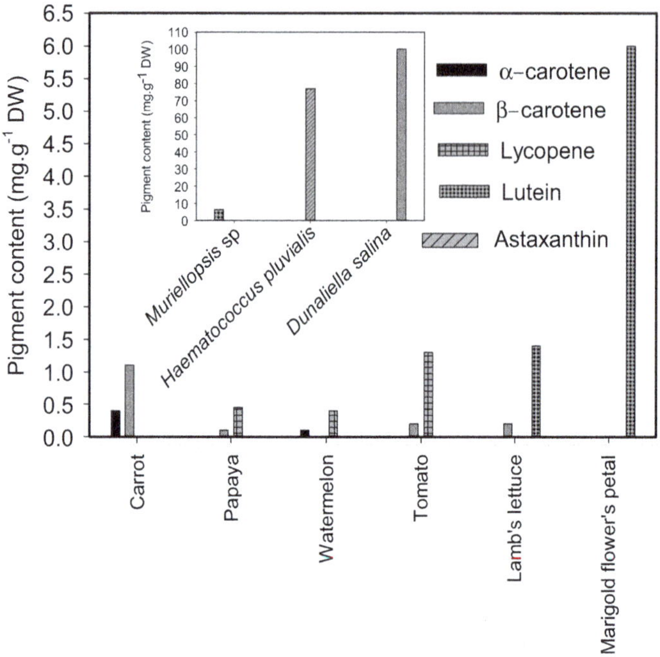

Fig. 3.2 Comparison of major pigment contents in some plants and microalgae (Del Campo et al. 2007b; Piccaglia et al. 1998; Mulders 2014)

3.1 Carotenoids

Carotenoids are derived from five-carbon isoprene units that are polymerized enzymatically to form highly conjugated 40-carbon structures, with up to 15 conjugated double bonds (Ambati et al. 2018). The use of carotenoids as a naturally derived food colorant has a recognizable impact and demand due to its striking color

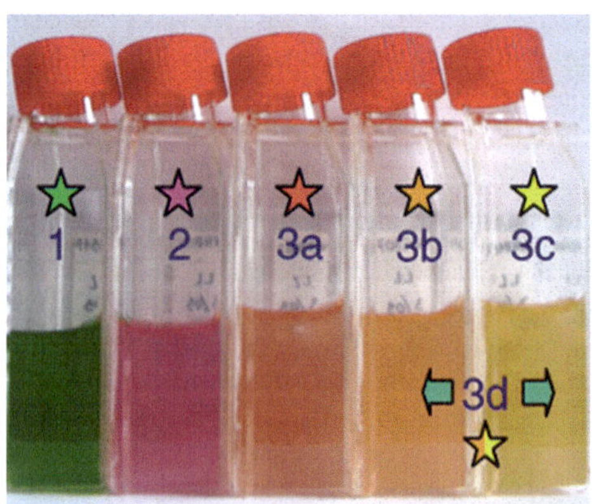

Fig. 3.3 Diversity of major pigment types (1–3) and subtypes (3a–c) of marine *Synechococcus* spp. Pigment subtype 3d represent the chromatic adaptation ability of the microalgae to modify their pigmentation from subtype 3b to 3c under different light spectral (colour) (Reproduced from Six et al. (2007) under Creative Commons attribution license)

Fig. 3.4 Commercial products derived from algal pigments as natural food colorant

characteristics and bioactivity as well as its well-documented antioxidative and preservative attributes (Rodriguez-Amaya 2019). For these reasons, carotenoids remain the best-studied algal pigment and represent the preferred choice of colorant by food producers, especially in the preparation of food with elevated content of fatty acids (e.g., margarine, butter, soft drinks, cakes, milk products). Carotenoids are lipid soluble in nature and can be obtained from photosynthetic vascular plants and algae and non-photosynthetic bacteria, fungi, and animals (Del Campo et al. 2007b). The global market for carotenoid was US$1.5 billion in 2017 and is projected to hit US$2.0 billion in 2022 (http://www.bccresearch.com). Carotenoids can be classified into two major groups: carotene (α-, β-, and ϒ-carotene) consisting of hydrogen and carbon atoms (true hydrocarbon) and xanthophylls containing oxo, hydroxyl, or epoxy groups. β-carotene is currently the most sought-after carotenoid pigment due to its wide applicability as a food colorant, a precursor to vitamin A synthesis in food and feed, an additive in multivitamin and cosmetic preparations, as well as an ingredient in functional foods. Generally, the color characteristics of carotenoids range from yellow, orange, and red and they have been shown to give colors to crustacean shell, fish skin (salmons), flowers, fruits, leaves, and so on.

Microalgae have been well recognized as one of the best producers of carotenoids (Martins et al. 2016). Microalgal carotenoids are stored in the chloroplasts in association with proteins as either a mixture of oil droplets or in a crystalline form which is specific for different microalgal class. For example, *Dunaliella* (Chlorophyceae) which is the most extensively studied β-carotene (E160a) producer has a β-carotene content ranging between 3.0 and 5.0% per dry weight of biomass. *D. salina*, a halotolerant green biflagellate, is cultured at commercial scale for the production of carotenoid (Fig. 3.2), boasting a remarkable yield of more than 400 mg β-carotene m^{-2} per cultivation area (Finney et al. 1984). β-carotenes derived from *Dunaliella* are also composed of other accessory carotenoids (in lower concentrations) which can exhibit additional nutritional benefits that cannot be found in synthetic alternatives. β-carotene is among the leading food colorants used globally, with a market value of US$253 million in 2009 (http://www.bccresearch.com). It is applied to many food products and beverages to enhance their attractiveness to customers. Among the food products that use β-carotene are margarine, cheese, juices, baked products, canned foods, confectioneries, health condiments, and dairy products (Table 3.2). Besides these applications to human foods, β-carotene is also used in pet food such as dogs, fish, birds, and cats to make them more appealing. Moreover, it is also used as animal feed to help improve the appearance (color) of birds, fish (e.g., flesh coloring of salmon), and crustaceans. In addition to its use as a colorant, β-carotene is also significantly exploited for its antioxidant properties, as well as a precursor to vitamin A. When used as a colorant in food, they can also provide additional medical benefits such as antihypertensive, anticancer, antidiabetic, and anti-inflammatory activities (Rodriguez-Amaya 2019). Furthermore, carotenoids sourced from microalgae are composed of both the *cis*- and *trans*-enantiomers which greatly improve its bioavailability and bio-efficacy that is not found in synthetic derivatives. It also possesses high bioactivity and anticancer property due to the presence of xanthophylls (oxygenated carotenoids). β-carotene derived from *Dunaliella* is marketed as β-carotene extracts, *Dunaliella* powder for human and dried *Dunaliella* for animal nutrition. The extracted and purified β-carotene is distributed in vegetable oil at concentrations ranging between 1 and 20% for coloring food products, as well as in softgels (5 mg/capsule) for personal

Table 3.2 Microalgal pigments with potential food colorant application

Colorant	E code	Pigment color	Uses	Major microalgal producers
Carotene (α-, β-, and trans-β-carotene)	160a	Red orange	Butter, margarine, cakes, dairy products, soft drinks	*Dunaliella salina*
Lutein	161b	Yellow orange	Dairy products, soft drinks, confections, salads	*Muriellopsis* sp.
Astaxanthin	161j	Red orange	Foods, nutraceuticals, and pharmaceuticals	*Haematococcus pluvialis*
Chlorophylls	140	Green	Beverages, fruit juices, pasta, dairy products, sweetener preparations, soups	*Chlorella* sp.
C-phycocyanin, allophycocyanin	n.a.	Blue	Food, nutraceuticals, pharmaceuticals, chewing gum, jellies	*Arthrospira platensis*
Phycoerythrin	n.a.	Red	Food, nutraceuticals, pharmaceuticals, chewing gum, jellies	*Porphyridium cruentum*
Phenolics	n.a.	Yellow, orange, red, dark purple	Beverages, dairy products, confectionary	*Arthrospira* sp.

n.a. not applicable

use. This purified β-carotene product is naturally associated with other carotenoids such as lutein, zeaxanthin, neoxanthin, violaxanthin, α-carotene, and cryptoxanthin derived from the *Dunaliella* cells.

Another high-value carotenoid of great commercial importance as a natural food colorant is astaxanthin (E161j), which is a ketocarotenoid from the xanthophyll family. Like β-carotene, astaxanthin has applications in animal and human nutrition, pharmaceuticals, cosmeceuticals, and nutraceuticals for the management of degenerative diseases including cancer prevention. Astaxanthin has a market value of US\$2500 kg^{-1} and a total market value of US\$257 million in 2009 (http://www.bccresearch.com). Despite the fact that the current market is dominated by synthetically derived astaxanthin, a significant increase in consumer push for natural products creates a promising opportunity for naturally sourced astaxanthin. Within the microalgae community, *Haematococcus pluvialis*, a freshwater alga, is the most abundant source of astaxanthin, accumulating this pigment at about 0.2–2.0% dry weight of biomass and currently cultivated on large-scale production (Borowitzka 1999).

3.2 Phycobiliproteins

Phycobiliproteins are accessory photosynthetic pigments that aggregate in algal cells to form the phycobilisomes, which are attached to the chloroplast's thylakoid membrane. Phycobiliproteins (PBPs) are made of chromophore-bearing polypeptides called the bilins. These bilins are synthesized from open-chain tetrapyrrole molecules covalently bonded to an apoprotein by thioether linkage. Two major classes are predominant and are typically known as the blue and red PBPs, called phycocyanin (C-phycocyanin, deep blue, and allophycocyanin, light blue) and phycoerythrin (red), respectively. Others in the phycobilin family are phycoviolobilin and phycourobilin, which are purple and yellow in color, respectively (Six et al. 2007). In addition to pharmaceutical and cosmetic applications, these water-soluble PBPs are widely sought-after as colorants in the food industry. The red or pinkish-red phycoerythrin is mostly produced from the microalga, *Porphyridium* sp. (e.g., *P. cruentum*, *P. purpureum*) with a yield of 200 mg L^{-1} and content of 15% dry weight (up to 30% under optimal condition) (Dufoss et al. 2005). This pigment can provide color in confectioneries, dairy products, and gelatin desserts at between 50 and 100 mg kg^{-1} of food (Table 3.2) (Mishra et al. 2008). When utilized as an ingredient in the preparation of dried food, its color is reported to be highly stable at 30 °C for 30 min, and when stored at low humidity condition at pH 6–7, the shelf life is seen to be extended (Dufoss et al. 2005). The application of red phycoerythrin from *Porphyridium* as a colorant in foodstuffs has received several patents (Dufoss et al. 2005). Besides food coloring, red phycoerythrin due to its inherent fluorescence properties can also be potentially used in foods, cake decoration, soft drinks, and alcoholic beverages that fluoresce under natural and ultraviolet illumination (Dufoss et al. 2005). However, the red phycoerythrin from *Porphyridium* sp. is yet to be approved by regulatory bodies for use in foodstuffs and cosmetics meant for human consumption.

A source of natural blue color from microalgae is the blue-green cyanobacterium, *Arthrospira* (*Spirulina*) sp. This cyanobacterium naturally synthesizes the blue PBP pigment called C-phycocyanin at elevated concentrations (Martelli et al. 2014). However, a red microalga, *Porphyridium aerugineum*, that lacks red phycoerythrin has also been reported to accumulate c-phycocyanin in large concentrations (Dufoss et al. 2005). In *Arthrospira*, the phycobilisome contains the c-phycocyanin (up to 20% dry weight) surrounding the allophycocyanin core peripherally. Unlike the c-phycocyanin from *Spirulina* which has already been certified safe and approved for use as a colorant in foods, that sourced from *P. aerugineum* is yet to be permitted for use in food and feed products by regulatory authorities. The current market value of C-phycocyanin is between the range of US$10 and 50 million annually, with the price of food-grade (OD_{620}/OD_{280} = <1.0) derivatives estimated at around US$500 kg^{-1} (Nwoba et al. 2019b). The amount of this blue pigment required for food coloring is between 140 and 180 mg kg^{-1} of food/drink (Dufoss et al. 2005). Blue colorants are commonly used in food products such as gelatin, ice cream, drinks, and confectioneries (Table 3.2). Blue colorants/pigments sourced from microalgae have been shown to be stable under ambient light, temperatures up to 60 °C for 40 min, and at pH 4–5 (Dufoss et al. 2005). Nevertheless, C-PC is highly sensitive to heat treatment or thermal processes (Chentir et al. 2018).

3.3 Lutein

Lutein represents one of the main carotenoids found in food products and human serum. In conjunction with zeaxanthin, they make up the vital component of the yellow spot pigment in the retina and lens of our eyes (Del Campo et al. 2007b; Whitehead et al. 2006). Lutein is currently utilized as a food colorant and also as a feed additive in fish and poultry farms. Previous studies have established the protective function of lutein in delaying the onset of chronic illnesses (Whitehead et al. 2006), resulting in its interest and need for large scale. The global market for lutein in 2009 was US$187 million, representing an annual growth of 6.9% from its value in 2004 (http://www.bccresearch.com).

Lutein is conventionally sourced from dark, leafy green vegetables (e.g., spinach and kale) and from food products of yellow color origin such as corn and egg yolk. The current main organic source for lutein in the market is from the crown petals of *Tagetes erecta* and *T. patula* (marigold) flowers (Del Campo et al. 2007b; Piccaglia et al. 1998). At least 95% of plant-derived lutein is esterified, in which almost half of the weight is composed of fatty acids. Therefore, processing of plant-derived lutein entails chemical saponification. In stark contrast to plant sources of lutein, microalgae accumulate lutein in the free non-esterified form (Del Campo et al. 2007a). Hence, microalgal-derived lutein is considered a more attractive and valuable alternative to plant-based lutein. The chlorophycean microalgae *Muriellopsis* and *Scenedesmus* spp. have been shown to accumulate large quantities of lutein in their biomass (Fig. 3.2). Furthermore, other microalgal species that can accumulate lutein in high concentration include *Chlorella protothecoides* and *C. zofingiensis* (Shi et al. 2006; Del Campo et al. 2004). Production of lutein from microalgae is yet to be commercialized when compared to other similar carotenoids such as β-carotene. However, pilot-scale outdoor production trials of lutein-enriched biomass from *Muriellopsis* and *Scenedesmus* spp. (Blanco et al. 2007; Del Campo et al. 2007b) have been established. *Muriellopsis* sp. cultivated in a 55 L closed tubular photobioreactor (surface area, 2.2 m^2) operated outdoor under a continuous culture system was shown to have a lutein content of 4.3 mg g^{-1} dry weight (DW) over biomass (Del Campo et al. 2007b). The biomass and lutein productivities obtained by Del Campo and colleagues were 40 g m^{-2} day $^{-1}$ and 180 mg m^{-2} day $^{-1}$, respectively. A lutein content of 4.5 mg g^{-1} and productivity of 290 mg m^{-2} day $^{-1}$ have been reported for *Scenedesmus almeriensis* cultivated in a 4000 L serpentine tubular photobioreactor positioned in a greenhouse (Del Campo et al. 2007b). Generally, the free lutein content of the biomass could range from 0.4 to 0.6% DW (e.g., *Muriellopsis* sp.), which is far higher than that of esterified lutein obtained from the petals of marigold plants.

3.4 Phenolics

Phenolic compounds are gaining attention as a promising group of naturally sourced food colorants. In addition to their coloring attributes, phenolics are widely recognized for their enhanced antioxidant, anti-inflammatory, health-promoting, and functional properties. More so, these compounds are generally responsible for the widespread variation in naturally derived food colors (Shahid and Mohammad 2013). However, the commercial production of phenolic compounds from plants is still limited by some limitations such as the challenge of employing good manufacturing practice regulations to cultivated plants, uncertainties due to the use of pesticide or herbicide deposits, and even other environmental contaminants (Kepekçi and Saygideger 2012). These challenges have sparked off a growing interest in using microalgae as a source of naturally derived phenolic compounds rather than vascular plants (Kepekçi and Saygideger 2012). Microalgae represent a promising source of phenolic compounds due to their ability to be cultivated in enclosed large-scale photobioreactors which reduces the risk of contamination and enhances the quality of the desired end product without the application of pesticides, herbicides, and toxic environmental pollutants. Harvesting and purification of the target compound are also much easier for microalgae when compared to vascular plants. However, all microalgal classes may not be suitable for the production of naturally derived phenolic compounds as some of them can be hindered by (a) possible toxin production, (b) low growth rates and consequently low yields, (c) difficulties in cultivation of some microalgal species, and (d) wide variation in the content of target compound (Kepekçi and Saygideger 2012; Li et al. 2007). The phenolic compounds often occur as simple phenols, tannins, lignin, phenolic acids, flavonoids, phenylpropanoids, and their derivatives. Generally, the safety, stability, and specificity of actions of these naturally sourced phenolic compounds still remain unclear (Carocho and Ferreira 2013), and hence they still do not have approved E codes (Table 3.2). Some examples of microalgae that have been previously investigated as potential sources of phenolic compounds are *Arthrospira* (*Spirulina*) *platensis*, *Phaeodactylum tricornutum*, *Chlorella vulgaris*, and *Tetraselmis suecica* (Batista et al. 2017). Among these species, the total phenolic content of *A. platensis* was found to be highest and ranged between 19 and 50 mg gallic acid equivalent per g of dry biomass (Batista et al. 2017; Kepekçi and Saygideger 2012).

3.5 Chlorophylls

Chlorophylls are considered the most abundant and common natural pigment which are currently sought after as both food and pharmaceutical colorants and functional food supplements. Besides their green colorant functions, chlorophylls have been also implicated as chemopreventive agents against the onset of degenerative diseases (Fernandes et al. 2007). Chemically, chlorophylls are porphyrin-containing compounds with macrocyclic tetrapyrrole nucleus. The tetrapyrrole backbones coordinate a centrally positioned magnesium atom bonded to the nitrogen atoms of the pyrrole molecules (da Silva Ferreira and Sant'Anna 2017). The pyrrole rings are linked to each other via methyne bridges with the double bonds forming a closed, conjugated loop (Fernandes et al. 2007). A long phytyl group derived from a 20-carbon isoprenoid alcohol molecule (phytol) is attached as the side chain, which confers the entire chlorophyll molecule its hydrophobic behavior. On the other hand, the closed loop of the conjugated double bonds forms the chromophore imparting light absorption ability to the molecule.

All plants and algae have chlorophylls as their primary pigment because of their vital role in photosynthesis. Recently, the commercial production of chlorophyll as a natural colorant in food (Table 3.2), feed, pharmaceutical, and cosmetic industries has been gaining immense attention (Christaki et al. 2015). Microalgae, being photosynthetic organisms, accumulate large amounts of chlorophyll in their biomass. Chlorophyll found in microalgae can be divided into various subclasses such as chlorophylls *a*, *b*, *c*, *d*, and *f*, in which *a* and *b* are the only green pigment found in most vascular plants. Chlorophyll *a* is abundantly present in all photosynthetic organisms and represents the main light-harvesting pigment required for photosynthesis to occur. Chlorophyll *b* (the second most abundant pigment) is distributed in all chlorophytes and their descendants and also in green plants, while chlorophyll *c* is only found in dinoflagellates, haptophytes, heterokonts, and cryptophytes (Table 3.1). Chlorophyll *d* is found in some red algae—rhodophytes (Schwartz and Lorenzo 1990), while chlorophyll *f* has been identified in certain cyanobacteria (da Silva Ferreira and Sant'Anna 2017). The different types of chlorophyll have been shown to slightly vary on the basis of their chemical structures, absorption spectra, and tonality (Christaki et al. 2015). Chlorophylls *a* and *b* are blue-green and brilliant green-colored pigments, respectively, whereas chlorophylls *c*, *d*, and *f* are somewhat of yellow-green, brilliant/forest green, and emerald green colors, respectively (Christaki et al. 2015). The major advantage of the commercial production of chlorophylls from microalgae is that it can be readily extracted and is economically viable without the need for its overproduction (da Silva Ferreira and Sant'Anna 2017).

Due to this inherent ability, microalgal species such as *Chlorella* with high growth rates can accumulate a total chlorophyll content of more than 45 mg g^{-1} DW under optimal culture conditions and represent an attractive candidate for commercial production (Christaki et al. 2015). Due to freshwater limitation around the globe, marine microalgal species represent a much more sustainable and attractive option for the commercial production of chlorophyll. It is to be noted that the content of chlorophyll in microalgal biomass can significantly decrease under so-called stress (unfavorable) conditions, in contrast to other bioactive compounds such as carotenoids whose overproduction is induced under unfavorable growth conditions (Markou and Nerantzis 2013).

4 Cultivation, Commercial-Scale Production, and Downstream Processing of Microalgal Pigments

Advances in the knowledge and improved awareness of the positive health benefits of naturally sourced pigments from microalgae have spurred the need for innovation in boosting their production. In order to meet the projected future demand of naturally sourced colorants, there is great need for large-scale production systems that are not only highly efficient but also cost-effective. Nonetheless, based on the inherent characteristics of each targeted pigment and its final application (purity) in the food industry, the choice of cultivation system may vary according to the type of microalgae and the targeted end product.

4.1 Current Cultivation Systems for the Production of Microalgae Pigments

Various configurations of different cultivation systems are currently employed for the large-scale production of multiple end products from microalgae. The choice of a particular cultivation system is only typically made after much consideration on the type of microalgae, its cultivation location, the type and use of the pigment, and, very importantly, the commercial value of the pigment. The cultivation systems currently employed for the production of useful metabolites by microalgae can be broadly classified into either open or closed systems (Nwoba et al. 2019a). A detailed description of these different systems including their advantages and disadvantages and also their potential applications is summarized in Table 3.3.

4.2 Cultivation Modes for the Production of Microalgae Pigments

Microalgae are capable of growth and the production of pigments under three different metabolic modes which are, namely, heterotrophic, photoautotrophic, and mixotrophic nutrition in which the choice of growth mode may vary depending on prevailing environmental conditions for some microalgae. These metabolic abilities and diversity have given microalgae a significant edge over other microorganisms and higher plants as a viable source of natural pigments. The different metabolic modes can be exploited for pigments and other metabolite production through the optimization of culture conditions as summarized in Table 3.4.

Table 3.3 Various types of systems used for cultivation of microalgae

Cultivation system	Description	Major characteristics
Natural waterbodies/ lakes/ponds	Natural lakes, lagoons, and other bodies of water that are exposed to direct solar illumination. They are enriched with nutrients and inoculated with the desired species of algae	They are very cheap, but productivity is usually very low because it is difficult to control culture conditions. Contamination is also a problem except for the growth of extremophiles. Mixing is usually by wind convection
Artificial ponds	These include various types of ponds such as circular, raceway, and rectangular ponds. They can be constructed of concrete or simply lined with polythene to prevent leakage. Mixing is usually by paddles of different designs. Large-scale ponds are usually exposed, but small- to medium-scale cultures can be under transparent roofs. The depth depends on the strain and solar light intensity in the area but usually not more than 30 cm	Productivity is usually higher than that of the natural bodies of water but much lower than that of closed photobioreactors. Contamination is also a problem. However, they are the most extensively used system for large-scale cultivation of microalgae, especially for the production of cell biomass and other high-volume, low-price products
Tubular photobioreactor	These are constructed of transparent tubes or glasses of various diameters and configurations, ranging from vertical, inclined, horizontal, and coils. Mixing is by air bubbling, but airlift systems are also very common. Illumination can be by solar irradiation, but indoor, tubular photobioreactors illuminated by LED, fluorescent lamps, halogen lamps, etc., are in use. LED lamps are now popular because of low energy consumption, low heat generation, longevity, and ease of control of the light wavelengths	The illumination surface-to-volume ratio is usually very high, and thus the productivity is much higher than that of ponds. The design can also be varied to maximize solar light interception, and the risk of contamination is lower. However, it is more expensive to construct and maintain. Cleaning of tubes can be technically challenging, and use of disposable tubes increases the production costs. They are suitable for small- to medium-scale production of relatively high-value products

(continued)

Table 3.3 (continued)

Cultivation system	Description	Major characteristics
Flat plate photobioreactors	These are hexagonal vessels with very small light paths. They vary in width (light path), length, and heights. They can be vertical or inclined at various angles. They are usually installed in units, and the spacing and orientation depend on the desired light intensity, which in turn depends on the strain, the product, and the prevailing solar light intensity. The higher the reactors, the higher will be an area of productivity, but the more difficult it is to fix them and avoid falling. Mixing is usually by air bubbling, and the lid can be closed or open	They are simpler to construct than tubular photobioreactors, and comparable productivities can be achieved. Since they are usually in units, the operation costs are usually higher than that of tubular photobioreactors. They are also suitable for small- to medium-scale cultivation of microalgae
Vertical column photobioreactors	These are cylindrical vertical transparent or opaque vessels that are mixed by bubbling and occasional paddle mixing (depending on the diameters). The small diameter units are mixed just by air bubbling, while the larger diameter units are mixed by both bubbling and intermittent paddle mixing. For large-scale production, many units are installed and operated independently. In such a case, the spacing is optimized to minimize shading and maximize solar light interception. The culture depth depends on the species, the prevailing solar light intensity, and the product	They are simpler to construct than flat plates, but the operation costs and productivities are comparable for most products. Some of them are used for on-site aquaculture feed production
Internally illuminated photobioreactors	These are conventional heterotrophic bioreactors that are internally illuminated by optical fibers, LED lamps, or fluorescent lamps	The productivities are much higher than other photobioreactors because the culture conditions (illumination, pH, temperature, mixing, etc.) can be precisely controlled just like the conventional heterotrophic bioreactors. However, they are very expensive, and it is still technically difficult to construct and operate very large-scale internally illuminated photobioreactors. They are thus suitable for only small-scale production of high-value products

Table 3.4 Culture modes for pigment production by microalgae

Cultivation mode	Description	Characteristics	References
Heterotrophic	Cultivation of microalgae in the dark, using organic carbon as the carbon and energy source	High biomass concentration can be achieved using conventional bioreactors. It is easy to control culture conditions and maintain monocultures. However, it applies to only strains with heterotrophic metabolism. For most strains, chlorophyll development and pigment production are suppressed under heterotrophic condition	Hu et al. (2018), Ogbonna and Tanaka (1998), Ogbonna and McHenry (2015)
Photoautotrophic	Cultivation of microalgae in light and the presence of inorganic carbon source but the absence of organic carbon source so that light is the energy source and inorganic carbon is used as a carbon source	Chlorophyll and pigment production is usually very high, but due to light limitation, the final biomass concentration is usually very low. Photobioreactors with good light supply (intensity and distribution) are required. Aside from open ponds which have contamination problems, it is technically challenging to construct efficient large-scale closed photobioreactors for phototrophy	Nwoba et al. (2019b)

(continued)

Table 3.4 (continued)

Cultivation mode	Description	Characteristics	References
Mixotrophic	Cultivation of microalgae in the presence of light and inorganic and organic carbon sources so that both heterotrophic and photoautotrophic metabolic activities take place simultaneously	Very high cell concentration with relatively high chlorophyll and pigment contents can be achieved. In some strains, both heterotrophic and photoautotrophic metabolic activities take place simultaneously and independently so that the growth rates and the final cell concentrations in mixotrophic cultures are the sums of the values obtained in heterotrophic and photoautotrophic cultures. However, in some strains, either the photoautotrophic or heterotrophic metabolic activities are inhibited or reduced in mixotrophic cultures. Depending on the type of bioreactor used, contamination can be a problem	Rezić et al. (2013)
Sequential heterotrophic/ photoautotrophic cultures	This is a two-phase cultivation method whereby the cells are first cultivated heterotrophically to high cell concentration, and the culture condition is switched to the photoautotrophic mode when the organic carbon source is exhausted or by harvesting the cells and suspending in photoautotrophic medium and illuminating the culture	Each phase can be optimized independently so that high cell concentrations with high pigment concentrations can be achieved. Only cells with heterotrophic metabolism can be employed, and for large-scale processes, two reactors – the conventional bioreactor and an efficient photobioreactors – are required	Ogbonna et al. (1997), Hata et al. (2001)

(continued)

Table 3.4 (continued)

Cultivation mode	Description	Characteristics	References
Sequential heterotrophic/ mixotrophic cultures	This is similar to sequential heterotrophic/ photoautotrophic cultures except that during the second phase, organic carbon is still present in the culture	The characteristics are similar to those of sequential heterotrophic/photoautotrophic cultures. However, since organic carbon is present in the second phase, contamination can be a problem. Relatively high cell concentration can be maintained in the second phase, but the cellular pigment concentrations are usually lower than that of the sequential heterotrophic/ photoautotrophic cultures	Bassi et al. (2014)
Cyclic heterotrophic/ photoautotrophic cultures	The cells are cultivated photoautotrophically during the day, but at night, a controlled amount of organic carbon source is added so that the cells grow heterotrophically	Night biomass loss which is usually experienced in outdoor photoautotrophic cultures is avoided, and cells grow continuously during the day and at night. Relatively high biomass and pigment concentrations can be achieved	Ogbonna and Tanaka (1998), Ogbonna et al. (2001), Mohsenpour and Willoughby (2013)
Cyclic mixotrophic/ heterotrophic cultures	This is outdoor mixotrophic cultivation of microalgae so that the cells grow mixotrophically during the day and heterotrophically during the night	Continuous cell growth during the day and night results in relatively high biomass concentration but pigment production is usually lower, and contamination can be a serious problem, depending on the type of photobioreactor used	Bouarab et al. (2004)

4.3 *Current Commercial Producers of Microalgae Pigments*

Currently, there are multiple existing companies that are engaged in the commercial production of pigments from microalgae. Despite much research and pilot-scale evaluation of indoor microalgal cultivation systems for the production of pigments, almost all of the current commercial producers employ outdoor systems that utilize natural solar light for the production of pigments (Table 3.5). In addition, most of them have also inclined toward the use of open ponds as their choice of cultivation system, while a very few have employed the use of tubular photobioreactors for the production of certain pigments such as astaxanthin.

Table 3.5 Commercial production of microalgae pigments

Company name	Microalgae	Product	Place/ country	Year of establishment	Production capacity (tons/ year)
Nutrex Hawaii	*Spirulina*, *Haematococcus pluvialis*	Hawaiian spirulina, BioAstin Hawaiian astaxanthin	Kailua-Kona, Hawaii	1990	n.a.
Jingzhou Natural Astaxanthin Inc.	*Haematococcus pluvialis*	Astaxanthin powder (1.5–3.0%), natural astaxanthin oleoresin, natural astaxanthin softgel (5–10%)	China	2003	18
Algatech	*Haematococcus pluvialis*, *Phaeodactylum tricornutum*, *Porphyridium cruentum*, *Nannochloropsis* sp.	AstaPure Astaxanthin, AstaPure Arava, FucoVital, Astaxanthin softgels, Astaxanthin beadlets	Israel, USA	1998	n.a.
Yaeyama Shokusan Co., Ltd.	*Chlorella*	Midori-no-Sachi Chlorella tablet	Okinawa, Japan	1975	420
Fuji Chemicals Industry Co., Ltd.	*Haematococcus pluvialis*	AstaReal, AstaTrol	Japan, Sweden	1946	1.6
Hydrolina Biotech Pvt. Ltd.	*Spirulina*	Vitalinaa-Spirulina capsules, tablets	Chennai, India	2003	n.a.
Mera Pharmaceuticals	*Haematococcus pluvialis*	Astaxanthin	Hawaii, USA	1983	6.6
Nikken Sohonsha Corporation	*Chlorella*, *Dunaliella*	β-carotene capsules and tablets	Japan	1975	n.a.
Sun Chlorella	*Chlorella*	Chlorella tablets and drinks	Osaka, Japan	1969	

(continued)

Table 3.5 (continued)

Company name	Microalgae	Product	Place/ country	Year of establishment	Production capacity (tons/ year)
Far East Microalgae Ind. Co., Ltd.	*Chlorella* and *Spirulina*	Organic spirulina and chlorella tablets, dietary supplements, and aquaculture feeds	Taiwan	1967	*Chlorella*—1000, *Spirulina*—200
Beijing Gingko Group	*Haematococcus pluvialis*	Pure astaxanthin	China	1995	0.9
Stone Forest Astaxanthin Biotech Co., Ltd.	*Haematococcus pluvialis*	Pure astaxanthin	China	n.a.	1.2
Yunnan Alphy Biotech Co., Ltd.	*Haematococcus pluvialis*	Pure astaxanthin	China	n.a.	0.6
Yunnan SGYJ Biotech Co., Ltd.	*Haematococcus pluvialis*	Pure astaxanthin	China	n.a.	0.4

n.a. information not available

4.4 Extraction and Further Downstream Processing/ Purification of the Pigments for Use as Food Colorants

The end application of microalgae pigments as a source of food colorant necessitates their isolation and availability in their intact form with high purity for maximum performance. The first step of pigment extraction from microalgae consists of harvesting/dewatering the biomass produced. There are various methods that are currently employed for harvesting microalgae which may include centrifugation, sedimentation using various flocculating agents including chemicals like alum (Gerchman et al. 2017) and bioflocculants such as *Moringa oleifera* seeds (Ogbonna and Edeh 2018) and chitosan (Chen et al. 2015), floatation and filtration method, or even a combination of some of these different methods. The choice and relative advantages of these harvesting methods depend on the type of microalgae as well as the scale of production. Continuous centrifugation is widely used for large-scale outdoor cultures, while sedimentation and filtration techniques are preferred for relatively small-scale systems.

Post harvesting, the algal biomass needs to be dried to reduce its water content and to preserve and make cell disruption much easier (brittle). There are various drying methods currently in use for pigment production such as hot air-drying and lyophilization. Targeted pigments can be located in various different organelles within algal cells; therefore, there is a need for efficient disruption or perforation of the cell walls for maximum extraction. The cell wall compositions of microalgae vary depending on its class, and thus the choice of disruption method according to its efficiency may vary according to the type of microalgae. Various mechanical and chemical methods including grinding, homogenization, ultrasound, or sonication (Hosikian et al. 2010) are used to break algal cell wall to facilitate solvent entry into the cell for pigment extraction. There are three basic criteria that must be taken into account in selecting a method for pigment extraction from microalgae. These considerations include the cell wall composition of the desired microalgae, the location of the targeted pigment(s), and the structure (stability) of the pigments. The composition of the cell wall determines the choice of the disruption technique, while the location of the pigment dictates the steps necessary for rupturing both the external and organelle walls.

The solvents used for extraction depend on the nature of the pigments to be extracted. For example, a mixture of organic solvents such as methanol and dichloroethane and N,N-hexane and ethanol is generally used to extract water-insoluble pigments such as astaxanthin. The choice of solvent for pigment extraction from microalgae is determined by the inherent ability of the solvent to dissolve and extract the pigment without interfering with the pigment structure and components (Soares et al. 2016). Any solvent selected for pigment extraction and purification must be able to isolate the target pigment with high efficiency and minimal impurities. It is therefore very pertinent to choose the right solvent(s) for each pigment of interest (Soares et al. 2016). The major requirements for solvents used for pigment extraction include non-toxicity, low cost, high efficiency in extracting the target pigment, and the ease with which the solvent can be recovered after extraction.

Current methods used for pigment extraction can be either grouped into conventional methods or non-conventional (electro-technological) methods (Table 3.6). Conventional methods involve the use of the individual or a combination of aqueous and organic solvents such as acetone, diethyl ether, ethanol, methanol, N,N-hexane, ethyl acetate, and chloroform for the extraction of pigments (Kumar et al. 2010). These solvents are used in various concentrations together with other treatment techniques such as saponification, repeated freezing and thawing, and heating. Aside from conventional solvents, supercritical fluid extraction (SFE) also represents a viable and efficient method for the extraction of various pigments. SFE excludes the need for organic solvent as it uses CO_2 as an extraction solvent. SFE is preferred to organic solvent extraction because the extracted products are usually intact and no organic solvents are required. The extracts are usually in their pure form, so less additional processing steps are required. The extraction temperature can be reduced to avoid extract breakdown. It is important to note that none of the conventional extraction methods discussed above fulfills all the requirements of a good extraction method that should include efficiency, rapidity, environmental friendliness, cost-effectiveness, and recoverability. Thus, there is still significant room for improvement that can be potentially achieved through the use of nonconventional procedures.

Table 3.6 Methods for pigment extraction from different species of microalgae

Microalgae	Pigment	Extraction method	References
Haematococcus pluvialis	Astaxanthin	Mix with dodecane for 48 h, saponify with 0.02 M NaOH, dissolve in MeOH, and leave to sediment in the dark at 4 °C for 12 h	Kang and Sim (2007)
Haematococcus pluvialis	Astaxanthin	Supercritical fluid extraction at 69.85 °C and 550 bar	Machmudah et al. (2006)
Haematococcus pluvialis	Astaxanthin	Digestion with 2 N HCl at 70 °C, followed by extraction with acetone for 1 h	Sarada et al. (2006)
Chlorococcum sp.	Astaxanthin	Use of a mixture of MeOH/dichloromethane in a ratio of 3:1, followed by saponification in darkness (50 mg NaOH in 100 ml MeOH) under 110 MPa. It is then filtered using a 0.45 μm filter	Ma and Chen (2001)
Dunaliella tertiolecta	Chlorophyll, β-carotene, fucoxanthin	Cold and hot soaking and ultrasound-assisted extraction, freeze-drying with acetone and ethanol	Pasquet et al. (2011)
Cylindrotheca closterium	Chlorophyll *a* and fucoxanthin	Since frustules present hinder solvent penetration, the biomass is irradiated for 3–15 min at 25–100 W with stirring at 56 °C before extraction	Pasquet et al. (2011)
Haematococcus pluvialis and *Dunaliella salina*	Astaxanthin and β-carotene	Solvent extraction, ultrasound-assisted extraction, microwave-assisted extraction, and supercritical fluid extraction	Saini and Keum (2018)
Desmodesmus sp.	Lutein, β-carotene, zeaxanthin, *trans*-lutein, chlorophyll	Ultrasound-assisted extraction with hexane, acetone, hexane/ethanol/acetone 10:6:7, and hexane/ethanol in various ratios. Extraction is done five times and concentrated with a rotary evaporator and dissolved in MeOH/dichloroethane mixture 1:1 (v/v), filtered and subjected to HPLC analysis	Soares et al. (2016)
Chlorella sp.	Chlorophyll and carotenoids	Biomass is ground with glass powder, preheating solvent and boiling with solvents. Acetone cold or hot extraction at 50 °C under shaking and centrifugation	Ilavarasi et al. (2012)
Acrochaete sp.			
Phormidium chlorinum,			
Jaaginema pseudgeminatum			
Botryococcus braunii UTEX LB572	Chlorophyll and carotenoids	Grinding with mortar and pestle. Pretreatment with 99.95% CO_2 rapidly, depressurization for 1 h at 21–49 °C and 6–13 MPa, followed by SFE at 40 °C and 30 MPa for 1 h	Uquiche et al. (2016)

(continued)

Table 3.6 (continued)

Microalgae	Pigment	Extraction method	References
Chlorella vulgaris	Lutein	Saponification with 10 M KOH and 2.5% ascorbic acid; incubate at 60 °C for 10 min at cool to room temperature, and extract with dichloromethane. Purify with a mixture of ethanol/water/dichloromethane or ethanol/water/hexane; washing with 30% aqueous ethanol removed water-soluble impurities	Li et al. (2002)
Haematococcus pluvialis	Astaxanthin	Dried biomass treated with 4 M HCl at 70 °C for 2 min, washed twice in deionized water, centrifuged and extracted with acetone by ultrasonication under ice for 20 min, or stepwise extraction with MeOH and acetone: ultrasonic extraction with 1 ml MeOH under ice, followed by 1 ml acetone extraction without light or soy oil extraction in which dried biomass is mixed with soy oil and agitated at room temperature for 2 h. It is filtered through a 0.22 μm cellulose filter	Dong et al. (2014)
Chlorococcum sp.	Astaxanthin	Saponification and organic solvent extraction to obtain crude astaxanthin	Li and Chen (2001)
Nannochloropsis gaditana	Chlorophyll and carotenoids	Acetone, ethanol, and methanol, freezing and thawing with liquid N_2 and in fridge at 60 °C, use of ultrasound-assisted extraction at 250 Hz	Henriques et al. (2007)

Non-conventional methods are innovative techniques which combine electrical power with or without the use of solvents. Among these are the use of electro-technologically assisted extraction processes such as pulsed electric field (PEF). This method of extraction does not generate heat. Here, the material (microalgae biomass) from which pigments are to be extracted is kept between two electrodes either in batches or in a continuous treatment chamber. The substrates are then subjected to a continual electric frequency that ranges from Hz to MHz with a strong electric field of about 0.1–80 kV cm^{-1} for less than a second. The pulses applied in PEF are either bi- or unipolar with exponential or square-shaped frequencies. The applied pulses create holes in the algal cell membranes and permit solvents to permeate into them. Controlling some of the parameters (intensity, strength, time) of the electric pulses applied can regulate the size of the holes on the cell membrane which may result in selective extraction of the pigment of interest (Töpfl 2006). Other electric-assisted extraction technologies include the use of a moderate electric field, high voltage electric discharges, supercritical and subcritical fluid extraction, pressurized liquid extraction, microwave-assisted extraction, ultrasound-assisted extraction, and high-pressure homogenization (Poojary et al. 2016). These upcoming technologies are fast (extraction can be completed within a short period) and may potentially eliminate the need of expensive solvents. Furthermore, since little or no solvents are used, a very little solvent is discarded into the environment. Thus, they are environmentally friendly. Most of these electro-technologies are very efficient, and therefore most of the target pigments can be extracted from the substrate. Furthermore, some of these methods can selectively target and isolate the pigment of interest with less impurity.

Post extraction, the crude pigment extracts often contain a mixture of different components that can consist of pigments, other metabolites, cell debris, and cellular components such as proteins and DNA, depending on the methods used for extraction, as well as residual solvents used in the extraction. The extracted pigments must therefore be separated and subjected to further purification steps before being formulated into desired end products. The choice of method and degree of purification highly depends on the intended use of the pigment. After extraction, rotary evaporators are typically used for evaporating the extraction solvent and concentrating the extracted pigment samples. Further separation and purification can be achieved by either filtration using membrane filter with different pore sizes, scanning electron photomicroscopy, reverse-phase high-performance liquid chromatography (HPLC), or photodiode array detector (Hosikian et al. 2010). The purified pigments are then formulated into powder, tablets, capsules, or even syrups. There are various methods used for pigment purification. These methods can either be used individually or in combinations based on the nature of the pigment, the type of contaminants, as well as the intended use. The first step in the purification of pigment extracts containing both fat-soluble pigments such as β-carotene and water-soluble pigments such as anthocyanin as impurities is to separate them using a biphasic system consisting of water-ethanol-dichloromethane at different ratios. Water-soluble impurities can be removed by washing in 30% aqueous ethanol, while fat-soluble impurities can be washed off through hexane extraction. This method has been shown to be successful for the recovery of lutein of about 90–98% purity (Li et al. 2002). Chromatographic separation methods are the most widely used method of pigment purification. The separation is based on the differences in the equilibrium distribution of the pigments and the contaminants and between the mobile and the stationary phases. For laboratory purifications (small scale), paper chromatography and thin-layer chromatographies (TLC) are used. For example, chlorophylls and carotenoids can be separated by first evaporating the extract using nitrogen gas and subsequently dissolving it in methanol/water/NH_3 solvent in the ratio of 90:10:0.0012 before separating it using a silica-based gel high-performance TLC (Tokarek et al. 2016).

Different chromatographic methods are commonly used for the large-scale purification of microalgae pigments. Adsorption chromatography involves the use of column packing materials that have great binding affinity to the targeted pigments which are selectively adsorbed to the column, while undesired contaminants are washed out using the mobile phase. The pigments are then eluted in pure form from the packing materials. In the case of ion-exchange chromatography, the column packing material is charged, and the relative adsorption of the pigment depends on the type and strength of their charge. This method is useful for the strong positively or negatively charged pigments. In such cases, negatively charged column packing materials are used for positively charged pigments, while positively charged packing materials are used for negatively charged pigments. Although anthocyanins are positively charged, most microalgae pigments such as carotenes and chlorophylls typically do not have a net electrical charge. Thus, the application of ion-exchange chromatography for purification of microalgae pigments is restricted. However, protein-based pigments such as B-phycoerythrin and R-phycoerythrin could be purified by relative ease through ion-exchange chromatography (Román et al. 2002). Size exclusion chromatography is commonly used when the molecular size of the pigment is different from those of the contaminants. Components with small molecular sizes enter the pores of the packing materials and thus have long residence time, while components with large molecular sizes are excluded from the pores of the packing material by virtue of their physical size and are carried along with the mobile phase so that they are eluted first. Affinity chromatography is another method widely used when the pigments or the contaminants have a high affinity for some specific ligands. By using ligands for which the pigments have high affinity, the pigments are retained on the column packing material (ligand), while all remaining contaminants are eluted. The retained pigments are later selectively eluted from the column using various buffered solutions containing the chosen ligands. Preparative high-speed countercurrent chromatography (HSCCC) with biphasic solvent system is an upcoming method that utilizes a mixed solution of n-hexane-ethanol-water in the ratio of 10:9:1 v/v for the purification of crude extract of canthaxanthin to a purity level of about 98.7% (Li et al. 2006). The same HSCCC with two-phase solvent system comprising n-hexane-ethyl acetate-ethanol-water in the ratio of 8:2:7:3 v/v has also been used to purify zeaxanthin to a purity level of up to 96.2% (Chen et al. 2005). Some protein pigments such as phycoerythrin require several subsequent purification steps even after the application of an initial chromatography method in order to obtain high-purity end products. For instance, the purification of crude extracts of phycoerythrin typically involves three distinguished chromatographic steps. Firstly, the crude extract of phycoerythrin is concentrated through ammonium sulfate precipitation. The concentrated pigment solution is then loaded onto a hydroxyapatite column and eluted with 100 mM phosphate buffer. The eluted fraction may have a purity ratio of about $A_{565}/A_{280} = 6.75$. The collected fraction is subsequently loaded onto Q-sepharose column that purifies it to $A_{565}/A_{280} = 15.48$ purity ratio. The third and final purification stage is done using a Sephacryl S-200 HR resin, and the resulting phycoerythrin has a purity ratio of $A_{565}/A_{280} = 17.3$ (Pumas et al. 2012). Other chromatographic techniques such as ion-exchange column can also be used in the purification of phycoerythrin. B-phycoerythrin represents another protein-based pigment that requires several stages of purification. For example, after obtaining a crude extract through osmotic shock, ultrafiltration is employed as the initial purification step for B-phycoerythrin. An additional purification using SOURCE 15Q exchange column is done, and the resulting B-phycoerythrin purity ratio was A_{545}/A_{280} of 5.1 (Tang et al. 2016). Open column chromatography (OCC) is another valuable form of chromatography that can be optimized for the fractionating of pigments for purification and identification purpose. The obtained fractions are usually confirmed by HPLC and then subjected to nuclear magnetic resonance (NMR) for structural elucidation (HPLC-NMR). For instance, Sivathanu and Palaniswamy (2012) used OCC to separate crude carotenoid pigments extracted from *Chlorococcum humicola* before subjecting the fractions to HPLC and NMR analysis for structural elucidation. Expanded bed adsorption (EBA) chromatography can also be used for phycoerythrin purification using dimethylaminoethyl cellulose (DEAE-c) as the resin. Bermejo et al. (2003) used EBA to purify crude phycoerythrin extracted from *Porphyridium cruentum*. The unbound contaminating proteins were washed with 50 mM acetic acid-sodium acetate buffer. The target protein (phycoerythrin) bound to the column was eluted with 250 mM acetic acid-sodium acetate buffer. Overall, chromatographic techniques are indispensable tools in pigment purification.

5 Factors Affecting the Production of Pigments in Microalgae

The inherent composition and physicochemical properties of microalgae are directly subjected to the growth and environmental conditions of which the algae is grown (Benavente-Valdés et al. 2016). Currently, there are two viable options in improving the yield of pigments in microalgae cells which can either involve the (a) manipulation of the growth/physical conditions of the microalgae cultures or (b) the alteration of metabolic pathways responsible for the production of pigments (Lamers et al. 2010; Mulders et al. 2014; Beer et al. 2009; Benavente-Valdés et al. 2016). The first option which is often termed as "stress conditions" typically involves drastic changes to the natural growth conditions of microalgae cultures that can either include (a) the use of subsaturation light conditions which have been shown to increase the concentration of primary pigments or (b) the application of unfavorable growth conditions (stress) to cultures for the increase of secondary pigments such as temperature, salinity, irradiance, salinity, and nutrient availability (Rao et al. 2007; Ördög et al. 2012; Benavente-Valdés et al. 2016).

5.1 Cultivation Condition

The type of cultivation can significantly impact the characteristics and composition of the desired microalgae. In general, most microalgae are autotrophic organisms which require light as a source of energy to fix inorganic carbon. Nevertheless, there are some species (heterotrophs) that are capable of exploiting organic sources of carbon as an energy and nutrient source for cell growth and development (Benavente-Valdés et al. 2016). An increase in the concentration of organic carbon under heterotrophic growth conditions has been found to favor the biosynthesis of secondary pigments in certain microalgae (Chen et al. 2009; Mojaat et al. 2008). In terms of direct comparison between the different cultivation conditions, photoautotrophic induction was found to be more efficient over heterotrophic induction methods for the production of astaxanthin in *Haematococcus pluvialis* (Kang et al. 2005).

5.2 Nutrient Limitation and Starvation

Nitrogen is a vital element for the survival of microalgae cells as it forms the major building blocks of protein and nucleic acids while also being essential for cellular division and growth (Benavente-Valdés et al. 2016). Algae cultures grown in the absence of nitrogen are typically under nitrogen starvation, while culture subjected to a low concentration of nitrogen are termed as nitrogen-limited (Bona et al. 2014). Changes in the concentration of nitrogen in the growth medium during the cultivation of microalgae can visibly affect a range of cellular metabolic activities. Cells grown under nitrogen limitation conditions are restricted in growth and generally respond by converting nitrogen-rich non-lipid cellular components into nitrogen for the formation of lipids (Benavente-Valdés et al. 2016). Under nitrogen-limiting conditions, photosynthesis of marine microalgae is negatively affected due to a decrease in chlorophyll content, while non-photosynthetic pigments such as carotenoids are found to increase (Berges et al. 1996; Benavente-Valdés et al. 2016). Chlorophylls are nitrogen-rich molecules that are generally degraded and serve as a nitrogen source to fuel further growth and biomass formation of cells under nitrogen-depleted conditions (Li et al. 2008). Despite its adverse effect on chlorophyll, a significant increase in carotenoids is observed for cells grown under nitrogen-limiting conditions (Fábregas et al. 1998; Benavente-Valdés et al. 2016). For example, nitrogen deficiency has been shown to be more efficient in increasing the astaxanthin content of *Haematococcus pluvialis* cultures than the increase brought forward by the changes in light irradiance (Fábregas et al. 1998).

Phosphorus represents another essential nutrient that can significantly impact the growth of microalgae through its key role in various metabolic processes such as respiration, cell division, and photosynthesis (Qu et al. 2008; John and Flynn 2000; Chen and Chen 2006). Phosphorus limitation has been observed to significantly limit the growth of microalgae while generally not affecting the content of primary pigments such as chlorophyll in algae cells (Kozłowska-Szerenos and Zieliński 2000; Kozłowska-Szerenos et al. 2004). In addition, the enhanced supplementation of other nutrients such as iron has also been shown to improve the synthesis of astaxanthin and β-carotene in microalgae (Cai et al. 2009; Mojaat et al. 2008).

5.3 Light Irradiance and Spectral Composition

Light (quantity and quality) is by far the main limiting factor affecting concentration of photosynthetic pigments in algae cells due to its direct impact on photosynthesis (Benavente-Valdés et al. 2016; Vadiveloo et al. 2016). In general, light irradiance has been inversely correlated with the cellular chlorophyll content of most green algae (Begum et al. 2016; Danesi et al. 2004; Vadiveloo et al. 2015). Microalgae cells grown under suboptimal or photo-limiting light irradiance have been shown to significantly increase the cellular concentration of primary pigments (photosystems) such as chlorophyll *a* (Begum et al. 2016; Chauhan and Pathak 2010; Danesi et al. 2004; Dubinsky and Stambler 2009). Such a response (photoacclimation) is typically brought forward by the algae cells as a mechanism to improve the efficiency of light absorption under these limiting conditions and has been typically observed in most microalgae (Nwoba et al. 2019a). Irradiance has been also shown to affect the accumulation of secondary pigments in algal cells; however, this response has only been typically observed in certain green microalgae (Chlorophytes) (Hejazi et al. 2004; Seyfabadi et al. 2011). In terms of photo-protective accessory pigments such as β-carotene and astaxanthin, high irradiance levels have been shown to significantly increase the concentration of these pigments (Fábregas et al. 1998; Hejazi et al. 2004). Oxidative stress brought forward by high irradiance is responsible for the enhanced production of secondary pigments (astaxanthin and β-carotene) for the protection of cells against oxidative damage (Ip and Chen 2005). On the contrary, low irradiance has been found to favor the production of accessory pigment such as phycobiliproteins in cyanobacteria (Grossman et al. 1995). It is important to note that quantity of light available to algae cells can be visibly affected by other confounding factors such as culture density, mixing regime, and shading which can result in the variation of pigment content in algal cells.

In addition, the spectral composition (color/wavelength) of incident light can also bring forward a significant change in the concentration of chlorophyll in algae cells by potentially acting as a photomorphogenic signal (Vadiveloo et al. 2017; Kagawa and Suetsugu 2007). Algal cultures subjected to wavelengths of light that are not efficiently absorbed by chlorophyll molecules (e.g., green and orange) have been shown to increase the cellular concentration of primary pigments (Vadiveloo et al. 2015; Vadiveloo et al. 2016). In terms of accessory pigments such as phycobiliproteins, either red or blue light has been reported to enhance its production in various cyanobacteria (Rodríguez et al. 1991). Other changes in light quality such as the effect of flashing light have also been found to significantly increase the production (at least fourfold) of secondary pigments in algal cells when compared to cultures grown in continuous light (Fábregas et al. 2001; Kim et al. 2006). Long-term adaptation of algae cultures under the absence of light (dark) can also significantly affect pigment composition (e.g., fucoxanthin, diatoxanthin, and diadinoxanthin) especially in diatoms (Veuger and van Oevelen 2011). The combination of ultraviolet radiation (UV-A) together with visible light (400–700 nm) has been reported to enhance the accumulation of carotenoids in algae cells when compared to cultures only subjected to visible light (Mogedas et al. 2009).

Overall, the quantity and quality of light are seen to affect the content of pigments in algal cells; however, most of these studies have only been performed in laboratory scales and might represent an expensive option for application during the mass outdoor cultivation of microalgae due to the need of additional artificial lighting based on the climatic condition of a locality.

5.4 Salinity

Changes in salinity or "salt stress" can be responsible for a wide range of bioenergetics and biochemical changes in microalgae through its effect on osmosis (excess concentration Na^+ and Cl^- ions) (Benavente-Valdés et al. 2016). In terms of pigment production, an increase in salinity above the tolerance threshold of microalgae cells is typically shown to increase the concentration of secondary pigments such as carotenoids while decreasing chlorophyll content (Ben-Amotz and Avron 1992; Wegmann 1986).

Freshwater microalgae such as *Chlorella* sp., *Haematococcus* sp., and *Scenedesmus* sp. have all been documented to significantly increase the content of their accessory pigments (e.g., astaxanthin and canthaxanthin) when grown under different saline conditions (Kobayashi et al. 1997; Li et al. 2009; Benavente-Valdés et al. 2016). Nonetheless, it is important to note that the continuous increase in salinity has been shown to negatively impact the growth of these microalgae species due to the imbalance in ion homeostasis, changes in osmotic pressure, and also the accumulation of reactive oxygen which ultimately results in programmed cell death (Affenzeller et al. 2009). Changes in salinity have also been observed to alter the pigment composition of marine microalgae. For example, the halotolerant green microalgae *Dunaliella salina* is reported to grow best at the optimum salinity range between 18 and 22% of NaCl (Borowitzka et al. 1984). However, the optimum salinity for the production of carotenoids has been highlighted to be above 27% NaCl, resulting in the accumulation of up to 14% of β-carotene over its biomass (Borowitzka et al. 1984). Nonetheless, Ishika et al. (2017) reported that halo-adaptation influenced fucoxanthin production in diatoms in which fucoxanthin content of diatoms has been found to be highest at their optimum salinities.

5.5 Temperature

Temperature is vital element responsible for the survival of all living organisms as it directly affects various metabolic pathways and the biochemical composition of cells (Fon Sing et al. 2011). The optimum growth temperature and tolerance of microalgae or the production of pigments generally vary according to individual species (Richmond 1986). For example, 25 °C, 35 °C, and 36 °C are reported to be the optimum temperature for the production of phycobiliproteins from *Spirulina platensis*, *Anabaena* sp., and *Nostoc* sp., respectively (Moreno et al. 1995). Overall, an increase in culture temperature has been shown to favor the accumulation of secondary carotenoids such as lutein and β-carotene (García-González et al. 2005). Nonetheless, any further increase in temperature above the tolerance threshold of a particular species can be detrimental as it can significantly reduce biomass productivity and result in the death of cells (Guedes et al. 2011).

5.6 *pH*

The pH of the inherent culture medium can also influence the pigment composition of microalgae. Microalgae are typically shown to produce maximum concentration of pigments at optimum pH which is species-specific, e.g., halophilic green microalgae *Dunaliella salina* produce β-carotene at pH 7.5 (García-González et al. 2005), green microalgae *Scenedesmus almeriensis* produce high amount of carotenoids at pH 8, whereas at pH 7 the production of carotenoids is highest in *Chlorococcum citriforme* and *Neospongiococcus gelatinosum* (Del Campo et al. 2000).

5.7 *Two-Stage Cultivation*

In most situations, an increase in the concentrations of pigments (primary and secondary) of microalgae cells is typically only achieved at the expense of growth and biomass formation. This outcome is highly unfavorable for the commercial production of algal biomass in which high-density cultures are required to improve production efficiency and its economics (Benavente-Valdés et al. 2016). As such, two-stage algae cultivation systems which involve the cultivation of the microalgae under two distinct different growth conditions (e.g., nutrients, temperature, salinity, and light) are preferred for the production of valuable pigments from microalgae.

Two-stage cultivation systems generally consist of high growth and biomass formation first phase, followed by a subsequent pigment induction phase that occurs separately. Such systems have been employed for improving the production yield of high-value compounds in microalgae cells. For example, the commercial production of astaxanthin from the green alga *Haematococcus* is typically carried out in two distinct stages which involves an initial phase consisting of the production of fast-growing and healthy green biomass under near-optimal growth conditions, followed by a second reddening phase in which astaxanthin accumulation is induced under environmental and nutrient stress conditions (Lorenz and Cysewski 2000).

Other strategies that are currently employed in two-stage cultivation systems for enhancing the accumulation of high-value products in microalgae cells can either include the combination and optimization of phototrophic, heterotrophic, or mixotrophic growth condition (Yen and Chang 2013; Ogbonna et al. 1997).

5.8 *Metabolic Engineering*

As highlighted earlier, the other prominent approach that can be exploited to improve the pigment content of microalgae is through the regulation of cellular metabolic pathways that can either be brought forward by (a) the up- or downregulation of enzymes responsible for the biosynthesis of pigments or (b) the formation of a metabolic sink (Mulders et al. 2014). Up- and downregulation of enzymes that catalyze the biosynthesis of pigments can specifically allow for the increase of the desired end pigment through the manipulation of the respective pigment metabolic activity (Mulders et al. 2014). In an ideal situation, this would only involve the overexpressing of enzymes that are directly responsible for the end production of the targeted pigment without affecting the concentration of other metabolites (Rosenberg et al. 2008). In addition, such approaches could also be exploited for the simultaneous increase of multiple pigment production such as chlorophylls and carotenoids (Estévez et al. 2001; Mulders et al. 2014). Nonetheless, in most situations, the end production of various metabolites is not generally governed by a single enzyme but shared by multiple enzymes (Kacser 1995). Thus, it is most likely that the overexpression of multiple enzymes is required to bring forward an increase in the production of an individual or group of pigments (Mulders et al. 2014). Downregulation of enzyme activity through the addition of enzyme inhibitors and gene knockout strategies can also contribute to the increased production of pigments (Mulders et al. 2014). Much different to upregulation strategies, downregulation involves the reduction of flux toward undesired side branches or end products. For example, an improved flux of the enzyme activity can be redirected to the formation of the desired pigment by blocking side pathways that result in the formation of undesired metabolites or pigments (Mulders et al. 2014). The major challenge associated with the overproduction of pigments is the storage and transport of these pigments within the photosynthetic apparatus which is absent in most microalgae species (Mulders et al. 2014). The use of specific enzymes has been proposed to introduce extra storage space outside of the photosystem and efficient transport of the pigment molecules (Mulders et al. 2014). In general, drastic changes to growth conditions (e.g., high light and salinity) have been typically observed to only increase the production of secondary carotenoids (e.g., β-carotene and astaxanthin). For the overproduction of other pigments found in microalgae, the use of metabolic engineering procedures such as the modification of enzyme expression seems to be promising. Nonetheless, the understanding behind the working mechanisms is limited, resulting in the end procedure being unpredictable. Further research and additional toolboxes are still required to improve current metabolic engineering procedures in microalgae, especially for pigment production.

5.9 Bioprospecting of New Strains

Currently, there are only a limited number of microalgae strains that have been commercially exploited for the production of pigments. Thus, there is significant room for further improvement through the screening and identification of new strains that are not only robust with high growth characteristics but also able to produce high concentrations of the targeted end products (Richmond 2000). Such screening and bioprospecting procedures must be comprehensive and exploit the physiological characteristics or physical properties of algae strains that can minimize or circumvent some of the core problems that are currently associated with the production of pigments from microalgae (Barclay and Apt 2013).

6 Conclusion

The significant increase in the preference and demand for natural colorants over synthetic ones by consumers necessitates the development and exploitation of available natural resources. Microalgae are ideal biofactories that clearly show potential to meet the demand for the production of food colorants as they are able to accumulate a wide range of pigments at high concentration coupled with their inherent fast growth rates. Commercial-scale pigment production from microalgae is currently a reality, although limited to a few microalgal species. *Haematococcus pluvialis* for astaxanthin, *Dunaliella salina* for β-carotene, *Chlorella vulgaris* for chlorophyll, and *Arthrospira platensis* for c-phycocyanin represent viable large-scale productions. The diversity and richness of pigment distribution in microalgae enhance its prospects for the development of other nonconventional colorants. Nonetheless, there are still some bottlenecks such as economic constraints and low technology readiness level that must be overcome and addressed before realizing its true potential. As such, the exploitation of microalgae for the production of natural colorants is highly favorable as it can increase profitability, it provides additional health benefits, and, most importantly, it can overcome sustainability challenges.

References

Affenzeller, M. J., Darehshouri, A., Andosch, A., Lütz, C., & Lütz-Meindl, U. (2009). Salt stress-induced cell death in the unicellular green alga *Micrasterias denticulata*. *Journal of Experimental Botany, 60*(3), 939–954.

Ambati, R. R., Gogisetty, D., Aswathanarayana, R. G., Ravi, S., Bikkina, P. N., Bo, L., et al. (2018). Industrial potential of carotenoid pigments from microalgae: Current trends and future prospects. *Critical Reviews in Food Science and Nutrition*, 1–22.

Bagchi, D. (2006). Nutraceuticals and functional foods regulations in the United States and around the world. *Toxicology, 221*(1), 1.

Barclay, W., & Apt, K. (2013). *Strategies for bioprospecting microalgae for potential commercial applications. Handbook of microalgal culture: Applied phycology and biotechnology* (2nd ed., pp. 69–79). Chichester: Wiley-Blackwell.

Bassi, A., Saxena, P., & Aguirre, A.-M. (2014). Mixotrophic algae cultivation for energy production and other applications. In R. Bajpal, A. Prokop, & M. Zappi (Eds.), *Algal biorefineries* (pp. 177–202). New York, NY: Springer.

Batista, A. P., Niccolai, A., Fradinho, P., Fragoso, S., Bursic, I., Rodolfi, L., Biondi, N., Tredici, M. R., Sousa, I., & Raymundo, A. (2017). Microalgae biomass as an alternative ingredient in cookies: Sensory, physical and chemical properties, antioxidant activity and in vitro digestibility. *Algal Research, 26*, 161–171.

Beer, L. L., Boyd, E. S., Peters, J. W., & Posewitz, M. C. (2009). Engineering algae for biohydrogen and biofuel production. *Current Opinion in Biotechnology, 20*(3), 264–271.

Begum, H., Yusoff, F. M., Banerjee, S., Khatoon, H., & Shariff, M. (2016). Availability and utilization of pigments from microalgae. *Critical Reviews in Food Science and Nutrition, 56*(13), 2209–2222.

Ben-Amotz, A., & Avron, M. (1992). *Dunaliella: Physiology, biochemistry, and biotechnology*. Boca Raton, FL: CRC press.

Benavente-Valdés, J. R., Aguilar, C., Contreras-Esquivel, J. C., Méndez-Zavala, A., & Montañez, J. (2016). Strategies to enhance the production of photosynthetic pigments and lipids in *chlorophyceae* species. *Biotechnology Reports, 10*, 117–125.

Berges, J. A., Charlebois, D. O., Mauzerall, D. C., & Falkowski, P. G. (1996). Differential effects of nitrogen limitation on photosynthetic efficiency of photosystems I and II in microalgae. *Plant Physiology, 110*(2), 689–696.

Bermejo, R., Acién, F. G., Ibáñez, M. J., Fernández, J. M., Molina, E., & Alvarez-Pez, J. M. (2003). Preparative purification of B-phycoerythrin from the microalga *Porphyridium cruentum* by expanded-bed adsorption chromatography. *Journal of Chromatography B, 790*(1–2), 317–325.

Blanco, A. M., Moreno, J., Del Campo, J. A., Rivas, J., & Guerrero, M. G. (2007). Outdoor cultivation of lutein-rich cells of *Muriellopsis* sp. in open ponds. *Applied Microbiology and Biotechnology, 73*(6), 1259–1266.

Bona, F., Capuzzo, A., Franchino, M., & Maffei, M. E. (2014). Semicontinuous nitrogen limitation as convenient operation strategy to maximize fatty acid production in *Neochloris oleoabundans*. *Algal Research, 5*, 1–6.

Borowitzka, M. A. (1999). Commercial production of microalgae: Ponds, tanks, tubes and fermenters. *Journal of Biotechnology, 70*(1), 313–321.

Borowitzka, M. A. (2013). High-value products from microalgae—Their development and commercialisation. *Journal of Applied Phycology, 25*(3), 743–756.

Borowitzka, L., Moulton, T., & Borowitzka, M. (1984). The mass culture of *Dunaliella salina* for fine chemicals: From laboratory to pilot plant. In C. J. Bird & M. A. Ragan (Eds.), *Eleventh international seaweed symposium* (pp. 115–121). Dordrecht: Springer.

Bouarab, L., Dauta, A., & Loudiki, M. (2004). Heterotrophic and mixotrophic growth of *Micractinium pusillum Fresenius* in the presence of acetate and glucose: Effect of light and acetate gradient concentration. *Water Research, 38*(11), 2706–2712.

Cai, M., Li, Z., & Qi, A. (2009). Effects of iron electrovalence and species on growth and astaxanthin production of *Haematococcus pluvialis*. *Chinese Journal of Oceanology and Limnology, 27*(2), 370.

Carocho, M., & Ferreira, I. C. (2013). A review on antioxidants, prooxidants and related controversy: Natural and synthetic compounds, screening and analysis methodologies and future perspectives. *Food and Chemical Toxicology, 51*, 15–25.

Carocho, M., Barreiro, M. F., Morales, P., & Ferreira, I. C. (2014). Adding molecules to food, pros and cons: A review on synthetic and natural food additives. *Comprehensive Reviews in Food Science and Food Safety, 13*(4), 377–399.

Carocho, M., Morales, P., & Ferreira, I. C. (2015). Natural food additives: Quo vadis? *Trends in Food Science & Technology, 45*(2), 284–295.

Chauhan, U., & Pathak, N. (2010). Effect of different conditions on the production of chlorophyll by *Spirulina platensis*. *Journal of Algal Biomass Utilization, 1*(4), 89–99.

Chen, G.-Q., & Chen, F. (2006). Growing phototrophic cells without light. *Biotechnology Letters, 28*(9), 607–616.

Chen, F., Li, H.-B., Wong, R. N.-S., Ji, B., & Jiang, Y. (2005). Isolation and purification of the bioactive carotenoid zeaxanthin from the microalga *Microcystis aeruginosa* by high-speed counter-current chromatography. *Journal of Chromatography A, 1064*(2), 183–186.

Chen, H., Jiang, J. G., & Wu, G. H. (2009). Effects of salinity changes on the growth of *Dunaliella salina* and its isozyme activities of glycerol-3-phosphate dehydrogenase. *Journal of Agricultural and Food Chemistry, 57*(14), 6178–6182.

Chen, W., He, Y., Zhou, Y., Shao, Y., Feng, Y., Li, M., & Chen, F. (2015). Edible filamentous fungi from the species Monascus: Early traditional fermentations, modern molecular biology, and future genomics. *Comprehensive Reviews in Food Science and Food Safety, 14*(5), 555–567.

Chentir, I., Hamdi, M., Li, S., Doumandji, A., Markou, G., & Nasri, M. (2018). Stability, bio-functionality and bio-activity of crude phycocyanin from a two-phase cultured *Saharian Arthrospira* sp. strain. *Algal Research, 35*, 395–406.

Christaki, E., Bonos, E., & Florou-Paneri, P. (2015). Innovative microalgae pigments as functional ingredients in nutrition. In S. K. Kim (Ed.), *Handbook of marine microalgae* (pp. 233–243). London: Elsevier.

da Silva Ferreira, V., & Sant'Anna, C. (2017). Impact of culture conditions on the chlorophyll content of microalgae for biotechnological applications. *World Journal of Microbiology and Biotechnology, 33*(1), 20.

Danesi, E. D. G., Rangel-Yagui, C. O., Carvalho, J. C. M., & Sato, S. (2004). Effect of reducing the light intensity on the growth and production of chlorophyll by *Spirulina platensis*. *Biomass Bioenergy, 26*(4), 329–335.

Del Campo, J. A., Moreno, J., Rodríguez, H., Vargas, M. A., Rivas, J., & Guerrero, M. G. (2000). Carotenoid content of chlorophycean microalgae: Factors determining lutein accumulation in *Muriellopsis* sp. (Chlorophyta). *Journal of Biotechnology, 76*(1), 51–59.

Del Campo, J., Rodriguez, H., Moreno, J., Vargas, M., Rivas, J., & Guerrero, M. (2004). Accumulation of astaxanthin and lutein in *Chlorella zofingiensis* (Chlorophyta). *Applied Microbiology and Biotechnology, 64*(6), 848–854.

Del Campo, J., Garcia-Gonzalez, M., & Guerrero, M. (2007a). Outdoor cultivation of microalgae for carotenoid production: Current state and perspectives. *Applied Microbiology and Biotechnology, 74*, 1163–1174.

Del Campo, J. A., García-González, M., & Guerrero, M. G. (2007b). Outdoor cultivation of microalgae for carotenoid production: Current state and perspectives. *Applied Microbiology and Biotechnology, 74*(6), 1163–1174.

Dias, M. I., Ferreira, I. C., & Barreiro, M. F. (2015). Microencapsulation of bioactives for food applications. *Food & Function, 6*(4), 1035–1052.

Dong, S., Huang, Y., Zhang, R., Wang, S., & Liu, Y. (2014). Four different methods comparison for extraction of astaxanthin from green alga *Haematococcus pluvialis*. *The Scientific World Journal, 2014*, 694305.

Dubinsky, Z., & Stambler, N. (2009). Photoacclimation processes in phytoplankton: Mechanisms, consequences, and applications. *Aquatic Microbial Ecology, 56*(2–3), 163–176.

Dufoss, L., Galaup, P., Yarnon, A., Arad, S. M., Blanc, P., Kotamballi, N. C., et al. (2005). Microorganisms and microalgae as source of pigments for use: A scientific oddity or an industrial reality? *Trends in Food Science and Technology, 16*, 389–406.

Estévez, J. M., Cantero, A., Reindl, A., Reichler, S., & León, P. (2001). 1-Deoxy-D-xylulose-5-phosphate synthase, a limiting enzyme for plastidic isoprenoid biosynthesis in plants. *Journal of Biological Chemistry, 276*(25), 22901–22909.

Fábregas, J., Domínguez, A., Álvarez, D. G., Lamela, T., & Otero, A. (1998). Induction of astaxanthin accumulation by nitrogen and magnesium deficiencies in *Haematococcus pluvialis*. *Biotechnology Letters, 20*(6), 623–626.

Fábregas, J., Otero, A., Maseda, A., & Domínguez, A. (2001). Two-stage cultures for the production of astaxanthin from *Haematococcus pluvialis*. *Journal of Biotechnology, 89*(1), 65–71.
Farré, G., Sanahuja, G., Naqvi, S., Bai, C., Capell, T., Zhu, C., & Christou, P. (2010). Travel advice on the road to carotenoids in plants. *Plant Science, 179*(1–2), 28–48.
Fernandes, T. M., Gomes, B. B., & Lanfer-Marquez, U. M. (2007). Apparent absorption of chlorophyll from spinach in an assay with dogs. *Innovative Food Science & Emerging Technologies, 8*(3), 426–432.
Finney, K., Pomeranz, Y., & Bruinsma, B. (1984). Use of algae Dunaliella as a protein supplement in bread. *Cereal Chemistry, 61*, 402–406.
Fon Sing, S., Isdepsky, A., Borowitzka, M., & Moheimani, N. (2011). Production of biofuels from microalgae. *Mitigation and Adaptation Strategies for Global Change*, 1–26. https://doi.org/10.1007/s11027-011-9294-x.
Gantt, E., & Cunningham Jr, F. X. (2001). Algal pigments. In: *Encyclopedia of life sciences*. John Wiley Publisher. https://doi.org/10.1038/npg.els.0000323.
García-González, M., Moreno, J., Manzano, J. C., Florencio, F. J., & Guerrero, M. G. (2005). Production of *Dunaliella salina* biomass rich in 9-cis-β-carotene and lutein in a closed tubular photobioreactor. *Journal of Biotechnology, 115*(1), 81–90.
Gerchman, Y., Vasker, B., Tavasi, M., Mishael, Y., Kinel-Tahan, Y., & Yehoshua, Y. (2017). Effective harvesting of microalgae: Comparison of different polymeric flocculants. *Bioresource Technology, 228*, 141–146.
Grossman, A. R., Bhaya, D., Apt, K. E., & Kehoe, D. M. (1995). Light-harvesting complexes in oxygenic photosynthesis: Diversity, control, and evolution. *Annual Review of Genetics, 29*(1), 231–288.
Guedes, A. C., Amaro, H. M., & Malcata, F. X. (2011). Microalgae as sources of carotenoids. *Marine Drugs, 9*(4), 625–644.
Hata, N., Ogbonna, J. C., Hasegawa, Y., Taroda, H., & Tanaka, H. (2001). Production of astaxanthin by *Haematococcus pluvialis* in a sequential heterotrophic-photoautotrophic culture. *Journal of Applied Phycology, 13*(5), 395–402.
Heer, K., & Sharma, S. (2017). Microbial pigments as a natural color: A review. *International Journal of Pharmaceutical Sciences and Research, 8*(5), 1913–1922.
Hejazi, M., Holwerda, E., & Wijffels, R. (2004). Milking microalga *Dunaliella salina* for β-carotene production in two-phase bioreactors. *Biotechnology and Bioengineering, 85*(5), 475–481.
Henriques, M., Silva, A., & Rocha, J. (2007). Extraction and quantification of pigments from a marine microalga: A simple and reproducible method. *Communicating Current Research and Educational Topics and Trends in Applied Microbiology Formatex, 2*, 586–593.
Hosikian, A., Lim, S., Halim, R., & Danquah, M. K. (2010). Chlorophyll extraction from microalgae: A review on the process engineering aspects. *International Journal of Chemical Engineering, 2010*, 11.
Hu, J., Nagarajan, D., Zhang, Q., Chang, J.-S., & Lee, D.-J. (2018). Heterotrophic cultivation of microalgae for pigment production: A review. *Biotechnology Advances, 36*(1), 54–67.
Humphrey, A. (1980). Chlorophyll. *Food Chemistry, 5*(1), 57–67.
Ilavarasi, A., Pandiaraj, D., Mubarakali, D., Ilyas, M., & Thajuddin, N. (2012). Evaluation of efficient extraction methods for recovery of photosynthetic pigments from microalgae. *Pakistan Journal of Biological Sciences, 15*, 883–888.
Ip, P.-F., & Chen, F. (2005). Production of astaxanthin by the green microalga *Chlorella zofingiensis* in the dark. *Process Biochemistry, 40*(2), 733–738.
Ishika, T., Moheimani, N. R., Bahri, P. A., Laird, D. W., Blair, S., & Parlevliet, D. (2017). Halo-adapted microalgae for fucoxanthin production: Effect of incremental increase in salinity. *Algal Research, 28*, 66–73. https://doi.org/10.1016/j.algal.2017.10.002.
Jacobson, M., & Kobylewski, S. (2010). Color us worried: Why synthetic food dyes should be banned. SAFE-FOOD REPORT. *Nutrition Action Healthletter, 37*(10).
Jeffrey, S. W., Wright, S. W., & Zapata, M. (2011). Microalgal classes and their signature pigments. In C. A. Llewellyn, E. S. Egeland, G. Johnsen, & S. Roy (Eds.), *Phytoplankton*

pigments: Characterization, chemotaxonomy and applications in oceanography. Cambridge environmental chemistry series (pp. 3–77). Cambridge: Cambridge University Press. https://doi.org/10.1017/CBO9780511732263.004.

John, E. H., & Flynn, K. J. (2000). Modelling phosphate transport and assimilation in microalgae; how much complexity is warranted? *Ecological Modelling, 125*(2–3), 145–157.

Kacser, H. (1995). *Recent developments beyond metabolic control analysis*. London: Portland Press Limited.

Kagawa, T., & Suetsugu, N. (2007). Photometrical analysis with photosensory domains of photoreceptors in green algae. *FEBS Letters, 581*(3), 368–374.

Kang, C. D., & Sim, S. J. (2007). Selective extraction of free astaxanthin from *Haematococcus* culture using a tandem organic solvent system. *Biotechnology Progress, 23*(4), 866–871.

Kang, C., Lee, J., Park, T., & Sim, S. (2005). Comparison of heterotrophic and photoautotrophic induction on astaxanthin production by *Haematococcus pluvialis*. *Applied Microbiology and Biotechnology, 68*(2), 237–241.

Kepekçi, R. A., & Saygideger, S. D. (2012). Enhancement of phenolic compound production in *Spirulina platensis* by two-step batch mode cultivation. *Journal of Applied Phycology, 24*(4), 897–905.

Kim, Z.-H., Kim, S.-H., Lee, H.-S., & Lee, C.-G. (2006). Enhanced production of astaxanthin by flashing light using *Haematococcus pluvialis*. *Enzyme and Microbial Technology, 39*(3), 414–419.

Kobayashi, M., Kurimura, Y., & Tsuji, Y. (1997). Light-independent, astaxanthin production by the green microalga *Haematococcus pluvialis* under salt stress. *Biotechnology Letters, 19*(6), 507–509.

Kozłowska-Szerenos, B., & Zieliński, P. (2000). Involvement of glycolate metabolism in acclimation of *Chlorella vulgaris* cultures to low phosphate supply. *Plant Physiology and Biochemistry, 38*(9), 727–734.

Kozłowska-Szerenos, B., Bialuk, I., & Maleszewski, S. (2004). Enhancement of photosynthetic O_2 evolution in *Chlorella vulgaris* under high light and increased CO_2 concentration as a sign of acclimation to phosphate deficiency. *Plant Physiology and Biochemistry, 42*(5), 403–409.

Kumar, P., Ramakritinan, C., & Kumaraguru, A. (2010). Solvent extraction and spectrophotometric determination of pigments of some algal species from the shore of Puthumadam, southeast coast of India. *International Journal of Oceans and Oceanography, 4*(1), 29–34.

Lamers, P. P., van de Laak, C. C. W., Kaasenbrood, P. S., Lorier, J., Janssen, M., De Vos, R. C. H., Bino, R. J., & Wijffels, R. H. (2010). Carotenoid and fatty acid metabolism in light-stressed *Dunaliella salina*. *Biotechnology and Bioengineering, 106*(4), 638–648.

Laokuldilok, N., Thakeow, P., Kopermsub, P., & Utama-Ang, N. (2016). Optimisation of microencapsulation of turmeric extract for masking flavour. *Food Chemistry, 194*, 695–704.

Li, H.-B., & Chen, F. (2001). Preparative isolation and purification of astaxanthin from the microalga *Chlorococcum* sp. by high-speed counter-current chromatography. *Journal of Chromatography A, 925*(1–2), 133–137.

Li, H.-B., Jiang, Y., & Chen, F. (2002). Isolation and purification of lutein from the microalga *Chlorella vulgaris* by extraction after saponification. *Journal of Agricultural and Food Chemistry, 50*(5), 1070–1072.

Li, H. B., Fan, K. W., & Chen, F. (2006). Isolation and purification of canthaxanthin from the microalga *Chlorella zofingiensis* by high-speed counter-current chromatography. *Journal of Separation Science, 29*(5), 699–703.

Li, H.-B., Cheng, K.-W., Wong, C.-C., Fan, K.-W., Chen, F., & Jiang, Y. (2007). Evaluation of antioxidant capacity and total phenolic content of different fractions of selected microalgae. *Food Chemistry, 102*(3), 771–776.

Li, Y., Horsman, M., Wang, B., Wu, N., & Lan, C. Q. (2008). Effects of nitrogen sources on cell growth and lipid accumulation of green alga *Neochloris oleoabundans*. *Applied Microbiology and Biotechnology, 81*(4), 629–636.

Li, Y., Huang, J., Sandmann, G., & Chen, F. (2009). High-light and sodium chloride stress differentially regulate the biosynthesis of astaxanthin in *Chlorella zofingiensis* (Chlorophyceae). *Journal of Phycology, 45*(3), 635–641.
Lorenz, R. T., & Cysewski, G. R. (2000). Commercial potential for *Haematococcus* microalgae as a natural source of astaxanthin. *Trends in Biotechnology, 18*(4), 160–167. https://doi.org/10.1016/S0167-7799(00)01433-5.
Ma, R. Y.-N., & Chen, F. (2001). Enhanced production of free trans-astaxanthin by oxidative stress in the cultures of the green microalga *Chlorococcum* sp. *Process Biochemistry, 36*(12), 1175–1179.
Machmudah, S., Shotipruk, A., Goto, M., Sasaki, M., & Hirose, T. (2006). Extraction of astaxanthin from *Haematococcus pluvialis* using supercritical CO_2 and ethanol as entrainer. *Industrial & Engineering Chemistry Research, 45*(10), 3652–3657.
Markou, G., & Nerantzis, E. (2013). Microalgae for high-value compounds and biofuels production: A review with focus on cultivation under stress conditions. *Biotechnology Advances, 31*(8), 1532–1542.
Martelli, G., Folli, C., Visai, L., Daglia, M., & Ferrari, D. (2014). Thermal stability improvement of blue colorant C-Phycocyanin from *Spirulina platensis* for food industry applications. *Process Biochemistry, 49*(1), 154–159.
Martins, N., Roriz, C. L., Morales, P., Barros, L., & Ferreira, I. C. (2016). Food colorants: Challenges, opportunities and current desires of agro-industries to ensure consumer expectations and regulatory practices. *Trends in Food Science & Technology, 52*, 1–15.
Mishra, S. K., Shrivastav, A., & Mishra, S. (2008). Effect of preservatives for food grade C-PC from *Spirulina platensis*. *Process Biochemistry, 43*(4), 339–345.
Mogedas, B., Casal, C., Forján, E., & Vílchez, C. (2009). β-Carotene production enhancement by UV-A radiation in *Dunaliella bardawil* cultivated in laboratory reactors. *Journal of Bioscience and Bioengineering, 108*(1), 47–51.
Mohsenpour, S. F., & Willoughby, N. (2013). Luminescent photobioreactor design for improved algal growth and photosynthetic pigment production through spectral conversion of light. *Bioresource Technology, 142*, 147–153.
Mojaat, M., Pruvost, J., Foucault, A., & Legrand, J. (2008). Effect of organic carbon sources and Fe^{2+} ions on growth and β-carotene accumulation by *Dunaliella salina*. *Biochemical Engineering Journal, 39*(1), 177–184.
Moreno, J., Rodríguez, H., Vargas, M. A., Rivas, J., & Guerrero, M. G. (1995). Nitrogen-fixing cyanobacteria as source of phycobiliprotein pigments. Composition and growth performance of ten filamentous heterocystous strains. *Journal of Applied Phycology, 7*(1), 17–23.
Mulders, K. J. M. (2014). *Phototrophic pigment production with microalgae*. Wageningen: Wageningen University.
Mulders, K. J. M., Lamers, P. P., Martens, D. E., & Wijffels, R. H. (2014). Phototrophic pigment production with microalgae: Biological constraints and opportunities. *Journal of Phycology, 50*(2), 229–242. https://doi.org/10.1111/jpy.12173.
Nwoba, E. G., Parlevliet, D. A., Laird, D. W., Alameh, K., & Moheimani, N. R. (2019a). Light management technologies for increasing algal photobioreactor efficiency. *Algal Research, 39*, 101433.
Nwoba, E. G., Parlevliet, D. A., Laird, D. W., Alameh, K., & Moheimani, N. R. (2019b). Sustainable phycocyanin production from *Arthrospira platensis* using solar-control thin film coated photobioreactor. *Biochemical Engineering Journal, 141*, 232–238.
Ogbonna, C. N. (2016a). Effects of carbon sources on pigment production by *Talaromyces purpurogenus* LC128689 in liquid surface cultures. *Bio-Research, 13*, 942–947.
Ogbonna, C. N. (2016b). Production of food colourants by filamentous fungi. *African Journal of Microbiology Research, 10*(26), 960–971.
Ogbonna, C. N., & Edeh, I. C. (2018). Harvesting *Chlorella variabilis*. Biomass using Moringa oleifera seed-induced sedimentation. *Journal of Advances in Biology and Biotechnology, 18*(4), 1–11.

Ogbonna, J. C., & McHenry, M. P. (2015). Culture systems incorporating heterotrophic metabolism for biodiesel oil production by microalgae. In N. R. Moheimani, M. P. McHenry, K. de Boer, & P. P. Bahri (Eds.), *Biomass and Biofuels from Microalgae* (pp. 63–74). Chem: Springer.

Ogbonna, J. C., & Tanaka, H. (1998). Cyclic autotrophic/heterotrophic cultivation of photosynthetic cells: A method of achieving continuous cell growth under light/dark cycles. *Bioresource Technology, 65*(1–2), 65–72.

Ogbonna, J. C., Masui, H., & Tanaka, H. (1997). Sequential heterotrophic/autotrophic cultivation–An efficient method of producing *Chlorella* biomass for health food and animal feed. *Journal of Applied Phycology, 9*(4), 359–366.

Ogbonna, J. C., Soejima, T., Ugwu, C. U., & Tanaka, H. (2001). An integrated system of solar light, artificial light and organic carbon supply for cyclic photoautotrophic-heterotrophic cultivation of photosynthetic cells under day–night cycles. *Biotechnology Letters, 23*(17), 1401–1406.

Ogbonna, C. N., Aoyagi, H., & Ogbonna, J. C. (2017). Isolation and identification of *Talaromyces purpurogenus* and preliminary studies on its pigment production potentials in solid state cultures. *African Journal of Biotechnology, 16*(13), 672–682.

Ördög, V., Stirk, W. A., Bálint, P., van Staden, J., & Lovász, C. (2012). Changes in lipid, protein and pigment concentrations in nitrogen-stressed *Chlorella minutissima* cultures. *Journal of Applied Phycology, 24*(4), 907–914.

Pasquet, V., Chérouvrier, J.-R., Farhat, F., Thiéry, V., Piot, J.-M., Bérard, J.-B., Kaas, R., Serive, B., Patrice, T., & Cadoret, J.-P. (2011). Study on the microalgal pigments extraction process: Performance of microwave assisted extraction. *Process Biochemistry, 46*(1), 59–67.

Piccaglia, R., Marotti, M., & Grandi, S. (1998). Lutein and lutein ester content in different types of *Tagetes patula* and *T. erecta*. *Industrial Crops and Products, 8*(1), 45–51.

Poojary, M., Barba, F., Aliakbarian, B., Donsì, F., Pataro, G., Dias, D., & Juliano, P. (2016). Innovative alternative technologies to extract carotenoids from microalgae and seaweeds. *Marine Drugs, 14*(11), 214.

Pumas, C., Peerapornpisal, Y., Vacharapiyasophon, P., Leelapornpisid, P., Boonchum, W., Ishii, M., & Khanongnuch, C. (2012). Purification and characterization of a thermostable phycoerythrin from hot spring cyanobacterium *Leptolyngbya* sp. KC45. *International Journal of Agriculture and Biology, 14*(1).

Qu, C.-B., Wu, Z.-Y., & Shi, X.-M. (2008). Phosphate assimilation by Chlorella and adjustment of phosphate concentration in basal medium for its cultivation. *Biotechnology Letters, 30*(10), 1735.

Rao, A. R., Dayananda, C., Sarada, R., Shamala, T., & Ravishankar, G. (2007). Effect of salinity on growth of green alga *Botryococcus braunii* and its constituents. *Bioresource Technology, 98*(3), 560–564.

Rezić, T., Filipović, J., & Šantek, B. (2013). Photo-mixotrophic cultivation of algae *Euglena gracilis* for lipid production. *Agriculturae Conspectus Scientificus, 78*(1), 65–69.

Richmond, A. (1986). Cell response to environmental factors. In A. Richmond (Ed.), *Handbook of microalgal mass culture* (Vol. 528, pp. 69–99). Boca Raton, FL: CRC Press.

Richmond, A. (2000). Microalgal biotechnology at the turn of the millennium: A personal view. *Journal of Applied Phycology, 12*, 441–451.

Rodríguez, H., Rivas, J., Guerrero, M. G., & Losada, M. (1991). Enhancement of phycobiliprotein production in nitrogen-fixing cyanobacteria. *Journal of Biotechnology, 20*(3), 263–270.

Rodriguez-Amaya, D. B. (2019). Natural food pigments and colorants. *Bioactive Molecules in Food*, 867–901.

Román, R. B., Alvarez-Pez, J., Fernández, F. A., & Grima, E. M. (2002). Recovery of pure B-phycoerythrin from the microalga *Porphyridium cruentum*. *Journal of Biotechnology, 93*(1), 73–85.

Rosenberg, J. N., Oyler, G. A., Wilkinson, L., & Betenbaugh, M. J. (2008). A green light for engineered algae: Redirecting metabolism to fuel a biotechnology revolution. *Current Opinion in Biotechnology, 19*(5), 430–436. https://doi.org/10.1016/j.copbio.2008.07.008.

Saini, R. K., & Keum, Y.-S. (2018). Carotenoid extraction methods: A review of recent developments. *Food Chemistry, 240*, 90–103.

Sarada, R., Vidhyavathi, R., Usha, D., & Ravishankar, G. (2006). An efficient method for extraction of astaxanthin from green alga *Haematococcus pluvialis*. *Journal of Agricultural and Food Chemistry, 54*(20), 7585–7588.

Schab, D. W., & Trinh, N.-H. T. (2004). Do artificial food colors promote hyperactivity in children with hyperactive syndromes? A meta-analysis of double-blind placebo-controlled trials. *Journal of Developmental & Behavioral Pediatrics, 25*(6), 423–434.

Schwartz, S. J., & Lorenzo, T. V. (1990). Chlorophylls in foods. *Critical Reviews in Food Science & Nutrition, 29*(1), 1–17.

Seyfabadi, J., Ramezanpour, Z., & Amini Khoeyi, Z. (2011). Protein, fatty acid, and pigment content of *Chlorella vulgaris* under different light regimes. *Journal of Applied Phycology, 23*(4), 721–726. https://doi.org/10.1007/s10811-010-9569-8.

Shahid, M., & Mohammad, F. (2013). Recent advancements in natural dye applications: A review. *Journal of Cleaner Production, 53*, 310–331.

Shi, X., Wu, Z., & Chen, F. (2006). Kinetic modeling of lutein production by heterotrophic *Chlorella* at various pH and temperatures. *Molecular Nutrition & Food Research, 50*(8), 763–768.

Shim, S.-M., Seo, S. H., Lee, Y., Moon, G.-I., Kim, M.-S., & Park, J.-H. (2011). Consumers' knowledge and safety perceptions of food additives: Evaluation on the effectiveness of transmitting information on preservatives. *Food Control, 22*(7), 1054–1060.

Sivathanu, B., & Palaniswamy, S. (2012). Purification and characterization of carotenoids from green algae *Chlorococcum humicola* by HPLC-NMR and LC-MS-APCI. *Biomedicine & Preventive Nutrition, 2*(4), 276–282.

Six, C., Thomas, J.-C., Garczarek, L., Ostrowski, M., Dufresne, A., Blot, N., Scanlan, D. J., & Partensky, F. (2007). Diversity and evolution of phycobilisomes in marine *Synechococcus* sp.: A comparative genomics study. *Genome Biology, 8*(12), R259.

Soares, A. T., Marques Júnior, J. G., Lopes, R. G., Derner, R. B., & Antoniosi Filho, N. R. (2016). Improvement of the extraction process for high commercial value pigments from *Desmodesmus* sp. microalgae. *Journal of the Brazilian Chemical Society, 27*(6), 1083–1093.

Spolaore, P., Joannis-Cassan, C., Duran, E., & Isambert, A. (2006). Commercial applications of microalgae. *Journal of Bioscience and Bioengineering, 101*(2), 87–96.

Tang, Z., Ju, B., Li, W., Wen, S., Pu, Y., & Qin, S. (2016). One-step chromatographic procedure for purification of B-phycoerythrin from *Porphyridium cruentum*. *Protein Expression and Purification, 123*, 70–74.

Thrane, J.-E., Kyle, M., Striebel, M., Haande, S., Grung, M., Rohrlack, T., & Andersen, T. (2015). Spectrophotometric analysis of pigments: A critical assessment of a high-throughput method for analysis of algal pigment mixtures by spectral deconvolution. *PLoS One, 10*(9), e0137645.

Timberlake, C., & Henry, B. (1986). Plant pigments as natural food colours. *Endeavour, 10*(1), 31–36.

Tokarek, W., Listwan, S., Pagacz, J., Leśniak, P., & Latowski, D. (2016). Column chromatography as a useful step in purification of diatom pigments. *Acta Biochimica Polonica, 63*(3).

Töpfl, S. (2006). Pulsed Electric Fields (PEF) for permeabilization of cell membranes in food-and bioprocessing–Applications, process and equipment design and cost analysis.

Uquiche, E., Antilaf, I., & Millao, S. (2016). Enhancement of pigment extraction from B. braunii pretreated using CO_2 rapid depressurization. *Brazilian Journal of Microbiology, 47*(2), 497–505.

USFDA. (2019). *Overview of food ingredients, additives and colors*. Retrieved March 25, 2019, from http://www.fda.gov/Food/IngredientsPackagingLabeling/FoodAdditivesIngredients/ucm094211.htm

Vadiveloo, A., Moheimani, N. R., Cosgrove, J. J., Bahri, P. A., & Parlevliet, D. (2015). Effect of different light spectra on the growth and productivity of acclimated *Nannochloropsis* sp.(Eustigmatophyceae). *Algal Research, 8*, 121–127.

Vadiveloo, A., Moheimani, N. R., Kosterink, N. R., Cosgrove, J. J., Parlevliet, D., Gonzalez-Garcia, C., & Lubián, L. M. (2016). Photosynthetic performance of two *Nannochloropsis* sp. under different filtered light spectra. *Algal Research, 19*, 168–177.

Vadiveloo, A., Moheimani, N. R., Cosgrove, J. J., Parlevliet, D., & Bahri, P. A. (2017). Effects of different light spectra on the growth, productivity and photosynthesis of two acclimated strains of *Nannochloropsis* sp. *Journal of Applied Phycology*, 1–10.

Veuger, B., & van Oevelen, D. (2011). Long-term pigment dynamics and diatom survival in dark sediment. *Limnology and Oceanography, 56*(3), 1065–1074.

Wang, Y., & Chen, T. (2008). The biosynthetic pathway of carotenoids in the astaxanthin-producing green alga *Chlorella zofingiensis*. *World Journal of Microbiology and Biotechnology, 24*(12), 2927–2932. https://doi.org/10.1007/s11274-008-9834-z.

Wang, C., Chen, D., Chen, M., Wang, Y., Li, Z., & Li, F. (2015). Stimulatory effects of blue light on the growth, monascin and ankaflavin production in Monascus. *Biotechnology Letters, 37*(5), 1043–1048.

Wegmann, K. (1986). Osmoregulation in eukaryotic algae. *FEMS Microbiology Reviews, 2*(1–2), 37–43.

Whitehead, A. J., Mares, J. A., & Danis, R. P. (2006). Macular pigment: A review of current knowledge. *Archives of Ophthalmology, 124*(7), 1038–1045.

Yen, H.-W., & Chang, J.-T. (2013). A two-stage cultivation process for the growth enhancement of *Chlorella vulgaris*. *Bioprocess and Biosystems Engineering, 36*(11), 1797–1801. https://doi.org/10.1007/s00449-013-0922-6.

Chapter 4
Algal Biotechnology: A Sustainable Route for Omega-3 Fatty Acid Production

B. S. Dhanya, Gandhi Sowmiya, J. Jeslin, Munusamy Chamundeeswari, and Madan L. Verma

Abstract The production of algal bioactive compounds, in particular, to omega-3 fatty acids, has been gaining increased attention, especially in nutraceuticals and aquaculture industries. Recently, marine algae have been regarded as a good source for producing polyunsaturated fatty acids (PUFAs). Algal cells vary in cell size, shape, cell wall structure, and characteristics; thus, it necessitates exploring various extraction methods for efficient recovery of such bioactive compounds. Effective recovery of omega-3 fatty acids from the microbial cells is subjected to the types of extraction and combination of various methods (mechanical disruption plus chemicals) employed in the process. Patent application related to microalgae ω-3's rich product is also discussed. The present chapter critically discussed about the biological techniques and their combinations used for the extraction of the ω-3 fatty acids.

Keywords Polyunsaturated fatty acids · Omega-3 fatty acids · Purification · Production · Solvent extraction

Dhanya B.S. and Gandhi Sowmiya contributed equally to this work.

B. S. Dhanya
Department of Biotechnology, Udaya School of Engineering, Nagercoil, India

G. Sowmiya · J. Jeslin · M. Chamundeeswari
Department of Biotechnology, St. Joseph's College of Engineering, Chennai, India

M. L. Verma (✉)
Department of Biotechnology, School of Basic Sciences, Indian Institute of Information Technology, Una, Himachal Pradesh, India
e-mail: madanverma@iiitu.ac.in

M. A. Alam et al. (eds.), *Microalgae Biotechnology for Food, Health and High Value Products*, https://doi.org/10.1007/978-981-15-0169-2_4

1 Introduction

Commercially and economically important bioproducts such as fatty acids, vitamins, and carotenoids can be synthesized using viable microalgae, a remarkable bio-organism with much gained attention worldwide (Verma et al. 2019a). The multiplicity effect of secondary metabolites produced by the microalgae such as antitumor, antifungal, antiviral, antimalarial, anti-inflammatory, antioxidant, antibacterial, and the presence of additional nutrients proves it to be the reasonably efficient research and development product in biopharma companies (Verma and Chandel 2019; Kumar et al. 2019; Verma et al. 2019b). Microalgae have the capacity to convert atmospheric carbon dioxide into valuable biomolecules such as lipids and carbohydrates that acts as a precious source in pharma companies (Thakur et al. 2019). Commercial use of microalgae faces several limitations, and it is quite challenging for upgrading to various technological methods (Alam and Wang, 2019).

The diverse microalgae in the environment renders a viable source for the production of economically highly valuable compounds, namely, fatty acids, antioxidative agents, polymers, pigments, as well as enzymes. Omega-3 fatty acids and carotenoid pigments serve as a highly valuable source among the other metabolites extracted from the microalgae (Markou and Nerantzis 2013). The 20-carbon backbone of long-chain polyunsaturated fatty acid has a number of desaturated double bonds that can be grouped as ω-3 or omega-6 fatty acids based on the double bond position from the methyl group end. The ω-3 fatty acids such as docosahexaenoic acid and eicosapentaenoic acid are the crucial components that are required for children and newborn babies for the effective maintenance of immune system (Voigt et al. 2000; Calder 2003). The therapeutic effects of ω-3 in diet can be used to treat cardiovascular disorders, metabolic disorders, obesity, and eczema (Navarro et al. 2000; Das 2002; Nugent 2004). The ω-3 fatty acid is the instinctively available polyunsaturated fatty acid (PUFA) that includes α-linolenic acid (ALA), eicosapentaenoic acid (EPA), docosapentaenoic acid (DPA), and docosahexaenoic acid (DHA) (Gupta et al. 2012). These fatty acids possess a great biomedical value in various disease treatments such as ventricular fibrillation, Alzheimer's disorder, and also in certain cancer treatments. It acts as an antithrombolytic agent and anti-inflammatory agent and is found to reduce the triglyceride deposit in the human body. Since it is an essential fatty acid, it can only be obtained through diet (Wen and Chen 2003; Ren et al. 2010; Xie et al. 2015). Traditionally, marine organisms serve as the major source for the commercial production of omega-3 fatty acid. The declination in the resources induces the exploration of microalgae for the increased omega-3 fatty acid production. The product differs by its type in terms of beverages, nutrient-enriched (fortified) food, neonate, as well as pet animal food (Credence Research 2016).

The market volume shares of ω-3 fatty acid and its necessity in 2025 are given in Fig.4.1. The future insight predicts the need for the large-scale production of ω-3 fatty acid. However, the yield is less with the traditional method of production from

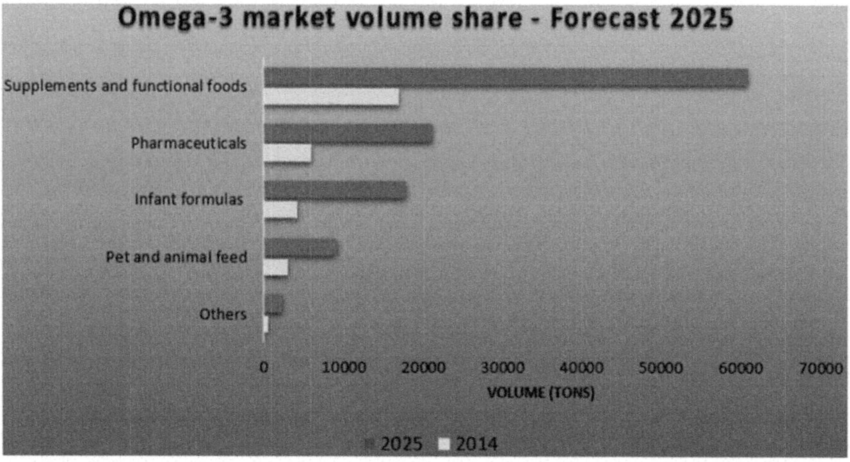

Fig. 4.1 Market share volume of ω-3 fatty acid in different industries and future perspective. (Adapted from Finco et al. 2016)

the oil extracted from fish and plants. The higher yield is obtained from the oil extracted from the microbes which is given in Fig. 4.2.

The present chapter discusses the production, extraction, and purification of omega-3 fatty acids with specific importance given to the newly emerged production using metabolic engineering.

2 Omega-3 Fatty Acid Extraction from Microalgae

2.1 Omega-3 Fatty Acids

Omega-3 fatty acids are polyunsaturated fatty acids with double bonds between third and fourth carbon atom which is important for the growth of eukaryotes (Ward and Singh 2005). Docosahexaenoic acid (DHA, 22:6) and eicosapentaenoic acid (EPA, 20:5) are nutritionally important omega-3 fatty acids. Both types of omega-3 fatty acids are proved to have many health-related benefits, specifically in controlling heart-based ailments such as stroke, arrhythmia, and hypertension (Liang et al. 2012). In addition to this, omega-3 fatty acids offer remedy to other health-related ailments such as rheumatoid arthritis, depression, and asthma (Schacky and Harris 2007).

Docosahexaenoic acid, a structural lipid, commonly found in the retina and brain is an acknowledged defense molecule, having a pivotal participation in minimizing the inflammation and reducing the formation of reactive oxygen species. Eicosapentaenoic acid, a fatty acid with long chain, is involved in managing

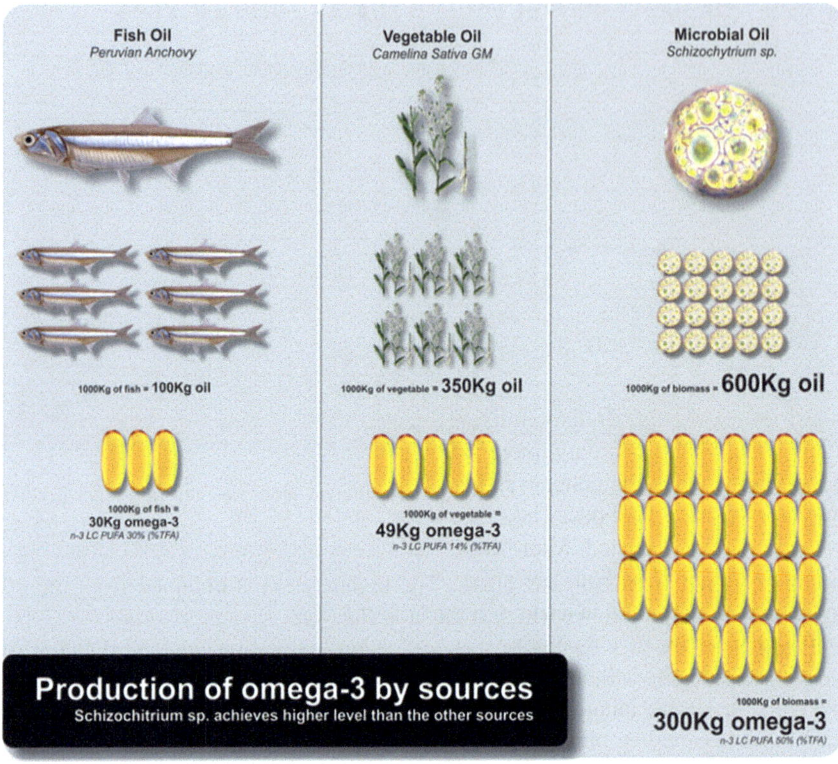

Fig. 4.2 Omega-3 fatty acid production and yield from different sources. (Adapted from Finco et al. 2016)

heart-related ailments (Winwood 2013). The structure of docosahexaenoic acid and eicosapentaenoic acid is shown in Fig. 4.3.

Microalgae have the capacity to produce docosahexaenoic and eicosapentaenoic acid, and it can survive in different culture conditions such as mixotrophic, autotrophic, or heterotrophic (Li et al. 2009). Microalgae that are heterotrophic in nature are an important source of docosahexaenoic acid (Van tol et al. 2009). About 26.7% DHA + EPA, 28% EPA, 45.1% DHA + EPA, 21.4% EPA, 27.7% DHA + EPA, 28% DHA + EPA, 23.4% EPA, 22.03% DHA + EPA, 39.9% EPA, 36% DHA + EPA, and 41.5% DHA + EPA are produced by *Nannochloropsis* sp. (Hu and Gao 2003), *Nannochloropsis salina* (Van Wagenen et al. 2012), *Thraustochytrium* sp. (Scott et al. 2011), *Dunaliella salina* (Bhosale et al. 2010), *Pavlova lutheri* (Carvalho and Malcata 2005), *Isochrysis galbana* (Yago et al. 2010), *Nannochloropsis oceanica* (Patil and Gogate 2015), *Pinguiococcus pyrenoidosus* (Sang et al. 2012), *Chlorella minutissima* (Yongmanitchai and Ward 1991), *Pavlova viridis* (Hu et al. 2008a, b), and *Pavlova lutheri* (Guiheneuf et al. 2009), respectively.

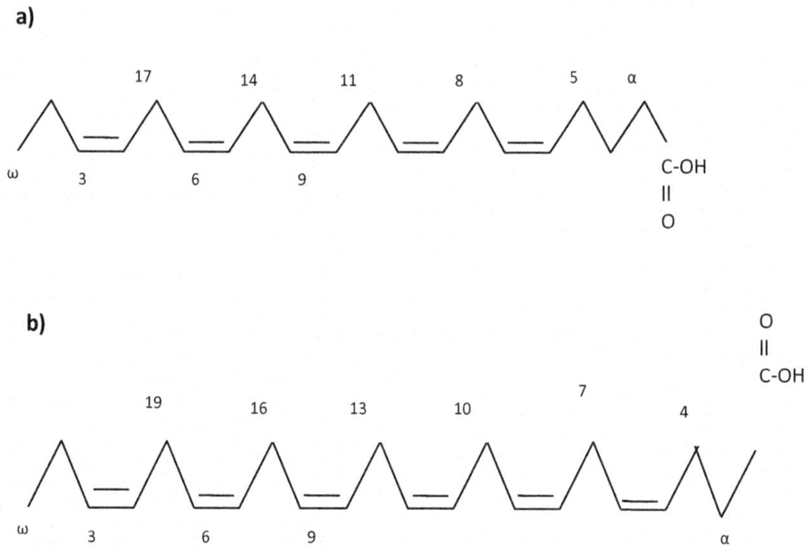

Fig. 4.3 Structure of (**a**) eicosapentaenoic acid and (**b**) docosahexaenoic acid

2.2 Extraction

During upstream processing, from the incubated agar plate, with the help of loop, about 30 mL culture is inoculated and incubated in a shaker at 18 °C for 4 days at 100 rpm. Further inoculation is done in a 95 ml culture for another consecutive 4 days. Carbon and nitrogen sources are the requirements during fermentation (Burja et al. 2006). The factors such as dissolved oxygen, pH, temperature, salinity, and light influence the omega-3 fatty acid production during fed batch process. Low temperatures favor the fatty acid production (Winwood 2013). In downstream processing, cell harvesting is executed using centrifugation process at 4500 rpm. Other than centrifugation, flocculation or filtration can also be used for harvesting (Ward and Singh 2005). After harvesting, extraction steps are followed. While extracting omega-3 fatty acids, it is necessary to switch off the oxidation process that may lead to rancidity. Figure 4.4 depicts the flow chart of extraction and purification of omega-3 fatty acid. The common extraction methods for omega-3 fatty acids are as follows:

- Mechanical pressing aided solvent extraction method.
- Supercritical fluid extraction.

Omega-3 fatty acids are found more in polar-based lipids—phospholipids and glycolipids, and so, nonpolar solvents are ineffective in the extraction process (Ryckebosch et al. 2012). Before extracting the oil, mechanical pressing is done. During pressing, the friction between the raw material and the expeller press

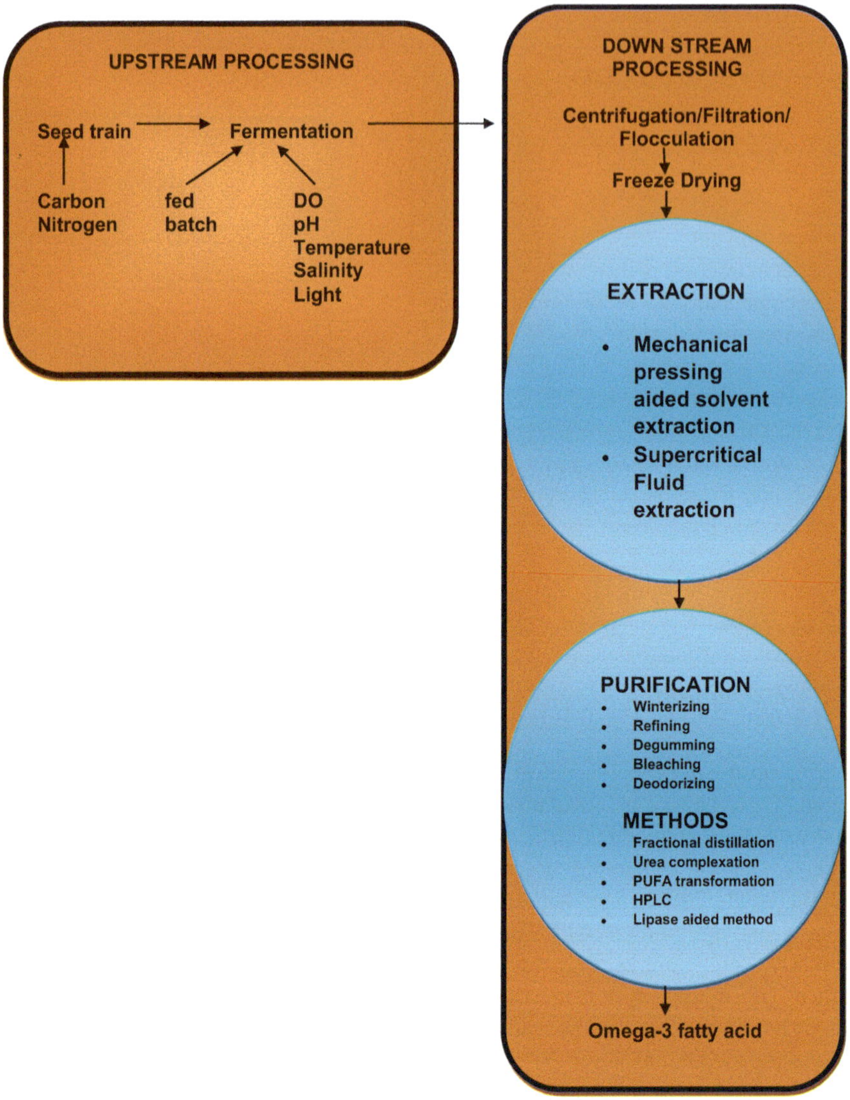

Fig. 4.4 Steps involved in omega-3 fatty acid production

generates heat which may be more than 120 F. A caged barrel-modeled cavity is present where the seeds are pressed for oil. The expeller press consists of an inlet where the algae are passed and an outlet where the pressed product is released. For compressing the algae, constant pressure and frictional force are applied. The appearance of algae is green in color with long fibers, and it is hard for these fibers to move through the screw; and hence, water is passed through the caged barrel (wetting the biomass) for easy passing. Through the small openings, the oil seeps and all the pressed algal biomass remains as cake after pressing, and hence, it must be removed from the expeller. While pressing, heat is generated and it ranges from 140 to 210 °F. At the final step, the oil is removed. For small-scale purposes, Bligh and Dyer method is used for extracting lipids (Bligh and Dyer 1959). In this method, methanol/chloroform solvent mixture is used for disrupting cell and extracting lipids. However, for large-scale purpose, hexane is applied as a solvent.

Extracting ω-3 fatty acids from food contents is applicable using the semicontinuous solvent extraction method that helps in attaining maximum isolation. Soxhlet extraction is the most commonly used semicontinuous method. Other than using organic liquids as solvent, supercritical carbon dioxide can be used for extraction since its usage is economically and environmentally feasible with seldom access to organic solvents. Supercritical fluid is a composite form of gas and liquid properties, and it exists above a critical temperature when heat is applied to the pressurized carbon dioxide. Supercritical fluid property of gas paves the way for its entry into the sample for proper extraction of large amount of omega-3 fatty acids. Supercritical carbon dioxide fluid can be mixed with the sample to be analyzed in a pressurized chamber with applied high amount of heat and results in extraction of omega-3 fatty acids.

For the extraction of microalgal compound and essential oils and decaffeination process, supercritical fluid extraction method is applied (Gil-Chavez et al. 2013). Supercritical fluid extraction put back the usage of solvent extraction, which involves the compounds to be separated at high temperature and pressure (Mercer and Armenta 2011). The density of supercritical fluid is the same as that of the fluids and the viscosity of supercritical fluid is similar to that of the gas. The major advantage of the supercritical fluid when compared to the conventional solvent extraction method lies in the time duration taken for extraction. For supercritical fluid extraction, the time consumed is less due to increased pressure applied that pushes the supercritical liquid to enter into the algal cells, enhancing the mass transfer effect. The diffusion rate of supercritical fluid and gas is the same. Further, extraction is purely selective and targetable in supercritical fluid extraction than the conventional solvent extraction method. Supercritical carbon dioxide is advantageous since it is chemically inert, safe, low cost, and nontoxic with apt pressure and temperature (Daintree et al. 2008). Involvement of chemical is nil in supercritical fluid extraction. In addition to this, the need of co-solvent is seldom important during the separation of extracted compounds from the solvent because of the gaseous nature of carbon dioxide. Carbon dioxide-rich flue gas acts as a low-cost source of carbon dioxide. Pressure and temperature control is necessary for execution of supercritical fluid extraction method. High cost of operation and increased infrastructure are required for supercritical fluid extraction method to work effectively.

2.3 *Purification*

To enhance the shelf-life, quality, and quantity of PUFA, processes such as deodorization, antioxidant addition, filtration, polishing, and bleaching must be done additionally. Purification steps include degumming, refining, bleaching, and deodorizing. The techniques used for purification include molecular/fractional distillation, molecular sieve technique, PUFA transformation/transesterification, urea complexation/fractionation, high-pressure liquid chromatography, and lipase-assisted method (Winwood 2013).

In relation to the structure of a molecule, the polyunsaturated fatty acids are isolated, and this is represented as fractionation process called urea complexation. Improvization of clarity, color, and odor is necessary before the final usage of the extracted omega-3 fatty acids. By subjecting it to chemical refining process, impurities such as phosphatides, trace metals, monoacylglycerol, free fatty acids, pigments, sterols, diacylglycerols, and waxes are removed. Caustic soda is used to remove the free fatty acid by undergoing saponification process, where the free fatty acid is removed as soap. The generated soap is removed by aqueous washing; consequently, physical separation of water and drying is done. Drying process removes the dissolved oxygen present (Winwood 2013).

Degumming, a process in which water is required to remove sterols and phosphatids, is performed after the neutralization process. Adding activated carbon or absorbent clay removes the pigments, trace metals, and other oxidation products. After removal, the added activated carbon is removed by normal filtration process. Especially, the unsaturated fatty acids can be removed from the total lipids by a method known as winterization or fractional (molecular) distillation, and the saturated lipids can be precipitated by oil temperature. Subsequently, dewaxing step/winterization process is performed to remove the wax, specifically triacylglycerols with saturated fatty acids. Dewaxing step provides clarity to the oil. The type of microalgal strain, the end product usage, and its physical nature such as cell wall properties and size determine the efficiency of extraction and harvesting techniques. Chilling is followed for the formation of wax crystals; further, centrifugation/filtration is done to remove the wax crystals. Deodorizing is done by passing the winterized oil to deodorizer/steam cleaner. Additional impurities can be removed by producing high-pressure steam and then cooling finally (Winwood 2013).

3 ω-3 Production in Microalgae

The stress environmental condition leads to the accumulation of lipids and carbohydrates in microalgae, thereby enabling the organism to withstand the adverse condition. In general, the oil content in microalgae is about 10–50%, and the total lipid content lies between 30 and 70% of its dry weight. The energy stockpile of microalgae, i.e., the accumulated fatty acids, favors it to adapt the undesired condition or to aid in the cell division process. The ω-3 is accumulated because of the excessive energy content and also by the necessary flow characteristics that are needed for the cellular functions (Tiez and Zeiger 2010; Cohen et al. 2000). Some of the higher ω-3-accumulating microalgae genera are *Schizochytrium*, *Phaeodactylum*, *Thraustochytrium*, and *Nannochloropsis* (Burja et al. 2006; Zhu et al. 2007) that are generally found to accumulate high percentage of docosahexaenoic acid and eicosapentaenoic acid. This high accumulating DHA and EPA are successfully commercialized by the production process in the optimal medium containing optimized carbon and nitrogen source as well as in the optimized external pH and temperature (Griffiths and Harrison 2009). For example, the high oil yielding *Schizochytrium* sp. growth in the optimized environmental condition accumulated higher DHA molecules with higher cell density (Ward and Singh 2005).

3.1 ω-3 Induction in the Autotrophic Microalgae

The growth condition in which the microalgae have grown highly influences the lipid accumulation. The microalgal lipid content can be enhanced by the instantaneous changes in their growth condition. Starch and lipid accumulated by the microalgae are the stress survival molecules that act as the growth-limiting agents during higher growth induction. The growth-inducing environmental condition includes a light source for the photosynthesis process, temperature change, UV irradiation, and nutrient stress (Singh et al. 2002; de Castro and Garcia 2005). These environmental conditions enhance the lipid accumulation up to two- to threefolds. For example, the microalga diatom *Phaeodactylum tricornutum* is induced to enhance its total lipid content to 81.2 mg/g with increased cell dry weight of about 168.5 mg/g by the changes in the growth-limiting factors. This lipid accumulation can also be enhanced by the addition of any trace elements or other nutrient supplements as well as by anaerobic sulfur depletion (Timmins et al. 2009).

The ω-3 induction can be induced by different environmental conditions likely salinity changes, UV irradiation, as well as lesser temperature. For instance, the EPA production by *Pavlova lutheri* is found to be enhanced to 30.3 M % from 20.3 M % by the changes in the growth temperature to 15 °C (Adarme-Vega et al. 2012). The flow properties of fatty acids that would maintain the cellular fluidity at lower temperature aid in the enhanced induction of these polyunsaturated fatty acids. The PUFA induction can also be influenced by the changes in the salinity during microalgal growth. For example, the presence of 9 g/L NaCl in the growth medium of *Crypthecodinium cohnii* ATCC 30556 results in the increased DHA content of about 56.9%. Similarly in *Phaeodactylum tricornutum*, the EPA accumulation is increased to 19.84% by UV irradiation (Liang et al. 2006). The PUFAs consist of the number of double bonds that would act as the antioxidant agent by eliminating free radicals.

4 Improved ω-3 Production by Metabolic Engineering

Metabolic engineering is the promising approach that helps in the induction of enhanced ω-3 in the microalgae apart from the environmental stress provided (Schuhmann et al. 2012). The biosynthesis of fatty acids in the microalgae has been less investigated, and the key enzymes involved in the fatty acid production are analyzed from the plant metabolism. The significant enzymes that influence the fatty acid induction is recognized from the *C. reinhardtii* (model organism), *Thalassiosira pseudonana*, *Phaeodactylum tricornutum*, and *Ostreococcus tauri* (Tonon et al. 2005; Xu et al. 2009; Wagner et al. 2010).

The *de novo* synthesis of fatty acids is found to occur in the chloroplast following acetyl-CoA carboxylation and condensation to give malonyl-CoA. The long-chain fatty acids are extended by the presence of the substrate called malonyl-ACP by the elongation process. Further, in the endoplasmic reticulum, these long-chain fatty acids are converted to triacylglycerol (TAG) by the influence of glycerol-3-phosphate by the presence of a metabolic intermediary, phosphatidic acid (Hu et al. 2008). Fatty acid biosynthesis has been shown in Fig. 4.5.

Several works are carried out to induce higher ω-3 in the plants and fungus by the regulation of Δ5, Δ6, and Δ12 desaturase enzymes. Research works have been initiated for the higher ω-3 induction in microalgae by the metabolic manipulation of

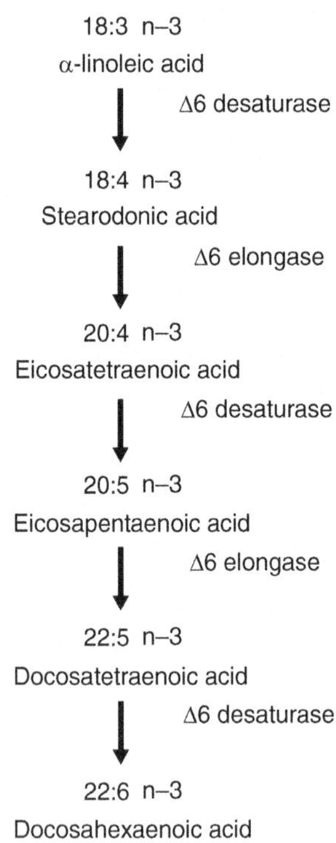

Fig. 4.5 The Δ6 pathway for DHA and EPA biosynthesis. (Adapted from Adarme-Vega et al. 2012)

significant enzymes. The higher expression of major enzymes elongases and desaturases through cisgenic strategies helps in the higher accumulation of DHA and EPA. The involvement of stress-responsive promoters in place of any constitutive promoters would prevent the routine cellular growth and function interference.

Furthermore, overcoming the PUFA degradation can also result in the enhancement of accumulation of ω-3 fatty acids. The desaturase enzymes are responsible for the formation of double bonds within the fatty acids before its degradation. Therefore, any changes in these enzymes would also hinder the higher fatty acid accumulation (Adarme-Vega et al. 2012).

5 Improved ω-3 Production from Microalgae

Microalgae, the potential organism for the enhanced production of high-value products, can be further modified for the increased production of bioproducts to meet the commercial need. One of the strategies is to induce a stress environment (such as low nitrogen and high saline condition) for the growth and development of microalgae, thereby critically altering or increasing the level of bioproduct such as the ω-3 production. The omics technology paves way to the efficient development of increased production by regulating gene and stress environment by means of genetic engineering techniques. Due to the depletion in the marine fish stocks for the production of ω-3 fatty acids, the cultivation of microalgae such as *Schizochytrium* sp. and *Crypthecodinium cohnii* commercially leads to the efficient production of DHA and certain marine microorganisms such as *Phaeodactylum* sp., *Nannochloropsis* sp., and *Nitzschia* spp. for the commercial formulation of EPA (Harwood and Guschina 2009). The large-scale industrial production of these fatty acids is highly technologically challenging and expensive to meet the global market.

The technological issues at the industrial production level of ω-3 fatty acids are meet by enhancing the acyl-CoA-dependent desaturase involvement in the production process (Venegas-Caleron et al. 2010). There are several studies carried out to use acyl-CoA-dependent Δ6 desaturase and metabolically engineer this enzyme in higher plants for the improved production of long-chain PUFA (Sayanova et al. 2012). However, Δ6 desaturation forms the rate-limiting step in the fatty acid production.

The alternative approach in the enhanced production of these fatty acids turns towards the genetic engineering of microalgae for the sustainable production of ω-3. For the commercial production of ω-3 fatty acid, *Phaeodactylum tricornutum*, is formerly recognized as a primary source for the higher yield of EPA. Genes encoding Δ5 and Δ6 desaturases have been successfully cloned for the enhanced EPA production. These microsomal enzymes are found to be contributing much in the fatty acid synthetic pathway for the sufficient production of ω-6 as well as ω-3 fatty acids, while DHA is synthesized from the EPA elongation by the enzyme called Δ5 elongase to form docosapentaenoic acid, which is then converted to EPA by the enzyme Δ6 desaturases (Arao and Yamada, 1994; Domergue et al. 2002).

Table 4.1 PUFA production from different commercially available and genetically engineered microalgae

PUFA	Microalgae	PUFA/L (%)	References
DGLA (dihomo-g-linolenic acid)	*Lobosphaera incisa* (genetically engineered)	21	Abu-Ghosh et al. (2015)
AA (arachidonic acid)	*Khawkinea quartana* SAG 1204-9	34.3	Lang et al. (2011), Tababa et al. (2012)
	Palmodictyon varium SAG 3.92	73.8	
	Parietochloris incisa	44	
	Rhabdomonas incurva SAG 1271–8	41.3	
EPA (eicosapentaenoic acid)	*Nannochloropsis oculata* strains	19	Olofsson et al. (2014), Ryckebosch et al. (2014)
	Nannochloropsis salina	23.6	Bellou and Aggelis (2012), Ryckebosch et al. (2014), Selvakumar and Umadevi (2014), Guiheneuf and Stengel (2015)
	Porphyridium cruentum	3.6	
	Porphyridium purpureum	15.8	
	Tetraselmis gracilis	14	
DHA (docosahexaenoic acid)	*Aurantiochytrium limacinum*	48.5	Li et al. (2015), Gong et al. (2015), Ling et al. (2015)
	Aurantiochytrium sp.	35	
	Crypthecodinium cohnii	25	
	Phaeodactylum tricornutum (genetically engineered)	12	Makri et al. (2011), Liu et al. (2014), Patil and Gogate (2015), Ling et al. (2015)
	Prorocentrum triestinum S2	20.4	
	Schizochytrium limacinum SR21	19.1	
	Schizochytrium sp.	39.6	
	Thraustochytriidae sp.	36.9	

Table 4.1 describes the varying production of PUFA from different microalgal species which can be further enhanced by genetic manipulation of the genes responsible for the efficient fatty acid production.

Generally, microalgae consist of a special arsenal that when genetically modified would result in the enhanced PUFA production. The overexpression of acyl-CoA D6 desaturase enzyme in *P. tricornutum* resulted in the slight increase in the DHA and EPA production. This confirms that this enzyme does not have any substantial effects in omega-3 fatty acid production, whereas the overexpression of this enzyme along with the C20 D5 elongase enzyme in the genetically modified *P. tricornutum* resulted in the increased DHA accumulation up to 8 times than the wild type, but subsequently reduced the EPA accumulation (Hamilton et al. 2014). The EPA accumulation is increased by the expression of D5 desaturase gene (PtD5b) in the prior organism (Peng et al. 2014).

6 Patent Application Related to Microalgae Omega-3's Rich Product

The patents related to microalgae can be categorized into three as follows:

(a) Patents based on microalgae strains such as *Chlamydomonas*, *Botryococcus*, and *Spirulina* and wild and genetically modified strains (engineered microorganisms, nucleic acid sequence, and biosynthetic pathway).
(b) Patents based on microalgae cultivation (photobioreactor, raceway, and heterotrophy) and extraction.
(c) Patents based on microalgae products and its applications as biodiesel, pharmaceutical composition, food feed and nutrition, cosmetic, and fatty acids.

The patent dataset segmentation is done based on the above categorized patterns, and it remains as the taxonomic first level. To construct a map, IPCs are used. The concepts are taken from the UK applications (from 1979), the PCT applications (from middle 2001), CN utility models, the US applications (15th March, 2001), FR applications, documents for WO (before 2000), the full official text in English of the EP applications (without euro-PCT from 1988), CN applications, documents for EP (before 1980s), and the issued US patents (1971–2000). The whole text consists of nominal phrases, and they are converted to standard rules and syntax. At last, each concept is arranged based on the field and the location where it happens. The concepts are the outcome of patent-related semantic contents.

The concept map is constructed in four steps: clustering, dimension, reduction, and drawing. Vector model is drawn. The concept map consists of clusters which are theoretically patents. Each cluster is designated with varied colors. The main page is filled with clusters which are further differentiated into families containing the concepts. The family title is visible if the cursor is pointed toward it. Each family represents three concepts. Distance and proportionality are followed keenly for better position of points. The constructed map is similar to topographic map. Main concepts are mentioned in a circular fashion and collocated into clusters, represented as shades of different intensity. The place (country and place) where the application is filed is known as office of first filing (OFF). After the initial application, the office of second filling (OSF) analysis is done within the families where the patent is protected. First filing shows the place where the technology evolved, and the second filing is related to the indirect market analysis, market location, and product development location. Table 4.2 represents the patents related to the production of omega-3 fatty acids from microalgae. The top-ten patent filing countries are shown in pie chart in Fig. 4.6.

Table 4.2 Patents on omega-3 fatty acids from microalgae

SI. no.	Patent title	Country of origin	Application claimed	Year	Microalgae used	Patent no.
1	Production of omega-3 fatty acids from *Pythium* species	USA	Production of eicosapentaenoic acid	2014	Pythium sp. Of alga	13/788,372
2	Transgenic microalgae with increased production of at least one omega-3 long-chain polyunsaturated fatty acid	USA	Enhancing the production of omega-3 fatty acids by using genetically modified microalgae	2018	Transgenic microalgae	0312888A1
3	Photoautotrophic growth of microalgae for omega-3 fatty acid production	USA	Microalgae Cultivation method (photoautotrophic) to produce omega-3 fatty acids	2013	*Chlorophyta*, *Thalassiosira* sp., *Chaetoceros* sp., *Prymnesiophyta*	8,603,488B2
4	Production of omega-3 fatty acids from crude glycerol	USA	Producing omega-3 fatty acid from microalgae using the substrate crude glycerol	2014	*Schizochytrium* sp., *Phaeodactylum* sp., *Thraustochytrid* sp., *Ulkenia* sp., *Labyrinthulea* sp.	015361
5	Production of omega-3 fatty acids in microflora of Thraustochytriales using modified media	EPO	Process of fermenting Thraustochythales	2007	*Ulkenia*, *Thraustochytrium*, *Schizochytrium*	2084290A1
6	Eicosapentaenoic acid containing oil and methods for its production	USA	Eicosapentaenoic acid production process	1996	Diatoms	5,567,732
7	Method of increasing omega-3 polyunsaturated fatty acids production in microalgae	EPO	Modification of microalgae for better production of EPA	2017	Recombinant microalgae	061347

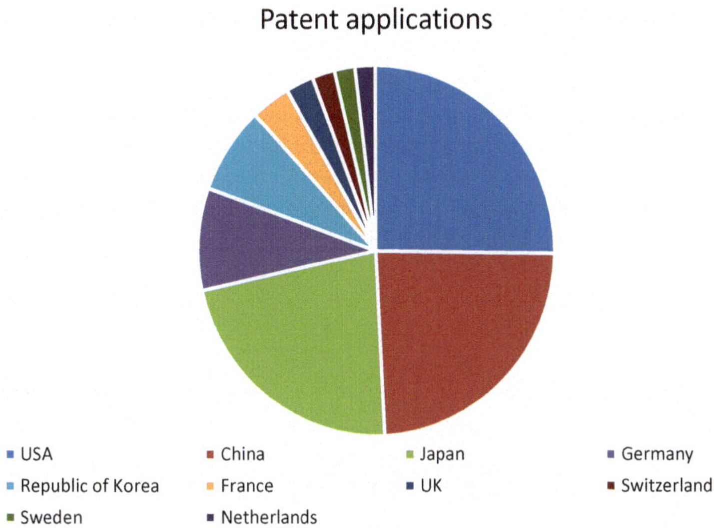

Fig. 4.6 Top 10 countries on patent filing—2019

7 Future Direction

Industrially and commercially, microalgae are important sources of health-related compounds. High selectivity, more efficiency, less extraction time, and minimum solvent usage make the nonconventional extraction methods effective than the traditional extraction methods. The cellular structure of microalgae and efficiency of extraction are interrelated. Microwave-assisted extraction method and ultrasound-assisted extraction methods are good in terms of fast extraction, and pressurized liquid extraction is good in terms of using very minimal solvent for extraction. Each technique has specific temperatures, and some specific temperatures will affect the carotenoids and degrade it since those carotenoids are thermolabile in nature. To overcome the disadvantages of nonconventional techniques, optimization is a better option. Cost reduction, minimizing the use of multiple instruments, and increasing yield are required, and they can be met by optimization. Most of the trials related in minimizing the cost are done at pilot level. Viability of all the techniques should be studied for its commercial success. Research developments are required to use carotenoids in the field of pharmaceutical, feed/food supplement, nutraceutical, and cosmeceuticals.

Life cycle analysis is done on raceway pond production and tubular reactor production. The analysis is related to the environmental rating of algae-based products. Among the two types of mass culturing, the raceway pond cultivation utilizes less energy and emits no greenhouse gas. Hence, the raceway method is environmentally sustainable. Using the energy and greenhouse gas balanced raceway method, EPA and DHA are extracted from *Nannochloropsis* and *Isochrysis.* Green labeling is done and nutrient starvation argument is effective as a marketing tool. Environmental appraisal is less for production in tubular photobioreactor since high amount of resource is utilized and it is expensive. Artificial lighting is required for biomass production. With the aid of genetic manipulation techniques, the production cost can be minimized by 50% with increased PUFA and carotenoid by 50–200%.

Upcoming technologies can be categorized into two: closed and open systems. However, closed type is preferred since there will be minimum risk of contamination in the algal culture. Yielding high amount of products and its sustainability are the challenges to be considered in the future.

8 Conclusion

Industrial applications of microalgae is widely recognized. Utilizing microalgae for manufacturing is quite valuable and practicable. The carbon dioxide and water are converted into biologically active components such as fatty acids and carotenoids in the presence of sunlight. Barriers exist in mass production of microalgae in terms of less efficiency of photosynthetic activity and productivity and the capacity of varied product lines. Sustainability is important for a production system.

Synthetic biology and systems biology are applied in finding solutions to many of the drawbacks pronounced. These novel approaches have the potential to develop photosynthetic cell factories efficiently. However, the availability of algal genome database and the advancement of recombinant DNA technology empower new species of algae to be designed and engineered. Such engineered algal species act as an effective tool to produce value-added products. Additionally, omics technology also leads to designing data-driven strain development. Innovation is a prerequisite for well-advanced production of powerful therapeutic compounds.

Microalgae act as a transparent cellular factory for various biomolecules, and it is considered as a global initiative to achieve sustainable agriculture, sustainable industry, and ecological economics. For large-scale manufacturing, the time consumed must be minimized with maximum production of bioactive compounds. Most of the algae available provided the status GRAS. This status runs in the utility of microalgae as a powerful cellular factory, thus generating large amount of useful compounds. Hence, microalgae rule industrially and commercially. Advancing to the gene-editing technologies such as CRISPR/Cas9, they are placed out of the GMO regulation cover, and no barriers exist in introducing such type of technologies.

If the end product composition is altered using genetic modification process, then the GRAS status is not valid. Microalgae gain high status commercially in the field of generating biofuels since they are of high demand than the biofuels derived from plants, thereby not utilizing fertile land. In a closed system of photobioreactor, genetically modified strains of microalgae can be grown, and this is an added advantage of microalgae. Further, open land is utilized for cultivating algae, and so, it is important to be noted while transgenic plants are used. Advantages in using microalgae are high with generation of value-added compounds, having application in various fields—nutritionally and pharmaceutically, making microalgae the most suitable entrant in developing cellular factories.

For last few decades, significance of biologically active compound extraction from microalgae has gained wide attention industrially and academically. The active compounds are located inside the microalgal cells, and its extraction is challenging. Also, some type of microalgae consists of thick cell walls, and its disruption needs hard treatment, and this hard treatment may affect the compounds to be extracted and also cause impurities in the extracted biomolecules. Mass cultivation at large-scale and compound extraction is quite challenging too. Innovative microalgal cultivation techniques, high quality, and efficient strain selection are needed to achieve high compound recovery that paves the way for a futuristic algal development-based industry.

References

Abu-Ghosh, S., Pal-Nath, D., Markovitch, D., Solovchenko, A., DidiCohen, S., Portugal, I., Khozin-Goldberg, I., Cohen, Z., & Boussiba, S. (2015). A novel source of dihomo- g-linolenic acid: Possibilities and limitations of DGLA production in the high-density cultures of the 5 desaturase-mutant microalga *Lobosphaera incisa*. *European Journal of Lipid Science and Technology, 117*, 760–766.

Adarme-Vega, T. C., Lim, D. K. Y., Timmins, M., Felicitas, V., Li, Y., & Schenk, P. M. (2012). Microalgal biofactories: A promising approach towards sustainable omega-3 fatty acid production. *Microbial Cell Factories, 11*(1), 96.

Alam, M. A., & Wang, Z. (Eds.). (2019). *Microalgae biotechnology for development of biofuel and wastewater treatment* (pp. 1–655). Singapore: Springer.

Arao, T., & Yamada, M. (1994). Biosynthesis of polyunsaturated fatty acids in the marine diatom, *Phaeodactylum tricornutum*. *Phytochemistry, 35*, 1177–1181.

Bellou, S., & Aggelis, G. (2012). Biochemical activities in *Chlorella* sp. and *Nannochloropsis salina* during lipid and sugar synthesis in a lab-scale open pond simulating reactor. *Journal of Biotechnology, 164*, 318–329.

Bhosale, R. A., Rajabhoj, M., Chaugule, B., et al. (2010). *Dunaliella salina Teod.* as a prominent source of eicosapentaenoic acid. *International Journal on Algae, 12*(2), 185–189.

Bligh, E. G., & Dyer, W. J. (1959). A rapid method of total lipid extraction and purification. *Canadian Journal of Biochemistry and Physiology, 37*, 911–917.

Burja, A. M., Radianingtyas, H., Windust, A., & Barrow, C. J. (2006). Isolation and characterization of polyunsaturated fatty acid producing *Thraustochytrium* species: Screening of strains and optimization of omega-3 production. *Applied Microbiology and Biotechnology, 72*(6), 1161–1169.

Calder, P. C. (2003). N-3 polyunsaturated fatty acids and inflammation: From molecular biology to the clinic. *Lipids, 38*, 343–352.

Carvalho, A. P., & Malcata, F. X. (2005). Optimization of ω-3 fatty acid production by microalgae: Crossover effects of CO_2 and light intensity under batch and continuous cultivation modes. *Marine Biotechnology, 7*(4), 381–388.

Cohen, Z., Khozin-Goldberg, I., Adlerstein, D., & Bigogno, C. (2000). The role of triacylglycerol as a reservoir of polyunsaturated fatty acids for the rapid production of chloroplastic lipids in certain microalgae. *Biochemical Society Transactions, 28*(6), 740–744.

Credence Research. (2016). *Algae products market by application*. Retrieved from http://www.credenceresearch.com/report/algae-products-market. Report code: 57848-11-18.

Daintree, L. S., Kordikowski, A., & York, P. (2008). Separation processes for organic molecules using SCF technologies. *Advanced Drug Delivery Reviews, 60*(3), 351–372.

Das, U. N. (2002). The lipids that matter from infant nutrition to insulin resistance. *Prostaglandins, Leukotrienes, and Essential Fatty Acids, 67*, 1–12.

De Castro, A. S., & Garcia, V. M. T. (2005). Growth and biochemical composition of the diatom *Chaetoceros* cf. wighamii Brightwell under different temperature, salinity and carbon dioxide levels. I. Protein, carbohydrates and lipids. *Aquaculture, 246*(4), 405–412.

Domergue, F., Lerchl, J., Zahringer, U., & Heinz, E. (2002). Cloning and functional characterization of *Phaeodactylum tricornutum* front-end desaturases involved in eicosapentaenoic acid biosynthesis. *European Journal of Biochemistry, 269*, 4105–4113.

Finco, A. M. D. O., Mamani, L. D. G., Carvalho, J. C., Pereira, G. V., Thomaz-Soccol, V., & Soccol, C. R. (2016). Technological trends and market perspectives for production of microbial oils rich in omega-3. *Critical Reviews in Biotechnology, 37*(5), 656–671.

Gil-Chavez, G. J., Villa, J. A., Ayala-Zavala, J. F., et al. (2013). Technologies for extraction and production of bioactive compounds to be used as nutraceuticals and food ingredients: An overview. *Comprehensive Reviews in Food Science and Food Safety, 12*(1), 5–23.

Gong, Y., Liu, J., Jiang, M., Liang, Z., Jin, H., Hu, X., Wan, X., & Hu, C. (2015). Improvement of omega-3 docosahexaenoic acid production by marine dinoflagellate *Crypthecodinium cohnii* using rapeseed meal hydrolysate and waste molasses as feedstock. *PLoS One, 10*, e0125368.

Griffiths, M. J., & Harrison, S. T. L. (2009). Lipid productivity as a key characteristic for choosing algal species for biodiesel production. *Journal of Applied Phycology, 21*(5), 493–507.

Guiheneuf, F., & Stengel, D. B. (2015). Towards the biorefinery concept: Interaction of light, temperature and nitrogen for optimizing the co-production of high-value compounds in Porphyridium purpureum. *Algal Research, 10*, 152–163.

Guiheneuf, F., Mimouni, V., Ulmann, L., et al. (2009). Combined effects of irradiance level and carbon source on fatty acid and lipid class composition in the microalga *Pavlova lutheri* commonly used in mariculture. *Journal of Experimental Marine Biology and Ecology, 369*(2), 136–143.

Gupta, A., Barrow, C. J., & Puri, M. (2012). Omega-3 biotechnology: Thraustochytrids as a novel source of omega-3 oils. *Biotechnology Advances, 30*, 1733–1745.

Hamilton, M. L., Haslam, R. P., Napier, J. A., & Sayanova, O. (2014). Metabolic engineering of *Phaeodactylum tricornutum* for the enhanced accumulation of omega-3 long chain polyunsaturated fatty acids. *Metabolic Engineering, 22*, 3–9.

Harwood, J. L., & Guschina, I. A. (2009). The versatility of algae and their lipid metabolism. *Biochemie, 91*, 679–684.

Hu, H., & Gao, K. (2003). Optimization of growth and fatty acid composition of a unicellular marine picoplankton, *Nannochloropsis* sp., with enriched carbon sources. *Biotechnology Letters, 25*(5), 421–425.

Hu, C., Li, M., Li, J., Zhu, Q., & Liu, Z. (2008a). Variation of lipid and fatty acid compositions of the marine microalga *Pavlova viridis* (Prymnesiophyceae) under laboratory and outdoor culture conditions. *World Journal of Microbiology and Biotechnology, 24*(7), 1209–1214.

Hu, Q., Sommerfeld, M., Jarvis, E., Ghirardi, M., Posewitz, M., Seibert, M., & Darzins, A. (2008b). Microalgal triacylglycerols as feedstocks for biofuel production: Perspectives and advances. *The Plant Journal, 54*(4), 621–639.

Kumar, P., Shaunak, I., & Verma, M. L. (2019). Biotechnological application of health promising bioactive molecules. In M. L. Verma & A. K. Chandel (Eds.), *Biotechnological production of bioactive compounds*. Amsterdam: Elsevier.

Lang, I., Hodac, L., Friedl, T., & Feussner, I. (2011). Fatty acid profiles and their distribution patterns in microalgae: A comprehensive analysis of more than 2000 strains from the SAG culture collection. *BMC Plant Biology, 11*(1), 124.

Li, Y., Qin, J. G., Moore, R. B., et al. (2009). *Perspectives of marine phytoplankton as a source of nutrition and bioenergy. Marine phytoplankton* (p. 14). New York, NY: Nova Science Pub Inc.

Li, J., Ruijie, L., Guifang, C., Xiangyu, L., Ming, C., Yuanfa, L., Qingzhe, J., & Xingguo, W. (2015). A strategy for the highly efficient production of docosahexaenoic acid by Aurantiochytrium limacinum SR21 using glucose and glycerol as the mixed carbon sources. *Bioresource Technology, 177*, 51–57.

Liang, Y., Beardall, J., & Heraud, P. (2006). Effect of UV radiation on growth, chlorophyll fluorescence and fatty acid composition of *Phaeodactylum tricornutum* and *Chaetoceros muelleri* (Bacillariophyceae). *Phycologia, 45*(6), 605–615.

Liang, Y., Zhao, X., Strait, M., et al. (2012). Use of dry-milling derived thin stillage for producing eicosapentaenoic acid (EPA) by the fungus *Pythium irregulare*. *Bioresource Technology, 1*, 1.

Ling, X., Guo, J., Liu, X., Zhang, X., Wang, N., Lu, Y., & Ng, I. S. (2015). Impact of carbon and nitrogen feeding strategy on high production of biomass and docosahexaenoic acid (DHA) by Schizochytrium sp. LU310. *Bioresource Technology, 184*, 139–147.

Liu, Y., Tang, J., Li, J., Daroch, M., & Cheng, J. J. (2014). Efficient production of triacylglycerols rich in docosahexaenoic acid (DHA) by osmoheterotrophic marine protists. *Applied Microbiology and Biotechnology, 98*, 9643–9652.

Makri, A., Bellou, S., Birkou, M., Papatrehas, K., Dolapsakis, N. P., Bokas, D., Papanikolaou, S., & Aggelis, G. (2011). Lipid synthesized by microalgae grown in laboratory- and industrial-scale bioreactors. *Engineering in Life Sciences, 11*, 52–58.

Markou, G., & Nerantzis, E. (2013). Microalgae for high-value compounds and biofuels production: A review with focus on cultivation under stress conditions. *Biotechnology Advances, 31*, 1532–1542.

Mercer, P., & Armenta, R. E. (2011). Developments in oil extraction from microalgae. *European Journal of Lipid Science and Technology, 113*(5), 539–547.

Navarro, E., Esteve, M., & Olive, A. (2000). Abnormal fatty acid pattern in rheumatoid arthritis. A rationale for treatment with marine and botanical lipids. *The Journal of Rheumatology, 27*, 298–303.

Nugent, A. P. (2004). The metabolic syndrome. *Nutrition Bulletin, 29*, 36–43.

Olofsson, M., Lamela, T., Nilsson, E., Berge´, J. P., Del Pino, V., Uronen, P., & Legrand, C. (2014). Combined effects of nitrogen concentration and seasonal changes on the production of lipids in Nannochloropsis oculata. *Marine Drugs, 12*, 1891–1910.

Patil, K. P., & Gogate, R. P. (2015). Improved synthesis of docosahexaenoic acid (DHA) using Schizochytrium limacinum SR21and sustainable media. *Chemical Engineering Journal, 268*, 187–196.

Peng, K. T., Zheng, C. N., Xue, J., Chen, X. Y., Yang, W. D., Liu, J. S., Bai, W., & Li, H. Y. (2014). Delta 5 fatty acid desaturase upregulates the synthesis of polyunsaturated fatty acids in the marine diatom Phaeodactylum tricornutum. *Journal of Agricultural and Food Chemistry, 62*, 8773–8776.

Ren, L. J., Ji, X. J., Huang, H., Qu, L., Feng, Y., Tong, Q., et al. (2010). Development of a stepwise aeration control strategy for efficient docosahexaenoic acid production by *Schizochytrium* sp. *Applied Microbiology and Biotechnology, 87*, 1649–1656.

Ryckebosch, E., Bruneel, A., Muylaert, K., et al. (2012). Microalgae as an alternative source of omega-3 long chain polyunsaturated fatty acids. *Lipid Technology, 24*(6), 128–130.

Ryckebosch, E., Bruneel, C., Termote-Verhalle, R., Goiris, K., Muylaert, K., & Foubert, I. (2014). Nutritional evaluation of microalgae oils rich in omega-3 long chain polyunsaturated fatty acids as an alternative for fish oil. *Food Chemistry, 160*, 393–400.

Sang, M., Wang, M., Liu, J., et al. (2012). Effects of temperature, salinity, light intensity, and pH on the eicosapentaenoic acid production of *Pinguiococcus pyrenoidosus*. *Journal of Ocean University of China*, 1–6. (English Edition).

Sayanova, O., Ruiz-Lopez, N., Haslam, R. P., & Napier, J. A. (2012). The role of A6-desaturase acyl-carrier specificity in the efficient synthesis of long-chain polyunsaturated fatty acids in transgenic plants. *Plant Biotechnology Journal, 10*, 195–206.

Schacky, C., & Harris, W. S. (2007). Cardiovascular benefits of ω-3 fatty acids. *Cardiovascular Research, 73*(2), 310–315.

Schuhmann, H., Lim, D. K. Y., & Schenk, P. M. (2012). Perspectives on metabolic engineering for increased lipid contents in microalgae. *Biofuels, 3*(1), 71–86.

Scott, S. D., Armenta, R. E., Berryman, K. T., et al. (2011). Use of raw glycerol to produce oil rich in polyunsaturated fatty acids by a *Thraustochytrid. Enzyme and Microbial Technology, 48*(3), 267–272.

Selvakumar, P., & Umadevi, K. (2014). Enhanced lipid and fatty acid content under photoheterotrophic condition in the mass cultures of Tetraselmis gracilis and Platymonas convolutae. *Algal Research, 6*, 180–185.

Singh, S. C., Sinha, R. P., & Hader, D. (2002). Role of lipids and fatty acids in stress tolerance in cyanobacteria. *Acta Protozoologica, 41*(4), 297–308.

Tababa, H. G., Hirabayashi, S., & Inubushi, K. (2012). Media optimization of Parietochlor is incisa for arachidonic acid accumulation in an outdoor vertical tubular photobioreactor. *Journal of Applied Phycology, 24*, 887–895.

Thakur, M., Bajaal, S., Rana, N., & Verma, M. L. (2019). Microalgal technology: A promising tool for waste water remediation. In P. Arora (Ed.), *Microbial technology for wastewater treatment and biodegradation*. Basel: Springer Nature.

Tiez, L., & Zeiger, E. (2010). *Plant physiology* (5th ed.). Sunderland: Sinauer Associates Inc, Publishers.

Timmins, M., Zhou, W., Lim, L., Thomas-Hall, S. R., Doebbe, A., Kruse, O., Hankamer, B., Marx, U. C., Smith, S. M., & Schenk, P. M. (2009). The metabolome of *Chlamydomonas reinhardtii* following induction of anaerobic H_2 production by sulphur deprivation. *The Journal of Biological Chemistry, 284*(35), 23415–23425.

Tonon, T., Sayanova, O., Michaelson, L. V., Qing, R., Harvey, D., Larson, T. R., Li, Y., Napier, J. A., & Graham, I. A. (2005). Fatty acid desaturases from the microalga *Thalassiosira pseudonana. The FEBS Journal, 272*(13), 3401–3412.

Van Tol, E. A. F., Willemsen, L. E. M., Koetsier, M. A., et al. (2009) Improvement of intestinal barrier integrity. EP patent 1,815,755.

Van Wagenen, J., Miller, T. W., Hobbs, S., et al. (2012). Effects of light and temperature on fatty acid production in *Nannochloropsis salina. Energies, 5*(3), 731–740.

Venegas-Caleron, M., Sayanova, O., & Napier, J. A. (2010). An alternative to fish oils: Metabolic engineering of oil-seed crops to produce omega-3 long chain polyunsaturated fatty acids. *Progress in Lipid Research, 49*, 108–119.

Verma, M. L., & Chandel, A. K. (2019). *Biotechnological production of bioactive compounds*. Amsterdam: Elsevier Press.

Verma, M. L., Jeslin, J., Koshy, A., & Chamundeeswari, M. (2019a). Recent progress in emerging microalgae technology for biofuel production. In N. Srivastava, M. Srivastava, P. K. Mishra, & P. W. Ramteke (Eds.), *Substrate analysis for effective biofuels production*. Basel: Springer Nature.

Verma, M. L., Kishor, K., Sharma, D., Kumar, S., & Sharma, K. D. (2019b). Microbial production of ω-3 polyunsaturated fatty acids. In M. L. Verma & A. K. Chandel (Eds.), *Biotechnological production of bioactive compounds*. Amsterdam: Elsevier.

Voigt, R. G., Jensen, C. L., Fraley, J. K., Rozelle, J. C., Brown, F. R., & Heird, W. C. (2000). Relationship between omega −3 long-chain polyunsaturated fatty acid status during early infancy and neurodevelopmental status at 1 year of age. *Journal of Human Nutrition and Dietetics, 15*, 111–120.

Wagner, M., Hoppe, K., Czabany, T., Heilmann, M., Daum, G., Feussner, I., & Fulda, M. (2010). Identification and characterization of an acyl-CoA: Diacylglycerol acyltransferase 2 (DGAT2) gene from the microalga *O. tauri. Plant Physiology and Biochemistry, 48*(6), 407–416.

Ward, O. P., & Singh, A. (2005). Omega-3/6 fatty acids: alternative sources of production. *Process Biochemistry, 40*(12), 3627–3652.

Wen, Z. Y., & Chen, F. (2003). Heterotrophic production of eicosapentaenoic acid by microalgae. *Biotechnology Advances, 21*, 273–294.

Winwood, R. J. (2013). Recent developments in the commercial production of DHA and EPA rich oils from micro-algae. *Oil Seeds and Fats Crops and Lipids, 20*(6), D604.

Xie, D., Jackson, E. N., & Zhu, Q. (2015). Sustainable source of omega-3 eicosapentaenoic acid from metabolically engineered *Yarrowia lipolytica*: From fundamental research to commercial production. *Applied Microbiology and Biotechnology, 99*, 1599–1610.

Xu, J., Zheng, Z., & Zou, J. (2009). A membrane-bound glycerol-3-phosphate acyltransferase from *Thalassiosira pseudonana* regulates acyl composition of glycerolipids. *Botany, 87*(6), 544–551.

Yago, T., Arakawa, H., Morinaga, T., Yoshie-Stark, Y., & Yoshioka, M. (2010). Effect of wavelength of intermittent light on the growth and fatty acid profile of the haptophyte isochrysis galbana. In H. J. Ceccaldi, I. Dekeyser, M. Girault, & G. Stora (Eds.), *Global change: mankind-marine environment interactions*. Dordrecht: Springer..

Yongmanitchai, W., & Ward, O. P. (1991). Growth of and omega-3 fatty acid production by *Phaeodactylum tricornutum* under different culture conditions. *Applied and Environmental Microbiology, 57*(2), 419–425.

Zhu, L., Zhang, X., Ji, L., Song, X., & Kuang, C. (2007). Changes of lipid content and fatty acid composition of *Schizochytrium limacinum* in response to different temperatures and salinities. *Process Biochemistry, 42*(2), 210–214.

Part II
Microalgae in Health Product Development

Chapter 5
Microalgae in Human Health and Medicine

Sajid Basheer, Shuhao Huo, Feifei Zhu, Jingya Qian, Ling Xu, Fengjie Cui, and Bin Zou

Abstract Microalgae contain various components those that have shown a great potential to be used for human health and medicine. The therapeutic properties of microalgae exhibit vast range of applications like cardiovascular health, anticancer, anti-inflammatory, anticoagulant, antiviral, antibacterial, antifungal, and others in human medicinal products. Microalgal components are used to enhance immune system and to reduce blood cholesterol and are effective against hypercholesterolemia. Microalgae contain effective components that can remove harmful elements from the human body and have properties of antitumor, stomach ulcer, and wound healing. The extract of microalgae enhances blood hemoglobin concentration and decrease blood sugar level. Some microalgal species are extensively used to form analgesic, broncholytic, and antihypertensive medicines. Large quantities of bioactive components obtained from microalgae have strong beneficial properties which reduce the production of inflammatory compounds, effective against muscle degradation. Microalgal bioactive components play a potential role in disease-inhibiting and health-promoting medicines like capsules, tablets, powders, and gels. This article also reviews the health risks regarding microalgae intake.

Keywords Microalgae · Medicine · Bioactive compounds · Human health

1 Introduction

In history, human has been dependent on microalgae to obtain nutrients. Documents proved that in China 2000 years ago *Nostoc* had been consumed by human and in other countries as well (Ciferri 1983). Microalgae use simple compounds present in

S. Basheer · S. Huo (✉) · J. Qian · L. Xu · F. Cui · B. Zou
School of Food and Biological Engineering, Jiangsu University, Zhenjiang, China
e-mail: huo@ujs.edu.cn

F. Zhu
Institute of Life Sciences, Jiangsu University, Zhenjiang, China

M. A. Alam et al. (eds.), *Microalgae Biotechnology for Food, Health and High Value Products*, https://doi.org/10.1007/978-981-15-0169-2_5

their environment to make complex compounds like carbohydrate, lipid, proteins, and other secondary metabolites of medicinal value (Hochman and Zilberman 2014). Some beneficial bioproducts are obtained from microalgae including antioxidant, natural dyes, polysaccharides, bioactive and functional pigments, eicosapentaenoic acid (EPA), and docosahexaenoic acid (DHA). Production of microalgae at commercial level has begun 50 years ago. In the 1960s, Japan presented the first commercial production of microalgae by producing *Chlorella* (Varfolomeev and Wasserman 2011). Microalgae contain several molecules, those that have shown potential important nutrients in human health. Micronutrients, polysaccharides, lipids, and proteins are among those molecules (Alam and Wang 2019). Microalgae have suggested nutritional supplements to meet nutritional requirements for the world population because of the unique composition. Adequate and balanced nutrients are required to be provided to the increasing human population worldwide for the better health and to avoid diseases; microalgae contain wide range of components, and it exhibits health benefits for human and can be utilized in pharmaceutical industry (Bishop and Zubeck 2012).

Minerals in microalgae are more bioavailable due to the lack of oxalate, phosphorus, and calcium contents that are similar to milk. Iron is provided to vegetarians by cereals. Moreover, the microalgae contain more iron than cereals. The nonexistence of oxalate and phytate made the absorption of phosphorus and calcium higher. For vegetarians, *Spirulina* is a better source of the minerals. All of these attributes made microalgae an important source of micro and macronutrients which are helpful to eliminate malnutrition to the mass of population that have not enough food to eat or to the people having physiological complications due to unusual food behavior (Nazih and Bard 2018). One of the main causes of death in the world is cardiovascular diseases. Atherosclerosis causes heart failure, coronary heart disease, peripheral artery disease, and stroke. Diabetes, hyperlipidemia, and hypertension are the major reasons of atherosclerosis. Many results showed that the risk factors can be reduced by microalgae; additionally, special composition of fatty acids may also inhibit these health problems (He et al. 2004). Microalgae contain functional ingredients which can optimize human health and prevent humans from chronic disorder (Smit 2004). Microalgae contain long-chain polyunsaturated fatty acids with omega 3 and 6 that are beneficial for cardiac and neurological development which support the human body against heart disease, cancer, hypertension, and cholesterol problem (Mata et al. 2010).

Microalgae similar to seaweed provide major elements (phosphorus, calcium, sodium, magnesium, sulfur, nitrogen) and other trace elements (zinc, manganese, iodine, copper, cobalt, selenium, and cobalt) along with the production of essential amino acids. Due to the source of nutritive components, it is demanded to be produced at larger scale to obtain huge quantities of those beneficial components (Kumari et al. 2010). Microalgae are the promising microorganisms that can provide valuable ingredients and bioactive nutrients important for human health which can be utilized in medicine production (Smit 2004). Nutrient deficiency is a worldwide problem; by taking one table spoon of edible microalgae per day, it can eliminate nutrient deficiency. The oral intake of microalgae regulates human body functions including respiratory, hormonal, immune, and nervous system (Chew and Park 2004; Kau et al. 2011). Due to its anti-inflammatory activity, microalgae are effective in wound and burn healing. Microalgae are effective in human health and well-being by providing solution to many serious conditions such as cardiovascular system and cognitive decline (Lordan et al. 2011; Riediger et al. 2009).

Microalgae contain various components that have shown a great potential to be used for human health. These bioactive components have been utilized in medicine and pharmaceutical products along with other applications. It exhibits vast range of applications like nutritional ingredient, high-value food, cardiovascular health, antioxidant, anticancer, anti-inflammatory, antimicrobial, anti-aging and skin protective and other medicinal properties. For instance, *Spirulina* is used to enhance immune system and to reduce cholesterol. Sulfated polysaccharides of *Spirulina* also have antiviral properties. There is a compound in *Chlorella* named β-1,3-glucan which lowers lipid in blood and stimulates immune system. In addition, it is effective in removing harmful elements from the human body, and it has properties of antitumor and stomach ulcer healing. Previously, it was reported that the extract of *Chlorella* enhances blood hemoglobin concentration and decreases blood sugar level. Oxidized carotenoids of *Dunaliella salina* have anticancer properties. Some microalgal species are extensively used to form analgesic, broncholytic, and antihypertensive medicines. Large quantities of astaxanthin obtained from *Haematococcus* have strong antioxidant properties which reduce the production of inflammatory elements, are effective against muscle degradation, and protect from oxidative stress. Tablets made by dry *Dunaliella* comprise of β-carotene which is precursor of vitamin A. Microalgal bioactive components play an important role in disease-inhibiting and health-promoting products. In this chapter, the importance of microalgae in human health and its utilization in pharmaceutical industry are summarized.

2 Commercial Production of Microalgae

Microalgal proteins are getting more interest in human health, although better nutrition and characteristics of microalgae to obtain biomass at large scale by cultivating it in conventional open ponds continue to be a challenge (Draaisma et al. 2013). In the 1960s, Japan firstly started cultivating *Chlorella* species at industrial scale for human consumption, as algae was proposed to be the solution to world food requirements (Burlew 1953). In the 1980s, production of microalgae was started in Asia, Australia, Israel, India, and the USA (Enzing et al. 2014). Recent advancement in technology made it possible to get high-valued nutrients from microalgae by making developments in culture systems and improvements on biotechnology of microalgae to obtain nutrients like fatty acids, polyhydroxyalkanoates, phycobilins, carotenoids, polysaccharides, and sterols (Borowitzka 2013).

Microalgal biomass is a plenteous protein source in quality and quantity; it is comparable with traditional protein sources like fish, egg, and soybean, making microalgae a better source of natural protein (Batista et al. 2013; Graziani et al. 2013). For instance, *Spirulina* sp. depending on the strain contains 50–70% of protein (Plaza et al. 2009). Although the biological value, utilization, protein efficiency ratio, and digestibility of microalgal protein are not higher than egg, casein and gold standard (Becker 2007).

Microalgae are source of fiber, enzymes, protein, oil, carbohydrate, and minerals (calcium, magnesium, iron, potassium, iodine) in several countries; they are considered significant source of food and are being used as nutritional supplement and human health products (Apt and Behrens 1999). Vitamin content of microalgae is also remarkable. Land-grown vegetables lack in vitamin B12, but microalgae are valuable source of this vitamin along with vitamins C, B1, and B2 (Martínez-Hernández et al. 2018). Brown microalgae are specifically abundant in a component called algin; it is represented by the name alginic acid. Alginates are derivatives of algin which are utilized as stabilizers and thickeners in industries like cosmetic, pharmaceutical, and food. Harmful components and heavy metals are removed from the human body because of higher chelating and exchange attribute of the alginic acid and the salts of alginic acid. Sodium alginate which is a soluble form can react with lead, and other metals convert it into insoluble chelates which are removed in feces from the human body. This attribute makes microalgae an important diet component especially for those people who are living in the contaminated environment (Zhao et al. 2018). Ordinary growth requirements and higher growth rate increase the growing interest of microalgae for many biotechnological implementations, and to obtain natural products, these properties all together make microalgae a suitable candidate. Precious pharmaceutical proteins like vaccines and antibodies can be obtained using microalgae; thus, this biotechnological application has attracted the industry and academia (Apt and Behrens 1999; Mayfield et al. 2007). Phycobiliproteins are obtained by microalgae which are Rhodophyta *Porphyridium* and cyanobacterium. These molecules are used as natural dyes, which are the common implementations of these molecules, and research has proved their biological attributes and huge applications in pharmaceutical industry. Phycocyanin is utilized in food items as coloring agent in ice lollies, chewing gums, soft drinks, candies, wasabi, and dairy products. In cosmetic products, other types of the natural pigments obtained from this chemical are also utilized (Carfagna et al. 2016). Due to the presence of micronutrients and macronutrients in microalgal biomass, it becomes suitable to be utilized as nutritional supplement and food since centuries. Huge protein content (50–60%) of microalgal dry biomass makes its use more interesting. *Spirulina* has attracted more due to its high-protein contents along with *Arthrospira maxima*, *Dunaliella bardawil*, *Dunaliella salina*, and *Chlorella* sp.; these species contain higher amounts of essential amino acids, which consist of nearly half part of protein (Belay et al. 1996). Modern research has shown that microalgal species such as *Chlorella*, *Dunaliella*, and *Scenedesmus* have potential to produce recombinant proteins that can be utilized in the formation of immunotoxins and antibodies useful in growth hormones, anticancer, gut therapy, vaccines, and therapeutic enzymes (Rasala and Mayfield 2015). For instance, in a research (Azabji-Kenfack et al. 2011), HIV-effected patients were given food supplements containing *Spirulina* and soya beans to compare their effectiveness in malnutrition. Both groups got weight remarkably after 12 weeks. Intervention period was 12 weeks; then it was observed that *Spirulina* group was more effective. CD4 count was increased, and viral load was also greatly reduced by *Spirulina* which was a better indicator of immune response. Plasma cholesterol was found to be decreased in rats when they were given *Porphyridium* sp., the red microalgae (Dvir et al. 2000). Moreover, cholesterol absorption was found to be decreased due to its inhibiting effect. *Spirulina maxima* found to have suppressive effect on lead in lipid metabolism (Ponce-Canchihuamán et al. 2010). The significance of commercial products from microalgae in human health and medicine was shown in Table 5.1.

3 Significant Effects of Microalgae in Human Health

As Fig. 5.1 shows, the significant effects of microalgae in human health mainly include anti-inflammatory effect, antiviral activity, antioxidant activity, antimicrobial activity, anticancer activity, and gut health.

3.1 *Anti-Inflammatory Effect*

The literature proves the anti-inflammatory properties of the extracts obtained from microalgae. In the research of analgesic and anti-inflammatory properties of methanol and aqueous *Phaeodactylum tricornutum* and *Chlorella tigmatophora* extracts, it was investigated that these kinds of functions have not seen in liposoluble fractions (Guzman et al. 2001).

3.2 *Antiviral Activity*

Several studies have already existed regarding antiviral properties of a number of polysaccharides obtained from several microalgal species which they release in cell medium. This activity has been proved in vitro in many viruses. In a study, antiviral properties for many microalgae like cyanobacteria, dinoflagellates, and rhodophytes have been discussed (Raposo et al. 2013). The antiviral activity of sulfated polysaccharides are among the most prominent ones, and at some stages, sulfate groups and calcium ion can form a molecular formation to have antiviral effect (Hayashi et al. 1996). Microalgal polysaccharides can block penetration and absorption of virus in the cell body (Hernández-Corona et al. 2002).

Table 5.1 Significance of commercial products in microalgae for human health and medicine

Microalgal species	Commercial compounds	Significance in human health and medicine	Reference
Spirulina sp., *Nostoc*, *porphyridium* sp.	Polysaccharides, nonlipid fraction phycocyanin	Cardiovascular health	Chai et al. (2015)
Arthrospira and *Chlorella*	Sulfated polysaccharides	Antimicrobial activity	Buono et al. (2014), Witvrouw and De (1997)
D. salina, *Chlorella* sp., *Haematococcus pluvialis*, *Nannochloropsis gaditana*	Carotenoids (lutein, astaxanthin, fucoxanthin), *Nannochloropsis gaditana*, PUFAs (EPA, DHA)	Anticancer	Chai et al. (2015), Liu et al. (2014), Peng et al. (2011)
Arthrospira platensis, *Spirulina*, *Chlorella* spp., *Arthrospira maxima*, *D. salina*	Docosahexaenoic acid (DHA) and eicosapentaenoic acid (EPA), protein, essential amino acids (cysteine and methionine), vitamins (lipophilic and hydrophilic)	Micronutrients and macronutrients	Santos-Sanchez et al. (2016)
Lyngbya majuscule, *Pharmaceuticals*, *Porphyridium cruentum*, *Odontella aurita*, *Nannochloropsis* sp., *Dunaliella salina*, *Chlorella* spp., *Arthrospira (Spirulina)*	Immune modulators, carotenoids (β-carotene and astaxanthin), fatty acids, EPA, polysaccharides, vitamin B12, protein, carbohydrate extract	Pharmaceuticals, healthy foods, nutritional supplements, antioxidant, immune system, atherosclerosis prevention, hypocholesterolemic	Cheong et al. (2010), de Jesús Paniagua-Michel et al. (2015), Pulz and Gross (2004)
Muriellopsis sp., *Chlorella zofingiensis*, astaxanthin, *Nannochloropsis*, *Haematococcus pluvialis*, *Dunaliella bardawil*	Carotenoids (lutein, zeaxanthin, β-carotene), protective antioxidants	Vision enhancement, liver health and obesity prevention, plasma cholesterol reduction, decrease liver inflammation	Ayelet et al. (2008), Garcíagonzález et al. (2005), Jin et al. (2010), Kleinegris et al. (2010), Kyle (2001), Lorenz and Cysewski (2000)
Dunaliella salina, *Chlorella vulgaris*, *Arthrospira platensis*, *Lyngbya majuscule*, *Phaeodactylum tricornutum*, *Porphyridium cruentum*	Beta-carotene, lutein, trace elements, minerals, phycocyanin, immune modulators, fucoxanthin, fatty acids, polysaccharides, PUFA (arachidonic acid)	Food supplements, Healthy food, Nutrition, Pharmaceuticals	Paniagua-Michel (2015)

(continued)

Table 5.1 (continued)

Microalgal species	Commercial compounds	Significance in human health and medicine	Reference
Euglena gracilis, *Euglena gracilis*, *Prototheca moriformis*, *a Chlorella* spp.	Vitamins (biotin, α-tocopherol, ascorbic acid)	Nutrition	Li et al. (2008)
Prorocentrum lima, *Gambierdiscus toxicus*	Okadaic acid	Therapeutics (antifungic), growth factors, secretion enhancement	Nagai et al. (1992), Pshenichkin and Wise (1995)

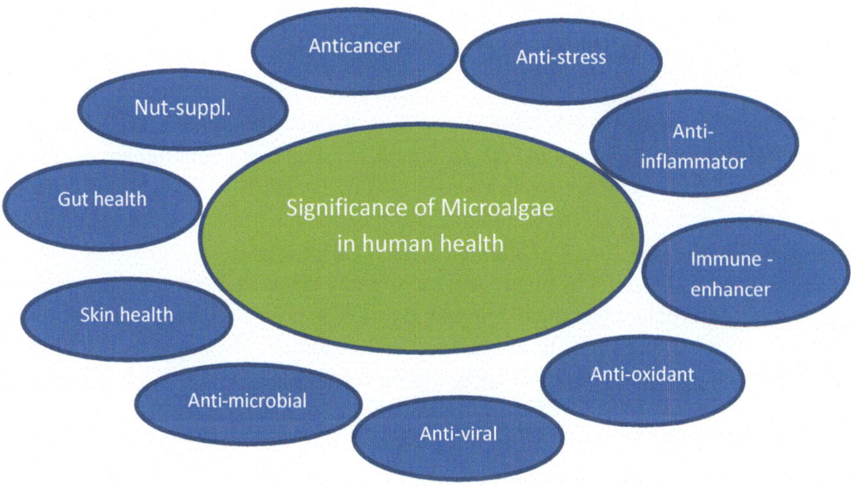

Fig. 5.1 The significant effects of microalgae in human health

3.3 Antioxidant Activity

Microalgae contain natural components and pigments (e.g., chlorophylls, phycobiliproteins, and carotenes). These components are very useful for human health since these cannot be produced by the human body (Sampath-Wiley et al. 2008). Some antioxidants are exclusively produced by microalgae, which have potential to reduce the risk of chronic disease development such as cardiovascular disease and cancer (Monego et al. 2017). Antioxidants minimize the damage caused by free radical in the human body, and these natural antioxidants can be obtained from microalgae (Chacón-Lee and González-Mariño 2010; Olivares et al. 2016).

3.4 Antimicrobial Activity

Antimicrobial properties have been found in extracts of huge number of microalgae; *Arthrospira* and *Chlorella* are most focused species in this property (Buono et al. 2014). Many microalgal species contain high amounts of sulfated polysaccharides that are found to have virus replication inhibitory property such as arenavirus, togavirus, flavivirus, herpesvirus, rhabdovirus, and orthopoxvirus families (Witvrouw and De 1997). It was found that microalgal extracts (alginate, laminarin, and fucoidan) and polysaccharides obtained from microalgae have the ability to resist viral diseases (Holdt et al. 2011).

3.5 Anticancer Activity

Worldwide, cancer is the second major reason of mortality (WHO 2016). Several studies related to molecular and cellular research found the antimalignant potential of microalgal-derived bioactive components (Kumar et al. 2013; Talero et al. 2015). Fucoxanthin, a carotenoid in microalgae, suppresses the genes which cause cancer and prevents the growth of malignant cells (Takahashi et al. 2015). In cancer research, there is more focus on preventive measures as well as proper treatment of the disease. Microalgal bioactive compounds such as polysaccharides, carotenoids, proteins, peptides, and lipids have been known due to their chemopreventive effects and beneficial properties against cancer (Talero et al. 2015).

3.6 Gut Health

Microalgae are effective in gut health because of prebiotic attribute and due to the presence of higher fiber and carbohydrate contents. The gut health plays a vital role in human health and provides protection against diseases due to containing microflora in human gut. The gut microflora gained much attraction due to scientific advancement and also because of its significance in human health (Rastall et al. 2010).

4 Important Microalgal Species

There are four groups of microalgae, chlorophytes (green), cyanobacteria (blue green), rhodophytes (red), and chromophytes (other kind of algae). There are hundreds of species in every group, and every species contains thousands of strains (Hochman and Zilberman 2014). Just a few microalgae have been investigated for its potential beneficial utilization. Chrysophyceae (golden algae), Chlorophyceae, Cyanophyceae, Bacillariophyceae (diatoms) are the most used microalgae for beneficial purpose (Pulz and Gross 2004). Now commercial production of microalgae at industrial scale has been grown much, and the major cultivated species are *Skeletonema*, *Schizochytrium*, *Nitzschia*, *Tetraselmis*, *Dunaliella*, *Spirulina*, and *Chlorella* (Lee 1997).

Since the last two decades, four major microalgal species have been focused biotechnologically: (1) *Spirulina* (*Arthrospira*), (2) *Dunaliella*, (3) *Chlorella vulgaris*, and (4) *Haematococcus pluvialis* (Table 5.2). Some of their characteristics, composition, and production process obtained from these species, *canthaxanthin* and *zeaxanthin*, are being used for medicinal purpose and to color chicken skin; they can be utilized as food coloring matter such as production of orange juice. Phycoerythrin and phycobiliproteins are utilized in cosmetics and food products, while phycocyanin is utilized as coloring agent in candy, ice cream, dietary food, beverage, medicine, and cosmetic production. Because of their higher fluorescence, photostability, and effectiveness in molecular absorption, these microalgae are highly utilized in immunological and clinical research laboratories (Anbuchezhian et al. 2015; Pulz and Gross 2004; Varfolomeev and Wasserman 2011). Extracts obtained from *Spirulina*, *Dunaliella*, *Chlorella*, and *Haematococcus* are utilized in the production of body lotions and face creams as well as in the production of hair masks, hair shampoos, and sun protection creams. Production of skin collagen is enhanced by *Chlorella vulgaris* extract. It is used to make skin surface wrinkle free and fiber synthesis. *Spirulina* extract slows skin aging due to its higher protein levels. Phycobiliprotein obtained from *Spirulina* is utilized as coloring agent, cosmetics, antioxidant, anti-inflammatory, and photodynamic to treat several tumors, cancers, production of fluorescent marker, and leukemia (de Jesús Paniagua-Michel et al. 2015; Patil et al. 2008). Many other species of microalgae are still being researched to be used as nutrition supplement. In the 1970s, large-scale production of *Scenedesmus* (green algae) was obtained, and wide research has been done to obtain single-cell protein. But the production process is not cost effective (Becker and Venkataraman 1980).

Table 5.2 Microalgal commercial products for human health and medicinal usage (Ansorena et al. 2013; Pulz and Gross 2004; Spolaore et al. 2006)

Microalgae	Products
Dunaliella salina	Powders
Chlorella	Powders, tablets, nectar
Aphanizomenon flos-aquae	Powders, crystals, capsules
Astaxanthin	Powder, liquids, gels, capsules
Spirulina (Arthrospira)	Powders, tablets, extracts

4.1 Spirulina

Immune system development and reduction in cholesterol can be achieved by using health beneficial properties of *Spirulina* (Belay et al. 1993). *Spirulina* contains higher amounts of pigments such as zeaxanthin, myxoxanthophyll, and phycocyanin, polyunsaturated fatty acid, minerals, essential amino acids, vitamins, and proteins. On dry weight basis, *Spirulina* has 4–9% of lipids, 8–16% of carbohydrates, and 46–71% of proteins (Zhu et al. 2014). Essential amino acids of *Spirulina*, that is, leucine, valine, and isoleucine, make it valuable. Higher amounts of provitamin A, K, and B12 and β-carotene also found in it. The polyunsaturated fatty acids of *Spirulina* are $\omega - 3$, $\omega - 6$, linolenic, and γ-linolenic acid. The amount of DHA is about 91% of all the fatty acids that *Spirulina* contains (Yukino et al. 2005). More than ten carotenoids are found in its antioxidant mixture. The amount of minerals present in *Spirulina* varies according to kind of water in which it has been grown. The content of magnesium, iron, and calcium available in *Spirulina* makes it highly nutritious (Belay et al. 1993; Hudek et al. 2014; Pulz and Gross 2004; Wu et al. 2005). It has been found that powder form of *Spirulina* contains (2.330 × 103 IU/kg) provitamin A, vitamin E (100 mg/100 g), β-carotene (140 mg/100 g), thiamin (3.5 mg/100 g), riboflavin (4.0 mg/100 g), niacin (14.0 mg/100 g), biotin (0.005 mg/100 g), inositol (64 mg/100 g), biotin (0.005 mg/100 g), vitamin B12 (0.32 mg/100 g), and vitamin K (2.2 mg/100 g) (Belay 1994; Wu et al. 2005). *Spirulina* is also known for its possible use to reduce malnutrition and hunger, as reported by FAO under the UN resolution (Habib 2008). *Spirulina* was tested with undernourished patient in clinical studies. In the 8-week study, 550 children below the age of 5 years which were undernourished were taken and given traditional meal with *Spirulina* (n = 170), only traditional meal (n = 40), Misola with *Spirulina* (n = 170), and only Misola (a mixture of millet, peanut, soya) (n = 170), with meals with *Spirulina* exhibited better results and found better in nutritional rehabilitation (Simpore et al. 2006). Mineral contents obtained from microalgae have potential to be utilized in human nutrition supplementation. *Spirulina* is considered to reserve calcium, phosphorus, and iron (Salmeán et al. 2015). Moreover, antioxidant activity of *Spirulina* is also a studied property which has preventive attribute from atherosclerosis. The effect was found against atherosclerotic mice (Ku et al. 2015).

In a clinical study about microalgae, hypolipidemic action was found on human. *S. maxima* was applied in oral administration in Mexican runners (Torres-Durán et al. 2012). Five gram *Spirulina* was taken by 41 runners; the treatment was applied for 15 days. A proper meal higher in fat before and after the *Spirulina* treatment used to be taken. After every 1.5 h, the fatty meal postprandial lipemia used to be measured before and after the *Spirulina* administration plasma triacylglycerol (TAG) was determined. It was found to be lowered after treatment from 71.47 to 57.06 mg/dL. Due to prevention of proliferation and adhesion of tumor cells, *Arthrospira* calcium spirulan prevents pulmonary metastasis (Saiki et al. 2004). The biomass of *Spirulina* used in extract form or processed with food products such as biscuit, pasta, or other food products assists to maintain healthy bacteria inside the intestine. *Spirulina* is found to enhance growth of lactobacilli species (Pulz and Gross 2004).

4.2 Chlorella

Species of *Chlorella* are spherical in shape, having a diameter of 2–10 μm, green, single cell, and photoautotrophic without flagella. *Chlorella* has chlorophyll a and b; hence, chloroplast, the green pigment, is able to carry photosynthesis. It increases in number quickly by getting minerals, water, CO_2, and sunlight. In circular open pounds, large circular tank, and photobioreactors, *Chlorella* has been grown commercially (Borowitzka 1999). After harvesting *Chlorella* by autoflocculation or centrifugation, the solid powder biomass is drum or spray dried to be further processed in tablet formation. In the composition by dry weight of *Chlorella*, it has protein 11–58%, lipid 2–46%, and carbohydrate 12–28% (Zhu et al. 2014). *Chlorella* reserves many kinds of vitamins like β-carotene (180 mg/100 g), provitamin A (55,500 IU/kg), thiamin (1.5 mg/100 g), riboflavin (4.8 mg/100 g), niacin (23.8 mg/100 g), vitamin B6 (1.7 mg/100 g), vitamin B12 (125.9 mg/100 g), inositol (165.0 mg/100 g), biotin (191.6 mg/100 g), pantothenic acid (1.3 mg/100 g), and folic acid (26.9 mg/100 g) (Belay 1994; Hudek et al. 2014).

The compound β-1,3-glucan is also reserved by *Chlorella*. It is valuable compound that lowers blood lipids and enhances immune system by capturing free radical. It is found to be effective in wound healing and removing harmful compounds from the body, and it is helpful against tumor and stomach ulcer. It is also effective in hypercholesterolemia. Extracts from *Chlorella* enhance hemoglobin concentration and lower blood sugar, and it is a hepatoprotective agent (de Jesús Paniagua-Michel et al. 2015; Varfolomeev and Wasserman 2011). *Chlorella* species are considered "healthy food" and play vital role in disease prevention like Alzheimer's disease (Caporgno and Mathys 2018). Due to the presence of vitamin B12, folate, and iron content, *Chlorella* was found to be effective against anemia for pregnant women (Nakano et al. 2010).

4.3 Dunaliella

The *Dunaliella* are single cell, extremophile green, flagellated, rich in nutrient, and edible microalgae. It is found in freshwater and marine habitats. *D. salina* has higher amount of antioxidant activity. Due to higher reserves of β-carotene as much as 14% of dry biomass weight, it is considered the ideal source for carotenoid β-fractions as compared to other β-carotene sources (Mobin and Alam 2017). *Dunaliella* retains protein 49–57%, lipids 6–8%, carbohydrate 4–32% by its dry weight composition (Zhu et al. 2014). There are several difficulties regarding *D. salina* harvesting: (1) in culture, cell density remains low; (2) cells are highly fragile because they do not contain cell wall, making it susceptible in harvesting; and (3) cells have the same densities as culture. After the harvest of *Dunaliella*, β-carotene can be extracted using solvent by hot oil and the biomass dried by drum or spray dried methods (Ruegg 1984). *Dunaliella* biomass contains the carotenoids; the oxidized form of *D. salina* is found to be effective against cancer. Antihypertensive, analgesic, and broncholytic drugs are being manufactured by these microalgae (Mobin and Alam 2017). β-Carotene is the precursor of vitamin A that is found in dried *Dunaliella* tablets (Anbuchezhian et al. 2015).

4.4 Haematococcus pluvialis

Haematococcus pluvialis is a biflagellate, freshwater, unicellular Chlorophyta microalga which is found all over the world. In stressful conditions, astaxanthin, the strong antioxidant, can be accumulated in huge quantities which are up to 2–3% of dry weight in the species. *H. pluvialis* is the major organism utilized commercially to obtain astaxanthin (de Jesús Paniagua-Michel et al. 2015). *H. pluvialis* is found to be ideal source of astaxanthin; it retains the valuable carotenoid pigment which consists of 5% of dry weight (Fig. 5.2) (Wayama et al. 2013). The arsenal pigment of *H. pluvialis* has strong biological attributes such as prevention from cancer and ulcer, immunomodulation, and antioxidant (de Jesús Paniagua-Michel et al. 2015). Astaxanthin has been a proven dietary supplement that can produce antibodies, which are associated with anti-inflammatory and antitumor which restrict liver, bladder, mammary, colon, and oral cancer. It decreases chance of Alzheimer's and Parkinson diseases and enhances cardiovascular health (Mata et al. 2010).

5 Bioactive Components and Their Benefits in Human Health

There are a number of bioactive components found in microalgae; some of those bioactive components of microalgae are being discussed here, having significant value in human health and medicine. Microalgae have the ability to be utilized in certain applications depending on products being manufactured either in non-food or food due to the capacity to use seawater, wastewater, residual nutrients, sunlight, and nonarable land which is the favored cultivation globally (Draaisma et al. 2013). Various bioactive metabolites that have the ability to be utilized in physiological problems such as oxidative stress, cancer, microbial infection, and hyperlipidemia can be obtained from microalga (da Silva Vaz et al. 2016; Plaza et al. 2008). Microalgae exist in complicated natural environments, and it can

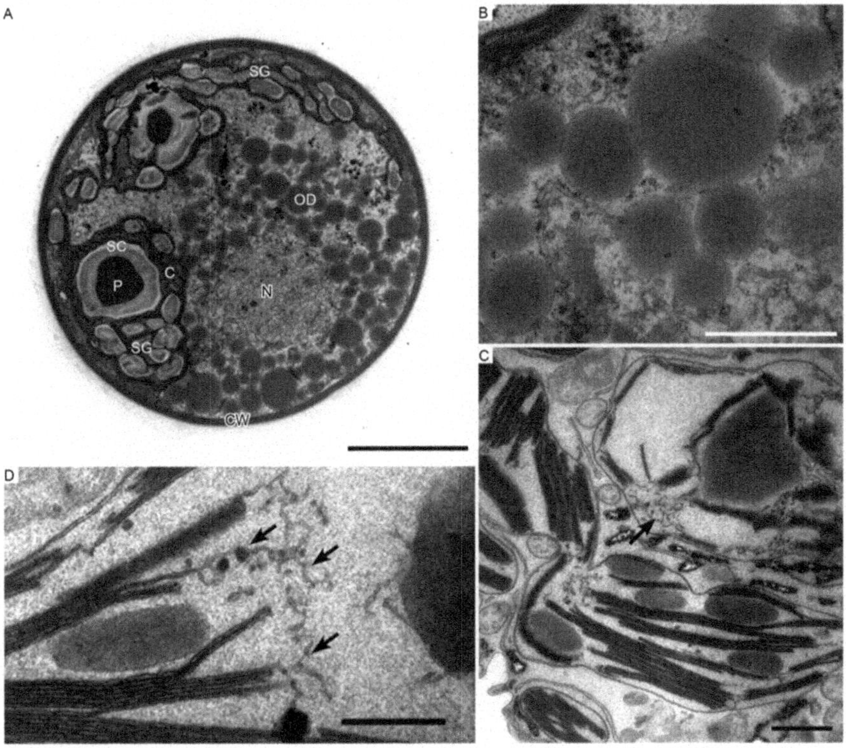

Fig. 5.2 Micrographs of *Haematococcus pluvialis* ((**a**) General ultrastructure. (**b**) Astaxanthin oil droplets. (**c**) Degradation of thylakoids. (**d**) Thylakoid degradation (high magnification)) (Peng et al. 2011)

adapt hard cultural environment such as nutrients, temperature, and variable salinity. As a result, they can make a number of effective and biologically active secondary metabolites with unique biological properties which are hardly seen in other living organisms (Anbuchezhian et al. 2015). Microalgae are being utilized in a number of industrial applications like nutritional components as nutraceuticals, pharmaceutical products, high-valued and healthy human food, food additives, cosmeceuticals, and antioxidants (Hudek et al. 2014; Liang et al. 2004; Mobin and Alam 2017; Mobin et al. 2001; Varfolomeev and Wasserman 2011). Anti-inflammatory, antioxidant, and anticancer properties have been shown by *Spirulina Porphyridium*, and *Arthrospira platensis* of red algae and phycobiliproteins obtained from cyanobacteria (Romay et al. 2003; Zheng et al. 2011). Microalgae produce many kinds of peptides that are considered anticancer agents. Dolastatins, a group of peptides, have been found in several cyanobacteria like *Symloca* and *Lyngbya* sp. (Fennell et al. 2003). Microalgal peptides with antitumor attributes in vitro are continuously increasing in number. Among those microalgae, there is polypeptide obtained from *Chlorella vulgaris*, *Chlorella pyrenoidosa*, and *S. platensis* (Wang and Zhang 2013).

5.1 Carotenoids

Bioactive compounds such as phycocyanin and β-carotene can also be extracted by microalgae. Astaxanthin, the powerful antioxidants, can be obtained in huge amounts from a *H. pluvialis* (Ambati et al. 2014; Liang et al. 2004). Carotenoids extracted from microalgae are much better in quality than the synthetic ones because of natural source (Lordan et al. 2011; Skjånes et al. 2013). The natural isomers of microalgal carotenoids exist in natural relationship. The β-carotene isomers obtained naturally have better quality to synthetically obtain trans isomers (Wang et al. 2015). Microalgae retain huge amounts of carotenoids, but the composition of exacts differs with the growth conditions and species. These molecules perform physiological benefits other than providing vitamin A (Nazih and Bard 2018).

5.2 Antioxidants

Haematococcus produces astaxanthin; exhibits the antioxidant attributes by protecting against oxidative stress, macular degradation, protein degradation; and suppresses the production of inflammatory compounds. In human, vitamin A formed by the conversion of β-carotene in the presence of that immune system performs its function effectively (Dufossé et al. 2005). There are 400 known carotenoids; among those, a few are commercially utilized such as β-carotene, lutein, lycopene, astaxanthin, bixin, and zeaxanthin. These are utilized as additives and natural food colorant. There are many carotenoid application in cosmetic production. The therapeutic and nutritional utility of a few carotenoids is that they are able to be converted to vitamin A. Moreover, carotenoids possess natural anti-inflammatory, antitumor, and chemopreventive activities (Fiedor and Burda 2014). Microalgae are the reservoir of the hydrophilic vitamins B1, B2, B3, B5, B6, B8 (biotin), B9, B12, and C along with lipophilic A and E vitamins. It attracted the vegetarian due to the presence of vitamin B12 because it is hard to meet daily required intake of the vitamin without consumption of meat (Nazih and Bard 2018). Antioxidant properties of microalgae have a number of health benefits such as protection against photooxidation, UV radiation, aging, immune response, liver functions, and eye and prostate health (Sheikhzadeh et al. 2012).

5.3 Amino Acids

All 20 protein amino acids are synthesized by microalgae, and it is an important traditional reservoir of all the essential amino acids that should be necessary in human nutrition (Spolaore et al. 2006). In the current era of world food security issues, there should be a sustainable source of bioactive peptides in the protein fraction of basic human food. There is great need of time to find a sustainable source of protein and to develop the technology to separate those and use in various applications. The marine microalgae are considered a sustainable and wonderful source of protein in order to obtain bioactive peptides (Udenigwe 2014). The functional peptides in microalgal substances gain attraction due to its revealed biological properties in provision of health benefits in certain conditions like oxidative stress, hypertension, immune disorder, diabetes, cancer, and inflammation (Nascimento 2015). Methionine and cysteine fraction in microalgae is less as compared to their presence in milk as casein or albumin, but these contents are still higher in amounts than in many known vegetable sources. *Spirulina* digestibility coefficient is comparatively higher (Devi and Venkataraman 1983).

5.4 Polysaccharides

Microalgal polysaccharides that retain sulfate esters are considered useful for industrial applications regarding their benefits toward human health (Markou et al. 2014). Sulfated polysaccharides found in microalgae called fucoidan contain high amounts of fucose and exhibit many useful activities for human health such as antiviral, anti-inflammatory, antiangiogenic, anticoagulant, antiadhesive, and immunomodulatory (Damonte et al. 2004). Half of the dry biomass of microalgae consists of polysaccharides; microalgae are found to be a good reservoir of polysaccharides. The contents of polysaccharides depend on species that can be found in cyanobacteria as glycogen and as hybrid glycogen or starch in some species (Plaza et al. 2009). Peptides and protein produced by microalgae are considered effective anticancer agents (Samarakoon and Jeon 2012). It was found that polysaccharides obtained from porphyridium cruentum exhibited higher antitumor properties. Sulfated polymer powerfully suppresses the tumor proliferation in vivo and in vitro of Graffi myeloid (Gardeva et al. 2009).

5.5 Fatty Acids

It incorporates a number of bioactive compounds which include polyunsaturated fatty acids that are comparable to traditional vegetable oil in the manufacture of human health products (Draaisma et al. 2013). Polyunsaturated fatty acids, soluble fibers, and sterol obtained from microalgae exhibited effectiveness in reduction of CVD (Plaza et al. 2008). It was found that α-linolenic and linolenic acids present in microalgae are in significant amounts, and more are present in soybean, sunflower, and rapeseed. Moreover, in several cases, the concentration of palmitic acid obtained from microalgae is found to be higher than other oils. The most attractive attribute of microalgal oil which makes it used as functional ingredient is the presence of docosahexaenoic acid (DHA, omega-3) and eicosapentaenoic acid (EPA, omega-3) long-chained polyunsaturated fatty acids that are being utilized functionally (Martins et al. 2013). Studies explain that intake of DHA and EPA provides health benefits by suppressing inflammation and reducing CVDs and helps in nervous system development in children and enhances brain functions (Endo and Arita 2016). Currently, marine fish, for example, cod, mullet, salmon, and mackerel, are the main provider of DHA and EPA. Fish oil is not considered better for usage as food ingredient due to its smell as it is undesirable for vegetarian. Moreover, due to the presence of contaminants like mercury in fish stocks gave the urge to look for other sources. Fungi, bacteria, and several plants can be possible alternative solutions to obtain commercial production of DHA and EPA. Fungi grow slowly and require organic carbon for its growth; the terrestrial plants need genetic modification to get long PUFAs. Microalgae are main producer of DHA and EPA under certain conditions such as mixotrophy, autotrophy, and heterotrophy (Ryckebosch et al. 2012). Lipids consist of 5–10% on dry mass basis and have importance in human health due to the presence of DHA and EPA in higher amounts (Santos-Sanchez et al. 2016). These polyunsaturated fatty acids are well known to suppress in CVD and inflammation (He et al. 2004).

5.6 Medicinal/Pharmaceutical Products

For the first time *Chlorella vulgaris* was found to produce useful compounds, and it was considered possible for utilization of microalgae in medicinal use (Pratt 1940). There are many microalgae that are utilized to produce antibiotics (bromophenols, alcohols, tannins, polysaccharide, fatty acids, and terpenoids). A number of neurotoxic and hepatotoxic compounds are produced by microalgae which can be utilized in pharmaceutical industry (Metting 1996). Toxins obtained by several blue green algae such as *Prymnesium parvum* and *Ochromonas* sp. have potential to be utilized in pharmaceutical production (Katircioglu et al. 2005).

Microalgal health products are available in the form of powder, capsule, and tablets; algal extracts are utilized to form other health products such as *Chlorella* drink containing growth factor, *Dunaliella* extract in capsule form, and *Spirulina* extract in (antioxidant) capsule form (Pulz and Gross 2004). Brands of *Spirulina* are commercially available and utilized to enhance health and effective against hypertension, high blood pressure, diabetes, and weight loss (Iwata et al. 1990). Dolastatin 1–16 are considered the prominent anticancer drug among others, and dolastatin 10 is recognized to suppress the assembly of microtubule; these medicines can also be obtained for utilization of microalgae (Costa et al. 2012).

5.7 *High-Value Food Products*

Microalgal species such as *Spirulina*, *Dunaliella*, and *Chlorella* are being utilized in food manufacturing. Microalgal biomass is being utilized in the formation of many health products such as capsules and tablets. Microalgae are utilized in the food products to increase their nutrition and health properties by utilizing as food additive in food items such as bread, noodles, biscuits, bean curd, ice cream, candies, and many other common food stuff (Finney et al. 1984; Liang et al. 2004; Villar et al. 1992). The *D. salina* protein is utilized in baking industry (Finney et al. 1984). Due to rich in nutrients, microalgal products are gaining more popularity in human health as healthy food and dominating the stores and market (Becker 2013). The nutrients obtained from microalgae have higher potential to be used in food products, but still their applications to develop human health and medicinal products are limited. Microalgal food products are categorized into two categories (Enzing et al. 2014). In the first category, microalgal products are available for customers to purchase and used as carbohydrates and protein sources to be further utilized in commodities; dried microalgae, in most of the cases the *Spirulina* and *Chlorella*, are used to obtain nutrients with higher nutrition values such as vitamins B12, D2, and C. The global distribution of microalgal companies for food and feed products is shown in Fig. 5.3. The products in the second category are more special microalgal-extracted components such as antioxidants, pigments, fatty acids (DHA and EPA), and proteins that are incorporated to enhance nutritional and health aspects of edible stuff (Vigani et al. 2015). Microalgal extracts are utilized in food products as stabilizer and thickeners such as agar (Bixler and Porse 2011). Figure 5.3 shows the quantity distribution of commercial microalgal products in the world, and Table 5.3 summarizes the research regarding microalgal effects on health and medicinal aspects.

6 Health Risks Regarding Microalgae Intake

Microcystin is the most studied and commonly found toxins whose 70 forms have already been discussed. Hepatocytes readily take cyclic heptapeptide; in this way, it makes structural damage restraining category 1 and 2A proteinous phosphatase which is compulsory in cell structural proteins. Human liver cancer also has linkage with microcystin. *Microcystis aeruginosa* strains are the main producers of microcystin; cyanobacteria *Oscillatoria agardhii* and *Nostoc* and *Anabaena flos-aquae* also contain this compound. *Aphanizomenon*, *Anabaena*, *Trichodesmium*, and *Oscillatoria* produce neurotoxins saxitoxins and anatoxins which are considered poisonous for humans. Cyanobacteria toxins effect the human health by skin contact inhalation and ingestion (Mulvenna et al. 2012). Microalgae contain huge amounts of nucleic acids (RNA and DNA); its components such as guanins, purines, and adenines are reduced by uric acid biochemical degradation. Uric acid

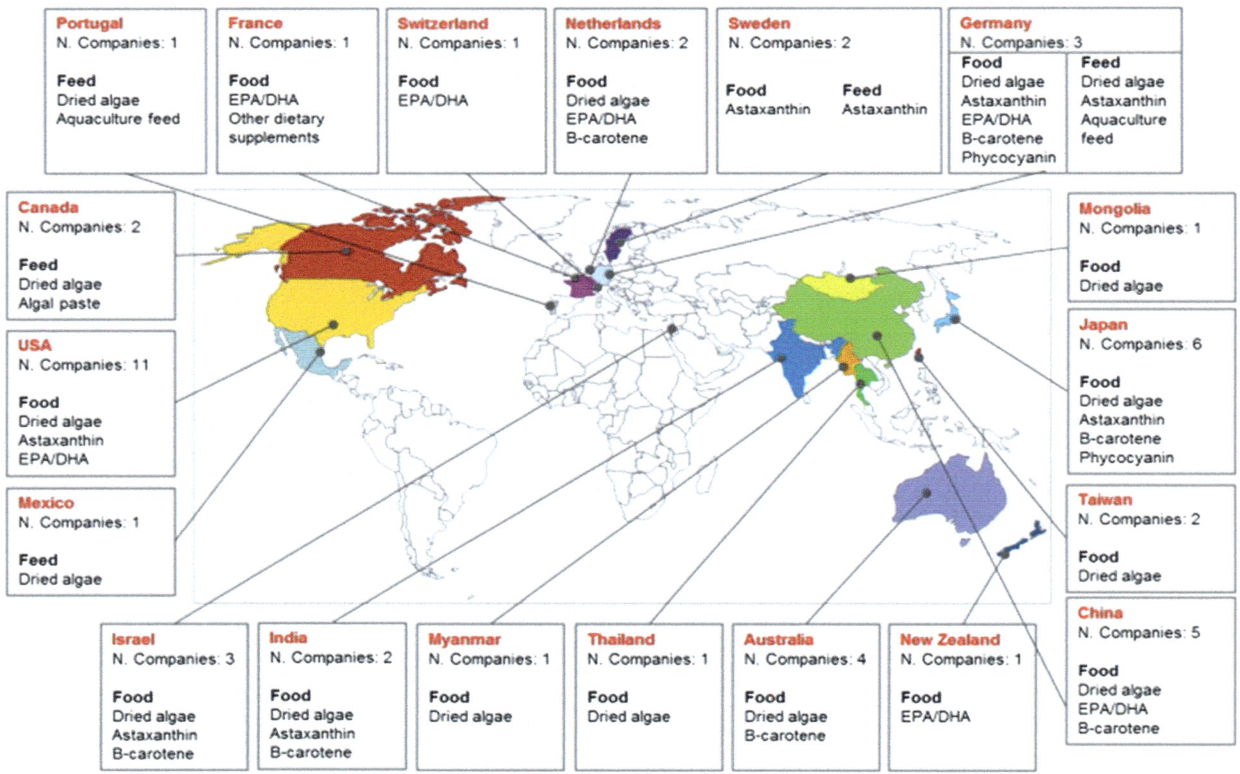

Fig. 5.3 Quantity distribution of commercial microalgal products in the world (Witvrouw and De 1997)

Table 5.3 Clinical studies regarding microalgal effects on health and medicinal aspects

Microalgae	Patients	Number of patients	Treatment duration (weeks)	Dose (g/day)	Healthy aspect	Reference
Dunaliella and β-carotene	Volunteer healthy men	60	12	0.56 g/day *Dunaliella* 40 mg/day β-carotene	Antioxidant protection and serum carotenoids increased Increase in retinol level in serum	Benamotz and Levy (1996)
Astaxanthin	Volunteer healthy men	20	1.5	6 mg/day	Blood circulation enhanced	Hiromi et al. (2008)
Arthrospira	Diabetic patients	–	12	8 g/day	Lower plasma lipids and blood pressure	Hee et al. (2008)

in high levels resulted in health problems like kidney stones and gout (Bux and Chisti 2016). Extensive intake of microalgae at higher concentrations could be injurious for human health (Spolaore et al. 2006).

7 Conclusion

Microalgae have been discovered to possess bioactive components such as carotenoids, polysaccharides, fatty acids, sterols, vitamins, and minerals that have huge potential in human health and their use as medicinal purposes, being excellent source of bioactive molecules. Microalgae can be utilized in the prevention of diseases by enhancing human immune system. Microalgae contain bioactive compounds which can be utilized in food stuff ultimately consumed by human and have impact on human health and could be considered a nutritional remedy. The medicinal properties of microalgal compounds such as antioxidant, anticancer, anti-inflammation, anti-aging, and antimicrobial have been utilized in medicinal purposes, and their application has been increasing in food and pharmaceutical industry. The production of microalgae is needed to enhance at large scale in order to meet its increasing demand due to potential use in vast products related to human health and medicine.

Acknowledgments This research was funded by the National Natural Science Foundation of China (21978120; 21506084), the Priority Academic Program Development of Jiangsu Higher Education Institutions, and the Training Project of the Young Core Instructor of Jiangsu University.

References

Alam, M. A., & Wang, Z. (Eds.). (2019). *Microalgae biotechnology for development of biofuel and wastewater treatment* (pp. 1–655). New York, NY: Springer.

Ambati, R., Phang, S.-M., Ravi, S., & Aswathanarayana, R. (2014). Astaxanthin: Sources, extraction, stability, biological activities and its commercial applications—A review. *Marine Drugs, 12*(1), 128–152.

Anbuchezhian, R., Karuppiah, V., & Li, Z. (2015). Prospect of marine algae for production of industrially important chemicals. In *Algal biorefinery: An integrated approach* (pp. 195–217). New York, NY: Springer.

Ansorena, D., Astiasarán, I., & Dominguez, H. (2013). *Development of nutraceuticals containing marine algae oils*. Cambridge: Woodhead Publishing Limited.

Apt, K. E., & Behrens, P. W. (1999). Commercial developments in microalgal biotechnology. *Journal of Phycology, 35*(2), 215–226.

Ayelet, H., Dror, H., Daniella, M., Hofit, C., Iris, B., Yehuda, K., Ayelet, G., Yariv, G., Ami, B. A., & Aviv, S. (2008). A 9-cis beta-carotene-enriched diet inhibits atherogenesis and fatty liver formation in LDL receptor knockout mice. *Journal of Nutrition, 138*(10), 1923.

Azabji-Kenfack, M., Dikosso, S. E., Loni, E., Onana, E., Sobngwi, E., Gbaguidi, E., Kana, A. N., Nguefack-Tsague, G., Von der Weid, D., & Njoya, O. (2011). Potential of *Spirulina platensis* as a nutritional supplement in malnourished HIV-infected adults in Sub-Saharan Africa: A randomised, single-blind study. *Nutrition and Metabolic Insights, 4*, 29.

Batista, A. P., Gouveia, L., Bandarra, N. M., Franco, J. M., & Raymundo, A. (2013). Comparison of microalgal biomass profiles as novel functional ingredient for food products. *Algal Research, 2*(2), 164–173.

Becker, E. (2007). Micro-algae as a source of protein. *Biotechnology Advances, 25*(2), 207–210.

Becker, E., & Venkataraman, L. (1980). Production and processing of algae in pilot plant scale experiences of the Indo-German project. In *Algae biomass: Production and use [sponsored by the National Council for Research and Development, Israel and the Gesellschaft fur Strahlen- und Umweltforschung (GSF), Munich, Germany]; editors, Gedaliah Shelef, Carl J. Soeder*. Amsterdam: Elsevier.

Becker, E. W. (2013). *Microalgae for human and animal nutrition*. London: Blackwell Science.

Belay, A. (1994). *Production of high quality spirulina at earthrise farms. Algal biotechnology in the Asia-Pacific region* (pp. 92–102). Kuala Lumpur: University of Malaya.

Belay, A., Kato, T., & Ota, Y. (1996). *Spirulina* (Arthrospira): Potential application as an animal feed supplement. *Journal of Applied Phycology, 8*(4-5), 303–311.

Belay, A., Ota, Y., Miyakawa, K., & Shimamatsu, H. (1993). Current knowledge on potential health benefits of *Spirulina*. *Journal of Applied Phycology, 5*(2), 235–241.

Benamotz, A., & Levy, Y. (1996). Bioavailability of a natural isomer mixture compared with synthetic all-trans beta-carotene in human serum. *American Journal of Clinical Nutrition, 63*(5), 729–734.

Bishop, W. M., & Zubeck, H. M. (2012). Evaluation of microalgae for use as nutraceuticals and nutritional supplements. *Nutrition & Food Sciences, 2*(5), 147.

Bixler, H. J., & Porse, H. (2011). A decade of change in the seaweed hydrocolloids industry. *Journal of Applied Phycology, 23*(3), 321–335.

Borowitzka, M. A. (1999). Commercial production of microalgae: Ponds, tanks, and fermenters. In *Progress in industrial microbiology* (Vol. 35, pp. 313–321). Amsterdam: Elsevier.

Borowitzka, M. A. (2013). High-value products from microalgae—Their development and commercialisation. *Journal of Applied Phycology, 25*(3), 743–756.

Buono, S., Langellotti, A. L., Martello, A., Rinna, F., & Fogliano, V. (2014). Functional ingredients from microalgae. *Food & Function, 5*(8), 1669–1685.

Burlew, J. S. (1953). *Algal culture from laboratory to pilot plant. Algal culture from laboratory to pilot plant*. Washington, DC: Carnegie Institution of Washington.

Bux, F., & Chisti, Y. (2016). *Algae biotechnology: Products and processes*. New York, NY: Springer.

Caporgno, M. P., & Mathys, A. (2018). Trends in microalgae incorporation into innovative food products with potential health benefits. *Frontiers in Nutrition, 5*, 58.

Carfagna, S., Salbitani, G., Bottone, C., & Vona, V. (2016). *Galdieria sulphuraria* as a possible source of food colorant. *Journal of Nutritional Ecology and Food Research, 3*(1), 67–70.

Chacón-Lee, T. L., & González-Mariño, G. E. (2010). Microalgae for "healthy" foods—Possibilities and challenges. *Comprehensive Reviews in Food Science and Food Safety, 9*(6), 655–675.

Chai, S. K., Kim, B., Pham, T. X., Yang, Y., Weller, C. L., Carr, T. P., Park, Y. K., & Lee, J. Y. (2015). Hypolipidemic effect of a blue-green alga (*Nostoc commune*) is attributed to its non-lipid fraction by decreasing intestinal cholesterol absorption in C57BL/6J mice. *Journal of Medicinal Food, 18*(11), 1214.

Cheong, S. H., Kim, M. Y., Sok, D. E., Hwang, S. Y., Kim, J. H., Kim, H. R., Lee, J. H., Kim, Y. B., & Kim, M. R. (2010). *Spirulina* prevents atherosclerosis by reducing hypercholesterolemia in rabbits fed a high-cholesterol diet. *Journal of Nutritional Science and Vitaminology, 56*(1), 34–40.

Chew, B. P., & Park, J. S. (2004). Carotenoid action on the immune response. *Journal of Nutrition, 134*(1), 257S.

Ciferri, O. (1983). *Spirulina*, the edible microorganism. *Microbiological Reviews, 47*(4), 551.

Costa, M., Costa-Rodrigues, J., Fernandes, M. H., Barros, P., Vasconcelos, V., & Martins, R. (2012). Marine cyanobacteria compounds with anticancer properties: A review on the implication of apoptosis. *Marine Drugs, 10*(10), 2181–2207.

da Silva Vaz, B., Moreira, J. B., de Morais, M. G., & Costa, J. A. V. (2016). Microalgae as a new source of bioactive compounds in food supplements. *Current Opinion in Food Science, 7*, 73–77.

Damonte, E. B., Matulewicz, M. C., & Cerezo, A. S. (2004). Sulfated seaweed polysaccharides as antiviral agents. *Current Medicinal Chemistry, 11*(18), 2399.

de Jesús Paniagua-Michel, J., Morales-Guerrero, E., & Soto, J. O. (2015). Microalgal biotechnology: Biofuels and bioproducts. In *Springer handbook of marine biotechnology* (pp. 1355–1370). New York, NY: Springer.

Devi, M., & Venkataraman, L. (1983). Supplementary value of the proteins of blue green algae *Spirulina platensis* to rice and wheat proteins. *Nutrition Reports International, 28*(5), 1029–1035.

Draaisma, R. B., Wijffels, R. H., Slegers, P. E., Brentner, L. B., Roy, A., & Barbosa, M. J. (2013). Food commodities from microalgae. *Current Opinion in Biotechnology, 24*(2), 169–177.

Dufossé, L., Galaup, P., Yaron, A., Arad, S. M., Blanc, P., Murthy, K. N. C., & Ravishankar, G. A. (2005). Microorganisms and microalgae as sources of pigments for food use: A scientific oddity or an industrial reality? *Trends in Food Science & Technology, 16*(9), 389–406.

Dvir, I., Chayoth, R., Sod-Moriah, U., Shany, S., Nyska, A., Stark, A. H., Madar, Z., & Arad, S. M. (2000). Soluble polysaccharide and biomass of red microalga *Porphyridium* sp. alter intestinal morphology and reduce serum cholesterol in rats. *British Journal of Nutrition, 84*(4), 469–476.

Endo, J., & Arita, M. (2016). Cardioprotective mechanism of omega-3 polyunsaturated fatty acids. *Journal of Cardiology, 67*(1), 22–27.

Enzing, C., Ploeg, M., Barbosa, M., & Sijtsma, L. (2014). *Microalgae-based products for the food and feed sector: An outlook for Europe. JRC scientific and policy reports* (pp. 19–37). Brussels: European Union.

Fennell, B., Carolan, S., Pettit, G., & Bell, A. (2003). Effects of the antimitotic natural product dolastatin 10, and related peptides, on the human malarial parasite Plasmodium falciparum. *Journal of Antimicrobial Chemotherapy, 51*(4), 833–841.

Fiedor, J., & Burda, K. (2014). Potential role of carotenoids as antioxidants in human health and disease. *Nutrients, 6*(2), 466–488.

Finney, K., Pomeranz, Y., & Bruinsma, B. (1984). Use of algae *Dunaliella* as a protein supplement in bread. *Cereal Chemistry, 61*, 402.

Garcíagonzález, M., Moreno, J., Manzano, J. C., Florencio, F. J., & Guerrero, M. G. (2005). Production of *Dunaliella salina* biomass rich in 9-cis-beta-carotene and lutein in a closed tubular photobioreactor. *Journal of Biotechnology, 115*(1), 81–90.

Gardeva, E., Toshkova, R., Minkova, K., & Gigova, L. (2009). Cancer protective action of polysaccharide, derived from red microalga *Porphyridium cruentum*—A biological background. *Biotechnology & Biotechnological Equipment, 23*(Suppl 1), 783–787.

Graziani, G., Schiavo, S., Nicolai, M. A., Buono, S., Fogliano, V., Pinto, G., & Pollio, A. (2013). Microalgae as human food: Chemical and nutritional characteristics of the thermo-acidophilic microalga *Galdieria sulphuraria*. *Food & Function, 4*(1), 144–152.

Guzman, S., Gato, A., & Calleja, J. (2001). Antiinflammatory, analgesic and free radical scavenging activities of the marine microalgae *Chlorella stigmatophora* and *Phaeodactylum tricornutum*. *Phytotherapy Research, 15*(3), 224–230.

Habib, M. A. B. (2008). *Review on culture, production and use of Spirulina as food for humans and feeds for domestic animals and fish*. Rome: Food and Agriculture Organization of the United Nations.

Hayashi, T., Hayashi, K., Maeda, M., & Kojima, I. (1996). Calcium spirulan, an inhibitor of enveloped virus replication, from a blue-green alga *Spirulina platensis*. *Journal of Natural Products, 59*(1), 83–87.

He, K., Song, Y., Daviglus, M. L., Liu, K., Van Horn, L., Dyer, A. R., & Greenland, P. (2004). Accumulated evidence on fish consumption and coronary heart disease mortality: A meta-analysis of cohort studies. *Circulation, 109*(22), 2705–2711.

Hee, L. E., Ji-Eun, P., Young-Ju, C., Kap-Bum, H., & Wha-Young, K. (2008). A randomized study to establish the effects of *Spirulina* in type 2 diabetes mellitus patients. *Nutrition Research and Practice, 2*(4), 295–300.

Hernández-Corona, A., Nieves, I., Meckes, M., Chamorro, G., & Barron, B. L. (2002). Antiviral activity of *Spirulina maxima* against herpes simplex virus type 2. *Antiviral Research, 56*(3), 279–285.

Hiromi, M., Jiro, T., Hiroki, T., & Isao, T. (2008). Effects of astaxanthin on human blood rheology. *Journal of Clinical Biochemistry and Nutrition, 43*(2), 69–74.

Hochman, G., & Zilberman, D. (2014). Algae farming and its bio-products. In *Plants and BioEnergy* (pp. 49–64). New York, NY: Springer.

Holdt, S. L., Kraan, S., Critchley, A., Oliveira, E. C., Cabellopasini, A., Weinberger, F., Hennequart, F., & Zuccarello, G. C. (2011). Bioactive compounds in seaweed: Functional food applications and legislation. *Journal of Applied Phycology, 23*(3), 543–597.

Hudek, K., Davis, L., Ibbini, J., & Erickson, L. (2014). *Commercial products from algae. Algal biorefineries*. New York, NY: Springer.

Iwata, K., Inayama, T., & Kato, T. (1990). Effects of *Spirulina platensis* on plasma lipoprotein lipase activity in fructose-induced hyperlipidemic rats. *Journal of Nutritional Science and Vitaminology, 36*(2), 165.

Jin, E., Feth, B., & Melis, A. (2010). A mutant of the green alga *Dunaliella salina* constitutively accumulates zeaxanthin under all growth conditions. *Biotechnology and Bioengineering, 81*(1), 115–124.

Katircioglu, H., Beyatli, Y., Aslim, B., Yüksekdag, Z., & Atici, T. (2005). Screening for antimicrobial agent production of some microalgae in freshwater. *Internet Journal of Microbiology, 2*, 1.

Kau, A. L., Ahern, P. P., Griffin, N. W., Goodman, A. L., & Gordon, J. I. (2011). Human nutrition, the gut microbiome and the immune system. *Nature, 474*(7351), 327–336.

Kleinegris, D. M. M., Es, M. A. V., Janssen, M., Brandenburg, W. A., & Wijffels, R. H. (2010). Carotenoid fluorescence in *Dunaliella salina*. *Journal of Applied Phycology, 22*(5), 645–649.

Ku, C. S., Kim, B., Pham, T. X., Yang, Y., Wegner, C. J., Park, Y.-K., Balunas, M., & Lee, J.-Y. (2015). Blue-green algae inhibit the development of atherosclerotic lesions in apolipoprotein E knockout mice. *Journal of Medicinal Food, 18*(12), 1299–1306.

Kumar, S. R., Hosokawa, M., & Miyashita, K. (2013). Fucoxanthin: A marine carotenoid exerting anti-cancer effects by affecting multiple mechanisms. *Marine Drugs, 11*(12), 5130–5147.

Kumari, P., Kumar, M., Gupta, V., Reddy, C. R. K., & Jha, B. (2010). Tropical marine macroalgae as potential sources of nutritionally important PUFAs. *Food Chemistry, 120*(3), 749–757.

Kyle, D. J. (2001). The large-scale production and use of a single-cell oil highly enriched in docosahexaenoic acid. *ACS Symposium, 788*, 92–107.

Lee, Y.-K. (1997). Commercial production of microalgae in the Asia-Pacific rim. *Journal of Applied Phycology, 9*(5), 403–411.

Li, Y., Horsman, M., Nan, W., Lan, C. Q., & Dubois-Calero, N. (2008). Biofuels from microalgae. *Biotechnology Progress, 24*(4), 815–820.

Liang, S., Liu, X., Chen, F., & Chen, Z. (2004). Current microalgal health food R & D activities in China. In *Asian pacific phycology in the 21st century: Prospects and challenges* (pp. 45–48). New York, NY: Springer.

Liu, J., Sun, Z., Gerken, H., Liu, Z., Jiang, Y., & Chen, F. (2014). *Chlorella zofingiensis* as an alternative microalgal producer of astaxanthin: Biology and industrial potential. *Marine Drugs, 12*(6), 3487–3515.

Lordan, S., Ross, R. P., & Stanton, C. (2011). Marine bioactives as functional food ingredients: Potential to reduce the incidence of chronic diseases. *Marine Drugs, 9*(6), 1056–1100.

Lorenz, R. T., & Cysewski, G. R. (2000). Commercial potential for *Haematococcus* microalgae as a natural source of astaxanthin. *Trends in Biotechnology, 18*(4), 160–167.

Markou, G., Vandamme, D., & Muylaert, K. (2014). Microalgal and cyanobacterial cultivation: The supply of nutrients. *Water Research, 65*, 186–202.

Martínez-Hernández, G. B., Castillejo, N., Carrión-Monteagudo, M. D. M., Artés, F., & Artés-Hernández, F. (2018). Nutritional and bioactive compounds of commercialized algae powders used as food supplements. *Food Science and Technology International, 24*(2), 172–182.

Martins, D., Custódio, L., Barreira, L., Pereira, H., Ben-Hamadou, R., Varela, J., & Abu-Salah, K. (2013). Alternative sources of n-3 long-chain polyunsaturated fatty acids in marine microalgae. *Marine Drugs, 11*(7), 2259–2281.

Mata, T. M., Martins, A. A., & Caetano, N. S. (2010). Microalgae for biodiesel production and other applications: A review. *Renewable and Sustainable Energy Reviews, 14*(1), 217–232.

Mayfield, S. P., Manuell, A. L., Chen, S., Wu, J., Tran, M., Siefker, D., Muto, M., & Marin-Navarro, J. (2007). *Chlamydomonas reinhardtii* chloroplasts as protein factories. *Current Opinion in Biotechnology, 18*(2), 126–133.

Metting, F. B. (1996). Biodiversity and application of microalgae. *Journal of Industrial Microbiology, 17*(5-6), 477–489.

Mobin, S., & Alam, F. (2017). Some promising microalgal species for commercial applications: A review. *Energy Procedia, 110*, 510–517.

Mobin, S., Kanai, K., & Yoshikoshi, K. (2001). Effects of feeding levels on the growth and survival of larval and juvenile Japanese flounder *Paralichthys olivaceus*. *Aquaculture Science, 49*(2), 207–218.

Monego, D. L., da Rosa, M. B., & do Nascimento, P. C. (2017). Applications of computational chemistry to the study of the antiradical activity of carotenoids: A review. *Food Chemistry, 217*, 37–44.

Mulvenna, V., Dale, K., Priestly, B., Mueller, U., Humpage, A., Shaw, G., Allinson, G., & Falconer, I. (2012). Health risk assessment for cyanobacterial toxins in seafood. *International Journal of Environmental Research and Public Health, 9*(3), 807–820.

Nagai, H., Murata, M., Torigoe, K., Satake, M., & Yasumoto, T. (1992). Gambieric acids, new potent antifungal substances with unprecedented polyether structures from a marine dinoflagellate Gambierdiscus toxicus. *The Journal of Organic Chemistry, 57*(20), 5448–5453.

Nakano, S., Takekoshi, H., & Nakano, M. (2010). *Chlorella pyrenoidosa* supplementation reduces the risk of anemia, proteinuria and edema in pregnant women. *Plant Foods for Human Nutrition, 65*(1), 25–30.

Nascimento, E. S. D. (2015). *Obtenção de hidrolisado proteico de sementes de quiabo Abelmoschus esculentus (L.) Moench e sua capacidade antioxidante*. Doctoral Dissertation. Universidade Federal da Paraíba.

Nazih, H., & Bard, J.-M. (2018). Microalgae in human health: interest as a functional food. In *Microalgae in health and disease prevention* (pp. 211–226). Amsterdam: Elsevier.

Olivares, H. G., Lagos, N. M., Gutierrez, C. J., Kittelsen, R. C., Valenzuela, G. L., & Lillo, M. E. H. (2016). Assessment oxidative stress biomarkers and metal bioaccumulation in macroalgae from coastal areas with mining activities in Chile. *Environmental Monitoring and Assessment, 188*(1), 1–11.

Paniagua-Michel, J. (2015). Chapter 16 – Microalgal nutraceuticals. In *Handbook of marine microalgae*. London: Academic Press.

Patil, V., Tran, K.-Q., & Giselrød, H. R. (2008). Towards sustainable production of biofuels from microalgae. *International Journal of Molecular Sciences, 9*(7), 1188–1195.

Peng, J., Yuan, J. P., Wu, C. F., & Wang, J. H. (2011). Fucoxanthin, a marine carotenoid present in brown seaweeds and diatoms: Metabolism and bioactivities relevant to human health. *Marine Drugs, 9*(10), 1806–1828.

Plaza, M., Cifuentes, A., & Ibáñez, E. (2008). In the search of new functional food ingredients from algae. *Trends in Food Science & Technology, 19*(1), 31–39.

Plaza, M., Herrero, M., Cifuentes, A., & Ibanez, E. (2009). Innovative natural functional ingredients from microalgae. *Journal of Agricultural and Food Chemistry, 57*(16), 7159–7170.

Ponce-Canchihuamán, J. C., Pérez-Méndez, O., Hernández-Muñoz, R., Torres-Durán, P. V., & Juárez-Oropeza, M. A. (2010). Protective effects of *Spirulina maxima* on hyperlipidemia and oxidative-stress induced by lead acetate in the liver and kidney. *Lipids in Health and Disease, 9*(1), 35.

Pratt, R. (1940). Influence of the size of the inoculum on the growth of chlorella vulgaris in freshly prepared culture medium. *American Journal of Botany, 27*(1), 52–56.

Pshenichkin, S. P., & Wise, B. C. (1995). Okadaic acid increases nerve growth factor secretion, mRNA stability, and gene transcription in primary cultures of cortical astrocytes. *The Journal of Biological Chemistry, 270*(11), 5994–5999.

Pulz, O., & Gross, W. (2004). Valuable products from biotechnology of microalgae. *Applied Microbiology and Biotechnology, 65*(6), 635–648.

Raposo, M., de Morais, R., & Bernardo de Morais, A. (2013). Bioactivity and applications of sulphated polysaccharides from marine microalgae. *Marine Drugs, 11*(1), 233–252.

Rasala, B. A., & Mayfield, S. P. (2015). Photosynthetic biomanufacturing in green algae; production of recombinant proteins for industrial, nutritional, and medical uses. *Photosynthesis Research, 123*(3), 227–239.

Rastall, R. A., Gibson, G. R., Gill, H. S., Guarner, F., Klaenhammer, T. R., Pot, B., Reid, G., Rowland, I. R., & Sanders, M. E. (2010). Modulation of the microbial ecology of the human colon by probiotics, prebiotics and synbiotics to enhance human health: An overview of enabling science and potential applications. *FEMS Microbiology Ecology, 52*(2), 145–152.

Riediger, N. D., Othman, R. A., Miyoung, S., & Moghadasian, M. H. (2009). A systemic review of the roles of n-3 fatty acids in health and disease. *Journal of the American Dietetic Association, 109*(4), 668–679.

Romay, C., Gonzalez, R., Ledon, N., Remirez, D., & Rimbau, V. (2003). C-phycocyanin: A biliprotein with antioxidant, anti-inflammatory and neuroprotective effects. *Current Protein and Peptide Science, 4*(3), 207–216.

Ruegg, R. (1984). *Extraction process for beta-carotene*. U.S. *Patent No. US4439629A*. Washington, DC: U.S. Patent and Trademark Office.

Ryckebosch, E., Bruneel, C., Muylaert, K., & Foubert, I. (2012). Microalgae as an alternative source of omega-3 long chain polyunsaturated fatty acids. *Lipid Technology, 24*(6), 128–130.

Saiki, I., Murata, J., Fujii, H., & Kato, T. (2004). Inhibition of tumor invasion and metastasis by calcium spirulan (Ca-SP), a novel sulfated polysaccharide derived from a blue-green alga *Spirulina platensis*. *Nutritional Sciences, 7*(3), 144–150.

Salmeán, G. G., Castillo, L. H. F., & Chamorro-Cevallos, G. (2015). Nutritional and toxicological aspects of *Spirulina* (Arthrospira). *Nutrición Hospitalaria, 32*(1), 34–40.
Samarakoon, K., & Jeon, Y.-J. (2012). Bio-functionalities of proteins derived from marine algae—A review. *Food Research International, 48*(2), 948–960.
Sampath-Wiley, P., Neefus, C. D., & Jahnke, L. S. (2008). Seasonal effects of sun exposure and emersion on intertidal seaweed physiology: Fluctuations in antioxidant contents, photosynthetic pigments and photosynthetic efficiency in the red alga *Porphyra umbilicalis* Kützing (Rhodophyta, Bangiales). *Journal of Experimental Marine Biology and Ecology, 361*(2), 83–91.
Santos-Sanchez, N., Valadez-Blanco, R., Hernandez-Carlos, B., Torres-Arino, A., Guadarrama-Mendoza, P., & Salas-Coronado, R. (2016). Lipids rich in ω-3 polyunsaturated fatty acids from microalgae. *Applied Microbiology and Biotechnology, 100*(20), 8667–8684.
Sheikhzadeh, N., Tayefi-Nasrabadi, H., Oushani, A. K., & Enferadi, M. H. (2012). Effects of *Haematococcus pluvialis* supplementation on antioxidant system and metabolism in rainbow trout (Oncorhynchus mykiss). *Fish Physiology and Biochemistry, 38*(2), 413–419.
Simpore, J., Kabore, F., Zongo, F., Dansou, D., Bere, A., Pignatelli, S., Biondi, D. M., Ruberto, G., & Musumeci, S. (2006). Nutrition rehabilitation of undernourished children utilizing Spiruline and Misola. *Nutrition Journal, 5*(1), 3.
Skjånes, K., Rebours, C., & Lindblad, P. (2013). Potential for green microalgae to produce hydrogen, pharmaceuticals and other high value products in a combined process. *Critical Reviews in Biotechnology, 33*(2), 172–215.
Smit, A. J. (2004). Medicinal and pharmaceutical uses of seaweed natural products: A review. *Journal of Applied Phycology, 16*(4), 245–262.
Spolaore, P., Joannis-Cassan, C., Duran, E., & Isambert, A. (2006). Commercial applications of microalgae. *Journal of Bioscience and Bioengineering, 101*(2), 87–96.
Takahashi, K., Hosokawa, M., Kasajima, H., Hatanaka, K., Kudo, K., Shimoyama, N., & Miyashita, K. (2015). Anticancer effects of fucoxanthin and fucoxanthinol on colorectal cancer cell lines and colorectal cancer tissues. *Oncology Letters, 10*(3), 1463.
Talero, E., García-Mauriño, S., Ávila-Román, J., Rodríguez-Luna, A., Alcaide, A., & Motilva, V. (2015). Bioactive compounds isolated from microalgae in chronic inflammation and cancer. *Marine Drugs, 13*(10), 6152–6209.
Torres-Durán, P. V., Ferreira-Hermosillo, A., Ramos-Jiménez, A., Hernández-Torres, R. P., & Juárez-Oropeza, M. A. (2012). Effect of *Spirulina maxima* on postprandial lipemia in young runners: A preliminary report. *Journal of Medicinal Food, 15*(8), 753–757.
Udenigwe, C. C. (2014). Bioinformatics approaches, prospects and challenges of food bioactive peptide research. *Trends in Food Science & Technology, 36*(2), 137–143.
Varfolomeev, S., & Wasserman, L. (2011). Microalgae as source of biofuel, food, fodder, and medicines. *Applied Biochemistry and Microbiology, 47*(9), 789–807.
Vigani, M., Parisi, C., Rodríguez-Cerezo, E., Barbosa, M. J., Sijtsma, L., Ploeg, M., & Enzing, C. (2015). Food and feed products from micro-algae: Market opportunities and challenges for the EU. *Trends in Food Science & Technology, 42*(1), 81–92.
Villar, R., Laguna, M., Calleja, J., & Cadavid, I. (1992). Effects of *Skeletonema costatum* extracts on the central nervous system. *Planta Medica, 58*(05), 398–404.
Wang, H.-M. D., Chen, C.-C., Huynh, P., & Chang, J.-S. (2015). Exploring the potential of using algae in cosmetics. *Bioresource Technology, 184*, 355–362.
Wang, X., & Zhang, X. (2013). Separation, antitumor activities, and encapsulation of polypeptide from *Chlorella pyrenoidosa*. *Biotechnology Progress, 29*(3), 681–687.
Wayama, M., Ota, S., Matsuura, H., Nango, N., Hirata, A., & Kawano, S. (2013). Three-dimensional ultrastructural study of oil and astaxanthin accumulation during encystment in the green alga *Haematococcus pluvialis*. *PLoS One, 8*(1), e53618.
WHO. (2016). *World health statistics 2016: Monitoring health for the SDGs sustainable development goals*. Washington, DC: Author.

Witvrouw, M., & De, C. E. (1997). Sulfated polysaccharides extracted from sea algae as potential antiviral drugs. *General Pharmacology, 29*(4), 497–511.

Wu, L.-C., Ho, J.-A. A., Shieh, M.-C., & Lu, I.-W. (2005). Antioxidant and antiproliferative activities of *Spirulina* and *Chlorella* water extracts. *Journal of Agricultural and Food Chemistry, 53*(10), 4207–4212.

Yukino, T., Hayashi, M., Inoue, Y., Imamura, J., Nagano, N., & Murata, H. (2005). Preparation of docosahexaenoic acid fortified *Spirulina platensis* and its lipid and fatty acid compositions. *Nippon Suisan Gakkaishi, 71*(1), 74–79.

Zhao, L., Wang, J., Zhang, P., Gu, Q., & Gao, C. (2018). Absorption of heavy metal ions by alginate. In *Bioactive seaweeds for food applications* (pp. 255–268). Amsterdam: Elsevier.

Zheng, L.-H., Wang, Y.-J., Sheng, J., Wang, F., Zheng, Y., Lin, X.-K., & Sun, M. (2011). Antitumor peptides from marine organisms. *Marine Drugs, 9*(10), 1840–1859.

Zhu, L., Hiltunen, E., Antila, E., Zhong, J., Yuan, Z., & Wang, Z. (2014). Microalgal biofuels: Flexible bioenergies for sustainable development. *Renewable and Sustainable Energy Reviews, 30*, 1035–1046.

Chapter 6
Astaxanthin Production from Microalgae

Thomas Butler and Yonatan Golan

Abstract Astaxanthin is commercially sold as a pigment for animal feed and as an antioxidant for the nutraceutical sector. Astaxanthin is predominantly manufactured synthetically from petrochemicals but is also obtained from the chlorophyte *Haematococcus pluvialis* (*Haematococcus lacustris*). The petrochemical-derived synthetic alternative has conventionally been used, attributable to its lower cost (\$1300–1800 kg^{-1}). However, it is inferior as an antioxidant, prohibited for direct human consumption, and may cause toxicity in the final product. Conventionally, astaxanthin from *H. lacustris* is produced in a two-stage production process, incorporating a green and red stage for maximising growth and astaxanthin production, respectively, but a one-stage process has been proposed. The *H. lacustris*-derived astaxanthin industry has been a commercial success, but several constraints have arisen including contamination issues, relatively low biomass and astaxanthin productivities, high downstream processing costs, and photobleaching issues in the red stage. These constraints need to be addressed for the production of astaxanthin from *H. lacustris* for the aquaculture sector. Alternatively, through the exploitation of an alternative life cycle stage, red motile macrozooids can be formed lacking the thick walls of aplanospores. It is envisaged that the red motile macrozooids could be harvested and fed as a whole-cell product directly to the aquaculture sector rich in astaxanthin and polyunsaturated fatty acids, bypassing the cell disruption and extraction steps to deliver bioavailable astaxanthin as a biobased feed.

Keywords Astaxanthin · Contamination · Biomass · Bioreactors · Extraction · *H. pluvialis*

T. Butler (✉)
Chemical and Biological Engineering Department, The University of Sheffield, Sheffield, UK
e-mail: tbutler2@sheffield.ac.uk

Y. Golan
CEO at Brevel Ltd., Lachish, Israel
e-mail: yonatang@brevel.co.il

M. A. Alam et al. (eds.), *Microalgae Biotechnology for Food, Health and High Value Products*, https://doi.org/10.1007/978-981-15-0169-2_6

1 Introduction

1.1 Carotenoids and Their Chemistry

Carotenoids are a family of greater than 600 naturally occurring pigments synthesised by higher plants, algae, fungi, and bacteria (Yaakob et al. 2014). Around 40 carotenoids are typically present in the human diet (BCC Research 2015). The chemical structure of carotenoids is derived from the carotenoid lycopene ($C_{40}H_{56}$). Carotenoids are mainly hydrocarbons which have two terminal rings joined by a conjugated double-bond chain or polyene system (Yuan et al. 2011). Two major groups of carotenoids have been characterised on the basis of their chemical structure: the carotenes (composed of carbon and hydrogen) and the xanthophylls (oxygenated derivatives). Astaxanthin, a xanthophyll, was found to be closely related to the other carotenoids, β-carotene, zeaxanthin, and lutein, and has many of the physiological and metabolic functions associated with carotenoids (Guerin et al. 2003). However, the presence of the hydroxyl and keto endings on each ionone ring reflect the unique properties, such as the ability to be esterified, a more polar configuration, and a higher antioxidant activity (inhibiting oxidation of other molecules) (Guerin et al. 2003). Each double bond from the polyene chain has been found to exist in two different configurations as geometric isomers *cis* or *trans*. *Cis* isomers are known to be thermodynamically less stable than *trans* isomers (Higuera-Ciapara et al. 2006) and in nature, most carotenoids predominate in the *trans* form (Stahl and Sies 2003).

Rodríguez-Sáiz et al. (2010) determined that astaxanthin contains two chiral centres and is present in three configurational isomers of the *trans* form (all-*E* isomer) (3R,3′R), (3R,3′S), and (3S,3′S). The (3S,3′S) form is the most abundant astaxanthin isomer in nature (Mont et al. 2010) and has been observed to be of the highest biotechnological value (Al-Bulishi 2015). Synthetic astaxanthin is generally composed of the three enantiomers (3R,3′R), (3R,3′S), and (3S,3′S) with a ratio of 1:2:1 and is unesterified, whereas astaxanthin from *H. pluvialis* is of the (3S,3′S stereoisomer), and 70% is in the monoester form, 10–15% in the diester form and 4–5% in the free form (Higuera-Ciapara et al. 2006; Ranga Rao et al. 2010; Young et al. 2017), which is also the main form in wild salmon (47.1–90%) (Young et al. 2017). The (3S,3′S) stereoisomer has been reported to impart a higher pigmentation in rainbow trout (*Oncorhynchus mykiss*) than other astaxanthin isomers and has been stated as the preferred additive for aquaculture (Choubert and Heinrich 1993). The 3S,3′S isomer has also been reported to have contributed to human health benefits, whereas the other forms have not been proven to have had positive biological effects (Capelli et al. 2013a; Guerin et al. 2003). Depending on their origin, astaxanthin can be found in association with other compounds such as proteins and biological lipids. In the case of *Haematococcus pluvialis*, up to 95% of astaxanthin molecules can be esterified with fatty acids (FAs) (commonly oleic, palmitic, and linoleic acid) (Lorenz and Cysewski 2000), with oleic acid as the major FA which is conjugated to astaxanthin molecules (Holtin et al. 2009). The synthetic form is found in the free, unesterified form as is astaxanthin derived from the yeast *Xanthophyllomyces dendrorhous* (Capelli et al. 2013a).

1.2 The Astaxanthin Market

The global carotenoid market reached US $1.5 billion in 2017 and is scheduled to reach US $2 billion by 2019 due to rising consumer awareness regarding the health benefits offered by the wide variety of carotenoids (BCC Research 2018). Panis and Rosales (2016) stated that in 2014 global astaxanthin production was 280 metric tons with a valuation of US $447, and the forecast for 2020 is US $1.5 billion (Panis and Rosales 2016; Allewaert et al. 2017; Molino et al. 2018). From industry reports, *H. pluvialis*-derived astaxanthin represents 5–8 tons (Pers. Com. Brevel Ltd.). Currently, 95% of the astaxanthin available in the market is generated synthetically from petrochemicals, <1% is produced from *H. pluvialis*, and the remainder is produced from the bacterium *Paracoccus carotinifaciens*, and the yeast *Xanthophyllomyces dendrorhous* (Koller et al. 2014; Panis and Rosales 2016; Shah et al. 2016). In 2009, 91% of commercial astaxanthin was used for animal feed pigments and 9.1% was used for nutraceuticals, with supply dominated by the synthetic form (Oilalgae 2015). It has been determined that the highest market share in 2016 (40%) was for the animal feed market (Market Watch 2019).

The market for astaxanthin has significantly grown from when it was first approved by the US Food and Drug Administration (FDA) in 1987 for its use as a feed additive in aquaculture and, over a decade later, when natural astaxanthin was subsequently approved to be used as a nutraceutical (Guerin et al. 2003). *H. pluvialis*-derived astaxanthin as a colour additive has been approved for salmonid feeds and additionally as a dietary supplement for human consumption in several European countries, the USA, and Japan (Yuan et al. 2011). To date, there is no European Food Safety Authority (EFSA) approval for the therapeutic application of *H. pluvialis*-derived astaxanthin. In line with EU 2015/2283, astaxanthin has been registered as a novel food and can be used to fortify foods equivalent to a maximum intake of 8 mg/day, but this is currently under review (https://www.efsa.europa.eu/sites/default/files/consultation/callsfordata/Callfordata-safetyassessmentofAstaxanthin.pdf). *H. pluvialis*-derived astaxanthin extracted using supercritical CO_2 has been granted Novel Food status by the UK Foods Standard Agency (FSA), and the US FDA has granted astaxanthin from *H. pluvialis* GRAS certified (Generally Recognised as Safe) (Shah et al. 2016). EU regulation 2015/1415 has limited synthetic astaxanthin to <100 ppm/kg of fish feed, whereas natural astaxanthin is widely accepted as safe (FDA GRAS Notice. No. GRN 000294). In the salmon farming industry, 10–15% of the total feed cost is attributable to astaxanthin (Mann et al. 2000; Nguyen 2013).

The major synthetic producers for astaxanthin are BASF, Royal DSM, and Zhejiang NHU Co. Ltd with a selling price of \$2000 (Koller et al. 2014), but current costs can be as low as \$1300 kg^{-1} for pure astaxanthin for the aquaculture industry (\$130 kg^{-1} for 10% astaxanthin) (Pers. Com. Brevel Ltd.). Natural astaxanthin can range from \$2500–7150 kg^{-1} (Kim et al. 2016; Koller et al. 2014), but from industrial reports, the price of nutraceutical grade pure astaxanthin from *H. pluvialis* is \$6000 kg^{-1} (Pers. Com. Brevel Ltd.). Currently, the estimated cost of production for synthetic astaxanthin is around \$1000 kg^{-1} compared to *H. pluvialis*-derived astaxanthin costing around \$3000–\$3600 kg^{-1} (Li et al. 2011). Concerns have been raised for synthetic astaxanthin use in human consumption due to it being derived from petrochemicals, making astaxanthin from natural sources a preferred choice (Li et al. 2011). Furthermore, there is concern that synthetically synthesised astaxanthin could be linked to cancer (Newsome 1986), but this has not been substantiated to date. Nevertheless, synthetic astaxanthin has not undergone safety testing for direct human use and has not been recorded to provide health benefits to humans; thus, it has not been registered with regulatory authorities for direct human use in any country (Capelli et al. 2013a), with the exception of DSM's AstaSana™. The level of synthetic food additives which are legally allowed into the market has steadily decreased due to suspected roles as promoters of carcinogenesis with additional claims of liver and renal toxicities (Guedes et al. 2011) creating stricter regulations for the human supplement market. Due to the high cost of production for *H. pluvialis*-derived astaxanthin and a requirement for low-cost astaxanthin in the animal feed sector, many *H. pluvialis* astaxanthin producers have targeted the higher value nutraceutical and pharmaceutical markets due to the numerous reported health benefits for natural astaxanthin (Guerin et al. 2003). However, *H. pluvialis*-derived astaxanthin has been found to be effective for animal feed through improved pigmentation of the flesh and skin, enhanced antioxidant potential, improved fish egg quality, increased growth, and survival of sea bream, rainbow trout, yellow croaker, and salmonid fry compared with the synthetic type (Li et al. 2014; Sheikhzadeh et al. 2012).

In the future, astaxanthin has the potential as a functional food, for example as a partial substitution of flour in cookies (Hossain et al. 2017). To be utilised in food matrices, further innovations in maintaining stability, preservation, encapsulation, and storage are required to avoid degradation and chemical changes (Martínez-delgado et al. 2017). The nutraceutical market today is dominated by astaxanthin from *H. pluvialis*, but one 'nature identical' synthetic product has entered the market, AstaSana™, manufactured by DSM. The Natural Algae Astaxanthin Association (NAXA) promotes *H. pluvialis*-derived astaxanthin (AlgaTechnologies Ltd., Cyanotech Corporation, Beijing Gingko Group (BGG) and Atacama Bio Natural), and they have been aiming to educate the public about the health benefits of natural astaxanthin and the major differences between the natural and synthetic form. A selection of the products are showcased in Fig. 6.1.

High demand by consumers has led to many new companies (at least 22 companies from 13 countries) entering the market in recent years (Fig. 6.2), and producers in China such as BGG are projected to become leaders in the market (Capelli 2018).

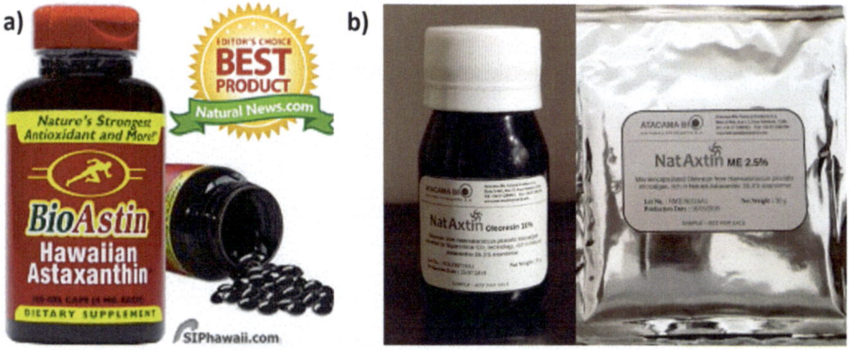

Fig. 6.1 Examples of products from (**a**) Cyanotech Corporation and (**b**) Atacama Bio Natural. Astaxanthin capsules are typically sold with 4–12 mg of astaxanthin as a 10% oleoresin with astaxanthin extracted from the dried powder and formulated in edible oils. (**a**) BioAstin® Hawaiian Astaxanthin®. (Courtesy of Nutrex Hawaii). (**b**) NatAxtin™ ME: microencapsulated oleoresin powder and NatAxtin™ oil: astaxanthin-rich 10% oleoresin from supercritical CO_2. (Courtesy of Atacama Bio Natural Products S.A., Chile). Astaxanthin from *H. pluvialis* cultivated in the Atacama Desert

Additionally, research is being undertaken on other sources of astaxanthin from natural sources such as *Chromochloris zofingiensis* (Chen et al. 2017), with the intention to compete with synthetic, yeast, bacterial and *H. pluvialis*-derived astaxanthin.

1.3 Astaxanthin Health Benefits

Astaxanthin has been reported to be one of the most potent compounds in terms of its antioxidant activity, 65 times more powerful than vitamin C; 10 times stronger than β-carotene, canthaxanthin, zeaxanthin, and lutein; and 100 times more effective than alpha-tocopherol (Capelli et al. 2013a; Miki 1991). The free radical quenching ability is attributable to the conjugated structure of astaxanthin which allows the molecule to intercalate within the phospholipid bilayers of the biological membranes and the terminal hydroxylated ring structures which remain exposed to the outer and inner surfaces of the membrane (Riccioni et al. 2012). The demand for natural astaxanthin as a nutraceutical has exponentially increased due to a growing clinical evidence base including early-stage human trials with many reported health benefits, most notably a high antioxidant potential (Guerin et al. 2003). Dr Joseph Mercola, one of the world's most followed physicians, had declared that astaxanthin was 'the #1 supplement you've never heard of that you should be taking'. Natural astaxanthin, with a molecular structure containing polar hydrophilic ends, can move throughout the entire body (Yuan et al. 2011) and cross the blood-brain barrier to bring antioxidant protection to the brain and eyes (Fig. 6.3), crossing the phospholipid bilayer, a unique characteristic, which only a few other carotenoids possess,

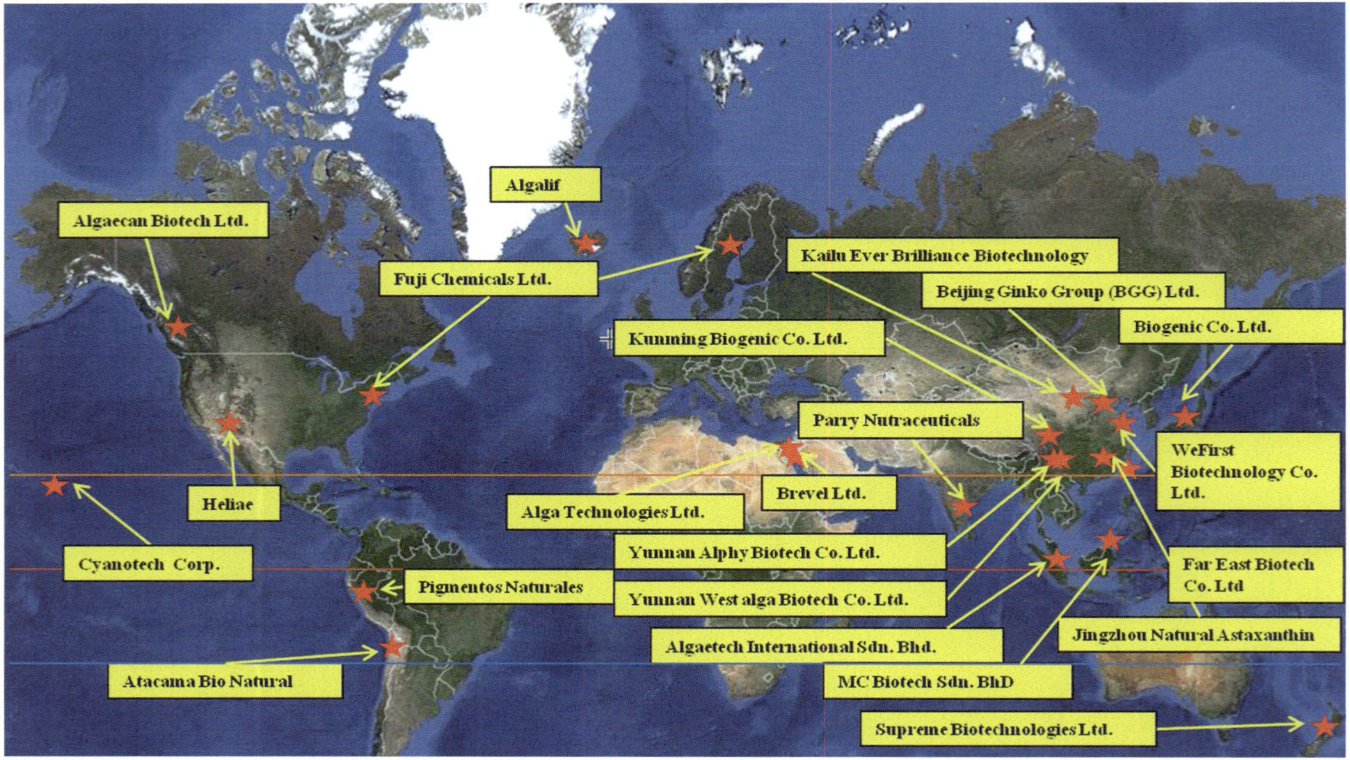

Fig. 6.2 Manufacturing locations of companies producing astaxanthin from *H. pluvialis*. The orange, red, and blue lines indicate the Tropic of Cancer, Equator, and the Tropic of Capricorn, respectively. There are 22 production companies with 77% of companies manufacturing outside of the tropics

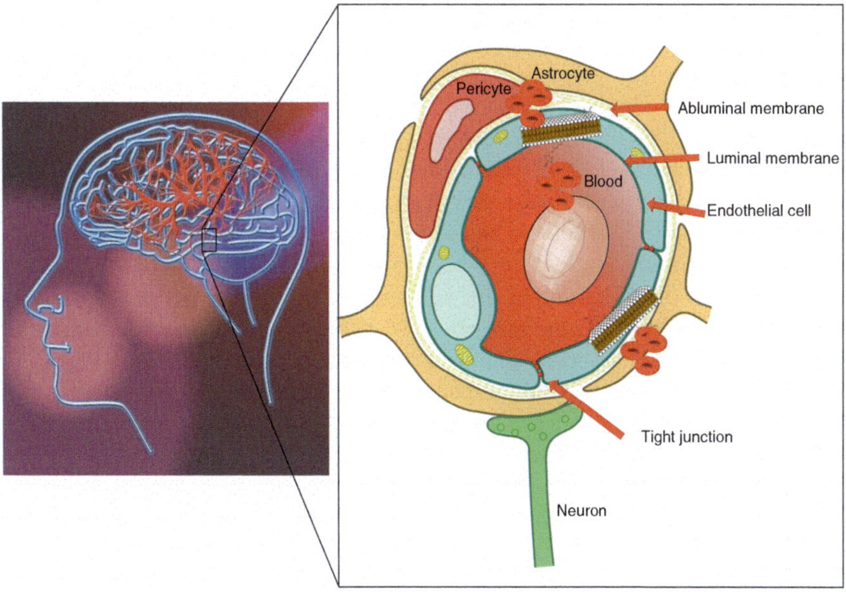

Fig. 6.3 Astaxanthin has the capacity to cross the blood-brain barrier to provide antioxidant protection to the brain and eyes. The blood-brain barrier is a semipermeable membrane which separates the blood from the cerebrospinal fluid forming a barrier to the passage of cells and large molecules but allowing the diffusion of hydrophobic and small polar molecules. Tight junctions create the barrier, and lipophilic substances can pass through the membrane

including lutein and zeaxanthin (Minatelli 2008). In rats, astaxanthin has been found to accumulate in the hippocampus and cerebral cortex (Manabe et al. 2018).

It has been clearly revealed that the source of astaxanthin can have a positive impact on health with extensive preclinical (in vitro and animal models) and clinical studies (Yuan et al. 2011). Natural *H. pluvialis*-derived astaxanthin has been showcased to be over 50 times more effective than synthetic astaxanthin in singlet oxygen quenching and 20 times more effective in free radical elimination (Capelli et al. 2013a). It has been found that the bioaccessibility of *H. pluvialis*-derived astaxanthin in supplements was higher than in aquaculture-derived salmon where synthetic astaxanthin is used for pigmentation (Chitchumroonchokchai and Failla 2017). Research by Nishida et al. (2007) revealed that astaxanthin has a singlet oxygen quenching capability 800 times greater than ubiquinone (antioxidant present in most cells in the body). These initial results have led to further studies, which have concluded antilipid peroxidation activities in vitro (Leite et al. 2010), anticancer properties in vitro and in vivo with rodent models (Tanaka et al. 2012), immune system boosting activity (Bolin et al. 2010), eye health maintenance (Guerin et al. 2003; Piermarocchi et al. 2012), alleviation of arthritic symptoms (Capelli et al. 2013b), and protection against cognitive decline (Katagiri et al. 2012; Satoh et al. 2009). In mice models, *H. pluvialis* astaxanthin has been shown to inhibit the growth of

Helicobacter pylori (common cause of peptic ulcers) and reduces bacterial load in the infected cells, but further research is needed to determine if this is the case in human patients (Kang and Kim 2017).

Most studies of *H. pluvialis*-derived astaxanthin have been in vitro and in animal models, and the efficacy has been well proven (Guerin et al. 2003; Visioli and Artaria 2017; Yuan et al. 2011). In addition, the health benefits of astaxanthin in human patients have been extensively reported: improvements in muscle endurance through reduced lactic acid and increased respiratory and sympathetic nervous system activities (Capelli et al. 2013b), antioxidant potential in bilateral cataract patients (Hashimoto et al. 2014), improved immune response and a reduction in inflammation and oxidative stress (Park et al. 2010), cognitive improvements with increased response time and accuracy of completing tasks (Katagiri et al. 2012; Satoh et al. 2009), cosmetic benefits through improvements in skin elasticity and a reduction in wrinkles (Tominaga et al. 2012), and improvements in semen quality with an associated increase in pregnancy rates (Elgarem et al. 2002; Comhaire et al. 2005). However, with regard to anticancer, cardiovascular health claims for the alleviation of oxidative stress in humans, and benefits in ocular health that have been reported in vitro and in vivo, no conclusive statements can be deduced, and further study is warranted.

The health benefits and published evidence suggest *H. pluvialis* astaxanthin is safe and orally bioavailable (Fassett and Coombes 2012), whilst having no provitamin A activity (which can lead to hypervitaminosis A) (Olaizola and Huntley 2003), and consequently, more clinical trials should be conducted (Fassett and Coombes 2012). The recommended dosage is 4–8 mg/day for normal health maintenance, and for athletes, 12 mg has been observed to be more effective (Capelli 2018). The FDA has approved *H. pluvialis*-derived astaxanthin for direct human consumption (up to 12 mg/day), and if taken for less than 30 days, 24 mg can be taken (Visioli and Artaria 2017). As the number of clinical studies and promotion of natural astaxanthin have increased, market demand has increased leading to a situation where demand is greater than the current ability to supply.

1.4 Sources of Astaxanthin

As outlined above, commercially, the main sources of astaxanthin are synthetically derived from natural astaxanthin *X. dendrorhous* and *H. pluvialis*. *H. pluvialis* is the best known natural producer of astaxanthin (Table 6.1). To date, only one higher plant species has produced astaxanthin (*Adonis annua*) with only 1% dry weight (DW) being observed in the petals (Renstrøm and Liaaen-Jensen 1981). However, despite the potential, the plant has relatively small flowers preventing suitable commercial production (Cunningham and Gantt 2011). *H. pluvialis*-derived astaxanthin is readily accepted for the human food market and of all living organisms has been reported to have the highest concentration of astaxanthin, with reports regularly revealing around 4% (Aflalo et al. 2007) and up to 7.7% at laboratory scale (Kang

Table 6.1 Microbial sources of astaxanthin: Natural vs. genetically modified (GM)

Species	Class	Astaxanthin content (% DW)	Reference
Haematococcus pluvialis NIES-144	Chlorophyceae	7.72	Kang et al. (2005)
Chromochloris zofingiensis ATCC30412	Chlorophyceae	0.71	Chen and Wang (2013)
Coelastrum sp. HA-1	Chlorophyceae	0.63	Liu et al. (2013)
Scenedesmus vacuolatus SAG 211/15	Chlorophyceae	0.27	Orosa et al. (2001)
Monoraphidium sp. GK12	Chlorophyceae	0.25	Fujii et al. (2006)
Scotiellopsis oocystiformis SAG 277/1	Chlorophyceae	1.09	Orosa et al. (2001)
Neochloris wimmeri CCAP 213/4	Chlorophyceae	1.92	Orosa et al. (2001)
Protosiphon botryoides SAG 731/1a	Chlorophyceae	1.43	Orosa et al. (2001)
Chlorococcum sp.	Chlorophyceae	0.57	Ma and Chen (2001)
Brevundimonas sp. N-5	Alphaproteobacteria	0.04	Asker (2017)
Agrobacterium aurantiacum	Alphaproteobacteria	0.01	Yokoyama et al. (1994)
Paracoccus carotinifaciens	Alphaproteobacteria	2.30	Ha et al. (2018)
Paracoccus kocurii	Alphaproteobacteria	1.10	Ha et al. (2018)
Xanthophyllomyces dendrorhous VKPM Y2476	Tremellomycetes	0.41	de la Fuente et al. (2010)
Xanthophyllomyces dendrorhous ATCC96594 (GM)	Tremellomycetes	0.97	Gassel et al. (2013)

et al. 2005). The purity of astaxanthin as total carotenoids in *H. pluvialis* is much higher than other microalgae and can be up to 95% of the total carotenoids (Harker et al. 1996), with most reports revealing an average of 85% (Capelli et al. 2013a; Dore and Cysewski 2003). Comparatively, *C. zofingiensis* has been reported to contain approximately 50% astaxanthin of the total carotenoids, the other main carotenoids being canthaxanthin and adonixanthin (Liu et al. 2014b). Interestingly, in a direct comparison on the basis of astaxanthin esterification, *C. zofingiensis* was observed to contain a higher percentage of astaxanthin diesters (76.3% of the total astaxanthin), but with a significant reduction in astaxanthin monoesters (18% of total astaxanthin) compared with *H. pluvialis* (35.5% of total astaxanthin as diesters and 60.9% as monoesters) (Yuan et al. 2011). For other microalgae to become commercially competitive, the extracts would need to be purified, adding cost to the process. The commercial sector producing *H. pluvialis*-derived astaxanthin is well established, and there have been no adverse effects associated with the administration of *H. pluvialis*-derived astaxanthin reported to date (Fassett and Coombes 2012; Satoh et al. 2009; Spiller and Dewell 2003).

1.5 Large-Scale Production of Astaxanthin

The commercial scale-up of *H. pluvialis* has been and currently is difficult, with cultures requiring strict environmental conditions in the green stage (Olaizola and Huntley 2003). The first large-scale study investigating astaxanthin production from *Haematococcus* in a commercial facility (500,000 L bioreactor, 4500 m^2) in California was in 1987 by Microbio Resources Inc. for the production of a powder (1% DW astaxanthin), marketed under the name Algaxan Red (Bubrick 1991). This dry algal powder was utilised for the aquaculture sector with a production cost of <US \$20 kg^{-1} for the astaxanthin biomass product and able to compete with the synthetic form on price US \$2000 kg^{-1} (Bubrick 1991). To date, there are several astaxanthin companies who have been unsuccessful including Fuji Chemicals with their BioDome™ system (Hawaii, USA), Aragreen (UK), and Maui Tropical Algae Farms (Hawaii, USA). Large-scale cultivation often results in low biomass densities susceptible to contamination issues, with high cell die-off (photobleaching) when transferred to the red stage under high light, and overall, it is a costly production process requiring extraction of the mechanically and chemically resistant, thick-walled aplanospores. Furthermore, there are issues with biofouling from cells forming in 'dead areas' with poor circulation which can lead to reduced light penetration and can cause considerable downtime, thus increasing the annual cost in large-scale cultivation. Companies such as Varicon Aqua Solutions Ltd. (PhycoFlow™) (http://www.variconaqua.com) along with individuals (Van De Ven and Van de Ven 2009) have patented technologies with automated self-cleaning systems, and these technologies offer the ability to reduce biofouling and downtime. The production capacity of *H. pluvialis* outdoors is further constrained by an intrinsically slow growth rate, low cell density, ease of contamination by other microorganisms, and susceptibility to adverse weather conditions (Ip et al. 2004). Most companies follow a production process similar to the schematic shown in Fig. 6.4.

Most companies have utilised *H. pluvialis* strains from culture collections which have been investigated in the literature such as CCAP 34/12, NIES-144, and SCCAP k-0084 (BCC Research 2015) that have been maintained in an artificial environment for a long period of time since their initial isolation from the natural environment. A few companies have used *H. pluvialis* strains, isolated directly from the natural environment that are unavailable in culture collections, including Cyanotech utilising H2B (Cifuentes et al. 2003) and MC Biotech utilising a local Brunei isolate. At a commercial level, astaxanthin is commonly produced using a two-stage culture system involving a green stage, for maximal biomass production, and a red stage, for maximising the product astaxanthin (Aflalo et al. 2007; Olaizola 2003; Olaizola and Huntley 2003), but a three-stage culture system (green, starvation, and red stage) is implemented by Algalif, Iceland. In the outdoor two-stage production process, astaxanthin productivities can reach 8–10 mg/L/day over a 10 day cycle (4 day green stage and 6 day red stage) with astaxanthin accounting for up to 4% DW under high light and nitrate-depleted conditions in the red stage (Aflalo et al. 2007). Alternatively, a continuous, one-stage production process has been demonstrated at pilot scale which produces astaxanthin from a mixed culture of motile macrozooids and palmelloids, resulting in almost twice the astaxanthin productivity (20.8 mg/L/

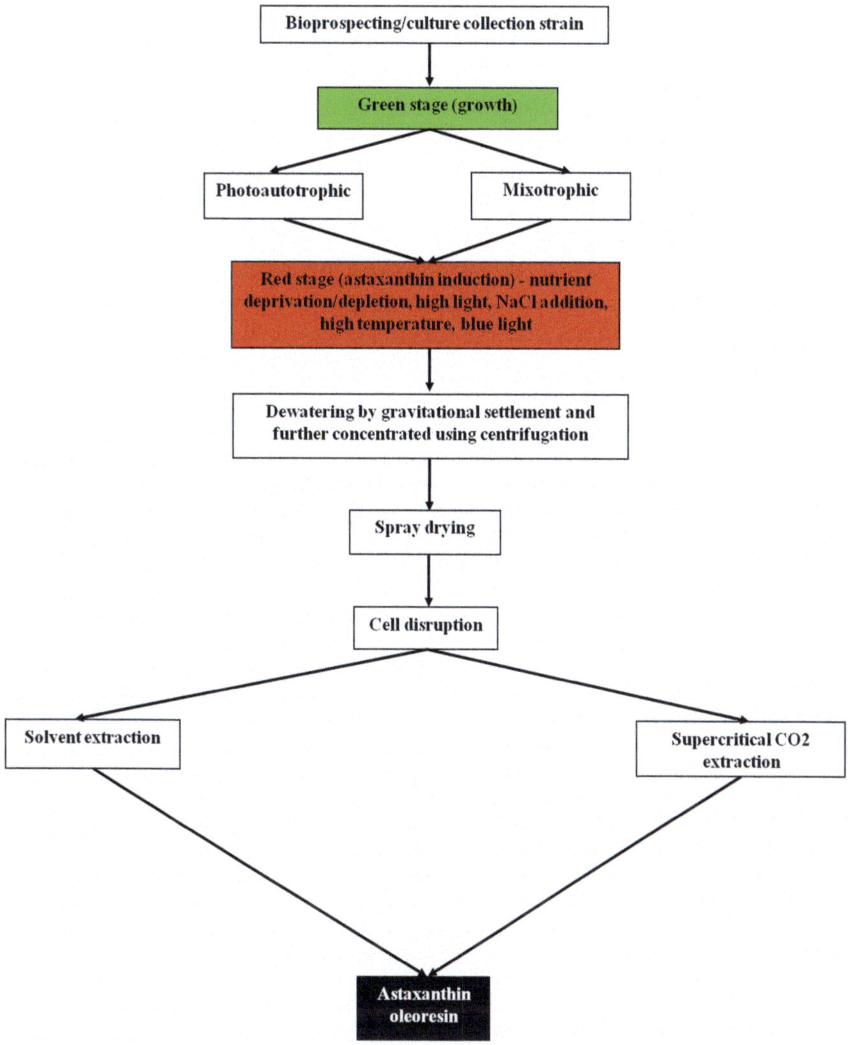

Fig. 6.4 Current process used on a commercial scale in the two-stage production process

day), formed under nitrate-deficient conditions (Del Río et al. 2008), but this process is yet to be adopted on a commercial scale.

Currently, the manufacture of astaxanthin is mainly conducted outdoors (Fig. 6.2), primarily due to the high light intensities and temperatures required for astaxanthin induction in the red stage which would be uneconomical indoors, but surprisingly most of the production occurs outside the tropics. In indoor cultivation, the red stage can account for 59% of the electricity costs, mainly due to the high light costs (Li et al. 2011). Only two companies employ a completely indoor production process to the author's knowledge, Fuji Chemicals and Algalif. Fuji Chemicals abandoned *H. pluvialis* culture outdoors in its BioDome™ system in

Hawaii due to contamination issues and subsequently continued with indoor mixotrophic culture in Sweden and Washington (Algae Industry Magazine 2015). Commercial production of *H. pluvialis*-derived astaxanthin in temperate zones is constrained due to unsuitable weather conditions for astaxanthin production outdoors, and consequently, only indoor culture is feasible. Aragreen in Gloucestershire was investigating astaxanthin production from *H. pluvialis*, but the company filed for bankruptcy in 2017 (Aragreen 2015).

Industrially, there are a wide range of cultivation methods, and most are aiming to utilise a more sustainable production process. In the case of Cyanotech, in the green stage *H. pluvialis* is cultivated indoors under strictly regulated culture conditions and then transferred outdoors for astaxanthin induction in the red stage in open ponds. AlgaTechnologies in Israel conducts their whole process outdoors in photobioreactors (PBRs) to exploit natural sunlight and utilises photovoltaic cells (Algae Industry Magazine 2015). Comparatively, Algalif utilises geothermal energy for an entire production process indoors using light-emitting diodes (LEDs). Most companies are focussing on phototrophic cultivation, but mixotrophic cultivation is being explored, for example, at Fuji Chemicals. Lorenz and Cysewski (2000) reported that astaxanthin induction can take 3–5 days, and during this stage, the encystment process results in the formation of aplanospores (cysts).

After cultivation and astaxanthin induction, the aplanospores are harvested by gravitational settlement and further concentrated by ultracentrifugation. The biomass is then dried, conventionally by spray drying as this is more economical than freeze drying and drum drying (Dore and Cysewski 2003; Shah et al. 2016). The dried thick-walled aplanospores are subjected to an extraction process to disrupt the cell walls and make the astaxanthin bioavailable. The walls of aplanospores resist digestion by animals (in feed applications) and by humans (nutraceutical applications), and therefore, the aplanospores must be disrupted for astaxanthin to become bioavailable (Olaizola and Huntley 2003). Sommer et al. (1991) observed that intact astaxanthin-rich aplanospores of *H. pluvialis* on ingestion do not result in pigmentation in salmonids. Care has to be taken in the extraction process to limit oxygen exposure and high temperatures which can damage the astaxanthin and result in losses in the process (Bustos-Garza et al. 2013). Extraction of astaxanthin on a commercial scale is most commonly by supercritical fluid extraction (SFE) with CO_2 ($ScCO_2$) (Shah et al. 2016). After extraction, the dried product is usually mixed with a preservative and shipped to feed manufacturers where it is incorporated into formulated feed (Olaizola and Huntley 2003). Alternatively, the astaxanthin is encapsulated and formulated for nutraceuticals which is discussed in Sect. 3.4.

2 Biology of *H. pluvialis/H. lacustris*

H. pluvialis Flotow belongs to the Chlorophyceae, order Volvocales, and family Haematococcaceae, and in the past, this species has been referred to as *Haematococcus lacustris* or *Sphaerella lacustris* (Shah et al. 2016). Currently,

H. pluvialis and *H. lacustris* are synonymous and the correct terminology is *H. lacustris*; therefore, this taxonomy will be followed hereafter (Buchheim et al. 2013; Mazumdar et al. 2018; Nakada and Ota 2016).

It has been determined that *Haematococcus* is non-monophyletic with two distinct *Haematococcus* lineages by using nuclear-encoded small (18S) and large (26S) subunit rRNA combined with internal transcribed spacer 2 (ITS2) genes (Buchheim et al. 2013). It has been determined that *H. pluvialis* (*H. lacustris*) is the only member of the *Haematococcus* genus (albeit with at least five distinct lineages A–E from bootstrap data), with motile macrozooids with 'delicate' cytoplasmic strands and the formation of aplanospores with copious amounts of astaxanthin. Buchheim et al. (2013) stated the other *Haematococcus* species (*H. buetschlii*, *H. capensis*, *H. zimbabwiensis*, and *H. droebakensis*) should be designated to the second lineage, the *Balticola* genus (cytoplasmic strands thickened at the base) as previously proposed by Droop (1956). Allewaert et al. (2015) reported three species of *Haematococcus* from European isolates (*H. pluvialis*, *H. rubens*, and *H. rubicundus*) with *H. pluvialis* having the lowest maximum growth rate. Mazumdar et al. (2018) have reported four *Haematococcus* lineages with five valid species: *H. lacustris*, *H. rubicundus*, *H. rubens*, *H. carocellus*, and *H. alpinus*. The *H. alpinus* species was recently isolated from an alpine zone in New Zealand and identified as a new species with no known relatives (Mazumdar et al. 2018).

H. lacustris is regarded as the 'birdbath' alga which is distinct from other species of *Haematococcus* due to its ability to produce a vegetative resting stage (cyst/haematocyst/aplanospore) and is known to accumulate high amounts of the carotenoid astaxanthin (Buchheim et al. 2013; Droop 1955). *H. lacustris* differs morphologically from other species of *Haematococcus*, having uniformly thin cytoplasmic strands compared to strands which are thickened at the base (Buchheim et al. 2013) and possessing three or more pyrenoids compared to only two in other species of the *Haematococcus* genera (Allewaert et al. 2015). *H. lacustris* is primarily a freshwater species, commonly found in ephemeral rain pools, natural and manmade ponds, and birdbaths (Burchardt et al. 2006). *H. lacustris* is circumglobal and has been found on every continent with the exception of Antarctica (Guiry 2010). The ability of *H. lacustris* to encyst allows this species to survive in extreme conditions: high light, temperature, and salinity (Proctor 1957a).

2.1 Life Cycle

The first reports in the literature on the life cycle of *H. lacustris* were written in the middle of the nineteenth century (Flotow 1844) (Peebles 1909; Elliot 1934), and there was a further surge of interest in the late 1950s (Droop 1956; Proctor 1957a, b). *H. lacustris* has a complex life cycle, typically encompassing four life cycle stages: with the green stage containing vegetative cells, green motile macrozooids (flagellates), and non-motile palmelloids (zoospores), the red stage containing non-motile aplanospores (haematocysts), and a gamete stage with microzooids (Elliot 1934;

Triki et al. 1997; Han et al. 2012). Triki et al. (1997) observed that microzooids were formed from aplanospores maintained in nitrate-starved conditions for over a month when transferred to a medium high in nutrients. Under favourable conditions, green motile macrozooids predominate (Triki et al. 1997), but when conditions become unfavourable, e.g. low nutrients, high light, and salinity stress, aplanospore formation occurs in conjunction with astaxanthin accumulation (Harker et al. 1996).

Wayama et al. (2013) reported that the life cycle of *H. lacustris* was more complex than originally perceived, describing the life cycle in more detail. It was reported that when *H. lacustris* aplanospores were resuspended in fresh medium, they germinated and released up to 32 green motile macrozooids by cytokinesis. After 3–5 days, these were noted to form green coccoid cells (palmelloids), and as aging progressed, the cells were transformed into intermediate cells (palmelloids) and aplanospores (Wayama et al. 2013). Even though a considerable body of work has been undertaken, the life cycle of *H. lacustris* is still not fully understood along with the morphotypes involved (Figs. 6.5 and 6.6). To date, little is known on what conditions contribute to palmelloid formation other than culture ageing and strain disposition (Allewaert et al. 2017). In addition, red motile macrozooids have also been observed in some strains, but it has not been fully elucidated why they are formed rather than aplanospores (Fig. 6.5) (Brinda et al. 2004; Butler et al. 2017; Del Río et al. 2005; Grünewald et al. 1997; Hagen et al. 2000; Tocquin et al. 2012).

Reproduction in *H. lacustris* is still a contentious issue as it is unknown whether *H. lacustris* undergoes sexual reproduction, and more direct evidence is still warranted (Chunhui et al. 2017). During division, sporocysts are formed which can contain 16 or 8 cells in the green and red stage, respectively (Figs. 6.7 and 6.8). *H. lacustris* is reported to be capable of sexual reproduction, but it is considered unusual (Triki et al. 1997). Triki et al. (1997) did not observe sexual reproduction in green motile macrozooids and reported that this was due to *H. lacustris* being heterothallic within culture collections, with populations derived from a single mating type (Droop 1956). Zheng et al. (2017) stated there is no convincing evidence that *H. lacustris* undergoes sexual reproduction, and it is unknown if Figs. 6.7d and 6.8b show the onset of sexual reproduction or whether the cells separate without fusing. For syngamy to occur, two comparable mating types have to be present (Triki et al. 1997). In the presence of a single clone, or numerous incompatible clones of *H. lacustris*, syngamy would not be possible (Triki et al. 1997). Determining if sexual reproduction occurs in *H. lacustris* would be of biotechnological interest because then mating trials could be conducted for selective breeding.

During the life cycle of *H. lacustris*, ultrastructural changes occur within the cell. These have been well documented by Wayama et al. (2013) and reviewed by Shah et al. (2016). The green motile macrozooids are surrounded by an extracellular matrix (Wayama et al. 2013). During the onset of encystment, *H. lacustris* cell walls thicken up to 2 μm, and the cells develop conspicuous pyrenoids with many starch grains located around the pyrenoids (Wayama et al. 2013). Circular oil droplets with various sizes containing astaxanthin are located around the

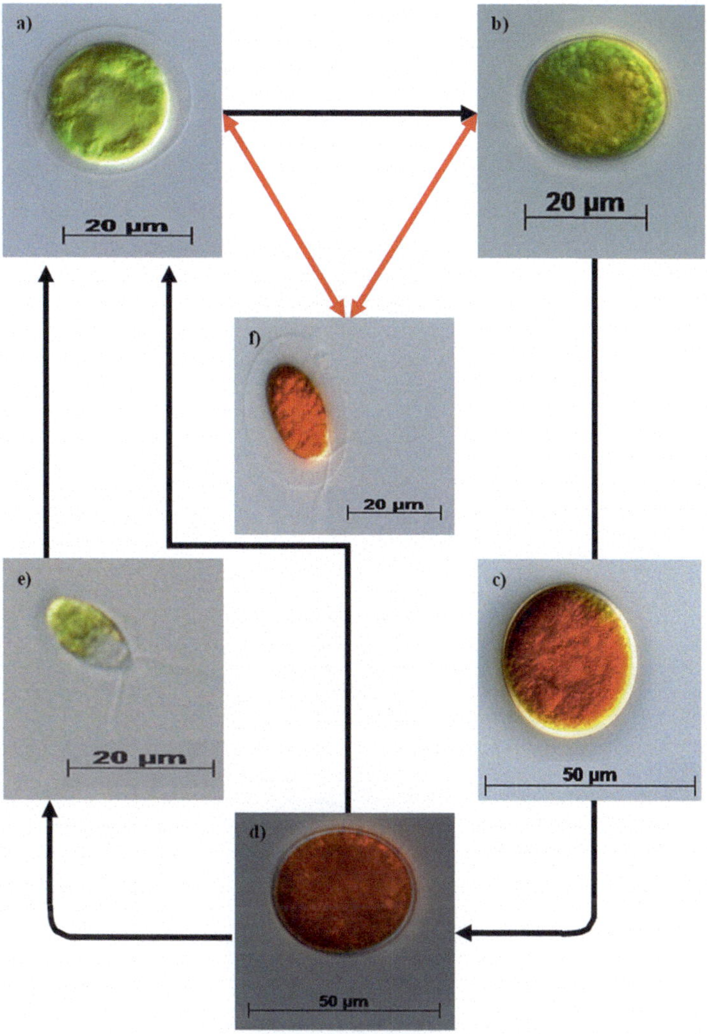

Fig. 6.5 Life cycle of *H. lacustris*: (**a**) green motile macrozooid, (**b**) early-stage palmelloids, (**c**) late-stage palmelloids, (**d**) aplanospore, (**e**) green motile microzooid, and (**f**) red motile macrozooid. The black lines indicate known interactions between life cycle stages, and the red lines indicate proposed interactions

nucleus (Wayama et al. 2013). As astaxanthin accumulates, the chloroplast reduces in volume, and the chloroplasts degenerate and are localised in the interspace between oil droplets, but the photosynthetic activity of the cell is maintained (Wayama et al. 2013). It is still unknown what happens to the ultrastructure of the red motile macrozooids and if it is the same as in aplanospores.

Cell type	Morphology		
Green motile macrozooid	Pyrenoid Anterior end Lipid bodies Flagellum	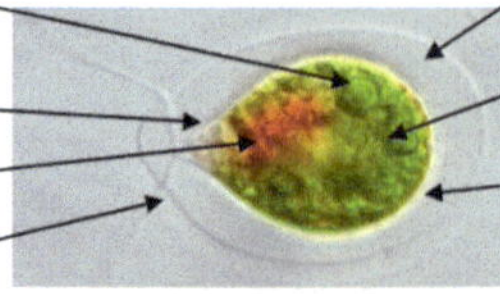	Extracellular matrix (0.64-0.8 µm) Chloroplast Faint cytoplasmic strands Cell weight: ~500 pg/cell Cell size: 18-25 µm
Early-stage palmelloid	Trilaminar sheath; algaenan		Pyrenoids (2-10) Cell wall up to 0.4 µm thick Cell weight: ~1000 pg/cell Cell size: 20-28 µm
Late-stage palmelloid	Trilaminar sheath; algaenan Secondary wall; mannose and cellulose	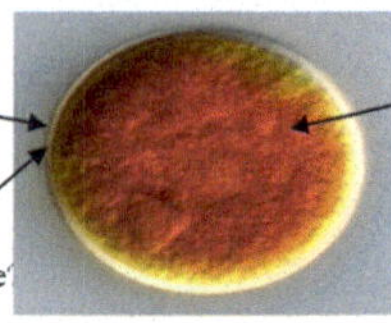	Astaxanthin accumulation Cell size: 26-34 µm
Aplanospore	Lipid bodies covering entire chloroplast Trilaminar sheath; algaenan Secondary wall; mannose and cellulose	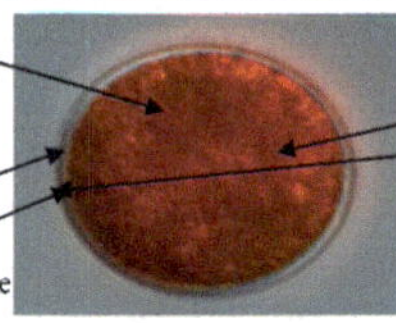	Astaxanthin Tertiary wall Decrease in thylakoid membrane number/volume Cell weight: ~3000 pg/cell Cell size: > 30-60 µm
Red motile macrozooid	Astaxanthin Lipid bodies covering entire chloroplast Flagella	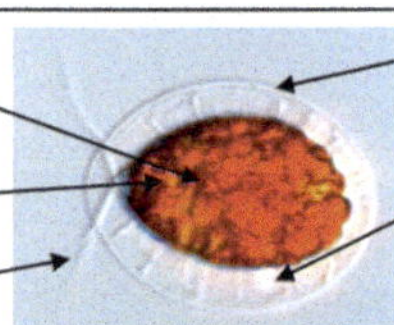	Extracellular matrix Cytoplasmic strands Cell size: 17-24 µm
Microzooid	Pyrenoid Vacuole Flagella	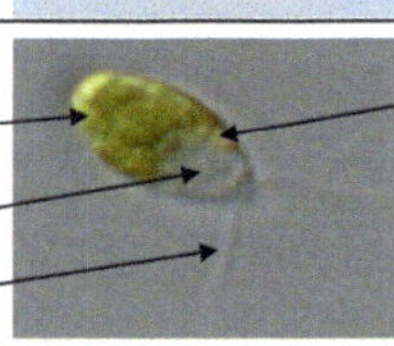	Lipid body Cell size: < 10 µm

Fig. 6.6 Morphological characteristics of life cycle stages. (Data for annotations obtained from García-Malea et al. 2006, Damiani et al. 2006, Wayama et al. 2013, and Chekanov et al. 2014)

2.2 *Biochemical Components: Proteins, Lipid, Carbohydrates, Pigments*

Morphogenesis from the green motile macrozooid to the aplanospore cell stage results in profound changes within the cell, including changes in cell wall structure which can be detected by electron microscopy and cytochemistry (Wayama et al. 2013). The green motile macrozooids exhibit an extracellular matrix (mainly

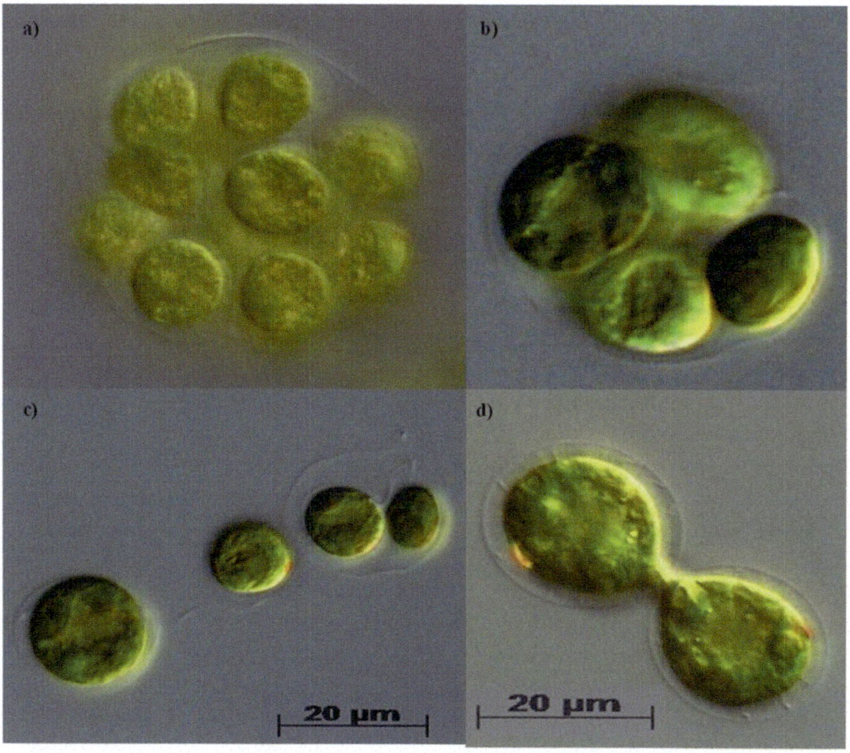

Fig. 6.7 Asexual reproduction in green motile macrozooids: (**a**) 16 daughter cells, (**b**) four daughter cells, (**c**) released green motile macrozooids after the breakdown the of extracellular matrix, and (**d**) pairing head to head in green motile macrozooids

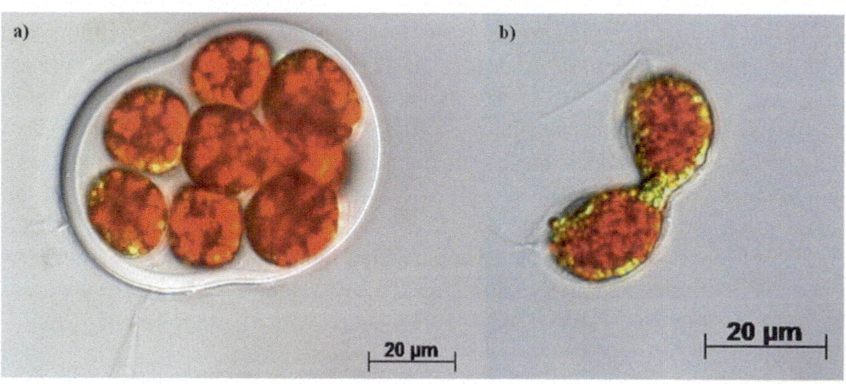

Fig. 6.8 (**a**) Asexual reproduction with eight daughter cells formed and (**b**) pairing head to head in red motile macrozooids

consisting of glycoproteins and lacking cellulose or acetolysis-resistant material) around the cell (Hagen et al. 2002). In ageing green motile macrozooids, a two-layered primary cell wall forms (containing β-1,4-glycosidic linkages), which is subsequently followed by a loss of motility and the development of palmelloid cells (Hagen et al. 2002). After completion of the primary cell wall, the formation of a trilaminar sheath is observed containing cellulose in the palmelloids (Damiani et al. 2006). During encystment, the *H. lacustris* cell develops a secondary cell wall containing algaenan, a sporopollenin-like material, which is highly resistant to chemical and mechanical breakage (Han et al. 2013a). Montsant et al. (2001) identified that the cell wall of aplanospores was two- to threefold thicker than green motile macrozooids using transmission electron microscopy (TEM). The composition of the aplanospores cell walls is 70% carbohydrate (89.4% mannose), 6% protein, 3% cellulose, and 3% acetolysis-resistant material (Hagen et al. 2002). The biochemical composition of *H. lacustris* varies depending on the life cycle stage and the environmental conditions.

H. lacustris in the green stage typically has a biochemical composition of 13.8–48.0% protein (higher protein related to a higher nitrogen content) (Gacheva et al. 2015; Sipaúba-Tavares et al. 2015), 39.0–64.2% carbohydrate, and 8.3–16.2% lipid (as a function of DW) (Lorenz 1999; Gacheva et al. 2015). The primary FAs in the green stage were linolenic (18:3 (n − 3)) and palmitic acid (16:0) (26.4% and 18.9% of the total FAs, respectively) (T. Butler, unpublished data). With regard to the pigment fraction, the green vegetative cells can contain chlorophylls a and b, and the carotenoids lutein (70%), neoxanthin (12%), violaxanthin (10%), β-carotene (8%) with zeaxanthin are also reported (Harker and Tsavalos 1996; Harker et al. 1996).

In varying stages of the red phase, *H. lacustris* had a proximate composition (on a dry basis): 14–26% crude protein, 2.6–26.3% lipid, 6.30–48.8% carbohydrate, 2.0–4.0% ash, and an approximate gross energy of 24.1 kJ/g (Boussiba and Vonshak 1991; Choubert and Heinrich 1993; Sarada et al. 2006; Kim et al. 2015; Molino et al. 2018). It was reported that the lipid was composed of 88.3% FAs with 48.20% as polyunsaturated fatty acids (PUFAs) (the main fatty acids were linoleic acid (18:2 (n − 6)), palmitic, and oleic acid (18:1) encompassing 74.25%) (Kim et al. 2015; Molino et al. 2018). The amino acid profile is comprised of glutamic acid, aspartic acid, leucine, and alanine with a total amino acid content of 10% DW and 46% of the amino acids in *H. lacustris* defined as essential (Kim et al. 2015). The monosaccharides were mostly glucose and mannose (46% and 40.9% of the composition, respectively) (Kim et al. 2015). In a commercial product, Cyanotech previously reported that spray-dried aplanospore biomass had an astaxanthin content of >1.5% with the biomass containing 20–30% protein, 30–40% carbohydrate, 5–15% ash (dry weight (DW)), and 4–9% moisture, with a particle size of 5–25 μm (Dore and Cysewski 2003).

Generally, it has been observed in the red stage that the carbohydrate content increases dramatically with up to 74% as starch observed (Boussiba and Vonshak 1991). When aplanospores are formed, cytoplasmic lipid droplets (FAs as mono- or diesters) can account for 40% DW and can contain 4% astaxanthin (Aflalo et al. 2007; Saha et al. 2013a). The neutral lipid fraction predominates in the green and red stages, and in the transition to the red stage, the neutral lipid fraction as triacylglycerides (TAGs) increases along with the glycolipid content (Damiani et al. 2010). In the red phase, the aplanospores are rich in palmitic (C16:0), linoleic (18:2), and linolenic (18:3), whereas in the red motile macrozooid stage, oleic acid (18:1) was also found to be abundant (Butler et al. 2017). The biochemical composition largely depends on several factors including cultivation conditions, e.g. light, temperature, nutrients, carbon dioxide, and the genetics of the cell. Nutrient starvation is well known to increase lipid content, and the FAs formed are suitable for biodiesel (Damiani et al. 2010; Saha et al. 2013a). In the red stage, the pigment composition undergoes a dramatic shift with astaxanthin comprising 80–99% of the total carotenoids (Dragos et al. 2010). The ratio of carotenoids to chlorophyll is around 0.2 in the green stage but increases to 2–9 in the red stage (Shah et al. 2016). In the aplanospores and red motile macrozooids, astaxanthin is not in the free form but is often esterified with mono- or diesters of palmitic (16:0), linoleic (18:2), or oleic acid (18:2) (Shah et al. 2016; Butler et al. 2017). In aplanospores, approximately 70% of the astaxanthin is monoesters, 25% diesters, and only 5% free astaxanthin (Johnson and Schroeder 1995; Solovchenko and Chekanov 2014), but in red motile macrozooids, 77% of the astaxanthin occurs as a monoester, 18% as diesters, and 1.4% as free astaxanthin (Butler et al. 2017).

2.3 Bioprospecting for Commercially Relevant Strains

H. lacustris is a ubiquitous freshwater green alga with members of this species found circumglobally, and to date, >150 strains have been isolated, with the majority from the northern hemisphere (Fig. 6.9). Of these unique strains, the location of isolation of at least 44% remains unknown (Alam and Wang 2019).

In the past, morphological traits have been used to determine the species and strains of *H. lacustris*. It has been identified that some culture collection strains predominantly grow as green motile macrozooids and others as predominantly green palmelloids (Han et al. 2012). However, morphology alone proves difficult to observe differences between strains, and genetics play an increasingly important role for differentiation, and therefore, a rapid DNA barcoding method is required for cataloguing new strains. Barcoding is important for identifying and cataloguing strains with desirable characteristics, including high growth rate and astaxanthin production and the capability to survive extremophilic conditions (an advantage in large-scale culture to avoid culture crashes from contaminants). Conventionally, microalgal cells are lysed using heating and cetyl trimethylammonium bromide (CTAB) (Doyle and Doyle 1990) to release DNA that is then amplified through the polymerase chain reaction (PCR), run on a gel, purified, and sequenced (Mostafa et al. 2011). As a pretreatment for DNA extraction, bead beating has been incorporated as it is notoriously difficult to extract high-yielding DNA (Peled et al. 2011). Recently, a simple colony PCR process has been established for *H. lacustris* with a simple heating step in a PCR machine for 10 min with PCR buffer, followed by subsequent PCR amplification (with improved band intensity with increasing cycle number) (Liu et al. 2014a).

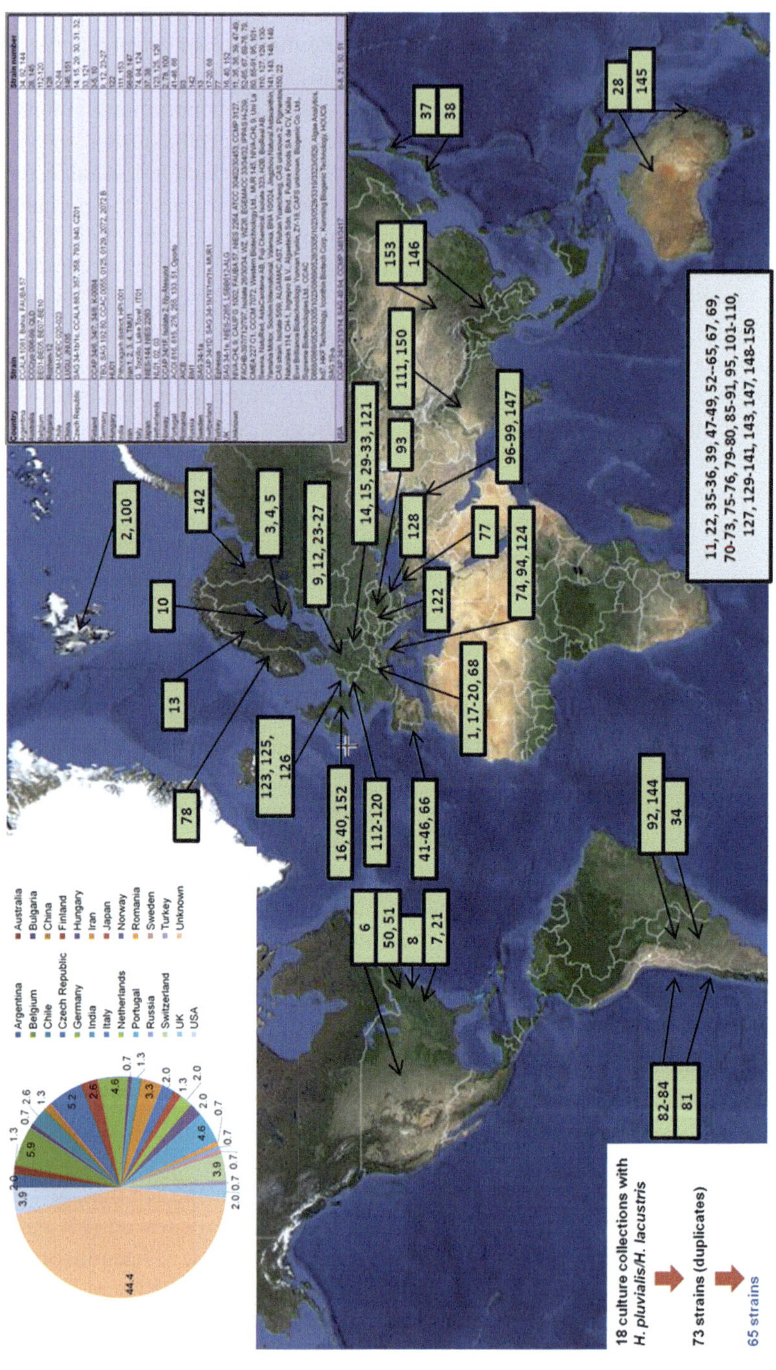

Fig. 6.9 Geographical mapping of 153 *H. pluvialis/H. lacustris* strains obtained from 18 culture collections and the published literature. From the 18 culture collections 65 unique strains were obtained. It was observed that for most of the strains the location of isolation was unknown (44.4%)

To date, no universal barcode has been established for eukaryotic microorganisms, but the hypervariable V4 region of the 18S rDNA was proposed as a prebarcode (Łukomska-Kowalczyk et al. 2016). Mostafa et al. (2011) used inter simple sequence repeat (ISSR) and random-amplified polymorphic DNA (RAPD) molecular markers to observe the genetic diversity between ten strains of *H. lacustris*, four sourced from Iran and the other six from CCAP which successfully produced a dendrogram showing the correct strains based on their geographical origin. Allewaert et al. (2015) used ITS2 and then complete ITS rDNA and rbcL molecular phylogenies (with ITS being more powerful for species/strain separations) to determine the relationship between European *H. lacustris* isolates (seven strains) and those in common culture collections. It was determined that six lineages could be resolved from the ITS rDNA phylogenetic data corresponding to three out of five of the ITS2 rDNA lineages (A, C, and E) reported by Buchheim et al. (2013). Denaturing gradient gel electrophoresis (DGGE) was also identified as a method for rapid identification of European temperate strains and has been highlighted as a method that could be used in the future (Allewaert et al. 2015).

Bioprospecting for new strains has been fairly commonplace for *H. lacustris* as the large, red aplanospore cells are easy to see, and isolates have been obtained globally from Temperate Zones over Europe (Allewaert et al. 2015) to Torrid Zones such as India (Prabhakaran et al. 2014) to Frigid Zones such as Svalbard (Klochkova et al. 2013). Many strains of *H. lacustris* in culture collections including the Culture Collection of Algae and Protozoa (CCAP) have been found in ephemeral pools (CCAP 2015). Other strains of *H. lacustris* have been found in extreme environments outside of their conventional niche. An isolate has been found at high altitude in India (Prabhakaran et al. 2014). Recently, a cold-tolerant strain of *H. lacustris* was isolated from the high Arctic in Blomstrand Halvøya, Svalbard, which has been found to exhibit growth between 4 and 15 °C and can produce astaxanthin at 4–10 °C (Kim et al. 2011; Klochkova et al. 2013). *H. lacustris* has even been recorded in samples from a nuclear-fuel storage pond in Sellafield, UK (Groben 2007). A thermophilic strain with the ability to grow at temperatures up to 41.5 °C has also been isolated (Gacheva et al. 2015). In one study that screened 30 natural isolates and compared them with culture collection strains, it was identified that the culture collection strains had a lower astaxanthin productivity which might have been attributed to a loss in photoprotective capacity during longer-term cultivation (Allewaert et al. 2017). Bioprospecting offers significant potential for identifying new strains of *H. lacustris* with desirable characteristics for biotechnological exploitation. Elucidating the diversity in *H. lacustris* species is essential for biotechnological applications as potential fast-growing and astaxanthin-hyperaccumulating strains can be identified, in conjunction with determining strains suitable to local climatic conditions.

An Arctic strain of biotechnological significance is BM1, found on coastal rocks off the coast of a Russian island, with the ability to tolerate salinities up to 25% (Chekanov et al. 2014). Typically, a salinity of 8% causes the cessation of growth in *H. lacustris* (Boussiba and Vonshak 1991). This strain could be cultivated in brackish water which would reduce the cost of production, would minimise the environmental burden, and would be suitable for areas with a limited supply of freshwater. In addition, astaxanthin accumulation was detected after only 10 days of cultivation (Chekanov et al. 2014). After 6 days of resuspension in distilled water, 27 °C, 480 μmol photons/m^2/s, and continuous light, the astaxanthin content reached 5–5.5% DW (99% of the total carotenoids).

Much of the scientific literature has been based on specific strains of *H. lacustris* from culture collections, including CCAP 34/7 (Harker and Tsavalos 1996; Mendes-Pinto et al. 2001; Mostafa et al. 2011; Rioboo et al. 2011), CCAP 34/8 (García-Malea et al. 2005, 2006, 2009), SCCAP k-0084 (Montsant et al. 2001; Peled et al. 2012; Wayama et al. 2013), and the highest number of publications on NIES-144 (Kobayashi et al. 1991, 1993, 1997a, b; Kang et al. 2005; Yoo et al. 2012; Wan et al. 2014a). There are few reports on other strains held in culture collections including CCAP strains 34/1D, 34/13, and 34/14. Published papers for these strains have mainly been restricted to the study of evolutionary relationships (Mostafa et al. 2011). Revisiting these strains could provide a quick method for identifying strains with suitable properties for commercial production of astaxanthin which could alleviate some of the issues currently found within the industry such as identifying red motile macrozooids with the highest reported astaxanthin content to date (2.74% DW) (Butler et al. 2017). To date, it has been observed that the highest astaxanthin content has been observed in NIES-144 and the highest astaxanthin productivity in CCAP 34/8.

To date, only two publications have comprehensively compared *H. lacustris* strains. From 25 strains from various culture collections, it was identified that CCAC 0125 was the optimal strain with a total biomass and astaxanthin yield of 91.2 g/m^2 and 1.4 g/m^2, respectively, with an associated astaxanthin content of 1.5% DW (Kiperstok et al. 2017). All 25 strains were capable of immobilised growth in a biofilm attaining biomass and astaxanthin yields of between 73 and 112 g/m^2, 0.74 g/m^2 and 2.1 g/m^2, respectively. Allewaert et al. (2017) undertook a screen of 30 strains which included recently isolated strains and those maintained in culture collections. It was concluded that these recently isolated strains generally had a higher astaxanthin productivity and there was a 15-fold difference, with BE02_09 having the highest astaxanthin productivity (4.59 mg/L/day). In future strain selections it would be beneficial to conduct a high-throughput screening method to identify astaxanthin-hyperproducing mutants. Fourier-transform infrared spectroscopy has been suggested rather than conventional high-performance liquid chromatography (HPLC) (Liu and Huang 2016).

2.4 PBR Development and Cultivation Mode

To date, *H. lacustris* has been cultivated in open ponds, plastic bags, fermentors, and PBRs (flat plate, horizontal/tubular, bubble columns, and airlift) and attached to a membrane system (Table 6.2). The highest astaxanthin productivity (20.8 mg/L/day) was obtained in an indoor system using a one-stage process (Del

Table 6.2 Biomass and astaxanthin productivities

Species/ strain	Cultivation vessel	Mode	Indoor/ outdoor	Scale	Cultivation time (day)	Biomass productivity green stage (g/L/day)	Biomass productivity red stage (g/L/ day)	Astaxanthin productivity (mg/L/day)	Astaxanthin productivity ($mg/m^2/day$)	Astaxanthin content (% DW)	Reference
CCAP 34/7	30 L airlift PBR	Photoautotrophic	Indoor	Small	90	–	0.02	0.44	30.00	2.50	Harker et al. (1996)
UTEX 16	3.7 L fermentor	Mixotrophic	Indoor	Small	20	–	0.14	3.22	–	2.35	Zhang et al. (1999)
Unknown origin: AQSE002	25,000 L Aquasearch Growth Modules (green stage) open pond (red stage)	Photoautotrophic	Outdoor. Hawaii, USA	Large pilot	9 months 19 (14 + 5)	0.04–0.05	–	–	–	0.70–3.40	Olaizola (2000)
NIES-144	2.3 L fermentor (green stage). PBR (red stage)	Heterotrophic-photoautotrophic	Indoor	Small	30 (20 + 8)	0.35	0.21	4.07	–	1.90	Hata et al. (2001)
NIES-144	250 mL Erlenmeyer flasks	Photoautotrophic	Indoor	Small	18	–	0.13	9.74	–	7.72	Kang et al. (2005)
CCAP 34/8	1.8 L bubble column PBR	Photoautotrophic	Indoor	Small	12	–	0.21	1.60 (carotenoids)	–	1.50 (carotenoids)	García-Malea et al. (2005)

(continued)

Table 6.2 (continued)

Species/ strain	Cultivation vessel	Mode	Indoor/ outdoor	Scale	Cultivation time (day)	Biomass productivity green stage (g/L/day)	Biomass productivity red stage (g/L/ day)	Astaxanthin productivity (mg/L/day)	Astaxanthin productivity (mg/m^2/day)	Astaxanthin content (% DW)	Reference
CCAP 34/8	2 L jacketed bubble column PBR (green and red stage)	Photoautotrophic	Indoor	Small	–	–	–	5.60	50.90	0.80	Del Río et al. (2005)
SAG 34/1b	1 L cylindrical airlift double-region PBR	Photoautotrophic	Indoor	Small	22 (12 + 10)	0.33 (fresh weight)	–	16.20	–	4.95	Suh et al. (2006)
CCAP 34/8	50 L airlift PBR with external-loop solar receiver	Photoautotrophic	Outdoor, Almeria, Spain.	–	3 months	0.58	–	–	–	–	García-Malea et al. (2006)
CCAP 34/8	55 L bubble column PBR	Photoautotrophic	Outdoor	Small	16	0.06	–	0.12	–	0.20	García-Malea López et al. (2006)
CCAP 34/8	55 L airlift tubular PBR	Photoautotrophic	Outdoor	Small	16	0.41	–	4.40	–	1.07	García-Malea López et al. (2006)
SCCAP k-0084	500 mL glass columns	Photoautotrophic	Indoor	Small	9 (5 + 4)	0.50	0.21	11.50	420.00	4.00	Aflalo et al. (2007)

Species/ strain	Cultivation vessel	Mode	Indoor/ outdoor	Scale	Cultivation time (day)	Biomass productivity green stage (g/L/day)	Biomass productivity red stage (g/L/ day)	Astaxanthin productivity (mg/L/day)	Astaxanthin productivity ($mg/m^2/day$)	Astaxanthin content (% DW)	Reference
SCCAP k-0084	500 L flat vertical panel PBR, 5 cm o.d. (green stage). 2000 L horizontal tubular type PBR, 5 cm o.d. (red stage)	Photoautotrophic	Outdoor, Sede-Boker campus, Israel	Medium	10 (4 + 6). Summer	0.37	0.27	10.10	–	3.80	Aflalo et al. (2007)
NIES-144	1 L Bubble column PBR (6.9 cm i.d.)	Photoautotrophic	Indoor	Small	34.58 (12.5 + 22.08)	0.40	0.19	14.46	–	7.46	Ranjbar et al. (2008)
NIES-144	1 L Airlift PBR (4.68 cm i.d.)	Photoautotrophic	Indoor	Small	30 (12.5 + 17.5)	–	0.38	20.00	–	5.21	Ranjbar et al. (2008)
CCAP 34/8	2 L bubble column PBR (6 cm o.d.)	Photoautotrophic	Indoor	Small	Unknown	–	1.90	20.80 (maximum 25.00)	–	1.10	Del Río et al. (2008)

(continued)

Table 6.2 (continued)

Species/ strain	Cultivation vessel	Mode	Indoor/ outdoor	Scale	Cultivation time (day)	Biomass productivity green stage (g/L/day)	Biomass productivity red stage (g/L/ day)	Astaxanthin productivity (mg/L/day)	Astaxanthin productivity (mg/m^2/day)	Astaxanthin content (% DW)	Reference
CCAP 34/8	50 L Airlift PBR with external-loop solar receiver (2.4 cm i.d.)	Photoautotrophic	Outdoor, Almeria, Spain	Small	Summer	–	0.70	8.00 (3.50 in winter)	107.20	1.30	García-Malea et al. (2009)
WBG 26	3 L open pond (30 cm o.d., 10 cm i.d.)	Photoautotrophic	Indoor	Small	12	–	0.15	4.26	202.60	2.79	Zhang et al. (2009)
NIES-144	1 L flat panel PBR (5 cm o.d.)	Photoautotrophic	Indoor	Small	13.5 (4.5 + 9)	0.33	0.44	14.10	–	4.80	Kang et al. (2010)
LB-16	5.2 airlift PBR (15 cm o.d.)	Photoautotrophic	Indoor	Small	30	–	0.14	3.33	–	2.38	Choi et al. (2011)
NIES-144	Plastic bag bubble column PBR, 60° V shaped bottom (10 cm o.d.)	Photoautotrophic	Indoor	Small	55	–	0.05	1.40	29.70	–	Yoo et al. (2012)

Species/strain	Cultivation vessel	Mode	Indoor/outdoor	Scale	Cultivation time (day)	Biomass productivity green stage (g/L/day)	Biomass productivity red stage (g/L/day)	Astaxanthin productivity (mg/L/day)	Astaxanthin productivity (mg/m^2/day)	Astaxanthin content (% DW)	Reference
K-0084	0.6 L bubble column PBR (i.d. 5 cm) (green stage), 0.6 L columns outdoors, east-west orientation (red stage)	Photoautotrophic	Outdoor, Mesa, Arizona, USA	Small	10 (July to December)	–	0.58	17.10	426.30	2.70	Wang et al. (2013)
NIES-144	3 L fermentor (green stage), 500 mL flasks (red stage)	Mixotrophic	Indoor	Small	38 (14 + 24)	0.18	0.32	15.80	–	4.90	Park et al. (2014)
NIES-144	1 L column PBR	Mixotrophic	Indoor	Small	12	–	–	–	33.90	1.10	Wan et al. (2014a)
NIES-144	Attached cultivation with algal disks (diameter of 33 ± 0.5 cm^2	Mixotrophic	Indoor	Small	12	–	–	–	65.80	1.30	Wan et al. (2014a)

(continued)

Table 6.2 (continued)

Species/strain	Cultivation vessel	Mode	Indoor/outdoor	Scale	Cultivation time (day)	Biomass productivity green stage (g/L/day)	Biomass productivity red stage (g/L/day)	Astaxanthin productivity (mg/L/day)	Astaxanthin productivity (mg/m²/day)	Astaxanthin content (% DW)	Reference
ZY-18	1 L column PBR	Heterotrophic-photoautotrophic	Indoor	Small	12	–	0.12	5.40	–	3.50	Wan et al. (2014b)
ZY-18	200 L raceway	Heterotrophic-photoautotrophic	Outdoor, Jiaxing City, Zhejiang Province, China	Medium	12	–	0.06	2.30	–	2.50	Wan et al. (2014a, b)
SAG 34-1b	0.7 L glass columns (5 cm i.d.) (green stage). Attached cultivation on laminar cultivation module (red stage)	Photoautotrophic	Outdoor	Small	–	–	–	–	164.50	2.20	Zhang et al. (2014)
SKLBE ZY-18	50 L fermentor (green stage), 1 L column PBRs (7 cm o.d.) (red stage)	Heterotrophic-photoautotrophic	Indoor	Small	27 (17 + 14)	1.58	0.05	6.40	–	4.60	Wan et al. (2015)

Species/ strain	Cultivation vessel	Mode	Indoor/ outdoor	Scale	Cultivation time (day)	Biomass productivity green stage (g/L/day)	Biomass productivity red stage (g/L/ day)	Astaxanthin productivity (mg/L/day)	Astaxanthin productivity ($mg/m^2/day$)	Astaxanthin content (% DW)	Reference
UTEX 2505	Column PBR (3.22 cm i.d.)	Photoautotrophic-mixotrophic	Indoor	Small	28 (8 + 20)	0.24	0.24	10.21	–	4.25	Pan-utai (2017)
JNU35	0.5 L column PBRs (6 cm i.d.) (green stage). 0.5 L column PBRs (diameter 3 cm i.d.) (red stage)	Photoautotrophic	Indoor	Small	30 (15 + 15)	1.34	0.91	18.10	–	2.70	Wang et al. (2019)

i.d. and o.d. for inner and outer diameters of the reactors, respectively. Cultivation time is represented as the total cultivation time with the cultivation times for the green and red stages, respectively, in brackets

Table 6.3 N/P ratio of common media used for *H. lacustris* culture

Medium	BBM	3N-BBM + V	OHM	FM:FB	NIES-C	BG-11	Basal medium	NSIII	Standard inorganic medium
Nitrate (mM)	2.94	8.82	4.05	2.70	1.9	17.65	0.1	9.99	10.17
Phosphate (mM)	1.72	1.72	0.21	4.60	0.16	0.23	0.087	1.76	0.37
N/P ratio	1.7	5.1	19.29	0.59	11.88	184.2	1.15	5.68	27.49

Río et al. 2008). The production process can range from 10 to 90 days, with biomass productivities in the green and red stages ranging from 0.04 to 1.58 g/L/day and 0.02–1.90 g/L/day, respectively, with astaxanthin productivities from 0.12 to 20.9 mg/L/day, and with an astaxanthin content of 0.20–7.72% DW (Table 6.3). It has previously been reported that tubular airlift systems are preferable over bubble column PBRs for the outdoor production of biomass and astaxanthin productivity due to an optimal average lighting irradiance of 130 μmol photons/m^2/s with nitrate decreasing to <5 mM for astaxanthin induction over the 16-day period (Table 6.3) (García-Malea López et al. 2006). In a direct comparison between a bubble column and an airlift PBR, the airlift PBR resulted in a 18% higher biomass concentration (4.8 g/L DW, 7×10^6 cells/mL) and a 16% higher astaxanthin yield (480 mg/L), attributable to the regular light/dark cycles and laminar flow in the downcomer of the airlift PBR (Ranjbar et al. 2008). In terms of the optimal light path, it has been observed that a 6 cm light path in the green stage, followed by a 3 cm light path in the red stage resulted in the highest astaxanthin productivity (20.1 mg/L/day) (Wang et al. 2019). In terms of the optimal bioreactor size, it has been determined that the smaller flat plate bioreactor system (17 L) resulted in a 97% higher cell density than in a 200 L system (Issarapayup et al. 2011) showcasing the difficulties of scaling up. Maximising hydrodynamic mixing through an optimised sparger and PBR shape is also critical (metal, 0.2 vvm, 1.3 cm diameter sparger, 60° V-shaped bottom) with a resultant 1.7-fold increase in astaxanthin productivity without the adherence of cells (Yoo et al. 2012). However, with 60+ years of PBR research (317 studies of algal reactors), it has been identified that there is little difference between the system used and the biomass productivity overall, but it was suggested that intermediate volume bioreactors with higher surface area-to-volume ratios could provide higher yields whilst simultaneously reducing the environmental footprint with lower energy consumption (Granata 2017).

Recently, there has been a large transition in the materials utilised for PBRs; traditionally, plastic tubes have been utilised due to their apparent low cost. Recently, many microalgal companies have partnered with Schott AG (e.g. A4F, Varicon Aqua Solutions Ltd., Ecoduna, and Heliae) and are replacing plastic PBRs with Schott glass due to higher resistance to UV radiation and chemicals, reductions in biofilm formation, and cost savings over a longer-term period (Schott 2019). It has been identified that over a 12-month period using a total tube length of 12 km, a 10%

higher biomass and astaxanthin productivity was attained using a wall thickness of 1.8 mm compared with 2.5 mm in Israel, attributable to higher sunlight penetration and more stable temperatures (Schott 2019).

Novel modes of cultivation have included attached cultivation on a membrane and utilising perfusion culture. In attached cultivation, the cells are cultivated in the water column and then seeded on a membrane to increase the light surface area, and this reduces the harvesting costs as the cells are already dewatered, resulting in up to a 90% reduction in water consumption (Zhang et al. 2014). Other benefits include overall energy savings from the lack of mixing/pumping and a reduction in contamination, particularly single-celled protozoa (Wan et al. 2014a). Furthermore, when attached cultivation is employed in the red stage, the astaxanthin induction is faster than in column PBRs (Wan et al. 2014a). Utilising a two-stage system with attached cultivation for the red stage, biomass and astaxanthin productivities of 3.7 g/m^2/day and 65.8 mg/m^2/day, respectively, were obtained utilising strain NIES-144 (12 day) (2.8- and 2.4-fold higher than conventional suspended bioreactor cultivation, respectively) (Wan et al. 2014a). The optimal temperatures for maximising biomass and astaxanthin production were conflicting, with the highest astaxanthin content (1.5% DW) at 15 °C but the maximum biomass and astaxanthin productivities at 25 °C (Wan et al. 2014a). Zhang et al. (2014) reported a slightly modified version of attached cultivation utilising strain SAG 34-1b with a different medium (BG-11 vs. NIES-N) with nitrate deprivation (1.8 mM) rather than depletion, resulting in an astaxanthin productivity of 164.5 mg/m^2/day. Through a strain screening process (25 strains) and utilising the optimal strain (CCAC 0125) for a twin-layer two-stage immobilisation system, a high biomass (19.4 g/m^2/day at 1015 μmol photons/m^2/s) and astaxanthin productivity (0.39 g/m^2/day at 1015 μmol photons/m^2/s) was obtained with 1% CO_2 supplementation, 14/10 photoperiod, and 28.5 °C (Kiperstok et al. 2017). Utilising a one-stage process resulted in similar biomass and astaxanthin productivities but in half the time (8 days) (Kiperstok et al. 2017).

Alternatively, perfusion culture in a fermentor has been demonstrated where the medium (NIES-C with 11.98 mM acetate, 2.58 mM nitrate, and 0.147 mM phosphate) is continuously replaced, removing inhibitory metabolites formed during cultivation and replenishing nutrients (Park et al. 2014). This process resulted in a biomass and astaxanthin yield of 12.3 g/L (0.18 g/L/day) and 602 mg/L, respectively, through stepwise increased light irradiance (150–450 μmol photons/m^2/s) (Park et al. 2014). However, this method required 54% more energy than a fed-batch stepwise photoautotrophic process and 24.5 additional days (Kang et al. 2010).

To date, most systems have focussed on phototrophic production using a two-stage strategy as the conditions for maximising growth and astaxanthin productivity are mutually exclusive (Table 6.2). The aim of the green stage is to maximise biomass productivity and the second red stage is to induce astaxanthin formation. Utilising this method, a biomass yield of almost 20 g/L could be attained after 21 days in the green stage using stepwise increases in irradiance (25–100–500 μmol photons/m^2/s) with an astaxanthin productivity of 11.5 mg/L/day (Aflalo et al. 2007). Wang et al. (2019) have obtained a biomass yield of 20.1 g/L (1.34 g/L/day) in the green stage and 27.3 g/L DW (0.91 g/L/day) after the red stage, the highest biomass yield to date in a *H. lacustris* photoautotrophic system. A single-stage astaxanthin production system has been devised (utilising an impinging irradiance of 1000 μmol photons/m^2/s, a specific average irradiance of 93.4 μmol photons/m^2/s, dilution rate of 0.9 μ/day, and 2.2 mM nitrate) which resulted in a biomass productivity after the red stage of 1.9 g/L/day and an astaxanthin productivity of 20.8 mg/L/day, the highest to date (Del Río et al. 2008). The technical feasibility of this process outdoors in summer (50 L tubular PBR) has been showcased with a biomass and astaxanthin productivity of 0.7 g/L/day and 8 mg/L/day, respectively, and further increases were believed to be attainable through increasing the availability of light (>53.45 μmol photons/m^2/s) to the cells (García-Malea et al. 2009). Furthermore, it has been proposed that night-time losses of biomass and astaxanthin could be reduced through identifying the optimised control temperature (Wan et al. 2014a). It was observed that 2.9 and 5-fold increases in biomass and astaxanthin productivities could be obtained with NIES-144 when the night temperature was maintained below 28 °C, but this will differ for each strain and the specific cultivation conditions (Wan et al. 2014a). However, it must be noted that several drawbacks of this system have been highlighted including the lower astaxanthin content compared to the two-stage process (0.9–1.1% vs. 3.8% DW), the requirement for artificial illumination at night which is economically unattractive, vulnerability to grazers, and difficulties in harvesting the heterogeneous cells (compared to gravitational settlement for the aplanospores) (Aflalo et al. 2007).

H. lacustris is capable of mixotrophic growth with sodium acetate and ribose appearing to be the most suitable substrates (Kobayashi et al. 1991; Pang and Chen 2017). It is well known that acetate is a suitable organic carbon source for maximum growth (4–12 mM) and for enhancing astaxanthin accumulation in the red stage (Cifuentes et al. 2003; Göksan et al. 2010; Gong and Chen 1997; Kang et al. 2007). Higher biomass and astaxanthin productivities (0.35 g/L/day and 4.54 mg/L/day, respectively) have been obtained when utilising NSII medium supplemented with 50 mM sodium acetate throughout the whole process, but with the addition of 100 mM potassium acetate in the red stage, a higher astaxanthin productivity could be attained (10.21 mg/L/day) with a 2.5–4.3% astaxanthin content, albeit with a reduced biomass productivity (0.24 g/L/day) (Pan-utai 2017). Using a fed-batch mode with sodium acetate results in a biomass yield of 1.77 g/L after 9 days (93.9% higher than a batch culture), and it was determined that the culture should be fed at night (Sun et al. 2015). Utilising ribose (9.66 mM) as a C5 carbon substrate has been suggested to prolong the green stage, increase the specific growth rate and biomass yields, and reduce the risk of contamination (Pang and Chen 2017). It has been stated that organic carbon feeding should be done at night under 16 °C, to minimise the loss of enzymatic activity (Sun et al. 2015).

H. lacustris has also been reported to be capable of heterotrophic growth using sodium acetate as a carbon source in the green stage, but the organism has slow metabolic growth (0.20–0.22 μ/day), and contamination issues have been encountered (Droop 1955; Hata et al. 2001). Comparatively, photoautotrophic production in the green stage results in a growth rate of 0.56 μ/day (García-Malea et al. 2005) and even up to 1.30 μ/day (Boussiba and Vonshak 1991). Heterotrophic growth (10–30 mM sodium acetate) offers the potential for producing high biomass yields in the green stage with Hata et al. (2001) obtaining a biomass yield of 7 g/L DW and Wan et al. (2015) attaining a biomass yield of 26 g/L DW with the highest biomass productivity (1.58 g/L/day) to date in the green stage (Table 6.2). Hata et al. (2001) revealed that after the third repeated fed batch, culture contamination became an issue (Hata et al. 2001), but Wan et al. (2015) did not report contamination issues. To date, there have been no confirmed reports of astaxanthin production in a commercial heterotrophic process; however, Kobayashi et al. (1997a, b) reported carotenoid formation in heterotrophic cultures of *H. lacustris* in the laboratory, but this was not confirmed to be astaxanthin, and to date, this remains unknown. A sequential heterotrophic-photoautotrophic production process has been suggested where astaxanthin formation was induced using nitrate deprivation, and subsequently, 5% CO_2 was supplied to the culture resulting in the highest cellular astaxanthin content to date of 7.72% DW (6.25 mg/L/day) (Kang et al. 2005). Commercially, some companies have been experimenting with heterotrophic cultivation of *H. lacustris*, but astaxanthin production in the red stage is currently insufficient for commercialisation. Utilising the alga's heterotrophic ability to produce high biomass yields in the green stage, followed by optimal induction in the red stage through photoautotrophic induction, could be a suitable method for attaining high astaxanthin productivities.

2.5 *Astaxanthin Induction*

Astaxanthin is accumulated outside the plastid in cytoplasmic lipid vesicles (Grunewald et al. 2001). Astaxanthin has been reported to have several functions within *H. lacustris* aplanospores, acting as a 'sunshade' protecting the photosynthetic apparatus, protection from photooxidative stress, and minimising oxidation of storage lipids (Han et al. 2012). To date, it is not fully known how astaxanthin acts to protect *H. lacustris*, and further studies need to be conducted to fully elucidate its role in the protection of *H. lacustris* cells in unfavourable conditions.

Astaxanthin synthesis was originally proposed to be induced by the cessation of cell division and only occur in the resting stage (Boussiba and Vonshak 1991; Kobayashi et al. 1997a, b). However, synthesis of astaxanthin has been demonstrated to be independent of cell division and can occur in vegetative cells (motile macrozooids and palmelloids) (Brinda et al. 2004; Butler et al. 2017; Del Río et al. 2005, 2008; Hagen et al. 2001). A range of factors have been investigated to explore astaxanthin synthesis by *H. lacustris*, including high light, high temperature, and nutrient deprivation/depletion (e.g. nitrate and phosphate). High light was suggested to have been one of the most effective factors in the stimulation of astaxanthin synthesis (Choi and Park 2002). High-temperature treatments have been reported to have resulted in greater levels of astaxanthin, but temperatures >30 °C have been found to reduce the biomass yield (Tjahjono et al. 1994).

Salinity stress by the addition of NaCl (0.1–0.5%) has also been used to increase astaxanthin levels, but concentrations of 0.6–0.8% NaCl can cause severe cell mortality (Cifuentes et al. 2003; Harker et al. 1996; Sarada et al. 2002).

It has been proposed and is well known that nitrate limitation is the critical factor inducing astaxanthin accumulation with high light and dilution rate as factors responsible for enhancing the astaxanthin content, but which alone are not inducers of the pigment itself (nitrate > dilution rate > light) (García-Malea et al. 2009). Christian et al. (2018) further validated the effect of light and identified that high light intensity alone had little effect on inducing astaxanthin production. It has been identified that nitrate limitation (<5–8 mM) results in the formation of astaxanthin with 2.2 mM nitrate being optimal for astaxanthin productivities (Ranjbar et al. 2008; Del Río et al. 2008). In the two-stage process under nitrate deprivation (4 mM), the astaxanthin content is 2.7% DW, but with nitrate depletion, the content is increased to 3.8% DW (Wang et al. 2013). In the case of urea as a nitrogen source, 3 mM resulted in the highest astaxanthin content (2.4% DW) (Wang et al. 2019). Other reports have stated that a nitrate concentration of 0.6 mM is the concentration necessary to induce astaxanthin accumulation, whilst avoiding culture washout (García-Malea et al. 2009). The specific concentration of nitrate required for induction is likely dependent on the PBR, cultivation conditions, and the strain employed.

2.6 *Astaxanthin Biosynthesis*

Astaxanthin is produced in the chloroplast, accumulates around the nucleus to protect the ultrastructures from reactive oxygen species (ROS), is esterified in the endoplasmic reticulum, and spreads into cytoplasmic lipids, with recent models revealing that astaxanthin and lipid biosynthesis are not synchronous, with lipid droplets accumulating faster than astaxanthin (Collins et al. 2011; Saha et al. 2013b; Solovchenko et al. 2013; Cheng et al. 2017a). It has been found that photorespiration can accelerate astaxanthin accumulation which is speculated to be through increasing glycerate-3-phosphate (PGA), a precursor to glyceraldehyde 3-phosphate (G3P) (Fig. 6.10) during the Calvin cycle (Zhang et al. 2016). It has been identified that isopentenyl pyrophosphate (IPP) is an essential intermediate of carotenoid synthesis and this molecule can originate from the mevalonate pathway (MVA) or non-mevalonate pathway (MEP) in the chloroplast (Lemoine and Schoefs 2010). In the MEP pathway, 1-deoxy-D-xylulose-5-phosphate is formed in the first stage. IPP then undergoes isomerisation to dimethylallyl diphosphate (DMAPP); however, it remains unknown which enzyme is responsible for this conversion (Shah et al. 2016). The isoprenoid chain is then elongated, initiated with DMAPP with a subsequent linear addition of three molecules of IPP, and is catalysed by geranylgeranyl pyrophosphate synthase (GGPS); then finally, geranylgeranyl pyrophosphate (GGPP) is formed, a C20 compound (Shah et al. 2016).

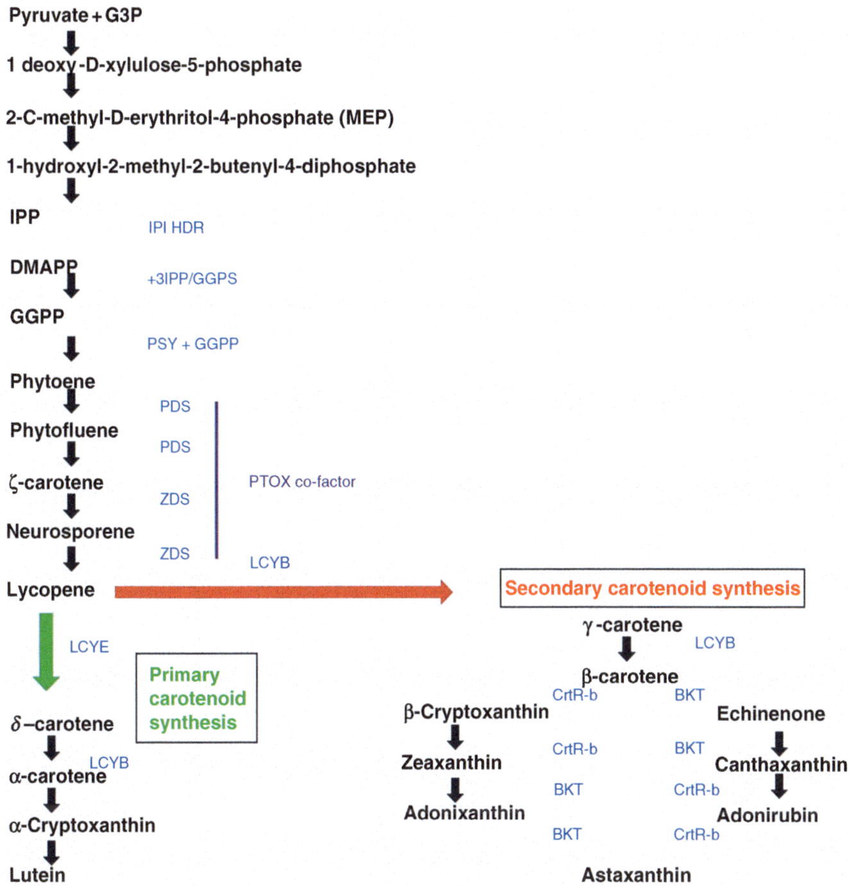

Fig. 6.10 Astaxanthin biosynthesis in *H. lacustris*. (Modified from Shah et al. 2016). The enzyme abbreviations are defined as follows: HDR, 4-hydroxy-3-methylbut-2-enyl diphosphate reductase; IPI, isopentenyl pyrophosphate isomerase; PSY, phytoene synthase; GGPS, geranylgeranyl pyrophosphate synthase; PDS, phytoene desaturase; ZDS, ζ-carotene desaturase; LCYE, lycopene ε-cyclase; LCYB, lycopene β-cyclase; CrtR-b, β-carotene 3,3′-hydroxylase; BKT, β-carotene ketolase

For the carotenoid synthesis, phytoene synthase (PSY) is the catalyst and initiates a head-to-tail condensation of two GGPP molecules to form the C40 compound, phytoene which acts as a precursor to astaxanthin (Cunningham and Gantt 2011). It is well known that PSY is upregulated in the transition from green to red stage cultures (Gwak et al. 2014). It has recently been proposed that a mutation in the PSY of *H. pluvialis* had an essential role in the evolution of hypercarotenogenesis (Pick et al. 2019). Lycopene is formed through four desaturation steps (increasing the number of conjugated carbon-carbon double bonds) involving two phytoene desaturases (PDS) and a ζ-carotene desaturase (ZDS) as the catalysts (Li et al. 2010; Nawrocki et al. 2015). Plastid terminal oxidases

(PTOX1 and PTOX2) are cofactors involved in carotenoid desaturation, and PTOX1 is coregulated with astaxanthin synthesis, but the function is still undefined (Wang et al. 2009). Following the desaturation stages, the two termini of lycopene undertake cyclisation through competing pathways regulating metabolic flux; primary carotenoid formation is catalysed by lycopene ε-cyclase (LCYE), and secondary carotenoids are synthesised by lycopene β-cyclase (LCYB) resulting in β-carotene formation (Gwak et al. 2014; Lao et al. 2017). It has been found that β-carotene is exported out of the chloroplast and is enzymatically converted to astaxanthin in the cytoplasm (Fig. 6.10) (Pick et al. 2019). The cyclisation of lycopene is an important regulatory branch in astaxanthin biosynthesis. Upregulation of LCYB at the transcriptional, proteomic, and metabolomic levels could result in elevated concentrations of astaxanthin. It has also been identified that the two final oxidation steps catalysed by β-carotene ketolase (BKT) and β-carotene hydroxylase (CrtR-b) are rate-limiting steps of astaxanthin synthesis (Shah et al. 2016).

Recent transcriptomic profiling on the effect of LEDs on the astaxanthin biosynthetic pathway has been conducted, and it has been revealed that blue LED irradiation results in an increased expression of the enzymes BKT and the carotenoid hydroxylase (CHY) compared to white light as the baseline, but comparatively, red LED irradiation results in downregulation (Lee et al. 2018). Upregulation of PSY, LCY, carotenoid ketolase (Crt-o), and CrtR-b was also observed when blue light was used for astaxanthin induction (Ma et al. 2018). The astaxanthin synthesis pathway is complex, and multiple regulatory mechanisms at the transcriptional, translational, and posttranslational level are involved with five key enzymes critical in the process: isopentenyl pyrophosphate isomerase (IPI), PSY, PDS, Crt-o, and CrtR-b (Li et al. 2010). It is essential for the pathway of astaxanthin to be understood before genetic engineering can be undertaken (Fig. 6.10).

2.7 Genetic Engineering

Genetic engineering of microalgae has been reported in over 30 species, but the toolbox available for *H. lacustris* is limited. To date, the chloroplast genome (1.35 Mb) has been sequenced by the Synthetic Genomics group for *H. lacustris* UTEX 2505 (close relative of UTEX 16, a descendent of NIES-2264), and it has been identified as the largest assembled chloroplast of any plant or alga to date, but more coverage is required for the nuclear genome to be sequenced (Bauman et al. 2018; Buchheim et al. 2013; Smith 2018). It was reported that the sequencing of the chloroplast genome leaves many unanswered questions; >90% of the DNA was non-coding, it has a non-standard genetic code, it only encodes 12 tRNAs (less than half of a typical plastome), and it is one of the few sequenced plastids that is not biased in adenine and thymine (Smith 2018).

Currently, most genetic improvements in *H. lacustris* have been limited to classic random mutagenesis due to the lack of an available nuclear genome and a poorly annotated chloroplast genome. To date, UV mutagenesis (Sun et al. 2008; Tripathi et al. 2001) and chemical mutagenesis using *N*-methyl-*N*-nitro-*N*-nitrosoguanidine (MNNG) (Hu et al. 2008) and ethyl methanesulphonate (EMS) (Sun et al. 2008; Tripathi et al. 2001) have been trialled for elevating astaxanthin content with the aim of inducing 85–95% mortality. Chemical mutagenesis has been more successful because of the ability of *H. lacustris* to tolerate light damage, and using MNNG has resulted in improvements of volumetric astaxanthin up to threefold (Hu et al. 2008). For screening these mutants, herbicides have typically been used such as nicotine and norflurazon (Shah et al. 2016). With astaxanthin mutants, the colonies are potentially easier to pick because of their brighter red colouration. A Chilean *H. lacustris* mutant (mutated with EMS) was cultivated in a commercial-sized open pond of 125,000 L, and the astaxanthin content was 30% greater than the wild-type strain on a DW basis and 72% greater on a per culture volume basis (Gómez et al. 2013). Irradiating FACHB-872 with 4000 Gy ^{60}Co-γ and then cultivation under high light with 15% CO_2 resulted in a 1.7-fold increase in astaxanthin compared with a wild-type strain, and importantly, 56% of the genes were significantly upregulated in the mutant cells including pyruvate kinase providing a feedstock for astaxanthin and phytoene synthase, lycopene beta-cyclase, and ZDS for β-carotene conversion to astaxanthin (Cheng et al. 2016, 2017a, b).

An emphasis has been on targeting rate-limiting steps in astaxanthin biosynthesis with the key enzymes being localized in the chloroplast, and it has been observed that PDS is a key target (Grünewald et al. 2001). A focus has been on nuclear transformations with conventional mutagenesis such as overexpression of a PDS with a point mutation for norflurazon resistance with transgenic lines possessing a 36% higher astaxanthin content after 2 days of high light induction (Steinbrenner and Sandmann 2006). Recently, the carotenoid biosynthesis pathway has been genetically modified through a plasmid transformation in the chloroplast with the endogenous PDS nuclear gene to overproduce astaxanthin (67% increase in astaxanthin content in transformed strains compared to the wild type) with induction under high light and nitrogen depletion without an adverse effect on growth or biomass productivity (Galarza et al. 2018).

Insertional mutagenesis has been investigated for producing high-yielding astaxanthin strains using *Agrobacterium*-mediated transformation (Kathiresan et al. 2009), biolistics (particle bombardment) (Steinbrenner and Sandmann 2006), and electroporation (Sharon-Gojman et al. 2015), but until recently they have lacked efficacy. Now stable transformation of the chloroplast and nuclear genome is possible with the introduction of two transgenes without the requirement of an additional antibiotic resistance gene (Gutiérrez et al. 2012; Sharon-Gojman et al. 2015). The PDS variant is used as a selection marker that confers resistance to the herbicide norflurazon with a single point mutation (L504A) (Shah et al. 2016). Enhancement of astaxanthin in *H. lacustris* could be achieved by upregulating the PSY and CrtR-b genes which are often defined as being rate limited for astaxanthin production and have previously been upregulated by high light (Han et al. 2013b; Li et al. 2008, 2010). To bring a disruptive change in astaxanthin yields from *H. lacustris*, advanced genetic engineering methods with a greater emphasis on targeted mutations are required, utilising homologous recombination and clustering regularly interspaced short palindromic repeat (CRISPR) technology, which is an important new platform for generating RNA-guided nucleases (RGNs), such as Cas9 which is an RNA-guided DNA nuclease employed to introduce targeted mutations into eukaryotic genomes (Brodie et al. 2017). This may become a reality once the *H. lacustris* nuclear genome has been sequenced with detailed annotations and more transformation tools are available. Future emphasis should be on developing a toolbox for genetically engineering *H. lacustris* targeting the astaxanthin biosynthetic pathway, potentially through upregulation of the rate-limiting enzymes BKT and CrtR-b (Shah et al. 2016).

Recently, a high-throughput method has been devised for screening and selecting hyperproducing mutants under UV mutagenesis (40 mJ/cm^2 with 32 min of UV exposure time) using 50 μm sodium azide to accelerate astaxanthin induction (Eui et al. 2018). Using a soybean oil-based extraction method combined with spectrophotometric analysis (OD 470 nm) enabled the detection of 31 strains (88.5% of the cells) with higher astaxanthin production than the wild type (NIES-144) with the M13 strain exhibiting an astaxanthin yield 1.59 times higher than the wild type (174.7 mg/L) (Eui et al. 2018). Utilising this high-throughput method can help identify transformants, and this information can be supported through transcriptomics, proteomics, and metabolomics to further target bottlenecks in the production of astaxanthin. If a genetically engineered strain is to be utilised for commercial production, caution is warranted because there are regulatory hurdles to overcome such as Directive 2001/18/EC in the European Union (EU) where the approval procedure can take 4–6 years and cost 7–15 million Euros (Hartung et al. 2014).

3 Commercial Constraints of Astaxanthin Production from *H. lacustris*

Commercial astaxanthin production has been successful, and a number of companies are successfully operating. However, the market is saturating and the price of astaxanthin is overinflated. As supply has increased, the price of astaxanthin has fallen. To date, the focal point of astaxanthin has been as a nutraceutical and functional food. However, the bulk of astaxanthin is used for aquaculture which is dominated by the synthetic form and by *X. dendrorhous*. In order to compete with these sources, the cost of production needs to decrease, and there is a requirement for challenges to be overcome in the production process, including improvements in biomass and astaxanthin productivity, mitigating contamination, and the requirement for green chemistry and engineering. There needs to be more collaboration between academia and industry to advance knowledge. More investigations are required to look into the commercial feasibility of astaxanthin from *H. lacustris*. Shah et al. (2016) identified that there are three key areas to target for further improvements: cultivation efficiency and cost, good cultivation practice with the control of predators, and extraction and purification of astaxanthin. In this section, the authors believe improving biomass/astaxanthin productivity, minimising cell die-off through photobleaching in the red stage, costly downstream processing methods, contamination, and the total cost of the process are critical problems to address.

3.1 Improved Biomass Productivity

Currently, the growth rate of *H. lacustris* is slow, and during cultivation as it is a flagellated form, it is vulnerable to shearing, attributable to hydrodynamic stress in the PBRs. There are difficulties in maintaining green motile macrozooids in the green stage without transitions to the palmelloid form which results in slower growing cells.

The optimal cultivation conditions have been determined from a range of studies. The optimal temperature for growth is strain dependent but is typically between 20 and 28 °C (Allewaert et al. 2015; Giannelli et al. 2015). Temperatures greater than 30 °C induce encystment and the formation of aplanospores within 2 days, resulting in the cessation of growth (Allewaert et al. 2015; Tjahjono et al. 1994). The optimal pH for *H. lacustris* has been reported in the range of 7.00–7.85 (Hata et al. 2001; Sarada et al. 2002). Typically, the optimal irradiance for *H. lacustris* is 70–177 μmol photons/m^2/s with a saturation index of 250 μmol photons/m^2/s (Zhang et al. 2014; Giannelli et al. 2015), but this has not been conclusive, and often lower irradiances have been investigated (Park et al. 2014). It is likely that other factors play a role in a multifactorial process. The impact of photoperiod has been preliminarily investigated with continuous illumination appearing to be optimal for higher-density cultures, but only a 12:12 and 24:0 photoperiod was compared (Domínguez-Bocanegra et al. 2007). The optimal light seems to be warm-white light, but this needs further investigation (Saha et al. 2013a). As an inorganic carbon source, CO_2 is commonly utilised to increase biomass productivity and ranges from 1% to 5% in the green stage (Kaewpintong et al. 2007). It has been determined that CO_2 supplied at 5% is beneficial for growth (3.3 g/L DW) but increases to 10% and results in a deterioration in growth, photosynthesis, and the assimilation of carbon, determined by PSII yield, NPQ activity, chlorophyll a content, and biomass yield (Chekanov et al. 2017). It has been observed that a γ-ray-irradiated mutant cultivated under CO_2 at 6% in conjunction with high light (108 μmol photons/m^2/s) had the highest biomass productivity (0.16 g/L/day) with a maximum growth rate of 0.6 μ/day (Cheng et al. 2016). Higher concentrations of CO_2 at 10% or 20% resulted in a decrease in growth (Chekanov et al. 2017). A further emphasis needs to be on carbon uptake and assimilation by the cells rather than the CO_2 input alone which has seldom been explored.

Biomass accumulation has been a major bottleneck in the two-stage process of astaxanthin production, and further optimisation of the growth media is required. It is clear that a wide range of media and their derivatives have been utilised for optimal *H. lacustris* cultivation (Table 6.3). Commonly, BBM or 3N-BBM has been used for *H. lacustris* culture (Oncel et al. 2011; Qinglin et al. 2007; Suh et al. 2006; Tocquin et al. 2012), along with BG-11 (Aflalo et al. 2007; Kiperstok et al. 2017; Torzillo et al. 2003; Zhang et al. 2009). It has been identified that BBM is more effective than 3N-BBM and BG-11 for attaining higher cell densities (Nahidian et al. 2018). Fábregas et al. (2000, 2001) utilised an optimised *Haematococcus* medium (OHM) with a final cell density of 3.77×10^5 cells/mL (3 times higher than BBM) with higher yields (1.62 g/L DW) attained by utilising high light (235 μmol photons/m^2/s). Through media optimisation, BBM with threefold the phosphate concentration (5.16 mM) and an N/P ratio of 1.71:1 resulted in an 86% increase in the growth rate, and supplementation with 0.185 mM boron or 0.046 mM iron appeared beneficial for growth (Nahidian et al. 2018). However, a conflicting study by Fábregas et al. (2000) identified that boron was a nonessential nutrient for the growth of *H. lacustris* along with iodine, zinc, and vanadium. The addition of vitamin B_{12} has also been reported to increase algal growth by up to 55% in comparison to media without (Kaewpintong et al. 2007).

Utilising commercially viable media has also been investigated. A commercial plant fertiliser (£0.24 ton^{-1}) with an N-P-K 20:20:20 resulted in 0.9 g/L DW after 12 days (Dalay et al. 2007). A common hydroponics fertiliser, FM:FB, was screened, and the optimised formulation resulted in a cell density of $>1 \times 10^6$ cells/mL (Tocquin et al. 2012). Tocquin et al. (2012) attributed the increased growth to its low N/P ratio of 0.6:1 (1.00 mM nitrate and 1.63 mM phosphate), much lower than other media that have been tested (Table 6.3). Reformulating the medium could further increase biomass productivities, and only certain studies have investigated this in detail (Fábregas et al. 2000; Tocquin et al. 2012; Tripathi et al. 1999; Wang et al. 2019). An emphasis should be on the macronutrients with the concentrations of carbon, nitrate, and phosphate and the C/N/P ratio.

With regard to the nitrogen source, sodium nitrate has been determined to be optimal (Sarada et al. 2002). Feng et al. (2017) deduced that according to the elevated activity of nitrate reductase in the *H. lacustris* metabolism, the concentrations of sodium nitrate, monopotassium phosphate, and sodium acetate could be determined (3.53 mM, 0.33 mM, and 13.16 mM, respectively) with nitrate being the main influencing factor. However, by utilising a newly isolated strain (JNU35), it was determined that BBM with sodium nitrate replaced by urea (18 mM) was optimal for biomass productivity, attributable to providing a nitrogen and carbon source (Wang et al. 2019). Furthermore, an interesting observation was noted where the strain was observed to have increased growth when resuspended in the red stage in nitrate-depleted BBM medium, attributed to the synthesis of storable nitrogen compounds in the green stage (Wang et al. 2019). This was also noted by Butler et al. (2017) with cells initially cultivated in BG-11, followed by resuspension in nitrate-depleted 3N-BBM + V.

3.2 Improved Astaxanthin Content

Astaxanthin induction factors have been well characterised, and nitrate deprivation has been determined as critical with high light enhancing the rate of astaxanthin accumulation (Christian et al. 2018; Del Río et al. 2008; García-Malea et al. 2009). To date, the highest astaxanthin content (7.72% DW) has been attained using a sequential heterotrophic-photoautotrophic production process where astaxanthin induction was initiated under nitrate deprivation and 5% CO_2 (Kang et al. 2005), but the highest astaxanthin productivity has been attained with the one-stage process, albeit with a low content of astaxanthin (1.1% DW) (Del Río et al. 2008). Utilising the one-stage process, the astaxanthin content (0.8–1.1% DW) is lower than the two-stage process (4% DW) with a lower astaxanthin proportion of the total carotenoid fraction (65% compared with 95%) (Aflalo et al. 2007). If a two-stage process is adopted, the initial biomass density in the red stage plays a critical role in increasing astaxanthin content with 0.8 g/L DW being optimal (Wang et al. 2013). With an initial density of 0.1 g/L, severe photoinhibition was observed, and using 2.7 g/L DW resulted in light limitation for astaxanthin production in outdoor cultivation (Wang et al. 2013).

Regarding other parameters, it has been identified that the optimal temperature for astaxanthin production is 27 °C (Evens et al. 2008). Increased temperatures have been hypothesised to result in the synthesis of astaxanthin through oxygen radical formation (Tjahjono et al. 1994). Abiotic stresses generally cause the generation of reactive oxygen species (ROS) within the cell and induce astaxanthin as a defence strategy (Eui et al. 2018). The addition of 0.45 mM Fe^{2+} in the form of ferrous sulphate can significantly increase the biosynthesis of carotenoids attributable to the formation of hydroxyl radicals (Kobayashi et al. 1993). Adding 0.45 mM Fe^{2+}, 2.25 mM sodium acetate, and high temperature (30 °C) results in a further increase in carotenoids (Kobayashi et al. 1993; Tjahjono et al. 1994). This has also been validated using transcriptome analysis where high light and sodium acetate addition (25 mM) resulted in expression of essential genes related to carotenoid biosynthesis and FA elongation, but this was not observed for Fe^{2+} (20 μM) which showed a decrease in gene expression related to photosynthesis-antenna proteins (He et al. 2018).

For astaxanthin induction in the red stage, it has been reported that *H. lacustris* cannot be cultivated heterotrophically in the dark and the production of astaxanthin should adopt a photosynthetic mode (Guedes et al. 2011). Using high light (950 μmol photons/m^2/s) in an indoor enclosed system for the red stage is costly and not economically feasible for indoor production (Olaizola 2000; Guedes et al. 2011). It has been determined that 300 μmol photons/m^2/s as an incident light intensity is the optimal for astaxanthin production with light saturation reached at 500 μmol photons/m^2/s, and 600 μmol photons/m^2/s resulted in a lower astaxanthin content with increased cell death (Evens et al. 2008; Giannelli et al. 2015; Li et al. 2010). It has been suggested that inducing astaxanthin in the red stage in a flat plate bioreactor (3 cm light path) would elevate the astaxanthin content to 5.6% after 15 days compared with 3.7% DW in a glass column (6 cm light path) (Wang et al. 2019). Alternatively utilising different wavelengths of light could reduce the light intensity required for astaxanthin induction. Blue light (380–470 nm) is well known to induce the transition to encystment and can be utilised as part of a two-stage strategy (red light for the green stage and blue light at a high light intensity for astaxanthin induction) with an astaxanthin content up to 5.5% DW within 12.5 days (Katsuda et al. 2004; Lababpour et al. 2004). Sun et al. (2015) determined that blue and white light (3:1) at 95 μmol photons/m^2/s increased the astaxanthin yield by 11.8% compared with blue light alone (91.8 mg/L) and reduced the encystment time.

Plant hormones have been investigated for elevating yields including jasmonic acid, abscistic acid, and methyl jasmonate and have been reviewed elsewhere (Shah et al. 2016). In addition, growth regulators such as salicylic acid, gibberellic acid, and 2,4-epibrassinolide were found promising for increasing astaxanthin content as reviewed by Shah et al. (2016). With these compounds, astaxanthin genes were upregulated (tenfold increase). The highest improvement of astaxanthin was with 50 mg/L salicylic acid under low light conditions (25 μmol photons/m^2/s) elevating the astaxanthin content sevenfold, but this yield was comparatively low compared to other reports. It was observed that higher levels of the hormones/growth regulators adversely affected growth and astaxanthin accumulation (Gao et al. 2012). In addition, micronutrients such as selenium have been observed to result in an increase in astaxanthin but have resulted in declines in the biomass yield (Zheng et al. 2017).

Utilising γ-ray-irradiated mutants cultivated under CO_2 at 6% in conjunction with high light (108 μmol photons/m^2/s) resulted in 2.4-fold higher astaxanthin content than with the wild-type strain (Cheng et al. 2016). The same mutant could be cultivated with 15% CO_2 resulting in 5.8 times higher astaxanthin accumulation than under aeration with astaxanthin induction occurring after 24 h (Li et al. 2017). Christian et al. (2018) observed that *H. lacustris* cells cultivated under 15% CO_2 and high light (300 μmol photons/m^2/s) turned orange within 1 day, and by day 2, astaxanthin was accumulated with a final astaxanthin content of 3.62% DW after 4 days, but a reduction in cell density was observed, presumably attributable to photobleaching.

To date, the most successful parameters for inducing astaxanthin in the red stage have been nitrate deprivation/depletion combined with high light and an increase in the C/N ratio. Further improvements have been proposed using PBR developmental work, blue light, elevated temperatures, plant hormones, and micronutrients and through genetic engineering. It is likely that these effects when utilised in combination will have a synergistic effect on the astaxanthin content. A particularly interesting aspect would be on developing a process with high astaxanthin accumulation in the red stage whilst the cells are still continuing to divide as revealed by Wang et al. (2019).

3.3 Minimising Cell Die-Off Through Photobleaching in the Red Stage

In the green stage, controlled culture conditions are provided with low light intensity for maximising growth as discussed above. Upon attainment of a suitable biomass concentration, the culture is subjected to unfavourable conditions, usually involving high light in the red stage to induce astaxanthin biosynthesis (Olaizola 2000). During the initial transition to the red stage, mass cell die-off (photobleaching) has been noted ranging from 20% to 80% depending on the strain, PBR, and induction conditions (Wang et al. 2013). After 24 h of high light, cells have been noted to lose their flagella, and after 48 h the cell density has been observed to decrease by 41% (Gu et al. 2014). The surviving cells have been observed to undergo profound biochemical and cellular changes, with transformations in the life cycle from the vegetative to the aplanospore stage (Wang et al. 2014). It has been observed that *H. lacustris* cells exposed to higher irradiance accumulated more astaxanthin, but exhibited higher cell mortality (Li et al. 2010). The exact cause of cell death when cells have been transferred from the green to the red stage remains unknown (Wang et al. 2014), but there is a necessity to reduce die-off between the green and the red stage to minimise losses in biomass productivity.

It has been suggested that optimising the initial cell concentration for the red stage (Wang et al. 2013) and applying palmelloid cells rather than green motile macrozooids to the red stage for the induction of astaxanthin may present a promising strategy for greater biomass and astaxanthin production (Wang et al. 2014). Another strategy is to use stepwise irradiance allowing the cells to acclimate (Park et al. 2014). Of the vegetative cells, the green motile macrozooid stage has been determined to have a higher susceptibility to photooxidative stress than palmelloids when transferred to the red stage (Han et al. 2012; Harker et al. 1996; Sarada et al. 2002). In most *H. lacustris* strains, there is an overall tendency to rapidly transform to palmelloid stages, and these morphotypes are often preferred over green motile macrozooids because they are believed to be more resistant to imposed stressors (Allewaert et al. 2017; Wang et al. 2013). Choi et al. (2011) determined that transferring these palmelloids (senescent cells) to the red stage resulted in a higher capacity to accumulate astaxanthin. However, there is no evidence that strains with more palmelloids will have a higher associated astaxanthin productivity (Allewaert et al. 2017).

3.4 Downstream Processing

Downstream processing encompasses harvesting, cell disruption, drying, extraction, encapsulation, and formulation of the compound. Depending on the application, downstream processing can account for 20–40% of the costs of the production process ('t Lam et al. 2018). *H. lacustris* aplanospores are large (>50 μm) and, therefore, can be harvested by gravitational settlement (6–8 h in a pond, then 12–24 h for tank sedimentation) resulting in a concentration factor of 5.33, and subsequently, centrifugation (disk-stack) is undertaken resulting in a total suspended solid content of 15% (Li et al. 2011; Olaizola and Huntley 2003; Panis and Rosales 2016). However, as the harvesting process progresses, the biochemical components including the astaxanthin can change affecting the product, especially in tropical locations with high light and heat; therefore, this stage requires further development work and optimisation.

For the cell disruption and extraction stage, this can either be conducted using a wet or dry method. To obtain dry biomass, freeze drying, spray drying, and drum drying have been employed (Kamath and Sarada 2007), but spray drying has been concluded to be the most appropriate in the case of *H. lacustris*-derived astaxanthin taking into account cost and recovery rate (95–100%), resulting in a 5% moisture content (Panis and Rosales 2016; Pérez-López et al. 2014). Care has to be taken during cell disruption and extraction not to damage the carotenoids which are vulnerable to thermal degradation and oxidation. Carotenoids such as astaxanthin are unstable due to their structural configuration (3-hydroxy, 4-keto end-group) (Mendes-Pinto et al. 2001). It has to be noted that drying biomass in conjunction with mechanical disruption results in enhanced extraction efficiency, but the energy burden increases along with the financial cost (Panis and Rosales 2016).

The aim of cell disruption and extraction is to release the astaxanthin from the thick trilaminar sheath and algaenan cell walls. Methods investigated for cell disruption and extraction of dried haematocysts are generally categorised as mechanical and nonmechanical (Fig. 6.11) (Mendes-Pinto et al. 2001; Kim et al. 2016; Shah et al. 2016; Molino et al. 2018; Liu et al. 2018; Kapoore et al. 2018). Bead milling has been suggested as the most effective and energy-efficient method for the extraction of astaxanthin, and dried biomass is used as the feedstock with a dried algal cake between 100 and 200 g/L being optimal (Greenwell et al. 2010). Alternative methods manipulating the life cycle of *H. lacustris* have been investigated by utilising cell germination (12–18 h), combined with ionic liquid treatment (Praveenkumar et al. 2015). Adding nitrate to the medium of formed aplanospores resulted in zooids being released without rigid cell walls, resulting in 19.2 pg astaxanthin/cell after 1 min ionic liquid extraction with 1-ethyl-3-methyllimidazolium ethylsulfate at room temperature (Praveenkumar et al. 2015). This method had several associated advantages including less-toxic solvent use, a lower energy input, and the avoidance of thermal stress, but future challenges to address are improving the germination rate and how to recover the expensive ionic liquids.

Today, there is a desire for natural and environmentally friendly methods of extraction to be developed because traditional solvent extraction requires large quantities of organic solvents and is labour intensive, and the labile pigments can be exposed to excessive light, heat, and oxygen (Denery et al. 2004). There is a requirement for a nontoxic method which is inexpensive in utilising a green chemistry approach. A successful alternative has been SFE, and several companies such as Phasex Corporation are now operating specifically in the SFE field for the extrac-

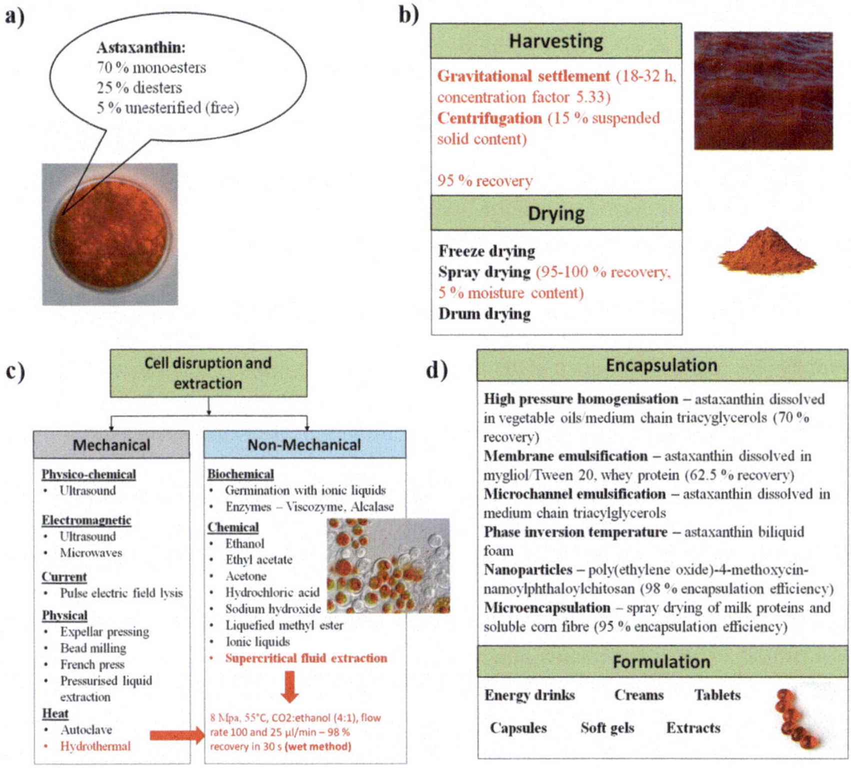

Fig. 6.11 Downstream processing for the production of astaxanthin from *H. lacustris*: (**a**) Aplanospore cells with desired astaxanthin product within the cell, (**b**) harvesting and drying, (**c**) cell disruption and extraction, and (**d**) encapsulation and formulation

tion of microalgae for the production of high-value nutraceuticals (Phasex Corporation 2015). SFE is the use of a substance such as CO_2 or water at a temperature and pressure above their critical point between the typical liquid and gas phase. Compared to most solvents, CO_2 has been found to be relatively cheap, nontoxic, chemically inert, and stable (Guedes et al. 2011). It has been found that $ScCO_2$ is industrially scalable and has been proven for coffee decaffeination, the extraction of hops, and for astaxanthin extraction from *H. lacustris* (Kwan et al. 2018).

Supercritical fluids have their own physicochemical properties similar to both gases and liquids such as high compressibility, high diffusibility, low viscosity, and a low surface tension allowing the fluid to easily diffuse through the natural extractant matrix, achieving higher-quality extractions compared to conventional liquid solvents (Pan et al. 2012). CO_2 is commonly used as the SFE solvent (Nobre et al. 2006), but CO_2 is not a good solvent for extraction with nonpolar molecules such as astaxanthin, and to improve the solvating power requires a cosolvent such as ethanol or olive oil to improve the polarity of CO_2 (Cheng et al. 2018; Wang et al. 2012). The challenge to overcome with $ScCO_2$ is to ensure high and consistent recovery rates of astaxanthin, but in some cases only 50% post cell disruption has been observed (Cheng et al. 2018; Nobre et al. 2006). Furthermore, $ScCO_2$ has a high investment cost (Kadam et al. 2013) and is a labour-intensive step of the production process requiring several hours for the extraction of astaxanthin from the biomass (Michalak and Chojnacka 2014). A way of mitigating this cost could be utilising a biorefinery approach with $ScCO_2$ for the extraction of astaxanthin and TAG simultaneously based on density sequential extraction (Kwan et al. 2018). Di Sanzo et al. (2018) extracted multiple products of interest with $ScCO_2$ with ball milling as a pretreatment with the maximum recovery of astaxanthin and lutein (98.6% and 52.3%, respectively) at 50 °C, 500 bar, and the maximum recovery of FA (93.2%) at 65 °C, 550 bar. Interestingly, at lower recovery rates, the purity of these compounds was higher (Di Sanzo et al. 2018).

To reduce the cost of the production process, wet processing methods have recently been explored. If a wet processing method is used, the extracted astaxanthin-containing biomass must be processed within a few hours to avoid spoilage (Shah et al. 2016). A new method of wet cell disruption has been developed involving hydrothermal disruption (200 °C, 10 min, 6 MPa) which has resulted in near-complete astaxanthin extraction from the biomass and is a more environmentally friendly method as water evaporation is avoided, reducing the energy input (Cheng et al. 2017b). However, it should be noted that although total astaxanthin content was similar to other treatments, this method did result in a loss of astaxanthin diesters, and changes to the stereoisochemistry were not investigated which warrants further investigation. Utilising a wet extraction method with hydrothermal disruption in combination with $ScCO_2$ extraction (8 MPa, 55 °C, with ethanol as a cosolvent, CO_2:ethanol (4:1), flow rate 100 μL/min and 25 μL/min) resulted in a 98% recovery rate of astaxanthin in a total of 30 s (Cheng et al. 2018). The high cost of cell disruption and extraction in conjunction with the process being time consuming has led to studies on cell wall-deficient mutants, MT 537 and MT 2978, which were obtained by chemical mutagenesis with reduced thickness of secondary cell walls to overcome the need for extraction, potentially reducing the costing of a cell wall disruption process (Wang et al. 2005), but further experimentation has not been undertaken, and more work is warranted before taking this research to pilot scale.

Although harvesting and extraction of astaxanthin from *H. lacustris* are well studied, the encapsulation and formulation of astaxanthin have received limited attention. The methods for encapsulation have been reviewed in detail by Khalid and Barrow (2018). Typical methods for encapsulation include high-pressure homogenisation, emulsification, phase inversion, nanoparticles, and microencapsulation (Fig. 6.11). Difficulties have been observed with maintaining the stability and functionality of astaxanthin during the final product formulation stage. There is a requirement for low energy methods for encapsulation, but scalability is the limiting factor in the success of emulsification technologies (Khalid and Barrow 2018). The stability of the encapsulated astaxanthin can be affected by the matrix composition, emulsifier type, and the stabilisers used, but the ingredients need to ensure functionality of the product and have the ability to satisfy regulatory requirements (Khalid and Barrow 2018). Further work in this area will result in a greater product range of astaxanthin products including beverages and creams (Fig. 6.11).

3.5 Contamination Threats

H. lacustris cultures obtained from the environment are often heavily contaminated by other organisms, including protozoa, other algae, fungi, and bacteria (Kim et al. 2011; González et al. 2009). Extensive successive isolation steps are required for obtaining an axenic culture which requires great expertise. To date, micropipetting, differential centrifugation, dilution, phototaxis, purification using UV light, and antibiotic treatment have been investigated (Andersen 2005; Allewaert et al. 2015). Cho et al. (2013) developed a comprehensive protocol for yielding axenic strains from environmental samples by using ultrasonication, cell sorting, and micropicking using agar, without the need for antibiotics or time-consuming micropipetting. Ultrasonication was found to reduce bacterial and fungal loading by detaching them from flocs of microalgae (Cho et al. 2013). Cho et al. (2013) reported that fluorescent-activated cell sorting (FACS) resulted in the removal of 99.5% of the bacteria, but it was difficult to completely remove the attached bacteria from all of the life stages of *H. lacustris*, especially palmelloids and aplanospores. FACS requires expensive equipment, is costly to maintain, and requires trained personnel; therefore, it is not suitable for small biotechnology companies.

H. lacustris is extremely susceptible to contamination in both the green and red stages, and predators can eliminate 90% of the *Haematococcus* biomass in <72 h (Bubrick 1991). It has been concluded that large-scale single-phase open-pond systems have proved unsatisfactory for the production of *H. lacustris* due to difficulties with contamination and control (Bubrick 1991; Margalith 1999). Contamination in microalgal cultures has become arguably the greatest threat to the industry (Day 2013; Han et al. 2013a, b). Proctor (1957c) found that of the microalgae investigated, *H. lacustris* was the most sensitive and *Chlamydomonas reinhardtii* was found to readily outcompete it within 3–5 days after inoculation, presenting a major threat of outcompeting *H. lacustris* when cultivated in open ponds in large-scale culture. Vampyrellids (common freshwater amoeba) may also present a threat to the *H. lacustris* industry, known to perforate algal cell walls to extract the cell contents (Carney and Lane 2014), increasing biomass loss and leakage of astaxanthin. The antibacterial/antiparasitic agent metronidazole has been investigated, and 3 μg/mL was determined to be effective for eliminating protozoa in *H. lacustris* cultures (Torres-Carvajal et al. 2017). To date, there have been limited publications on the issues of contamination in the large-scale production of *H. lacustris* (Poonkum et al. 2015; Torzillo et al. 2003). It has been argued that Cyanotech's red stage process of astaxanthin induction in open ponds for 5–6 days is too short for a contaminant to impact the system and the culture conditions are unsuitable for the growth of any possible contaminant (Olaizola and Huntley 2003). In industry, cases of contamination remain unreported, with the exception of Fuji Chemicals BioDome™ system in Hawaii (Algae Industry Magazine 2015). This may be due to insufficient monitoring of stocks, and commercial sensitivity definitely has a role. It has been identified in a commercial sample from AlgaeLabs Ltd. that the main microalgal contaminant was the Chlorophyte *Coelastrella* sp. on the basis of ITS ½ fragment sequencing (Dawidziuk et al. 2017).

Since 2008, there have been several publications on a new chytrid contaminant of *H. lacustris* cultures (Gutman et al. 2009; Hoffman et al. 2008), described as the most serious hurdle responsible for reductions in astaxanthin productivities and frequent culture collapse (Han et al. 2013a, b). The source of the *H. lacustris* chytrid remains unknown, but it has been identified as *Paraphysoderma sedebokerensis* and is closely related to the plant pathogen *Physoderma* (Gutman et al. 2009). Glucose in the medium was found to increase susceptibility to infection (Gutman et al. 2009; Hoffman et al. 2008), revealing a potential disadvantage of mixotrophic production. An interesting characteristic of the chytrid is the high level of resistance against desiccation, with the chytrid remaining viable even after being placed in an evacuated desiccator for 3 weeks which may allow it to spread to *H. lacustris* cultures in the air (Hoffman et al. 2008). Hoffman et al. (2008) reported that when the chytrid was placed in cultures of various strains of *H. lacustris* (20 strains tested), it infected them all in the palmelloid and aplanospore stages, including NIES-144, SAG 34/1b, SAG 192.80, CCAP 34/19, and SCCAP k-0084.

To date, only *H. lacustris* SCCAP k-0084 has been investigated in depth for its susceptibility to infection by *P. sedebokerensis* (Gutman et al. 2011, 2009; Hoffman et al. 2008). Green *H. lacustris* cultures infected by the chytrid will turn dark brown and will clump (Hoffman et al. 2008). Hoffman et al. (2008) observed that the chytrid infected palmelloids and aplanospores, but motile macrozooids remained unaffected. Agitation of cultures resulted in infection rates of up to 100% after 3 days in the green stage (Hoffman et al. 2008). Gutman et al. (2009) revealed that after 4 days of incubation, all *H. lacustris* cells in the aplanospores stage had *P. sedebokerensis* attached. *H. lacustris* cells were thought to be more susceptible to infection when the cultures were nitrogen starved (Gutman et al. 2009) and to be a representative of an unfavourable environment for *H. lacustris* when aplanospore formation occurs. Gutman et al. (2009) revealed that of the 13 Chlorophytes studied, the chytrid appeared to be *Haematococcus* specific with infection occurring both in the green and red stages. Information on the pathogen's life cycle is scarce, and little data exists on its nutritional requirements and mode of infection (Strittmatter et al. 2016). It is hypothesised that *P. sedebokerensis* transitions between a vegetative and resting phase depending on favourable or unfavourable growth conditions (Strittmatter et al. 2016). Future research on dissemination of the amoeboid and flagellated propagules of *P. sedebokerensis* warrants investigation as these have been identified as the most vulnerable to adverse environmental conditions, but the life cycle is complex and has not been fully elucidated (Strittmatter et al. 2016). Further emphasis should be on an early prevention or elimination strategy to avoid significant culture collapses of this important commercial alga. Currently, the pathogen's nutritional requirements are being determined through integrated metabolomics and transcriptomics, and *H. lacustris* cultures are being screened for random clones resistant to infection and with high astaxanthin productivities (A4F 2015).

To date, there is a lack of effective solutions to prevent or treat microbial contamination of mass cultures on a commercial scale, and most methods are reliant on microscopy and staining for early detection. Other methods have included flow cytometry, but this is a high capital investment for small microalgal biotechnology companies and cannot be used to identify microalgal contamination (Carney and Lane 2014). More recently, a high-resolution melting (HRM) analysis has been developed to detect fungal and microalgal contaminants and can identify contaminants with low levels of DNA in 5 h (2.5 ng/mL for fungi and 1.25 ng/mL for microalgae) (Dawidziuk et al. 2017). Natural, algal-mediated chemicals exist such as abscisic acid, and chemical agents have been observed to be effective against chytrids such as copper sulphate and Triton-N (Carney and Lane 2014). To date, five patents have been filed in relation to *P. sedebokerensis*, but the methods either rely on an early detection method using qPCR or epifluorescence microscopy which is not easily accessible in the field or easily available to small biotechnology companies, or through the use of fungicides, but the efficacy is unknown (WO2013127280A1, CN106755393A, AU2013353154B2, CN103857785A, CN202519240U). Developing methods of control which could lead to the eradication of this 'pest' are critical. In addition, crop protection strategies against other grazers and predators are also required which remain environmentally inert.

3.6 *The Question of Economics*

Producing astaxanthin from *H. lacustris* is more costly than other species of algae such as *Spirulina* due to the necessity for PBRs in the green stage, high electricity consumption, and the requirement for extraction of astaxanthin from the thick-walled cells, adding to the overall production cost (Issarapayup et al. 2011; Li et al. 2011). As outlined above in terms of overall economics, *H. lacustris*-derived astaxanthin is more costly to produce than the synthetic form, with production costs for *H. lacustris*-derived astaxanthin at large companies such as Cyanotech, Alga Technologies, and Mera Pharmaceuticals estimated at up to \$3600 kg^{-1} (Li et al. 2011).

Several studies have produced detailed techno-economic assessments for the economical production of astaxanthin in different locations: Europe, the Middle East, and the Far East. When *H. lacustris* was cultivated in a flat plate airlift PBR, it was determined that the reactor size and the cost of production were directly correlated, but a biomass reduction was simultaneously observed (Issarapayup et al. 2011). Issarapayup et al. (2011) identified that one of the major costs of cultivation of *H. lacustris* was the high electricity costs (40% of total operating costs); however, it was determined that artificial illumination results in a 107% higher productivity. Using a life cycle assessment (LCA) in Ireland, it was determined that electricity dominated the environmental burdens (Pérez-López et al. 2014). In Ireland, astaxanthin can only be produced outdoors for 5 months of the year with indoor production essential for the remainder of the calendar year (Pérez-López et al. 2014). Lower light intensities and changes in bioreactor design could aid in reducing the cost of the production process and limit the environmental burden. Reusing the medium was not an effective option for reducing costs because the productivity of the system decreased by 30% (Issarapayup et al. 2011), but utilising wastewater could offer a cost-effective alternative.

A process design and an economic assessment were conducted for astaxanthin production in Lebanon, and it was determined that at a market price higher than US \$1500 kg^{-1}, a production process could be economically feasible, and if the astaxanthin market price was US \$6000 kg^{-1}, a payback period of 1 year and a return on capital employed (ROCE) of 113% were possible (Zgheib et al. 2018). This was based on an annual production of 2592 kg of astaxanthin taking into account the harvesting costs (gravity sedimentation and disk-stack centrifuge), cell disruption (bead miller), drying (spray dryer), extraction (supercritical CO_2), and the fact that the residual biomass would be fed into an anaerobic digester and used for biogas and as a biofertiliser (Zgheib et al. 2018). The cost of the upstream processing was not taken into account and, therefore, represents an extreme underestimate.

Based on a modelling approach simulating large-scale production of astaxanthin, it has been proposed that the cost of producing astaxanthin in Livadeia, Greece, and Amsterdam, Netherlands, could be €1536 kg^{-1} astaxanthin (426 kg/year for a 2 ha site) and €6403 kg^{-1} (143 kg/year for a 2 ha site), respectively (Panis and Rosales 2016). For the upstream process, horizontal tubular PBRs (5 cm diameter) were chosen for the green stage (pH 7.5) and open raceway ponds for the red stage (pH 8.0) with an assumption that each system occupied 1 ha. For the CO_2 supply, flue and flaring gases were selected (2.2 mg/L CO_2—5–10% CO_2). For the downstream process, harvesting encompassed gravitational settlement followed by subsequent disk-stack centrifugation. The biomass was then spray dried resulting in a moisture content of 5%, and subsequent supercritical fluid extraction was applied (60 °C, 30 MPa with ethanol as a cosolvent (9.4%)) resulting in a 10–20% astaxanthin oleoresin. It was elucidated from the model that June–August were the most productive months in terms of astaxanthin production. It was determined that temperature was the most sensitive parameter for astaxanthin productivity. It was found that using the residual biomass as a biofertiliser (\$30–60 kg^{-1}) could be a more economical process. Water consumption was found to be high, and water recycling in the red stage has been suggested, resulting in an improvement in the economics and footprint (Panis and Rosales 2016). Tubular PBR cooling was observed to be the most energy-intensive process (Panis and Rosales 2016). It was observed that cultivation in the PBRs was the highest cost of the process with tubular PBR cooling being the major costing. Even with these cost savings, it was concluded that *H. lacustris*-derived astaxanthin in Greece and the Netherlands could not compete with synthetic production of astaxanthin (€880 kg^{-1}) for the feed market (Panis and Rosales 2016). It was further concluded that cultivation in equatorial regions such as Hawaii and Israel was more favourable (Panis and Rosales 2016).

For *H. lacustris*-derived astaxanthin to be cost competitive with the synthetic form, significant developments in the streamlining of the upstream (culturing) and downstream processing (dewatering, cell disruption, and extraction) will be required. Currently, the easiest way to reduce overall cost is to transfer production to a low-cost site, which is why a large number of companies are operating in China such as BGG. Li et al. (2011) published a report on the potential production cost of *H. lacustris*-derived astaxanthin outdoor operation in Kunming, China, using airlift tubular bioreactors and a raceway pond. It was assumed 33 tons of biomass per year (2.5% astaxanthin DW) could be produced equating to 914 kg of astaxanthin, and assuming a 10-year depreciation on fixed capital cost, the direct production cost for biomass and astaxanthin was estimated at \$18 kg^{-1} and £718 kg^{-1}, respectively.

In addition to the production process being economical, it also needs to be environmentally sustainable reducing the carbon footprint. A key way to do this is through energy reduction, utilising renewable energy sources, and through reductions in water consumption. To date, only one LCA has been published on *H. lacustris*-derived astaxanthin, and electricity represented the major contributor to the environmental burden, particularly in the green stage for cultivation (Pérez-López et al. 2014). Companies such as AlgaTechnologies have implemented solar power as the primary source of energy (specified at 250 W, 15% conversion efficiency, 1.65 m^2) (AlgaTechnologies 2015; Panis and Rosales 2016). In terms of water consumption, industrial cultivation of *H. lacustris* for astaxanthin requires 1000–1500 tons of freshwater for the production of 1 ton of *H. lacustris* biomass, but attached cultivation in the red stage has already been showcased to be effective for astaxanthin induction, and the water consumption is 30% of that in an open pond (Wan et al. 2014a).

Utilising a biorefinery approach using a 'high-value product first' principle where astaxanthin, phytosterols, and PUFAs are all produced (Bilbao et al. 2016), with the residual biomass being used as a protein source or a biofertiliser, could offer a future sustainable production process. Utilising waste streams such as carbon dioxide from flue gases and carbonates from the soft drink industry could offer cost reductions as well as environmentally sustainable solutions. It has been identified that if flue gases are used, a buffering system is required to the conventional and expensive HEPES, and Choi et al. (2017) determined that utilising a bicarbonate and phosphate buffer could be suitable using 10 mM KOH and 0.1 mM phosphate enabling a pH of 7 to be maintained.

4 Exploiting an Alternative *H. lacustris* Life Cycle Stage and the Future Direction for the Astaxanthin Industry

To target the global aquaculture market, around 130–1000 tons of astaxanthin is required annually to feed the salmonids with a diet containing 50 mg astaxanthin/kg (Zhang et al. 2009; Solovchenko and Chekanov 2014). Cerón et al. (2007) stated that the evaluation of the quality of the *H. lacustris* biomass should take into account both astaxanthin content and the FA profile. It has been identified that aquaculture-derived salmon contains ten times less astaxanthin than wild salmon and the bioaccessibility of astaxanthin in wild salmon is greater (Chitchumroonchokchai and Failla 2017). It has been found that during in vitro digestion of salmon flesh, >80% of the astaxanthin is recovered in the human body (Chitchumroonchokchai and Failla 2017).

Sommer et al. (1991) had clearly revealed that *H. lacustris* intact cells failed to achieve satisfactory pigmentation in salmonids and had to be disrupted which was further validated by Mendes-Pinto et al. (2001). Cell disruption in a scalable pressure treatment system (5000 psi for bioavailable astaxanthin) was deemed effective for lysing *H. lacustris* cells, and ethoxyquin was added to minimise oxidation (Young et al. 2017). It was shown that the lysed cells should be spray dried, incorporated into the feed, and frozen at −20 °C before being fed to salmonids for commercially acceptable pigmentation (Young et al. 2017). However, it is well known that the cell disruption and extraction costs are high. To successfully incorporate *H. lacustris*-derived astaxanthin into aquaculture feeds, a series of developments need to be made. One idea is to increase the biomass and astaxanthin productivities in the two-stage system through PBR development, optimised cultivation, and induction parameters in conjunction with operating in a low-cost locality for an economically favourable process. In addition, a biorefinery process could be implemented where the PUFAs are sold in an aquaculture feed in conjunction with a defatted microalga for partial fishmeal protein replacement (12.5% has been showcased in shrimp feeding trials) (Shah et al. 2018).

Another alternative is exploiting the accumulation of astaxanthin in red motile macrozooids. Synthesis of astaxanthin is not induced by the cessation of cell division and is independent of aplanospore formation (Butler et al. 2017; Hagen et al. 2000). Red motile macrozooids have been formed that contain up to 2.74% DW astaxanthin (78.4% of total carotenoids) and are rich in PUFAs; thus, it is envisioned that they could be directly fed to aquaculture species' (Butler et al. 2017). This is a process that is being explored at Brevel Ltd., Israel (Fig. 6.12). Currently, the biomass yield is low, and therefore, this could be incorporated into the one-stage process as devised by Del Río et al. (2008). Adding value to the *H. lacustris* biomass through the development of an oral vaccine such as against furunculosis in salmonids and immune-boosting supplements to reduce the mortality rates (which can be as high as 27%) (Overton et al. 2018) could increase the supply of *H. lacustris* to the aquaculture sector. Selenium is an example of an immune-boosting supplement and is incorporated

Fig. 6.12 Brevel Ltd. has a technology based on indoor internally illuminated photobioreactors which have a significantly lower surface area and are thus less prone to contamination and have better monitor and control processes. Light is either natural sunlight concentrated and transported into the cultivation tank via optical fibres or based on artificial lighting. Natural sunlight is spectrally filtered at the source in order to reduce heat and damage to cells by IR and UV, and thus, temperature control requirements are significantly reduced. The enclosed and fully controlled environment makes the possibility of mixotrophic cultivation viable and thus could further significantly increase yields and reduce cultivation costs. Under current estimates, Brevel will be capable of producing algal astaxanthin for the salmonid market by 2020 at a comparable retail price to the current synthetic product

into selenoproteins that have antioxidant and anti-inflammatory effects, and selenium is also highly protective against mercury which has been found to bioaccumulate in fish (Ralston et al. 2014; Rayman 2012). Micronutrients such as selenium could be added directly to the medium to incorporate into the cell, and decreases in biomass have not been observed at 3 mg/L (17.3 μM), and total selenium could accumulate up to 646 μg/g with 380 μg/g organic selenium (Zheng et al. 2017). Many developments have occurred in increasing the biomass/astaxanthin productivities and cost reductions from *H. lacustris*-derived astaxanthin, and the future bodes well for the replacement of synthetic astaxanthin by this biobased source.

References

't Lam, G. P., Vermuë, M. H., Eppink, M. H. M., Wijffels, R. H., & Van Den Berg, C. (2018). Multi-product microalgae biorefineries: From concept towards reality. *Trends in Biotechnology, 36*, 216–227.

A4F. (2015). *Project 1: Towards controlling chytrid pathogens in industrial cultures of Haematococcus sp*. Retrieved May 1, 2019, from https://msc-alff.org/about/alff-students/noreen-hiegle/.

Aflalo, C., Meshulam, Y., Zarka, A., & Boussiba, S. (2007). On the relative efficiency of two- vs. one-stage production of astaxanthin by the green alga *Haematococcus pluvialis*. *Biotechnology and Bioengineering, 98*, 300–305.

Alam, M. A., & Wang, Z. (Eds.). (2019). *Microalgae biotechnology for development of biofuel and wastewater treatment* (pp. 1–655). New York, NY: Springer.
Al-Bulishi, M. S. M. (2015). Health aspects of astaxanthin: A review. *Canadian Journal of Clinical Nutrition, 3*, 71–78.
Algae Industry Magazine. (2015). *Algae Industry Magazine*. Retrieved October 10, 2015, from http://www.algaeindustrymagazine.com/?s=astaxanthin.
Algatechnologies. (2015). *About Us*. Retrieved June 17, 2015, from http://www.algatech.com/about.asp?cat=004.
Allewaert, C., et al. (2015). Species diversity in European *Haematococcus pluvialis* (Chlorophyceae, Volvocales). *Phycologia, 54*, 583.
Allewaert, C. C., Vanormelingen, P., Daveloose, I., Verstraete, T., & Vyverman, W. (2017). Intraspecific trait variation affecting astaxanthin productivity in two *Haematococcus* (Chlorophyceae) species. *Algal Research, 21*, 191–202.
Andersen, R. A. (Ed.). (2005). *Algal culturing techniques*. Amsterdam: Elsevier.
Aragreen. (2015). *Aragreen*. Retrieved May 20, 2015, from http://www.aragreen.com/home.asp.
Asker, D. (2017). Isolation and characterization of a novel, highly selective astaxanthin-producing marine bacterium. *Journal of Agricultural and Food Chemistry, 65*, 9101–9109.
Bauman, N., Akella, S., Hann, E., Morey, R., Schwartz, A. S., Brown, R., & Richardson, T. H. (2018). Next-generation sequencing of *Haematococcus lacustris* reveals an extremely large 1.35-megabase chloroplast genome. *Genome Announcements, 6*, e00181–e001818.
BCC Research. (2015). *Boom in astaxanthin supplements boosting global carotenoid market, according to BCC Research*. Retrieved July 23, 2015, from http://www.bccresearch.com/press-room/fod/boom-in-astaxanthin-supplements-boosting-global-carotenoid-market-according-to-bcc-research.
BCC Research. (2018). *The global market for carotenoids*. Retrieved April 23, 2019, from https://www.bccresearch.com/market-research/food-and-beverage/the-global-market-for-carotenoids.html.
Bilbao, P. G. S., Damiani, C., Salvador, G. A., & Leonardi, P. (2016). *Haematococcus pluvialis* as a source of fatty acids and phytosterols: Potential nutritional and biological implications. *Journal of Applied Phycology, 28*, 3283–3294.
Bolin, A. P., Macedo, R. C., Marin, D. P., Barros, M. P., & Otton, R. (2010). Astaxanthin prevents in vitro auto-oxidative injury in human lymphocytes. *Cell Biology and Toxicology, 26*, 457–467.
Boussiba, S., & Vonshak, A. (1991). Astaxanthin accumulation in the green alga *Haematococcus pluvialis*. *Plant & Cell Physiology, 32*, 1077–1082.
Brinda, B. R., Sarada, R., Sandesh Kamath, B., & Ravishankar, G. A. (2004). Accumulation of astaxanthin in flagellated cells of *Haematococcus pluvialis* - Cultural and regulatory aspects. *Current Science, 87*, 1290–1295.
Brodie, J., Chan, C. X., De Clerck, O., Cock, J. M., Coelho, S. M., Gachon, C., Grossman, A. R., Mock, T., Raven, J. A., Smith, A. G., Yoon, H. S., & Bhattacharya, D. (2017). The algal revolution. *Trends in Plant Science, 22*, 726–738.
Bubrick, P. (1991). Production of astaxanthin from Haematococcus. *Bioresource Technology, 38*, 237–239.
Buchheim, M. A., Sutherland, D. M., Buchheim, J. A., & Wolf, M. (2013). The blood alga: Phylogeny of *Haematococcus* (Chlorophyceae) inferred from ribosomal RNA gene sequence data. *European Journal of Phycology, 48*, 318–329.
Burchardt, L., Balcerkiewicz, S., Kokocinski, M., Samardakiewicz, S., & Adamski, Z. (2006). Occurrence of *Haematococcus pluvialis* Flotow emend. Wille in a small artificial pool on the university campus of the collegium biologicum in Poznan (Poland). *Resource Conservation, 1*, 163–166.
Bustos-Garza, C., Yáñez-Fernández, J., & Barragán-Huerta, B. E. (2013). Thermal and pH stability of spray-dried encapsulated astaxanthin oleoresin from *Haematococcus pluvialis* using several encapsulation wall materials. *Food Research International, 54*, 641–649.

Butler, T., McDougall, G., Campbell, R., Stanley, M., & Day, J. (2017). Media screening for obtaining *Haematococcus pluvialis* Red motile macrozooids rich in astaxanthin and fatty acids. *Biology (Basel), 7*, 2.

Capelli, B. (2018). *AstaZine™ natural astaxanthin: The supplement you can feel.*

Capelli, B., Bagchi, D., & Cysewski, G. R. (2013a). Synthetic astaxanthin is significantly inferior to algal-based astaxanthin as an antioxidant and may not be suitable as a human nutraceutical supplement. *Nutrafoods, 12*, 145–152.

Capelli, B., Jenkins, U., & Cysewski, G. R. (2013b). Role of astaxanthin in sports nutrition. In *Nutrition and enhanced sports performance* (pp. 465–471). New York, NY: Academic Press.

Carney, L. T., & Lane, T. W. (2014). Parasites in algae mass culture. *Frontiers in Microbiology, 5*, 1–8.

CCAP. (2015). *Haematococcus pluvialis*. Retrieved September 10, 2015, from http://www.ccap.ac.uk/results2014.php.

Cerón, M. C., García-Malea, M. C., Rivas, J., Acien, F. G., Fernandez, J. M., Del Río, E., Guerrero, M. G., & Molina, E. (2007). Antioxidant activity of *Haematococcus pluvialis* cells grown in continuous culture as a function of their carotenoid and fatty acid content. *Applied Microbiological Biotechnology, 74*, 1112–1119.

Chekanov, K., Lobakova, E., Selyakh, I., Semenova, L., Sidorov, R., & Solovchenko, A. (2014). Accumulation of astaxanthin by a new *Haematococcus pluvialis* strain BM1 from the white sea coastal rocks (Russia). *Marine Drugs, 12*, 4504–4520.

Chekanov, K., Schastnaya, E., Solovchenko, A., & Lobakova, E. (2017). Effects of CO_2 enrichment on primary photochemistry, growth and astaxanthin accumulation in the chlorophyte *Haematococcus pluvialis*. *Journal of Photochemistry and Photobiology B: Biology, 171*, 58–66.

Chen, T., & Wang, Y. (2013). Optimized astaxanthin production in *Chlorella zofingiensis* under dark condition by response surface methodology. *Food Science and Biotechnology, 22*, 1–8.

Chen, J., Liu, L., & Wei, D. (2017). Enhanced production of astaxanthin by *Chromochloris zofingiensis* in a microplate-based culture system under high light irradiation. *Bioresource Technology, 245*, 518–529.

Cheng, J., Li, K., Yang, Z., Zhou, J., & Cen, K. (2016). Enhancing the growth rate and astaxanthin yield of *Haematococcus pluvialis* by nuclear irradiation and high concentration of carbon dioxide stress. *Bioresource Technology, 204*, 49–54.

Cheng, J., Li, K., Zhu, Y., Yang, W., Zhou, J., & Cen, K. (2017a). Transcriptome sequencing and metabolic pathways of astaxanthin accumulated in *Haematococcus pluvialis* mutant under 15% CO_2. *Bioresource Technology, 228*, 99–105.

Cheng, X., Riordon, J., Nguyen, B., Ooms, M. D., & Sinton, D. (2017b). Hydrothermal disruption of algae cells for astaxanthin extraction. *Green Chemistry, 19*, 106–111.

Cheng, X., Qi, Z., Burdyny, T., Kong, T., & Sinton, D. (2018). Low pressure supercritical CO_2 extraction of astaxanthin from *Haematococcus pluvialis* demonstrated on a micro fluidic chip. *Bioresource Technology, 250*, 481–485.

Chitchumroonchokchai, C., & Failla, M. L. (2017). Bioaccessibility and intestinal cell uptake of astaxanthin from salmon and commercial supplements. *Food Research International, 99*, 936–943.

Cho, D. H., Ramanan, R., Kim, B. H., Lee, J., Kim, S., Yoo, C., Choi, G. G., Oh, H. M., & Kim, H. S. (2013). Novel approach for the development of axenic microalgal cultures from environmental samples. *Journal of Phycology, 49*(4), 802–810.

Choi, Y. E., & Park, J. M. (2002). Evaluation of factors promoting astaxanthin production by a unicellular green alga, *Haematococcus pluvialis*, with fractional factorial design. *Biotechnology Progress, 18*, 1170–1175.

Choi, Y. E., Yun, Y. S., Park, J. M., & Yang, J. W. (2011). Determination of the time transferring cells for astaxanthin production considering two-stage process of *Haematococcus pluvialis* cultivation. *Bioresource Technology, 102*, 11249–11253.

Choi, Y., Joun, J., Lee, J., Hong, M., Pham, H., Chang, W., & Sim, S. (2017). Development of large-scale and economic pH control system for outdoor cultivation of microalgae *Haematococcus pluvialis* using industrial flue gas. *Bioresource Technology, 244*, 1235–1244.

Choubert, G., & Heinrich, O. (1993). Carotenoid pigments of the green alga *Haematococcus pluvialis*: Assay on rainbow trout, *Oncorhynchus mykiss*, pigmentation in comparison with synthetic astaxanthin and canthaxanthin. *Aquaculture, 112*, 217–226.

Christian, D., Zhang, J., Sawdon, A. J., & Peng, C. (2018). Enhanced astaxanthin accumulation in *Haematococcus pluvialis* using high carbon dioxide concentration and light illumination. *Bioresource Technology, 256*, 548–551.

Chunhui, Z., Jianguo, L. I. U., & Litao, Z. (2017). Cell cycles and proliferation patterns in *Haematococcus pluvialis*. *Chinese Journal of Oceanology and Limnology, 35*, 1205–1211.

Cifuentes, A. S., González, M. A., Vargas, S., Hoeneisen, M., & González, N. (2003). Optimization of biomass, total carotenoids and astaxanthin production in *Haematococcus pluvialis* Flotow strain Steptoe (Nevada, USA) under laboratory conditions. *Biological Research, 36*, 343–357.

Collins, A. M., Jones, H. D. T., Han, D., Hu, Q., Beechem, T. E., & Timlin, J. A. (2011). Carotenoid distribution in living cells of *Haematococcus pluvialis* (Chlorophyceae). *PLoS One, 6*, 1–7.

Comhaire, F. H., El Garem, Y., Mahmoud, A., Eertmans, F., & Schoonjans, F. (2005). Combined conventional/antioxidant "Astaxanthin" treatment for male infertility: A double blind randomized trial. *Asian Journal of Andrology, 7*, 257–262. https://doi.org/10.1111/j.1745-7262.2005.00047.x.

Cunningham, F. X., & Gantt, E. (2011). Elucidation of the pathway to astaxanthin in the flowers of *Adonis aestivalis*. *Plant Cell, 23*, 3055–3069.

Dalay, M. C., Imamoglu, E., & Demirel, Z. (2007). Agricultural fertilizers as economical alternative for cultivation of *Haematococcus pluvialis*. *Journal of Microbiology and Biotechnology, 17*, 393–397.

Damiani, M. C., Leonardi, P. I., Pieroni, O. I., & Cáceres, E. J. (2006). Ultrastructure of the cyst wall of *Haematococcus pluvialis* (Chlorophyceae): Wall development and behaviour during cyst germination. *Phycologia, 45*, 616–623.

Damiani, M. C., Popovich, C. A., Constenla, D., & Leonardi, P. I. (2010). Lipid analysis in *Haematococcus pluvialis* to assess its potential use as a biodiesel feedstock. *Bioresource Technology, 101*, 3801–3807.

Dawidziuk, A., Popiel, D., Luboinska, M., & Grzebyk, M. (2017). Assessing contamination of microalgal astaxanthin producer *Haematococcus* cultures with high-resolution melting curve analysis. *Journal of Applied Genetics, 58*, 277–285.

Day, J. G. (2013). Grazers: The overlooked threat to the sustained production of future algal biofuels. *Biofuels, 4*, 459–461.

Del Río, E., Acién, F. G., García-Malea, M. C., Rivas, J., Molina-Grima, E., & Guerrero, M. G. (2005). Efficient one-step production of astaxanthin by the microalga *Haematococcus pluvialis* in continuous culture. *Biotechnology and Bioengineering, 91*, 808–815.

Del Río, E., Acién, F. G., García-Malea, M. C., Rivas, J., Molina-Grima, E., & Guerrero, M. G. (2008). Efficiency assessment of the one-step production of astaxanthin by the microalga *Haematococcus pluvialis*. *Biotechnology and Bioengineering, 100*, 397–402.

Denery, J. R., Dragull, K., Tang, C. S., & Li, Q. X. (2004). Pressurized fluid extraction of carotenoids from *Haematococcus pluvialis* and *Dunaliella salina* and kavalactones from *Piper methysticum*. *Analytica Chimica Acta, 501*, 175–181.

Di Sanzo, G., Mehariya, S., Martino, M., Larocca, V., Casella, P., Chianese, S., Musmarra, D., Balducchi, R., & Molino, A. (2018). Supercritical carbon dioxide extraction of astaxanthin, lutein, and fatty acids from *Haematococcus pluvialis* microalgae. *Marine Drugs, 16*, 1–18.

Domínguez-Bocanegra, A. R., Ponce-Noyola, T., & Torres-Muñoz, J. A. (2007). Astaxanthin production by Phaffia rhodozyma and *Haematococcus pluvialis*: A comparative study. *Applied Microbiology and Biotechnology, 75*, 783–791.

Dore, J. E., & Cysewski, G. R. (2003). *Haematococcus algae meal as a source of natural astaxanthin for aquaculture feeds*. Kailua-Kona: Cyanotech Corporation.

Doyle, J. J., & Doyle, J. (1990). Isolation of plant DNA from fresh tissue. *Focus, 12*, 13.

Dragos, N., Bercea, V., Bica, A., Druga, B., Nicoara, A., & Coman, C. (2010). Astaxanthin production from a new strain of *Haematococcus pluvialis* grown in batch culture BT. *Annals of the Romanian Society for Cell Biology, XV*, 353–361.

Droop, M. R. (1955). Some factors governing encystment in *Haematococcus pluvialis*. *Archiv für Mikrobiologie, 21*, 267–272.

Droop, M. R. (1956). *Haematococcus pluvialis* and its allies. I. The sphaerellaceae. *Revue Algologique, 2*, 53–71.

Elgarem, Y., Lignell, A., & Comhaire, F. H. (2002). Supplementation with astaxanthin (astacarox) improves semen quality in infertile men. In *Proceedings of the 13th International Carotenoid Symposium, Honolulu, HI* (pp. 180–197).

Elliot, A. M. (1934). Morphology and life history of *Haematococcus pluvialis*. *Archiv für Protistenkunde, 82*, 250–272.

Eui, M., Il, H., Seok, H., Hwang, S., & Joon, Y. (2018). Rapid selection of astaxanthin-hyperproducing *Haematococcus* mutant via azide-based colorimetric assay combined with oil-based astaxanthin extraction. *Bioresource Technology, 267*, 175–181.

Evens, T. J., Niedz, R. P., & Kirkpatrick, G. J. (2008). Temperature and irradiance impacts on the growth, pigmentation and photosystem II quantum yields of *Haematococcus pluvialis* (Chlorophyceae). *Journal of Applied Phycology, 20*, 411–422.

Fábregas, J., Dominguez, A., & Regueiro, M. (2000). Optimization of culture medium for the continuous cultivation of the microalga *Haematococcus pluvialis*. *Applied Microbiology and Biotechnology, 53*, 530–535.

Fábregas, J., Otero, A., Maseda, A., & Domínguez, A. (2001). Two-stage cultures for the production of Astaxanthin from *Haematococcus pluvialis*. *Journal of Biotechnology, 89*, 65–71.

Fassett, R. G., & Coombes, J. S. (2012). Astaxanthin in cardiovascular health and disease. *Molecules, 17*, 2030–2048.

Feng, L. H., Xiaonan, L., Xuecheng, Z., Bangxiang, Z., & Yating, H. (2017). Cloning and transcription analysis of the nitrate reductase gene from *Haematococcus pluvialis*. *Biotechnology Letters, 39*, 589–597.

de la Fuente, J. L., Rodríguez-Sáiz, M., Schleissner, C., Díez, B., Peiro, E., & Barredo, J. L. (2010). High-titer production of astaxanthin by the semi-industrial fermentation of *Xanthophyllomyces dendrorhous*. *Journal of Biotechnology, 148*, 144–146.

Fujii, K., Imazato, E., Nakashima, H., Ooi, O., & Saeki, A. (2006). Isolation of the non-fastidious microalga with astaxanthin-accumulating property and its potential for application to aquaculture. *Aquaculture, 261*, 285–293.

Gacheva, G., Dimitrova, P., & Pilarski, P. (2015). New strain *Haematococcus cf. pluvialis* Rozhen-12 - Growth, biochemical characteristics and future perspectives. *Genetics and Plant Physiology, 5*, 29–38.

Galarza, J. I., Gimpel, J. A., Rojas, V., & Arredondo-vega, B. O. (2018). Over-accumulation of astaxanthin in *Haematococcus pluvialis* through chloroplast genetic engineering. *Algal Research, 31*, 291–297.

Gao, Z., Meng, C., Zhang, X., Xu, D., Miao, X., Wang, Y., Yang, L., Lv, H., Chen, L., & Ye, N. (2012). Induction of salicylic acid (SA) on transcriptional expression of eight carotenoid genes and astaxanthin accumulation in *Haematococcus pluvialis*. *Enzyme and Microbial Technology, 51*, 225–230.

García-Malea López, M. C., Del Río Sánchez, E., Casas López, J. L., Acién Fernández, F. G., Fernández Sevilla, J. M., Rivas, J., Guerrero, M. G., & Molina Grima, E. (2006). Comparative analysis of the outdoor culture of *Haematococcus pluvialis* in tubular and bubble column photobioreactors. *Journal of Biotechnology, 123*, 329–342.

García-Malea, M. C., Brindley, C., Del Río, E., Acién, F. G., Fernández, J. M., & Molina, E. (2005). Modelling of growth and accumulation of carotenoids in *Haematococcus pluvialis* as a function of irradiance and nutrients supply. *Biochemical Engineering Journal, 26*, 107–114.

García-Malea, M. C., Acién, F. G., Fernández, J. M., Cerón, M. C., & Molina, E. (2006). Continuous production of green cells of *Haematococcus pluvialis*: Modeling of the irradiance effect. *Enzyme and Microbial Technology, 38*, 981–989.

García-Malea, M. C., Acién, F. G., Del Río, E., Fernández, J. M., Cerón, M. C., Guerrero, M. G., & Molina-Grima, E. (2009). Production of astaxanthin by *Haematococcus pluvialis*: Taking the one-step system outdoors. *Biotechnology and Bioengineering, 102*, 651–657.

Gassel, S., Schewe, H., Schmidt, I., Schrader, J., & Sandmann, G. (2013). Multiple improvement of astaxanthin biosynthesis in *Xanthophyllomyces dendrorhous* by a combination of conventional mutagenesis and metabolic pathway engineering. *Biotechnology Letters, 35*, 565–569.

Giannelli, L., Yamada, H., Katsuda, T., & Yamaji, H. (2015). Effects of temperature on the astaxanthin productivity and light harvesting characteristics of the green alga *Haematococcus pluvialis*. *Journal of Bioscience and Bioengineering, 119*, 345–350.

Göksan, T., Ak, I., & Gökpinar, Ş. (2010). An alternative approach to the traditional mixotrophic cultures of *Haematococcus pluvialis* flotow (Chlorophyceae). *Journal of Microbiology and Biotechnology, 20*, 1276–1282.

Gómez, P. I., Inostroza, I., Pizarro, M., & Pérez, J. (2013). From genetic improvement to commercial-scale mass culture of a Chilean strain of the green microalga *Haematococcus pluvialis* with enhanced productivity of the red ketocarotenoid astaxanthin. *AoB Plants, 5*, 1–7.

Gong, X., & Chen, F. (1997). Rapid detection of heterotrophic growth of *Haematococcus pluvialis* using indirect conductimetry. *Biotechnology Techniques, 11*, 841–844.

González, M. A., Cifuentes, A. S., & Gómez, P. I. (2009). Growth and total carotenoid content in four Chilean strains of *Haematococcus pluvialis* flotow, under laboratory conditions. *Gayana Botánica, 66*, 58–70.

Granata, T. (2017). Dependency of microalgal production on biomass and the relationship to yield and bioreactor scale-up for biofuels: A statistical analysis of 60+ years of algal bioreactor data. *Bioenergy Research, 10*, 267–287.

Greenwell, H. C., Laurens, L. M. L., Lovitt, R. W., Flynn, K. J., & Shields, R. J. (2010). Placing microalgae on the biofuels priority list: A review of the technological challenges. *Journal of The Royal Society Interface, 7*, 703–726.

Groben, R. (2007). *Phytoplankton and nutrient analysis of a nuclear fuel-storage pond at Sellafield.*

Grünewald, K., Hagen, C., & Braune, W. (1997). Secondary carotenoid accumulation in flagellates of the green alga *Haematococcus lacustris*. *European Journal of Phycology, 32*, 387–392.

Grunewald, K., Hirschberg, J., & Hagen, C. (2001). Ketocarotenoid biosynthesis outside of plastids in the unicellular green alga *Haematococcus pluvialis*. *The Journal of Biological Chemistry, 276*, 6023–6029.

Gu, W., Li, H., Zhao, P., Yu, R., Pan, G., Gao, S., Xie, X., Huang, A., He, L., & Wang, G. (2014). Quantitative proteomic analysis of thylakoid from two microalgae (*Haematococcus pluvialis* and *Dunaliella salina*) reveals two different high light-responsive strategies. *Scientific Reports, 4*, 6661.

Guedes, A. C., Amaro, H. M., & Malcata, F. X. (2011). Microalgae as sources of carotenoids. *Marine Drugs, 9*, 625–644.

Guerin, M., Huntley, M. E., & Olaizola, M. (2003). *Haematococcus* astaxanthin: Applications for human health and nutrition. *Trends in Biotechnology, 21*, 210–216.

Guiry, M. D. (2010). *AlgaeBase*. Retrieved June 12, 2015, from http://www.algaebase.org.

Gutiérrez, C. L., Gimpel, J., Escobar, C., Marshall, S. H., & Henríquez, V. (2012). Chloroplast genetic tool for the green microalgae *Haematococcus pluvialis* (chlorophyceae, volvocales). *Journal of Phycology, 48*, 976–983.

Gutman, J., Zarka, A., & Boussiba, S. (2009). The host-range of *Paraphysoderma sedebokerensis*, a chytrid that infects *Haematococcus pluvialis*. *European Journal of Phycology, 44*, 509–514.

Gutman, J., Zarka, A., & Boussiba, S. (2011). Evidence for the involvement of surface carbohydrates in the recognition of *Haematococcus pluvialis* by the parasitic blastoclad *Paraphysoderma sedebokerensis*. *Fungal Biology, 115*, 803–811.

Gwak, Y., Hwang, Y.-S., Wang, B., Kim, M., Jeong, J., Lee, C.-G., Hu, Q., Han, D., & Jin, E. (2014). Comparative analyses of lipidomes and transcriptomes reveal a concerted action of multiple defensive systems against photooxidative stress in *Haematococcus pluvialis*. *Journal of Experimental Botany, 65*, 4317–4334.

Ha, P. T., Hoang, N. H., Lien, N. T. K., Phuong, N. T. D., Huy, N. Q., & Thoa, N. K. (2018). Selection of bacterial strains belonging to the astaxanthin producing *Paracoccus* genus. *Vietnam Journal of Biotechnology, 16*(3), 565–572.

Hagen, C., Grünewald, K., Schmidt, S., & Müller, J. (2000). Accumulation of secondary carotenoids in flagellates of *Haematococcus pluvialis* (Chlorophyta) is accompanied by an increase in per unit chlorophyll productivity of photosynthesis. *European Journal of Phycology, 35*, 75–82.

Hagen, C., Grünewald, K., Xyländer, M., & Rothe, E. (2001). Effect of cultivation parameters on growth and pigment biosynthesis in flagellated cells of *Haematococcus pluvialis*. *Journal of Applied Phycology, 13*, 79–87.

Hagen, C., Siegmund, S., & Braune, W. (2002). Ultrastructural and chemical changes in the cell wall of *Haematococcus pluvialis* (Volvocales, Chlorophyta) during aplanospore formation. *European Journal of Phycology, 37*, 217.

Han, D., Wang, J., Sommerfeld, M., & Hu, Q. (2012). Susceptibility and protective mechanisms of motile and non motile cells of *Haematococcus pluvialis* (Chlorophyceae) to photooxidative stress1. *Journal of Phycology, 48*, 693–705.

Han, D., Li, Y., & Hu, Q. (2013a). Astaxanthin in microalgae: Pathways, functions and biotechnological implications. *Algae, 28*, 131–147.

Han, D., Li, Y., & Hu, Q. (2013b). Biology and commercial aspects of *Haematococcus pluvialis*. In *Handbook of microalgal culture: Applied phycology and biotechnology* (2nd ed., pp. 388–405). Oxford: Wiley-Blackwell.

Harker, M., & Tsavalos, A. J. (1996). Autotrophic growth and carotenoid production of *Haematococcus pluvialis* in a autotrophic growth and carotenoid production of *Haematococcus pluvialis* in a 30 liter air-lift photobioreactor. *Journal of Fermentation and Bioengineering, 82*, 113.

Harker, M., Tsavalos, A. J., & Young, A. J. (1996). Factors responsible for astaxanthin formation in the chlorophyte *Haematococcus pluvialis*. *Bioresource Technology, 55*, 207–214.

Hartung, F., Schiemann, J., & Quedlinburg, D. (2014). Precise plant breeding using new genome editing techniques: Opportunities, safety and regulation in the EU. *The Plant Journal, 78*, 742–752.

Hashimoto, H., Arai, K., Hayashi, S., Okamoto, H., Takahashi, J., Chikuda, M., & Obara, Y. (2014). Effects of astaxanthin on antioxidation in human aqueous humor. *Journal of Clinical Biochemistry and Nutrition, 54*, 86–89.

Hata, N., Ogbonna, J. C., Hasegawa, Y., Taroda, H., & Tanaka, H. (2001). Production of astaxanthin by *Haematococcus pluvialis* in a sequential heterotrophic-photoautotrophic culture. *Journal of Applied Phycology, 13*, 395–402.

He, B., Hou, L., Dong, M., Shi, J., Huang, X., Ding, Y., Cong, X., Zhang, F., Zhang, X., & Zang, X. (2018). Transcriptome analysis in *Haematococcus pluvialis*: Astaxanthin induction by high light with acetate. *International Journal of Molecular Sciences, 19*, 175–193.

Higuera-Ciapara, I., Félix-Valenzuela, L., & Goycoolea, F. M. (2006). Astaxanthin: A review of its chemistry and applications. *Critical Reviews in Food Science and Nutrition, 46*, 185–196.

Hoffman, Y., Aflalo, C., Zarka, A., Gutman, J., James, T. Y., & Boussiba, S. (2008). Isolation and characterization of a novel chytrid species (phylum Blastocladiomycota), parasitic on the green alga *Haematococcus*. *Mycological Research, 112*, 70–81.

Holtin, K., Kuehnle, M., Rehbein, J., Schuler, P., Nicholson, G., & Albert, K. (2009). Determination of astaxanthin and astaxanthin esters in the microalgae *Haematococcus pluvialis* by LC-(APCI) MS and characterization of predominant carotenoid isomers by NMR spectroscopy. *Analytical and Bioanalytical Chemistry, 395*, 1613–1622.

Hossain, A. K. M. M., Brennan, M. A., Mason, S. L., Guo, X., Zeng, X. A., & Brennan, C. S. (2017). The effect of astaxanthin-rich microalgae "*Haematococcus pluvialis*" and wholemeal flours incorporation in improving the physical and functional properties of cookies. *Food, 6*, E57.
Hu, Z., Li, Y., Sommerfeld, M., Chen, F., & Hu, Q. (2008). Enhanced protection against oxidative stress in an astaxanthin-overproduction *Haematococcus* mutant (Chlorophyceae). *European Journal of Phycology, 43*, 365–376.
Ip, P. F., Wong, K. H., & Chen, F. (2004). Enhanced production of astaxanthin by the green microalga *Chlorella zofingiensis* in mixotrophic culture. *Process Biochemistry, 39*, 1761–1766.
Issarapayup, K., Powtongsook, S., & Pavasant, P. (2011). Economical review of *Haematococcus pluvialis* culture in flat-panel airlift photobioreactors. *Aquacultural Engineering, 44*, 65–71.
Johnson, E., & Schroeder, W. (1995). Microbial carotenoids. *Advances in Biochemical Engineering/Biotechnology, 53*, 119–178.
Kadam, S. U., Tiwari, B. K., & O'donnell, C. P. (2013). Application of novel extraction technologies for bioactives from marine algae. *Journal of Agricultural and Food Chemistry, 61*, 4667–4675.
Kaewpintong, K., Shotipruk, A., Powtongsook, S., & Pavasant, P. (2007). Photoautotrophic high-density cultivation of vegetative cells of *Haematococcus pluvialis* in airlift bioreactor. *Bioresource Technology, 98*, 288–295.
Kamath, S., Sarada, R. (2007). *Biotechnological production of microalgal carotenoids with reference to astaxanthin and evaluation of its biological activity*. Doctoral Dissertation, University of Mysore.
Kang, H., & Kim, H. (2017). Astaxanthin and β-carotene in *Helicobacter pylori* -induced gastric inflammation: A mini-review on action mechanisms. *Journal of Cance R Prevention, 22*, 57–61.
Kang, C. D., Lee, J. S., Park, T. H., & Sim, S. J. (2005). Comparison of heterotrophic and photoautotrophic induction on astaxanthin production by *Haematococcus pluvialis*. *Applied Microbiology and Biotechnology, 68*, 237–241.
Kang, C. D., Lee, J. S., Park, T. H., & Sim, S. J. (2007). Complementary limiting factors of astaxanthin synthesis during photoautotrophic induction of *Haematococcus pluvialis*: C/N ratio and light intensity. *Applied Microbiology and Biotechnology, 74*, 987–994.
Kang, C. D., Han, S. J., Choi, S. P., & Sim, S. J. (2010). Fed-batch culture of astaxanthin-rich *Haematococcus pluvialis* by exponential nutrient feeding and stepwise light supplementation. *Bioprocess and Biosystems Engineering, 33*, 133–139.
Kapoore, R. V., Butler, T. O., Pandhal, J., & Vaidyanathan, S. (2018). Microwave-assisted extraction for microalgae: From biofuels to biorefinery. *Biology, 7*(1), 18.
Katagiri, M., Satoh, A., Tsuji, S., & Shirasawa, T. (2012). Effects of astaxanthin-rich *Haematococcus pluvialis* extract on cognitive function: A randomised, double-blind, placebo-controlled study. *Journal of Clinical Biochemistry and Nutrition, 51*, 102–107.
Kathiresan, S., Chandrashekar, A., Ravishankar, G. A., & Sarada, R. (2009). *Agrobacterium*-mediated transformation in the green alga *Haematococcus pluvialis* (chlorophyceae, volvocales). *Journal of Phycology, 45*, 642–649.
Katsuda, T., Lababpour, A., Shimahara, K., & Katoh, S. (2004). Astaxanthin production by *Haematococcus pluvialis* under illumination with LEDs. *Enzyme and Microbial Technology, 35*, 81–86.
Khalid, N., & Barrow, C. J. (2018). Critical review of encapsulation methods for stabilization and delivery of astaxanthin. *Journal of Food Bioactives, 1*, 104–115.
Kim, G., Klochkova, T. A., Won Han, J., Kang, S. H., Gu Choi, H., Wha Chung, K., & Ja Kim, S. (2011). Freshwater and terrestrial algae from Ny-Ålesund and Blomstrandhalvøya island (svalbard). *Arctic, 64*, 25–31.
Kim, J. H., Affan, M. A., Jang, J., Kang, M. H., Ko, A. R., Jeon, S. M., Oh, C., Heo, S. J., Lee, Y. H., Ju, S. J., & Kang, D. H. (2015). Morphological, molecular, and biochemical characterization

of astaxanthin-producing green microalga *Haematococcus* sp. KORDI03 (*Haematococcaceae, chlorophyta*) isolated from Korea. *Journal of Microbiology and Biotechnology, 25*, 238–246.

Kim, D. Y., Vijayan, D., Praveenkumar, R., Han, J. I., Lee, K., Park, J. Y., Chang, W. S., Lee, J. S., & Oh, Y. K. (2016). Cell-wall disruption and lipid/astaxanthin extraction from microalgae: *Chlorella* and *Haematococcus*. *Bioresource Technology, 199*, 300–310.

Kiperstok, A. C., Sebestyén, P., Podola, B., & Melkonian, M. (2017). Biofilm cultivation of *Haematococcus pluvialis* enables a highly productive one-phase process for astaxanthin production using high light intensities. *Algal Research, 21*, 213–222.

Klochkova, T. A., Kwak, M. S., Han, J. W., Motomura, T., Nagasato, C., & Kim, G. H. (2013). Cold-tolerant strain of *Haematococcus pluvialis* (Haematococcaceae, Chlorophyta) from Blomstrandhalvøya (Svalbard). *Algae, 28*, 185–192.

Kobayashi, M., Kakizono, T., & Nagai, S. (1991). Astaxanthin production by a green alga, *Haematococcus pluvialis* accompanied with morphological changes in acetate media. *Journal of Fermentation and Bioengineering, 71*, 335–339.

Kobayashi, M., Kakizono, T., & Nagai, S. (1993). Enhanced carotenoid biosynthesis by oxidative stress in acetate-induced cyst cells of a green unicellular alga, *Haematococcus pluvialis*. *Applied and Environmental Microbiology, 59*, 867–873.

Kobayashi, M., Kakizono, T., Nishio, N., Nagai, S., Kurimura, Y., & Tsuji, Y. (1997a). Antioxidant role of astaxanthin in the green alga *Haematococcus pluvialis*. *Applied Microbiology and Biotechnology, 48*, 351–356.

Kobayashi, M., Kurimura, Y., & Tsuji, Y. (1997b). Light-independent, astaxanthin production by the green microalga *Haematococcus pluvialis* under salt stress. *Biotechnology Letters, 19*, 507–509.

Koller, M., Muhr, A., & Braunegg, G. (2014). Microalgae as versatile cellular factories for valued products. *Algal Research, 6*, 52–63.

Kwan, T. A., Kwan, S. E., Peccia, J., & Zimmerman, J. B. (2018). Selectively biorefining astaxanthin and triacylglycerol co-products from microalgae with supercritical carbon dioxide extraction. *Bioresource Technology, 269*, 81–88.

Lababpour, A., Hada, K., Shimahara, K., Katsuda, T., & Katoh, S. (2004). Effects of nutrient supply methods and illumination with blue light emitting diodes (LEDs) on astaxanthin production by *Haematococcus pluvialis*. *Journal of Bioscience and Bioengineering, 98*, 452–456.

Lao, Y. M., Jin, H., Zhou, J., Zhang, H. J., & Cai, Z. H. (2017). Functional characterization of a missing branch component in *Haematococcus pluvialis* for control of algal carotenoid biosynthesis. *Frontiers in Plant Science, 8*, 1–15.

Lee, C., Woo, J., Jin, A., Kim, B., Young, J., & Yoon, K. (2018). Comparative transcriptome analysis of *Haematococcus pluvialis* on astaxanthin biosynthesis in response to irradiation with red or blue LED wavelength. *World Journal of Microbiology and Biotechnology, 34*, 1–14.

Leite, M. F., De Lima, A., Massuyama, M. M., & Otton, R. (2010). In vivo astaxanthin treatment partially prevents antioxidant alterations in dental pulp from alloxan-induced diabetic rats. *International Endodontic Journal, 43*, 959–967.

Lemoine, Y., & Schoefs, B. (2010). Secondary ketocarotenoid astaxanthin biosynthesis in algae: A multifunctional response to stress. *Photosynthesis Research, 106*, 155–177.

Li, Y., Sommerfeld, M., Chen, F., & Hu, Q. (2008). Consumption of oxygen by astaxanthin biosynthesis: A protective mechanism against oxidative stress in *Haematococcus pluvialis* (Chlorophyceae). *Journal of Plant Physiology, 165*, 1783–1797.

Li, Y., Sommerfeld, M., Chen, F., & Hu, Q. (2010). Effect of photon flux densities on regulation of carotenogenesis and cell viability of *Haematococcus pluvialis* (Chlorophyceae). *Journal of Applied Phycology, 22*, 253–263.

Li, J., Zhu, D., Niu, J., Shen, S., & Wang, G. (2011). An economic assessment of astaxanthin production by large scale cultivation of *Haematococcus pluvialis*. *Biotechnology Advances, 29*, 568–574.

Li, M., Wu, W., Zhou, P., Xie, F., Zhou, Q., & Mai, K. (2014). Comparison effect of dietary astaxanthin and *Haematococcus pluvialis* on growth performance, antioxidant status and immune response of large yellow croaker Pseudosciaena crocea. *Aquaculture, 434*, 227–232.

Li, K., Cheng, J., Ye, Q., He, Y., Zhou, J., & Cen, K. (2017). In vivo kinetics of lipids and astaxanthin evolution in *Haematococcus pluvialis* mutant under 15% CO_2 using Raman microspectroscopy. *Bioresource Technology, 244*, 1439–1444.

Liu, J., & Huang, Q. (2016). Screening of astaxanthin-hyperproducing *Haematococcus pluvialis* using fourier transform infrared (FT-IR) and Raman microspectroscopy. *Applied Spectroscopy, 70*, 1639–1648.

Liu, Z., Liu, C., Hou, Y., Chen, S., Xiao, D., Zhang, J., & Chen, F. (2013). Isolation and characterization of a marine microalga for biofuel production with astaxanthin as a co-product. *Energies, 6*, 2759–2772.

Liu, J., Gerken, H., & Li, Y. (2014a). Single-tube colony PCR for DNA amplification and transformant screening of oleaginous microalgae. *Journal of Applied Phycology, 26*, 1719–1726.

Liu, J., Sun, Z., Gerken, H., Liu, Z., Jiang, Y., & Chen, F. (2014b). *Chlorella zofingiensis* as an alternative microalgal producer of astaxanthin: Biology and industrial potential. *Marine Drugs, 12*, 3487–3515.

Liu, Z., Zeng, X., & Cheng, J. (2018). The efficiency and comparison of novel techniques for cell wall disruption in astaxanthin extraction from *Haematococcus pluvialis*. *International Journal of Food Science & Technology, 53*, 2212–2219.

Lorenz, R. T. (1999). *A technical review of Haematococcus algae* (NatuRose™ technical bulletin #060) (pp. 1–12). Kailua-Kona: Cyanotech Corporation.

Lorenz, R. T., & Cysewski, G. R. (2000). Commercial potential for *Haematococcus* microalgae as a natural source of astaxanthin. *Trends in Biotechnology, 18*, 160–167.

Łukomska-Kowalczyk, M., Karnkowska, A., Krupska, M., Milanowski, R., & Zakryś, B. (2016). DNA barcoding in autotrophic euglenids: Evaluation of coi and 18s rDNA. *Journal of Phycology, 52*, 951–960.

Ma, R. Y., & Chen, F. (2001). Enhanced production of free trans -astaxanthin by oxidative stress in the cultures of the green microalga *Chlorococcum* sp. *Process Biochemistry, 36*, 1175–1179.

Ma, R., Thomas-Hall, S. R., Chua, E. T., Alsenani, F., Eltanahy, E., Netzel, M. E., Netzel, G., Lu, Y., & Schenk, P. M. (2018). Gene expression profiling of astaxanthin and fatty acid pathways in *Haematococcus pluvialis* in response to different LED lighting conditions. *Bioresource Technology, 250*, 591–602.

Manabe, Y., Komatsu, T., Seki, S., & Sugawara, T. (2018). Dietary astaxanthin can accumulate in the brain of rats. *Bioscience, Biotechnology, and Biochemistry, 8451*, 1–4.

Mann, V., Harker, M., Pecker, I., & Hirschberg, J. (2000). Metabolic engineering of astaxanthin production in tobacco flowers. *Nature Biotechnology, 18*, 888–892.

Margalith, P. Z. (1999). Production of ketocarotenoids by microalgae. *Applied Microbiology and Biotechnology, 51*, 431–438.

Market Watch. (2019). *Astaxanthin market projected to witness vigorous growth of 800 million by 2024*. Retrieved May 10, 2019, from https://www.marketwatch.com/press-release/astaxanthin-market-projected-to-witness-vigorous-growth-of-800-million-by-2024-2019-04-24.

Martínez-delgado, A. A., Khandual, S., Josefina, S., & Rodríguez, V. (2017). Chemical stability of astaxanthin integrated into a food matrix: Effects of food processing and methods for preservation. *Food Chemistry, 225*, 23–30.

Mazumdar, N., Gopalakrishnan, K. K., Visnovsky, G., & Novis, M. (2018). A novel alpine species of *Haematococcus* (Chlamydomonadales: Chlorophyta) from New Zealand. *New Zealand Journal of Botany, 56*, 216.

Mendes-Pinto, M. M., Raposo, M. F. J., Bowen, J., Young, A. J., & Morais, R. (2001). Evaluation of different cell disruption processes on encysted cells of *Haematococcus pluvialis*: Effects on astaxanthin recovery and implications for bio-availability. *Journal of Applied Phycology, 13*, 19–24.

Michalak, I., & Chojnacka, K. (2014). Algal extracts: Technology and advances. *Engineering in Life Sciences, 14*, 581–591.

Miki, W. (1991). Biological functions and activities of animal carotenoids. *Pure and Applied Chemistry, 63*, 141–146.

Minatelli, J. (2008). Astaxanthin and eye health: The newest carotenoid dietary supplement solution. *Nutraceutical Business & Technology*, 18–19.

Molino, A., Rimauro, J., Casella, P., Cerbone, A., Larocca, V., Chianese, S., Karatza, D., Mehariya, S., & Ferraro, A. (2018). Extraction of astaxanthin from microalga *Haematococcus pluvialis* in red phase by using generally recognized as safe solvents and accelerated extraction. *Journal of Biotechnology, 283*, 51–61.

Mont, L., Juc, A., Pedrosa, C., Seabra, J., Federal, U., Grande, R., & Cordeiro, G. G. (2010). Astaxanthin: Structural and e funcionais. *Revista de Nutrição, 23*, 1041–1050.

Montsant, A., Zarka, A., & Boussiba, S. (2001). Presence of a nonhydrolyzable biopolymer in the cell wall of vegetative cells and astaxanthin-rich cysts of *Haematococcus pluvialis* (chlorophyceae). *Marine Biotechnology, 3*, 515–521.

Mostafa, N., Omar, H., Tan, S. G., & Napis, S. (2011). Studies on the genetic variation of the green unicellular alga *Haematococcus pluvialis* (chlorophyceae) obtained from different geographical locations using ISSR and RAPD molecular marker. *Molecules, 16*, 2599–2608.

Nahidian, B., Ghanati, F., Shahbazi, M., & Soltani, N. (2018). Effect of nutrients on the growth and physiological features of newly isolated *Haematococcus pluvialis* TMU1. *Bioresource Technology, 255*, 229–237.

Nakada, T., & Ota, S. (2016). What is the correct name for the type of *Haematococcus* Flot. *Taxon, 65*, 343–348.

Nawrocki, W. J., Tourasse, N. J., Taly, A., Rappaport, F., & Wollman, F.-A. (2015). The plastid terminal oxidase: Its elusive function points to multiple contributions to plastid physiology. *Annual Review of Plant Biology, 66*, 49–74.

Newsome, R. (1986). Food colors. *Food Technology, 40*, 49–56.

Nguyen, K. D. (2013). Astaxanthin: A comparative case of synthetic vs. natural production. *Chemical and Biomolecular Engineering Publications and Other Works, 1*, 1–9.

Nishida, Y., Yamashita, E., & Miki, W. (2007, September). *Comparison of Astaxanthin's singlet oxygen quenching activity with common fat and water soluble antioxidants*. Paper presented at the 21st Annual Meeting on Carotenoid Research held at Osaka, Japan (Vol. 6).

Nobre, B., Marcelo, F., Passos, R., Beirão, L., Palavra, A., Gouveia, L., & Mendes, R. (2006). Supercritical carbon dioxide extraction of astaxanthin and other carotenoids from the microalga *Haematococcus pluvialis*. *European Food Research and Technology, 223*, 787–790.

Oilalgae. (2015). *Astaxanthin*. Retrieved March 30, 2015, from http://www.oilgae.com/non_fuel_products/astaxanthin.html.

Olaizola, M. (2000). Commercial production of astaxanthin from *Haematococcus pluvialis* using 25,000-liter outdoor photobioreactors. *Journal of Applied Phycology, 12*, 499–506.

Olaizola, M. (2003). Commercial development of microalgal biotechnology: From the test tube to the marketplace. *Biomolecular Engineering, 20*, 459.

Olaizola, M., & Huntley, M. E. (2003). Recent advances in commercial production of astaxanthin from microalgae. In *Recent advances in marine biotechnology. Volume 9. Biomaterials and bioprocessing* (Vol. 9, pp. 143–164). New Hampshire: Science Publishers.

Oncel, S. S., Imamoglu, E., Gunerken, E., & Sukan, F. V. (2011). Comparison of different cultivation modes and light intensities using mono-cultures and co-cultures of *Haematococcus pluvialis* and *Chlorella zofingiensis*. *Journal of Chemical Technology and Biotechnology, 86*, 414–420.

Orosa, M., Valero, J. F., Herrero, C., & Abalde, J. (2001). Comparison of the accumulation of astaxanthin in *Haematococcus pluvialis* and other green microalgae under N -starvation and high light conditions. *Biotechnology Letters, 23*, 1079–1085.

Overton, K., Samsing, F., Oppedal, F., Stien, H., & Dempster, T. (2018). Lowering treatment temperature reduces salmon mortality: A new way to treat with hydrogen peroxide in aquaculture. *Pest Management Science, 74*, 535.

Pan, J. L., Wang, H. M., Chen, C. Y., & Chang, J. S. (2012). Extraction of astaxanthin from *Haematococcus pluvialis* by supercritical carbon dioxide fluid with ethanol modifier. *Engineering in Life Sciences, 12*, 638–647.

Pang, N., & Chen, S. (2017). Effects of C5 organic carbon and light on growth and cell activity of *Haematococcus pluvialis* under mixotrophic conditions. *Algal Research, 21*, 227–235.

Panis, G., & Rosales, C. J. (2016). Commercial astaxanthin production derived by green alga *Haematococcus pluvialis*: A microalgae process model and a techno-economic assessment all through production line. *Algal Research, 18*, 175–190.

Pan-utai, W. (2017). Effect of inducing agents on growth and astaxanthin production in *Haematococcus pluvialis*: Organic and inorganic. *Biocatalysis and Agricultural Biotechnology, 12*, 152–158.

Park, J. S., Chyun, J. H., Kim, Y. K., Line, L. L., & Chew, B. P. (2010). Astaxanthin decreased oxidative stress and inflammation and enhanced immune response in humans. *Nutrition & Metabolism (London), 7*, 18.

Park, J. C., Choi, S. P., Hong, M. E., & Sim, S. J. (2014). Enhanced astaxanthin production from microalga, *Haematococcus pluvialis* by two-stage perfusion culture with stepwise light irradiation. *Bioprocess and Biosystems Engineering, 37*, 2039–2047.

Peebles, F. (1909). The life history of *Sphaerella lacustris* (*Haematococcus pluvialis*) with reference to the nature and behavior of the zoospores. *Zentralblatt fuer Bakteriologie Jena Abt II, 2*(24), 511–521.

Peled, E., Leu, S., Zarka, A., Weiss, M., Pick, U., Khozin-Goldberg, I., & Boussiba, S. (2011). Isolation of a novel oil globule protein from the green alga *Haematococcus pluvialis* (chlorophyceae). *Lipids, 46*, 851–861.

Peled, E., Pick, U., Zarka, A., Shimoni, E., Leu, S., & Boussiba, S. (2012). Light-induced oil globule migration in *Haematococcus pluvialis* (Chlorophyceae). *Journal of Phycology, 48*, 1209–1219.

Pérez-López, P., González-García, S., Jeffryes, C., Agathos, S. N., McHugh, E., Walsh, D., Murray, P., Moane, S., Feijoo, G., & Moreira, M. T. (2014). Life cycle assessment of the production of the red antioxidant carotenoid astaxanthin by microalgae: From lab to pilot scale. *Journal of Cleaner Production, 64*, 332–344.

Phasex Corporation. (2015). *Phasex* Corporation. Retrieved October 9, 2015, from http://www.phasex4scf.com/.

Pick, U., Zarka, A., Boussiba, S., & Davidi, L. (2019). A hypothesis about the origin of carotenoid lipid droplets in the green algae *Dunaliella* and *Haematococcus*. *Planta, 249*, 31–47.

Piermarocchi, S., Saviano, S., Parisi, V., Tedeschi, M., Panozzo, G., Scarpa, G., Boschi, G., & Lo Giudice, G. (2012). Carotenoids in Age-related maculopathy Italian study (CARMIS): Two-year results of a randomized study. *European Journal of Ophthalmology, 22*, 216–225.

Poonkum, W., Powtongsook, S., & Pavasant, P. (2015). Astaxanthin induction in microalga *H. pluvialis* with flat panel airlift photobioreactors under indoor and outdoor conditions. *Preparative Biochemistry & Biotechnology, 45*, 1.

Prabhakaran, M., Elumalai, S., Santhose, B. I., & Kanna, G. R. (2014). Collection, isolation and identification of *Haematococcus pluvialis* Flotow from high altitude region of Pithoragarh district, Uttarakhand, India. *Golden Research Thoughts, 3*, 1–7.

Praveenkumar, R., Lee, K., Lee, J., & Oh, Y. (2015). Breaking dormancy: An energy-efficient means of recovering astaxanthin from microalgae. *Green Chemistry, 17*, 1226–1234.

Proctor, V. W. (1957a). Some controlling factors in the distribution of *Haematococcus pluvialis*. *Ecology, 38*, 457–462.

Proctor, V. W. (1957b). Preferential assimilation of nitrate ion by *Haematococcus pluvialis*. *American Journal of Botany, 44*, 141.

Proctor, V. W. (1957c). Studies of algal antibiosis using *Haematococcus* and *Chlamydomonas*. *Limnology and Oceanography*, *2*(2), 125–139.
Qinglin, D., Xueming, Z., Xiangying, X., Jianzhong, H., & Jixian, G. (2007). Concomitant NH_4^+ secretion during astaxanthin synthesis in *Haematococcus pluvialis* under high irradiance and nitrogen deficient conditions. *Chinese Journal of Chemical Engineering, 15*, 162–166.
Ralston, N. V. C., Azenkeng, A., Ralston, C. R., Blackwell, J. L., & Raymond, L. J. (2014). Selenium-health benefit values as seafood safety criteria. In *Seafood science: Advances in chemistry, technology and applications* (pp. 433–457). Boca Raton, FL: CRC Press.
Ranga Rao, A., Raghunath Reddy, R. L., Baskaran, V., Sarada, R., & Ravishankar, G. A. (2010). Characterization of microalgal carotenoids by mass spectrometry and their bioavailability and antioxidant properties elucidated in rat model. *Journal of Agricultural and Food Chemistry, 58*, 8553–8559.
Ranjbar, R., Inoue, R., Katsuda, T., Yamaji, H., & Katoh, S. (2008). High efficiency production of astaxanthin in an airlift photobioreactor. *Journal of Bioscience and Bioengineering, 106*, 204–207.
Rayman, M. P. (2012). Selenium and human health. *Lancet, 379*, 1256–1268.
Renstrøm, B., & Liaaen-Jensen, S. (1981). Fatty acid composition of some esterified carotenols. *Comparative Biochemistry and Physiology B, 69*, 625–627.
Riccioni, G., Speranza, L., Pesce, M., Cusenza, S., D'Orazio, N., & Glade, M. J. (2012). Novel phytonutrient contributors to antioxidant protection against cardiovascular disease. *Nutrition, 28*, 605–610.
Rioboo, C., González-Barreiro, Ó., Abalde, J., & Cid, Á. (2011). Flow cytometric analysis of the encystment process induced by paraquat exposure in *Haematococcus pluvialis* (Chlorophyceae). *European Journal of Phycology, 46*, 89–97.
Rodríguez-Sáiz, M., De La Fuente, J. L., & Barredo, J. L. (2010). *Xanthophyllomyces dendrorhous* for the industrial production of astaxanthin. *Applied Microbiology and Biotechnology, 88*, 645–658.
Saha, S. K., Hayes, J., Moane, S., & Murray, P. (2013a). Tagging of biomolecules with deuterated water (D_2O) in commercially important microalgae. *Biotechnology Letters, 35*, 1067–1072.
Saha, S. K., McHugh, E., Hayes, J., Moane, S., Walsh, D., & Murray, P. (2013b). Effect of various stress-regulatory factors on biomass and lipid production in microalga *Haematococcus pluvialis*. *Bioresource Technology, 128*, 118–124.
Sarada, R., Bhattacharya, S., & Ravishankar, G. A. (2002). Optimization of culture conditions for growth of the green alga *Haematococcus pluvialis*. *World Journal of Microbiology and Biotechnology, 18*, 517–521.
Sarada, R., Vidhyavathi, R., Usha, D., Ravishankar, G. A., Sarada, R., Vidhyavathi, R., Usha, D., & Ravishankar, G. A. (2006). An efficient method for extraction of astaxanthin from green alga *Haematococcus pluvialis*. *Journal of Agricultural and Food Chemistry, 54*, 7585–7588.
Satoh, A., Tsuji, S., Okada, Y., Murakami, N., Urami, M., Nakagawa, K., Ishikura, M., Katagiri, M., Koga, Y., & Shirasawa, T. (2009). Preliminary clinical evaluation of toxicity and efficacy of a new astaxanthin-rich *Haematococcus pluvialis* extract. *Journal of Clinical Biochemistry and Nutrition, 44*, 280–284.
Schott. (2019). Schott. Retrieved May 1, 2019, from https://pbr-blog.schott.com/2017/06/06/free-guide-the-five-main-reasons-why-you-should-focus-your-attention-on-selecting-the-right-glass-components-for-your-pbr/.
Shah, M. M. R., Liang, Y., Cheng, J. J., & Daroch, M. (2016). Astaxanthin-producing green microalga *Haematococcus pluvialis*: From single cell to high value commercial products. *Frontiers in Plant Science, 7*, 531.
Shah, M. R., Lutzu, G. A., Alam, A., Sarker, P., Kabir Chowdhury, M. A., Parsaeimehr, A., Liang, Y., & Daroch, M. (2018). Microalgae in aquafeeds for a sustainable aquaculture industry. *Journal of Applied Phycology, 30*, 197–213.
Sharon-Gojman, R., Maimon, E., Leu, S., Zarka, A., & Boussiba, S. (2015). Advanced methods for genetic engineering of *H. pluvialis*. *Algal Research, 10*, 8–15.

Sheikhzadeh, N., Panchah, I. K., Asadpour, R., Tayefi-Nasrabadi, H., & Mahmoudi, H. (2012). Effects of *Haematococcus pluvialis* in maternal diet on reproductive performance and egg quality in rainbow trout (*Oncorhynchus mykiss*). *Animal Reproduction Science, 130*, 119–123.

Sipaúba-Tavares, L. H., Berchielli-Morais, F. A., & Scardoeli-Truzzi, B. (2015). Growth of *Haematococcus pluvialis* flotow in alternative media. *Brazilian Journal of Biology, 75*, 796–803.

Smith, D. R. (2018). *Haematococcus lacustris*: The makings of a giant-sized chloroplast genome. *AoB Plants, 10*, 1–7.

Solovchenko, A., & Chekanov, K. (2014). Production of carotenoids using microalgae cultivated in photobioreactors. In *Production of biomass and bioactive compounds using bioreactor technology* (pp. 63–91). New York, NY: Springer.

Solovchenko, A., Aflalo, C., Lukyanov, A., & Boussiba, S. (2013). Nondestructive monitoring of carotenogenesis in *Haematococcus pluvialis* via whole-cell optical density spectra. *Applied Microbiology and Biotechnology, 97*, 4533–4541.

Sommer, T. R., Potts, W. T., & Morrissy, N. M. (1991). Utilization of microalgal astaxanthin by rainbow trout (*Oncorhynchus mykiss*). *Aquaculture, 94*(1), 79.

Spiller, G. A., & Dewell, A. (2003). Safety of an astaxanthin-rich *Haematococcus pluvialis* algal extract: A randomized clinical trial. *Journal of Medicinal Food, 6*, 51–56.

Stahl, W., & Sies, H. (2003). Antioxidant activity of carotenoids. *Molecular Aspects of Medicine, 24*, 345–351.

Steinbrenner, J., & Sandmann, G. (2006). Transformation of the green alga *Haematococcus pluvialis* with a phytoene desaturase for accelerated astaxanthin biosynthesis. *Applied and Environmental Microbiology, 72*, 7477–7484.

Strittmatter, M., Guerra, T., Silva, J., & Gachon, C. M. M. (2016). A new flagellated dispersion stage in Paraphysoderma sedebokerense, a pathogen of *Haematococcus pluvialis*. *Journal of Applied Phycology, 27*, 1553.

Suh, I. S., Joo, H. N., & Lee, C. G. (2006). A novel double-layered photobioreactor for simultaneous *Haematococcus pluvialis* cell growth and astaxanthin accumulation. *Journal of Biotechnology, 125*, 540–546.

Sun, Y., Liu, J., Zhang, X., & Lin, W. (2008). Strain H2-419-4 of *Haematococcus pluvialis* induced by ethyl methanesulphonate and ultraviolet radiation. *Chinese Journal of Oceanology and Limnology, 26*, 152–156.

Sun, H., Kong, Q., Geng, Z., Duan, L., Yang, M., & Guan, B. (2015). Enhancement of cell biomass and cell activity of astaxanthin-rich *Haematococcus pluvialis*. *Bioresource Technology, 186*, 67–73.

Tanaka, T., Shnimizu, M., & Moriwaki, H. (2012). Cancer chemoprevention by carotenoids. *Molecules, 17*, 3202–3242.

Tjahjono, A. E., Hayama, Y., Kakizono, T., Terada, Y., Nishio, N., & Nagai, S. (1994). Hyper-accumulation of astaxanthin in a green alga *Haematococcus pluvialis* at elevated temperatures. *Biotechnology Letters, 16*, 133–138.

Tocquin, P., Fratamico, A., & Franck, F. (2012). Screening for a low-cost *Haematococcus pluvialis* medium reveals an unexpected impact of a low N/P ratio on vegetative growth. *Journal of Applied Phycology, 24*, 365–373.

Tominaga, K., Hongo, N., Karato, M., & Yamashita, E. (2012). Cosmetic benefits of astaxanthin on humans subjects. *Acta Biochimica Polonica, 59*, 43–47.

Torres-carvajal, L. K., González, Á. D., Francisco, U., & Santander, D. P. (2017). Astaxanthin production from *Haematococcus pluvialis*: Effects of light wavelength and salinity. *Contemporary Engineering Sciences, 10*, 1739–1746.

Torzillo, G., Goksan, T., Faraloni, C., Kopecky, J., & Masojidek, J. (2003). Interplay between photochemical activities and pigment composition in an outdoor culture of *Haematococcus pluvialis* during the shift from the green to red stage. *Journal of Applied Phycology, 15*, 127–136.

Triki, A., Maillard, P., & Gudin, C. (1997). Gametogenesis in *Haematococcus pluvialis* Flotow (Volvocales, Chlorophyta). *Phycologia, 36*, 190–194.

Tripathi, U., Sarada, R., Rao, S. R., & Ravishankar, G. A. (1999). Production of astaxanthin in *Haematococcus pluvialis* cultured in various media. *Bioresource Technology, 68*, 197–199.
Tripathi, U., Venkateshwaran, G., Sarada, R., & Ravishankar, G. A. (2001). Studies on *Haematococcus pluvialis* for improved production of astaxanthin by mutagenesis. *World Journal of Microbiology and Biotechnology, 17*, 143–148.
Van de Ven, M., & Van de Ven, J. J. M. F. (2009). Photobioreactor with a cleaning system and method for cleaning such a reactor. PCT Patent Application WO2009/051478. Accessed 16 Nov 2019.
Visioli, F., & Artaria, C. (2017). Astaxanthin in cardiovascular health and disease: mechanisms of action, therapeutic merits, and knowledge gaps. *Food & Function, 8*, 39–63.
Wan, M., Hou, D., Li, Y., Fan, J., Huang, J., Liang, S., Wang, W., Pan, R., Wang, J., & Li, S. (2014a). The effective photoinduction of *Haematococcus pluvialis* for accumulating astaxanthin with attached cultivation. *Bioresource Technology, 163*, 26–32.
Wan, M., Zhang, J., Hou, D., Fan, J., Li, Y., Huang, J., & Wang, J. (2014b). The effect of temperature on cell growth and astaxanthin accumulation of *Haematococcus pluvialis* during a light–dark cyclic cultivation. *Bioresource Technology, 167*, 276–283.
Wan, M., Zhang, Z., Wang, J., Huang, J., Fan, J., Yu, A., Wang, W., & Li, Y. (2015). Sequential heterotrophy–dilution–photoinduction cultivation of *Haematococcus pluvialis* for efficient production of astaxanthin. *Bioresource Technology, 198*, 557–563.
Wang, S. B., Chen, F., Sommerfeld, M., & Hu, Q. (2005). Isolation and proteomic analysis of cell wall-deficient *Haematococcus pluvialis* mutants. *Proteomics, 5*, 4839–4851.
Wang, J., Sommerfeld, M., & Hu, Q. (2009). Occurrence and environmental stress responses of two plastid terminal oxidases in *Haematococcus pluvialis* (Chlorophyceae). *Planta, 230*, 191–203.
Wang, L., Yang, B., Yan, B., & Yao, X. (2012). Supercritical fluid extraction of astaxanthin from *Haematococcus pluvialis* and its antioxidant potential in sun flower oil. *Innovative Food Science & Emerging Technologies, 13*, 120–127.
Wang, J., Han, D., Sommerfeld, M. R., Lu, C., & Hu, Q. (2013). Effect of initial biomass density on growth and astaxanthin production of *Haematococcus pluvialis* in an outdoor photobioreactor. *Journal of Applied Phycology, 25*(1), 253.
Wang, B., Zhang, Z., Hu, Q., Sommerfeld, M., Lu, Y., & Han, D. (2014). Cellular capacities for high-light acclimation and changing lipid profiles across life cycle stages of the green alga *Haematococcus pluvialis*. *PLoS One, 9*, e106679.
Wang, F., Gao, B., Wu, M., Huang, L., & Zhang, C. (2019). A novel strategy for the hyper-production of astaxanthin from the newly isolated microalga *Haematococcus pluvialis* JNU35. *Algal Research, 39*, 101466.
Wayama, M., Ota, S., Matsuura, H., Nango, N., Hirata, A., & Kawano, S. (2013). Three-dimensional ultrastructural study of oil and astaxanthin accumulation during encystment in the green alga *Haematococcus pluvialis*. *PLoS One, 8*, e53618.
Yaakob, Z., Ali, E., Zainal, A., Mohamad, M., & Takriff, M. S. (2014). An overview: Biomolecules from microalgae for animal feed and aquaculture. *Journal of Biological Research, 21*, 1–10.
Yokoyama, A., Izumida, H., & Miki, W. (1994). Production of astaxanthin and 4-ketozeaxanthin by the marine bacterium, *agrobacterium aurantiacum*. *Bioscience, Biotechnology, and Biochemistry, 58*, 1842–1844.
Yoo, J. J., Choi, S. P., Kim, B. W., & Sim, S. J. (2012). Optimal design of scalable photo-bioreactor for phototropic culturing of *Haematococcus pluvialis*. *Bioprocess and Biosystems Engineering, 35*, 309–315.
Young, A. J., Pritchard, J., White, D., & Davies, S. (2017). Processing of astaxanthin-rich *Haematococcus* cells for dietary inclusion and optimal pigmentation in Rainbow trout, *Oncorhynchus mykiss* L. *Aquaculture Nutrition, 23*, 1304–1311.
Yuan, J.-P., Peng, J., Yin, K., & Wang, J.-H. (2011). Potential health-promoting effects of astaxanthin: A high-value carotenoid mostly from microalgae. *Molecular Nutrition & Food Research, 55*, 150–165.

Zgheib, N., Saade, R., Khallouf, R., & Takache, H. (2018). Extraction of astaxanthin from microalgae: Process design and economic feasibility study. *IOP Conference Series Materials Science and Engineering, 323*, 1–12.

Zhang, X. W., Gong, X. D., & Chen, F. (1999). Dynamics and stability analysis of the growth and astaxanthin production system of *Haematococcus pluvialis*. *Journal of Industrial Microbiology and Biotechnology, 23*(2), 133–137.

Zhang, B. Y., Geng, Y. H., Li, Z. K., Hu, H. J., & Li, Y. G. (2009). Production of astaxanthin from *Haematococcus* in open pond by two-stage growth one-step process. *Aquaculture, 295*, 275–281.

Zhang, W., Wang, J., Wang, J., & Liu, T. (2014). Attached cultivation of *Haematococcus pluvialis* for astaxanthin production. *Bioresource Technology, 158*, 329–335.

Zhang, C., Zhang, L., & Liu, J. (2016). The role of photorespiration during astaxanthin accumulation in *Haematococcus pluvialis* (Chlorophyceae). *Plant Physiology and Biochemistry, 107*, 75–81.

Zheng, Y., Li, Z., Tao, M., Li, J., & Hu, Z. (2017). Effects of selenite on green microalga *Haematococcus pluvialis*: Bioaccumulation of selenium and enhancement of astaxanthin production. *Aquatic Toxicology, 183*, 21–27.

Chapter 7
Microalgae Nutraceuticals: The Role of Lutein in Human Health

M. Vila Spinola and E. Díaz-Santos

Abstract Lutein is a carotenoid compound belonging to the xanthophyll family whose more attractive bioactivity is its antioxidant capacity. This carotenoid is mainly distributed in vegetables and fruits and is present within the macula lutea as a pigment responsible of the yellow hue. Lutein has been widely found in the pigmentation of animal tissues as well as considered as an important nutraceutical and used for the coloration of foods, drugs, and cosmetics. Recently, lutein has been found to be effective in the prevention of age-related macular degeneration, cataracts, cardiovascular diseases, and certain types of cancer, having attracted thus great attention in relation to human health. At this time, the main source for an industrial-scale production of lutein is marigold oleoresin although, each time more, continuous reports concerning lutein-producing microalgae pose the question if those microorganisms could become a feasible alternative. In fact, several microalgae strains, such as *Scenedesmus almeriensis*, *Chlorella zofingiensis*, or *Muriellopsis* sp., have higher lutein content than most marigold cultivars and have been shown to yield productivities hundreds of times higher than marigold crops on a per square meter basis, suggesting that, in the current state of the art, microalgae could compete with marigold or other lutein producers. The potential of the lutein as nutraceutical and its role in metabolic functions related to human health as well as its production from microalgae are reviewed in this chapter.

Keywords Lutein · Antioxidant · Human health · Macular degeneration · Microalgae

M. V. Spinola
Laboratory of Biochemistry, Faculty of Experimental Sciences, Marine International Campus of Excellence (CEIMAR), University of Huelva, Huelva, Spain
e-mail: marta.vila@dqcm.uhu.es

E. Díaz-Santos (✉)
Laboratory of Biology and Biotechnology of Cyanobacteria, Institute for Integrative Biology of the Cell (I2BC), CEA, CNRS, Université Paris-Sud, Université Paris-Saclay, Gif-sur-Yvette Cedex, France

M. A. Alam et al. (eds.), *Microalgae Biotechnology for Food, Health and High Value Products*, https://doi.org/10.1007/978-981-15-0169-2_7

1 Microalgae

Microalgae are photosynthetic unicellular organisms, which can be found in most ecosystems, due to their great ability to adapt to different environments. They are microorganisms with important ecological interest because they are the basis of the trophic chain of marine ecosystems and responsible for fixing almost 80% of atmospheric CO_2. For more than a decade, they have aroused an enormous interest in biotechnology due to the high variety of compounds they are able to synthesize and the interest that these compounds present from the biotechnological point of view (Alam and Wang 2019). Not only for using in aquaculture but also for the elaboration of functional feeds, in the pharmacological industry, in the food industry, and increasingly in medicine, as nutraceuticals, they serve as a source of energy to produce hydrogen or biodiesel. This biotechnological interest presented by microalgae is due to its ability to synthesize carotenoids, which are potent antioxidants, in addition to PUFAs and other compounds of interest.

Carotenoids are isoprenoid molecules with conjugated double bonds. They can be synthesized only by plants, microalgae, and some bacteria. Up to date, more than 700 have been described. The most important include β-carotene, α-carotene, lycopene, lutein, zeaxanthin, β-cryptoxanthin, α-cryptoxanthin, γ-carotene, neurosporene, ζ-carotene, phytofluene, and phytoene.

They have certain beneficial pharmaceutical effects on human health due to its strong antioxidant properties. Carotenoids could be rapidly oxidized by a series of oxidants, which greatly reduces the availability of free radicals to react with other cellular components, such as unsaturated lipids, protein, and DNA (Woodall et al. 1997).

In this chapter, we will talk about lutein, a carotenoid that at the present time is of huge interest for its beneficial properties for health, of its production, and of the future prospects of its use.

2 Lutein and Its Importance in Human Health

Lutein is a lipophilic tetraterpene belonging to the family of carotenoids known as xanthophylls. Differently than carotenes, the other group of carotenoids, the xanthophylls possess different oxygenated groups in the chemistry structure (OH-) (Fig. 7.1). In the particular case of lutein and its stereoisomer zeaxanthin, the oxygen is presented as hydroxyl groups conferring primarily to these carotenoids a high chemical reactivity with oxygen species (ROS) and consequently the properties of removal singlet oxygen particles (Koushan et al. 2013; Perrone et al. 2016; Buscemi et al. 2018). Furthermore, lutein has the capacity of filtering the light absorption spectrum absorbing majority blue light (higher oxidative damage induction than orange light) and preventing the photoreceptor damage produced by it (Koushan et al. 2013). In addition, as is described for the other carotenoids, the presence of

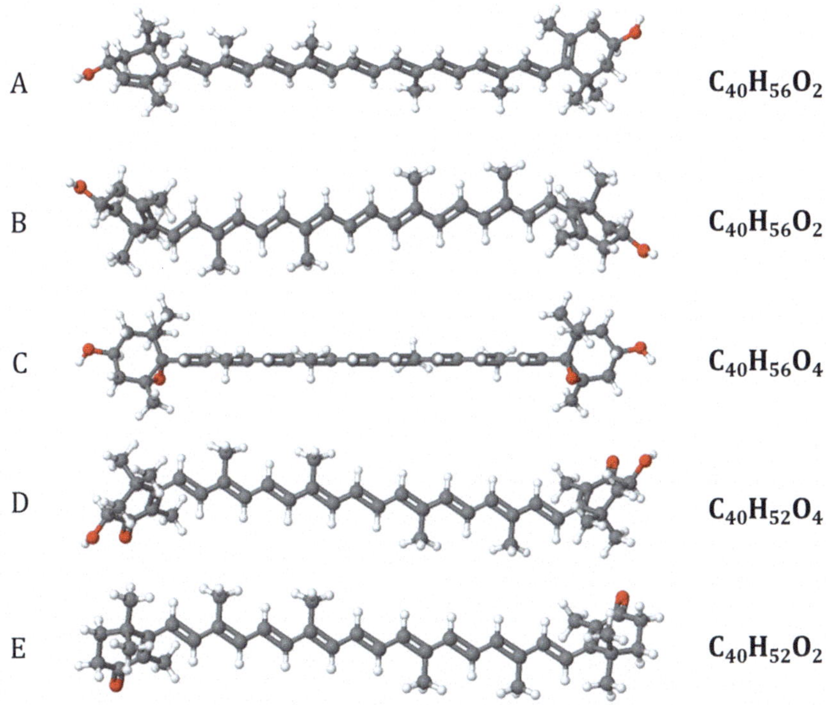

Fig. 7.1 Chemical and molecular structure of the different xanthophylls. (**a**) Lutein, (**b**) Zeaxanthin, (**c**) Violaxanthin, (**d**) Astaxanthin and (**e**) Canthaxanthin

conjugated double bonds along the carbon chain of this kind of molecules determines the photochemical and reactive properties, giving to carotenoids, carotenes, and xanthophylls their main biological function: the antioxidant activity.

Lutein can be found in certain foods of the human diet such as orange-red fruits, eggs, veggies, and especially in leafy dark greens (Sommerburg et al. 1998). In the human body, lutein (with zeaxanthin) is mainly distributed in the eye structures, especially abundant in the macula lutea, and to a lesser extent found in the skin, cervix, brain, or breasts (Granado et al. 2003; Perrone et al. 2016). Since it is known that lutein cannot be synthesized by humans and due to its high antioxidant properties and therefore its potential effect on human health (Sun et al. 2015), many efforts have focused on the market to obtain an alternative source of lutein food supplements. Traditionally, marigold, due to its high lutein content, has been considered the conventional source of this marketed carotenoid which is harvested periodically and meticulously from the petals of these plants (Jian-Hao et al. 2015). Therefore, over the last few years, microalgae have become a group of very attractive microorganisms for the obtainment of nutraceutical lutein, since certain species of them such as *Muriellopsis* sp., *Scenedesmus almeriensis*, *Chlorella protothecoides*, or *Chlorella zofingiensis* among others (Fernández-Sevilla et al. 2010) have been reported as naturally producers of high lutein content, being able to become a viable alternative to marigold (Jian-Hao et al. 2015).

Effects of lutein as a nutraceutical have been well studied mainly in the problems related to the vision and structure of the eye (age-related macular degeneration, cataracts, or retinopathies) because of the high content of this carotenoid in this part of the human body. In another hand, although it is evident that there is controversy in this regard, some studies go beyond these eye diseases and have focused on the investigation of antioxidant and anti-inflammatory potential of lutein in the prevention of other chronic diseases as cardiopathies, cancer risk, diabetes, or cognitive disorders (Granado et al. 2003; Buscemi et al. 2018) (Fig. 7.2).

2.1 *Eye-Related Diseases*

2.1.1 Age-Related Macular Degeneration (AMD)

The AMD is a progressive and degenerative eye disease affecting the macula lutea, a yellow ocular region located in the posterior central part of the retina, with approximately 5 mm diameter, formed mainly by lutein and zeaxanthin, and involved in visual acuity and vision of details (Pennington and DeAngelis 2016). Recent genetic studies suggest that the expression of the gene *HTRA1* in the chromosome 10q26 is associated with the risk of AMD (Liao et al. 2017). The development of this disease is caused by a continuous exposure of photoreceptors to the oxidative stress of light radiation, over the years. AMD is the principle cause of blindness in old-aged people, and it is considered a multifactor risk disease depending of light exposition duration, diet, age, genetic, or daily lifestyle (Koushan et al. 2013). Lutein and

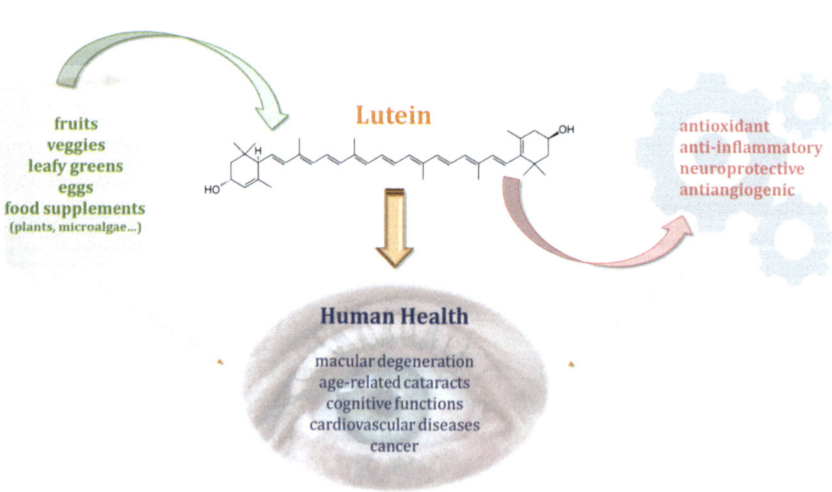

Fig. 7.2 Scheme. Lutein in human health

zeaxanthin, considered the major pigments in the main structures of the eye, play important roles in the development of AMD due to their capacity to absorb UV light, thus becoming photoprotective pigments against the damage caused by the reactive oxygen species (Sathasivam and Ki 2018).

The first clinical case-control study that was carried out in AMD indicating a positive influence of lutein in the decrease of the risk of suffering AMD is reported in the 1990s (Seddon et al. 1994). To date, numerous epidemiological studies have been carried out in which it is gathered that a supplementation of lutein in the human diet reports beneficial effects for the prevention and development of age-related macular degeneration and even indicating the possible influence in the partial recuperation of the amount of the carotenoids existing in the macula lutea (Koushan et al. 2013; Buscemi et al. 2018).

2.1.2 Cataracts

Cataracts are the opacity or the clouding of the lens that form part of the eye caused by the precipitation of proteins on the lens by ROS damage and preventing a clear vision (Maci and Santos 2015). Most cataracts are age related (ARC) and considered the main cause of blindness worldwide, along with AMD. The fact that oxidative damage is the most important factor in the development of cataracts makes lutein and zeaxanthin, xanthophyll concentrated and involved in eye structures, good candidates for fighting ARC (Manayi et al. 2016).

Several clinical studies indicate the relation between an intake of lutein and other carotenoids and the risk of cataracts or other age-related eye diseases. Many of them suggest that a supplementation of diet with lutein has a possible beneficial effect in the prevention and slow development of cataracts although a few studies describe that in the same subtypes of ARC the results are inconsistent or failed. Thus, a major precise understanding of cataract mechanisms and an increment number of clinical studies in which ARC patients are involved becomes necessary (Trumbo and Ellwood 2006; Buscemi et al. 2018).

2.1.3 Retinopathies and Other Eye-Related Diseases

Although, to a much lesser extent, the implication of lutein and its possible beneficial effects as antioxidants has been studied for other diseases related to the eye, diabetic retinopathy and retinopathy of prematurity, retinal detachment, retinitis pigmentosa, or uvea diseases are also being focused. In all these cases, there are few clinical studies with human patients, and many of them are in preliminary experimental phases in model animals (Koushan et al. 2013; Buscemi et al. 2018).

2.2 Cardiovascular Health

Since the anti-inflammatory properties of carotenoids and especially of lutein are known, several medical and epidemiologic research focused on the importance of lutein in diet to prevent primary and secondary cardiovascular diseases (CVD) (Maria et al. 2015). Lutein possesses the ability of decreasing the transcription and expression of the genes involved in the synthesis of the inflammatory response factors such as interleukins, cytokines, prostaglandins, or macrophage proteins that sum up to the protection against ROS stress suggesting the potential beneficial effect of lutein in cardiovascular health. Many studies are focused on the investigation of these aspects, and it is concluded that a higher concentration of lutein and other carotenoids is directly associated to the prevention of hypertension and, what is more, the resolution of chronic inflammation of CVD, the protection against atherosclerosis, and the benefit in cardiovascular complications (Maria et al. 2015; Chung et al. 2017). In addition, in other studies it is demonstrated that lutein could be involved with the reduction of LDL cholesterol and triglycerides and in the length of the telomers which could have effect in cardiovascular strokes (Buscemi et al. 2018).

2.3 Cancer Risk

Although the implications of lutein and different carotenoids, as nutraceuticals, in the prevention of cancer must be taken with great caution, in recent years there are already studies in which the role of antioxidant and anti-inflammatory molecules, that is, supplement in human diet, is beginning to be suggested. Considering that cancer is a multifactorial disease in which uncontrolled cell growth and a dysfunctional anti-inflammatory response are involved and can be localized in all different organs, the investigations in this aspect and the results are differently found. Among them, the most important can be found in colon, pancreas, non-Hodgkin lymphoma, or esophageal cancer (Buscemi et al. 2018). In addition, in 2018 Xiaoming Gong et al. reported a novel mechanism by which lutein could inhibit selectively the growth of breast cells, thus opening the way to increase the clinical research concerning this kind of diseases.

2.4 Other Human Diseases

Recent studies have been focused on other human diseases in which lutein could be involved. Among the most important is the accumulation of lutein in the brain conferring an improvement of the cognitive function in older people (Perrone et al. 2016). Few evidences involve lutein in lung or bone health as well as indicate the possible beneficial effects over diseases in pregnancy. Furthermore, some studies show the protection function of lutein in skin against the damage induced by solar UV radiations (Grether-Beck et al. 2017; Zielińska et al. 2017).

3 Production of Lutein

The elevated number of different lutein esters found in *Tagetes* is related to the structure of these xanthophylls. Lutein displays an asymmetric structure with terminal β- and ε-rings hydroxylated, respectively, at 3- and 3′-position that can lead to the formation of monoesters and diesters. Either lutein monoacylation or diacylation by two different fatty acids (FA) yields distinct regioisomers (Breithaupt et al. 2002a, b). To illustrate, considering the acylation of the (all-E)-form of this xanthophyll only with saturated FA containing chain length between 10 and 18 carbons, theoretically, 35 different (all-E)-lutein esters can be possibly formed. In addition, lutein can be often found as 9, 9′, 13, and 13′(Z)-isomers and further acylated with different saturated or unsaturated FA in diverse combinations, leading to a quite high number of different structures with similar characteristics. Mono- and dihydroxylated xanthophylls other than lutein can also be found esterified with FA in the most diverse combinations, further enhancing the variability of carotenoid structures (Rodrigues et al. 2019).

Lutein's structure is an extended conjugated double-bond system. These polyene chains exist in either a cis or trans conformation (predominant), giving rise to a large number of possible mono-cis and poly-cis isomers. Under light, oxygen heat, or pH stress, the geometric isomerization can convert all-trans carotenoids into a cis configuration. Since xanthophylls that accumulate in the lens and macular region of the retina are primarily all-trans lutein, the bioavailability of all-trans lutein may be higher than that of cis-lutein in the human body (Chitchumroonchokchai et al. 2004).

3.1 *Lutein from Microalgae*

Biotechnological production of lutein by microalgae exhibits several advantages including high lutein contents, fast growth rates, and high biomass and can be harvested throughout the year (i.e., season-independent harvest) (Chen et al. 2017; Fernández-Sevilla et al. 2010).

All carotenoids are formed from a C5 building block, common precursor of all isoprenoids, and the isopentenyl pyrophosphate (IPP), via the plastidial 2-C-methyl-D-erythriol 4-phosphate (MEP) pathway (Schwender et al. 2001). Differently from higher plants, where both the cytosolic mevalonate (MVA) and the plastidial MEP, also known as non-MVA, pathways supply IPP to chlorophyte microalgae, the cytosolic MVA pathway appears to be absent (Capa-Robles et al. 2009).

The first committing step of carotenoid biosynthesis is the condensation of two molecules of geranylgeranyl pyrophosphate (GGPP) to yield the first uncolored carotenoid, phytoene, catalysed by phytoene synthase (PSY). PSY is one of the most important regulatory enzymes in the pathway (Fig. 7.3). From this step on, phytoene is subjected to four sequential desaturations catalysed by phytoene (PDS) and ζ-carotene (ZDS) desaturases, resulting in the formation of prolycopene, which is isomerized by a specific isomerase (CRTISO) to all-*trans* lycopene, the first colored carotenoid (Varela et al. 2015).

At the level of lycopene, the pathway splits into two branches. In one branch, lycopene is cyclized at both ends by lycopene cyclase β (LCYB), yielding β-carotene with two β-ionone end groups. These can be further hydroxylated by a non-heme diiron hydroxylase, β-carotene hydroxylase (BCH), to yield zeaxanthin. In the other branch, the concerted action of LCYB and ε-cyclases (LCYE) results in the formation of α-carotene. Hydroxylation of α-carotene catalysed by two heme-containing cytochrome P450 monoxygenases, a carotene β-hydroxylase (CYP97A5) and a carotene ε-hydroxylase (CYP97C3), leads to the formation of lutein (Varela et al. 2015).

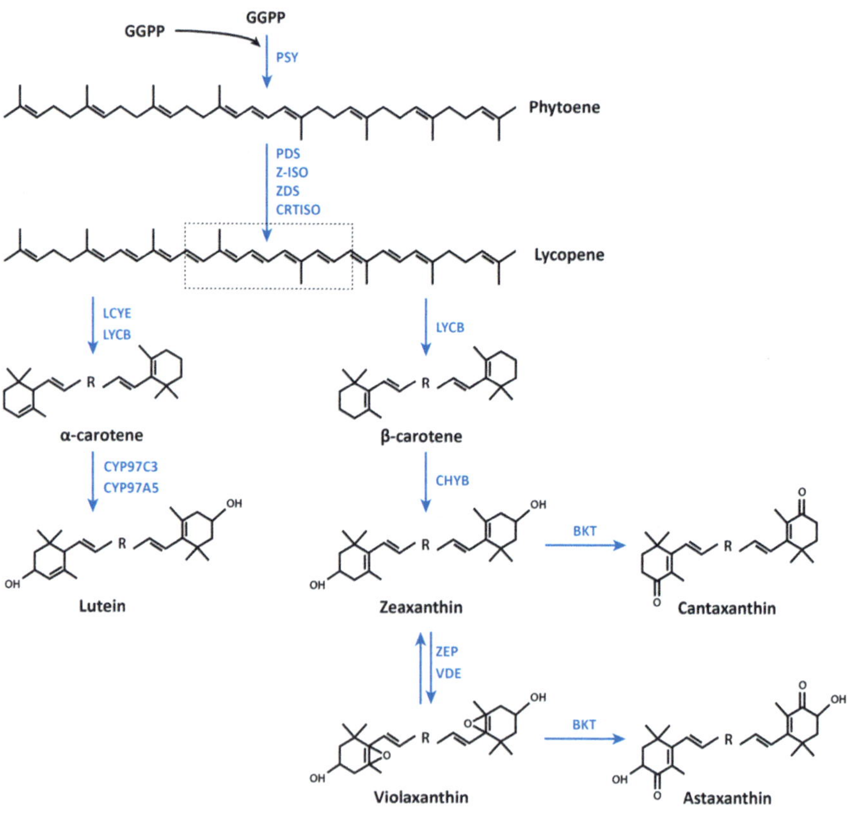

Fig. 7.3 Lutein biosynthesis in microalgae

3.2 Marigold Versus Microalgae

Over the years, lutein for human consumption has been obtained primarily from marigold petals in an arduous procedure. Although well established and developed, it requires a thorough extraction and production process, mainly carried out by hand, which means that in this sense, the research and use of other microorganisms rich in lutein has increased, including microalgae, such as an alternative via the production of lutein for food supplements. Lutein content of marigold flowers (e.g., *T. erecta*, *T. patula*, and *C. officinalis*) is estimated based on fresh flowers (80% moisture content) or granules or powder (10% moisture content), while that of microalgae is mostly based on dry products (Lin et al. 2015).

No direct comparisons of lutein content in marigold flowers and in microalgae have been made. For instance, the lutein content of *T. erecta* was in the range of 0.216–0.976 g/kg based on the weight of fresh flowers, which equals 0.02–0.1 wt% (Bosma et al. 2003; Liang et al. 2007), or 0.829–27.946 g/kg (0.08–2.8 wt%) for dried powders (Francisco and Octavio 1996). The lutein content of *T. patula* ranged at 0.597–12.31 g/kg based on the weight of dried powders (Natchigal et al. 2012). Additionally, based on dried powder weight, lutein content of *C. officinalis* ranged at 0.04–0.301 g/kg (Pintea et al. 2003; Natchigal et al. 2012). A reasonable estimate for lutein contents in microalgae biomass is 20 g/kg, mostly in lutein ester form. Del Campo et al. (2000) reported the lutein contents (converting to g/kg biomass) for various microalgal species as follows: *Chlorella fusca* (4.2–4.7), *Chlorococcum citroforme* (7.4), *Coelastrum proboscideum* (3.4–5.0), *Muriella aurantiaca* (2.6), *Muriella decolor* (0.5), *Neospondiococcum gelatinosum* (4.4), *Tetracysis aplanosporum* (5.9), *Tetracystis intermedia* (3.5), *Tetracystis tetrasporum* (4.4), and *Chlorella zofingiensis* (2.4–2.8). Other species with high lutein contents were reported by Shi et al. (2002) (*Chlorella protothecoides*), Ho et al. (2014) (*Scenedesmus* sp.), and Xie et al. (2013) (*Desmodesmus* sp.). A reasonable estimate for lutein contents in microalgae biomass is 5 g/kg, mostly in free lutein form. The main interest of microalgae as potential producers of carotenoids, mostly in certain species, lutein, lies in the greater ease of extracting the said pigment directly from the processing of microalgal biomass, the possible obtaining of other high-added value byproducts at the same time in the lutein extraction procedure, and its high lutein content, reaching up to 1.2% of its total dry weight (Lin et al. 2015). Moreover, other advantages of microalgae as lutein producers are remarkable (Fernandez-Sevilla et al. 2010; Yen et al. 2013):

1. Microalgae are a cheap and effective bioresource that can be used to produce value-added compounds, including chemicals, vitamins, carotenoids, and polysaccharides.
2. Microalgae growth rate is 5–10 times that that of higher plants.
3. Microalgae, which can be cultivated in seawater or brackish water and on nonarable land, do not compete for resources with conventional agriculture.
4. Microalgal biomasses can be harvested all year.

However, obtaining lutein from microalgae still has some notable limitations that could be studied, in order to achieve a fully profitable microalgal commercialization of this carotenoid. Among these, the most important ones would be the harvesting of microalgae biomass and the subsequent industrial steps of extraction and purification, especially at the energy level, requiring a redesign of the entire procedure to achieve a highly energetic efficient, environmentally friendly, and sustainable circular economy (Lin et al. 2015).

3.3 Improving Lutein Production from Microalgae

To make the production of lutein from microalgae profitable, it is necessary to increase the production of lutein in microalgae. Currently, there are many studies on the use of microalgae for the production of lutein, which optimizes not only the production but also the extraction and purification of the compound. In this chapter, we will deal with the production of lutein in microalgae by modifying the culture medium and conditions (Table 7.1) and also the genetic manipulation of microalgae to improve the synthesis of lutein.

3.3.1 Improving Lutein Production by Modifying Culture Conditions

It is well described that the conditions of stress or nutritional deficiency, as well as other types of stress (pH, light, temperature, etc.), cause in the microalgae an increase of the synthesis of carotenoids and therefore, an increase in the production of lutein, in order to avoid oxidative stress. This fact has served as the basis for many studies, in which the different forms and conditions of cultivation are studied to increase the productivity of microalgae.

Photoautotrophic Cultivation

The current productivity of microalgal biomass in photoautotrophic cultivation ranges from 0.055 to 0.061 g/L/day at the laboratory scale and is much lower at the industrial scale and, therefore, cannot meet the demands of the global lutein market (Rodolfi et al. 2009). Although heterotrophic fermentation can dramatically enhance the algal biomass productivity and provide large amounts of protein or oil resources (Gao et al. 2008), the low content of pigments greatly limits the economics of lutein extraction and results in a severe bottleneck for its commercial production in microalgae (Xiao et al. 2018).

The heterotrophic-photoautotrophic transition culture mode was investigated in *Auxenochlorella protothecoides*, for lutein accumulation, changing from organic carbon to increase biomass in dark fermentation to irradiation under nitrogen-rich conditions. This strategy increased the lutein content 10 times along with chloroplast

Table 7.1 Productivity and production of lutein from different microalgae

Microalga	Culture conditions	Productivity/ production	Reference
Auxenochlorella protothecoides	Heterotrophic-photoautotrophic transition culture	12.36 mg/L/day	Yibo Xiao et al. (2018)
Chlorella sorokiniana mutant	Semicontinuous cultivation	6.24 mg/L/day	Chen et al. (2019)
Chlorella sorokiniana mutant	Outdoor cultivation conditions (temperature of 35 °C/25 °C for a 12 h/12 h light/dark cycle)	3.34 mg/L/day	Chen et al. (2019)
Chlorella protothecoides	Add 0.01 mmol/L H_2O_2 and 0.5 mmol/L NaClO to the culture medium	31.4 mg/L/day	Wei et al. (2008)
Chlorella protothecoides	Add to the culture medium: 0.1 mmol/L H_2O_2 and 0.01 mmol/L NaClO plus 0.5 mmol/L Fe^2	29.8 mg/L/day	Wei et al. (2008)
Chlamydomonas strain JSC4	Temperature of 20–25 °C and ratio of 3:1 (white light-blue light)	3.25 mg/L/day	Zhao et al. (2019)
Chlamydomonas strain JSC4	High light irradiation of 625 μmol/ m^2/s	5.08 mg/L/day	Zhao et al. (2019)
Chlamydomonas JSC4	Salinity gradient	1.92 mg/L/day	Zhao et al. (2019)
Chlamydomonas strain JSC4	Light intensity to 750 μmol/m^2/s	1821.5 mg/L/day	Ma et al. (2019)
Scenedesmus obliquus FSP-3	Using a TL5 fluorescent lamp at a light intensity of 300 μmol/m^2/s	4.08 mg/L/day	Ho et al. (2014)
Coccomyxa onubensis	Continuous illumination with PAR (photosynthetically active radiations) + UVA (ultraviolet A, 8.7 W/m^2)	7.07 mg/g DW	Bermejo et al. (2018)
Coccomyxa onubensis	The addition of 100 mM NaCl	7.80 mg/g DW	Bermejo et al. (2018)
Desmodesmus sp. F51	Light intensity and initial nitrate concentration were 600 μmol/m^2/s and 2.2 mM, respectively	3.56 ± 0.10 mg/L/ day	Xie et al. (2013)
Desmodesmus sp. F51	Bicarbonate-C/ammonium-N ratio and ammonium-N concentration were 1:1 and 150 mg/L, respectively	5.22 mg/L/day	Xie et al. (2017)
Mutant MR-16 of *C. sorokiniana*	By random mutagenesis using MNNG	7.00 mg/g DW	Cordero et al. (2011)

regeneration and little biomass loss in 48 h. The highest lutein productivity and production in the heterotrophic-photoautotrophic transition culture reached 12.36 mg/L/day and 34.13 mg/L, respectively, within 7 days (Xiao et al. 2018).

Semicontinuous Cultivation

Chen et al. (2019) have shown that using semicontinuous cultivation, in a *Chlorella sorokiniana* mutant, with a medium replacement ratio of 75% resulted in a higher lutein productivity and lutein concentration of 6.24 mg/L/day and 50.6 mg/L, respectively, which were markedly higher than those obtained from batch and fed-batch cultivation. Also, they show that cultivation of these microalgae under simulated outdoor cultivation conditions (i.e., temperature of 35 °C/25 °C for a 12 h/12 h light/dark cycle) could achieve the highest lutein productivity and lutein concentration of 3.34 mg/L/day and 30.8 mg/L, respectively.

Oxidative Stress

The addition of 0.1 mmol/L H_2O_2 and 0.01 mmol/L NaClO plus 0.5 mmol/L Fe^{2+} to the culture enhanced the lutein content from 1.75 to 1.90 mg/g and 1.95 mg/g, respectively, in *Chlorella protothecoides*. The lutein content further increased to 1.98 mg/g when 0.01 mmol/L H_2O_2 and 0.5 mmol/L NaClO were added. The maximum yield of lutein (28.5, 29.8 and 31.4 mg/L) and a high biomass concentration (15.0, 15.3 and 15.9 g/L) were also achieved through the above treatments (Wei et al. 2008).

Light and Temperature

Environmental conditions, including light quality, temperature, and light wavelength mixing ratio, were individually altered to enhance the cell growth rate and lutein production in *Chlamydomonas strain JSC4*. Zhao et al. (2019) showed that optimal cell growth was obtained under white light and a temperature of 35 °C, while the optimal lutein content was obtained under blue light and a lower temperature of 20–25 °C. The best lutein production occurred when using a mixing ratio of 3:1 (white light-blue light). Among them, the two-stage strategy proved to be effective markedly improving lutein content from 2.52 to 4.24 mg/g and resulting in the highest lutein productivity of 3.25 mg/L/day. In addition, the light intensity was manipulated, in the same microalga, to enhance cell growth and lutein production. High lutein productivity (5.08 mg/L/day) was achieved under high light irradiation of 625 μmol/m^2/s. Further increase in light intensity to 750 μmol/m^2/s enhanced the biomass productivity to 1821.5 mg/L/day, but led to a decrease in lutein content (R. Ma et al. 2019).

In the microalga *Scenedesmus obliquus FSP-3*, the results demonstrate that using white LED resulted in better lutein production efficiency when compared to the other three monochromatic LEDs (red, blue, and green). The optimal lutein productivity of 4.08 mg/L/day was obtained when using a TL5 fluorescent lamp at a light intensity of 300 μmol/m^2/s, and this performance is better than that reported in most related studies. Moreover, the time-course profile of lutein accumulation in the microalga shows that the maximal lutein content and productivity were obtained at the onset of nitrogen depletion (Ho et al. 2014).

Under continuous illumination with PAR (photosynthetically active radiations) + UVA (ultraviolet A, 8.7 W/m^2), the microalga *Coccomyxa onubensis* showed a growth rate of 0.40 day^{-1} and produced 226.3 mg/L/day biomass, containing lipids (487.26 mg/g DW) and lutein (7.07 mg/g DW) (Bermejo et al. 2018).

Culture Medium

The medium types, nitrate N, and sea salt concentration are individually investigated to promote the cell growth rate and lutein production of marine microalga *Chlamydomonas JSC4* (Y. Xie et al. 2019). It has clearly demonstrated that salinity is a significant inducer of lutein accumulation by strain *JSC4* and that lutein production can be successfully optimized using the salinity-gradient strategy, which is beneficial for the outdoor large-scale lutein production in the future. An innovative salinity-gradient strategy is operated to dramatically enhance biomass productivity (560 mg/L/day) and lutein content (3.42 mg/g), resulting in the optimal lutein productivity (1.92 mg/L/day).

In the eukaryotic microalga *Coccomyxa onubensis*, the addition of 100 mM NaCl improved the growth rate (from 0.30 to 0.54 day^{-1}), biomass productivity (from 122.50 to 243.75 mg/L/day), and lipid accumulation (from 300.39 to 416.16 mg/g DW) and lutein production from 5.30 to 6.70 mg/g DW. However, when 200–500 mM salt was added, its growth was inhibited, but there was a significant induction of lutein (up to 7.80 mg/g DW) (Bermejo et al. 2018). In the cases of mixotrophic cultures grown on urea, *C. onubensis* accumulated up to 3.55 mg/g of lutein, showing that mixotrophic cultivation on urea efficiently enhances growth and productivity (Casal et al. 2011).

In the microalga *Desmodesmus* sp. *F51*, the medium composition, nitrate concentration, and light intensity were manipulated to improve the phototrophic growth and lutein production. It was found that a nitrogen-sufficient condition was required for lutein accumulation, while a high light intensity enhanced cell growth but caused a decrease in the lutein content. The best cell growth and lutein production occurred when the light intensity and initial nitrate concentration were 600 μmol/m^2/s and 8.8 mM, respectively. The highest lutein productivity (3.56 ± 0.10 mg/L/day) and content (5.05 ± 0.20 mg/g) were obtained when pulse feeding of 2.2 mM nitrate was employed (Xie et al. 2013).

Also, the type and concentration of inorganic carbon and nitrogen sources were manipulated in *Desmodesmus* sp. *F51* to improve cell growth and lutein productivity. Using nitrate as nitrogen source, the better cell growth and lutein accumulation were obtained under 2.5% CO_2 supply when compared to the addition of $NaHCO_3$

or Na_2CO_3. The highest biomass productivity (939 mg/L/day) and lutein productivity (5.22 mg/L/day) were obtained when the bicarbonate-C/ammonium-N ratio and ammonium-N concentration were 1:1 and 150 mg/L, respectively. The lutein productivity of 5.22 mg/L/day is the highest value ever reported in the literature using batch phototrophic cultivation (Xie et al. 2017).

3.3.2 Improving Lutein Production by Genetic Manipulation

Genetic engineering can open up the possibility of enhancing the productivity of commercial carotenoids. If it is certain that there are few microalgae that today have been genetically manipulated in a stable manner (Díaz-Santos 2019), many researchers work in this way, which *Chlorella sorokiniana* has been selected for lutein production, since it showed both a high content in this carotenoid and a high growth rate. Cordero et al. (2011) have obtained high lutein-yielding mutant of *C. sorokiniana* by random mutagenesis, using *N*-methyl-*N'*-nitro-nitrosoguanidine (MNNG) as a mutagen, and selecting mutants by their resistance to the inhibitors of the carotenogenic pathway, nicotine and norflurazon. The mutant MR-16 exhibited a two-fold higher volumetric lutein content than that of the wild type, attaining values of 42.0 mg/L, and mutants DMR-5 and DMR-8 attained a lutein cellular content of 7.0 mg/g DW (Cordero et al. 2011). Another *Chlorella zofingiensis* mutant (CZ-bkt1) was found to consist of a dysfunctional carotenoid ketolase, leading to the accumulation of zeaxanthin rather than to its downstream ketocarotenoid astaxanthin. CZ-bkt1 accumulated zeaxanthin up to 7.00 ± 0.82 mg/g when induced by high light irradiation and nitrogen deficiency and up to 36.79 ± 2.23 mg/L by additional feeding with glucose. Furthermore, in addition to zeaxanthin, CZ-bkt1 also accumulated high amounts of β-carotene (7.18 ± 0.72 mg/g or 34.64 ± 1.39 mg/L) and lutein (13.81 ± 1.23 mg/g or 33.97 ± 2.61 mg/L) (Huang et al. 2018).

The overexpression of CzPSY in *Chlamydomonas reinhardtii*, by nuclear transformation, has led to an increase in the corresponding CzPSY transcript level as well as in the content of the carotenoids violaxanthin and lutein which were 2.0- and 2.2-fold higher than in untransformed cells. Phytoene synthase gene from the green microalga *Chlorella zofingiensis* (CzPSY), involved in the first step of the carotenoid biosynthetic pathway, has been performed (Cordero et al. 2011).

4 Lutein Extraction and Purification from Microalgae Biomass

Although lutein is currently separated and purified from marigold flowers by a saponification–extraction–recrystallization method (F. Khachik 1999), no industrial processes have been proposed considering microalgae biomass as raw material. Some work has been done with microalgae on the optimization of separate operations as disruption of algal cells (W. Farrow and Tabenkin 1966; M. Ruane 1977; A. M. Nonomura 1987; Mendes-Pinto et al. 2001), extraction of carotenoids with organic solvents (T. Roukas and Mantzouridou 2001; M. A. Hejazi et al. 2002; H. Li et al. 2002; G. An and Cho 2003; P. K. Park et al. 2007), and/or saponification of vegetable biomass (M. Kimura et al. 1990; R. Fernandez et al. 2000; F. Granado et al. 2001; E. Larsen and Christensen 2005). On the other hand, the few studies dealing with the overall process of lutein recovery from microalgae biomass are developed at a very small scale because they are oriented to analytical purposes (D. R. Kull and Pfander 1997; H. Li et al. 2002). These methods require much time and high volumes of solvents and disregard the importance of cell disruption in the yield of the process. The development of a new single-step method that skips drying and combines extraction, saponification, and purification approach may save both time and solvent. Previously, in 2016, Wang et al. developed a procedure for a combined method of lutein extraction from marigold flowers, but similar studies for more microalgal lutein extraction and purification are rare. It is of great interest to reduce the operating units and to investigate the kinetics of this process, in order to minimize the time and cost of free lutein recovery from microalgae. In the study shown by Gong and co-workers in 2017, it has been described that (1) the development and process analysis of a single-step extraction, saponification, and purification method was conducted for extraction of lutein from wet microalgae biomass; (2) the extraction kinetics of microalgal lutein extraction were first monitored under different conditions for a better understanding and optimization of the process; (3) the experimental data was fitted using mathematical modelling; and (4) the diffusion coefficients were determined and analyzed for different conditions to determine the extraction rate. Moreover, different methods have been described depending on the microalga. In 2008, Cerón and co-workers developed a complete process for the recovery of lutein from the lutein-rich new strain *S. almeriensis*. The method takes into account the existence of a hard cell wall in this strain, and thus, a cell disruption step is included. In addition, an alkaline treatment is done to complete cell disruption and help remove ionizable lipids. The last stage is a multistep solvent extraction procedure, which is then optimized to minimize the number of extractions and therefore the amount of solvent used. The optimized method allows the obtainment of a final lutein-rich extract that could be used for commercial purposes. Li HB and co-workers have optimized a simple and efficient method for the isolation and purification of lutein from the microalga *Chlorella vulgaris*. Crude lutein was obtained by extraction with dichloromethane from the microalga after saponification. Partition values of lutein in the two-phase system of ethanol–water–dichloromethane at different ratios were measured by HPLC so as to assist the determination of an appropriate condition for washing water-soluble impurities in the crude lutein. Partition values of lutein in another two-phase system of ethanol–water–hexane at different ratios were also measured by HPLC for determining the condition for removing fat-soluble impurities. The water-soluble impurities in the crude lutein were removed by washing with 30% aqueous ethanol, and the fat-soluble impurities were removed by extraction with hexane. The final purity of lutein obtained was 90–98%, and the yield was 85–91%.

5 Lutein Market

Commercial lutein production is concerned with all-trans lutein content, not total lutein content. Plant materials contain all-trans isomer of lutein; nevertheless, cis isomers of lutein are generated, apart from other agents, also by the actions of light and temperature. Lutein esters must be deesterified before they are absorbed by the body since the in vivo hydrolysis of lutein esters into lutein occurs with an efficacy of less than 5% (Breithaupt et al. 2002a, b; Granado et al. 2002). Although the claims on benefits of lutein products differentiate between free lutein and lutein esters, evidence proving these claims is lacking. Medical research that used lutein esters instead of free lutein improved patients with macular disease (Landrum et al. 1997; Koh et al. 2004). Other data (Subagio et al. 2001) showed that esterification had no effect on the antioxidant activity of lutein, because lutein esters have higher bioavailability. Finding about certain antioxidant reactions is forthcoming; however, they may not completely identify which is better: free lutein or lutein esters (Lin et al. 2015).

Currently, the commercialization of lutein is carried out mainly in the form of capsules (pure lutein) or as a mixture with other carotenoids in suspension being part of corn and sunflower oils. Some of these commercialized products are FloraGLO® from Kemin Foods or Xangold® from Cognis (Fig. 7.4).

The growing demand of health supplements that contain lutein has been largely driven by the lutein market, over the past few years. Lutein is ranked as the second most high-value commercial carotenoids after β-carotene with a global market value (März, 2015). The market for lutein, in terms of value, is estimated to be USD 263.8 million in 2017 and is projected to reach USD 357.7 million by 2022, at a CAGR of 6.3% from 2017 (A. McWilliams 2018). The Asia Pacific region is pro-

Fig. 7.4 Different lutein commercial products

jected to be the fastest-growing market for lutein over the next 5 years. In this region, countries such as India, one of the fastest-growing markets for lutein in the Asia Pacific region China, and Japan hold a major share of more than 60% of the Asia Pacific lutein market.

6 Future Prospect and Concluding Remarks

Since the potential of lutein as antioxidant and anti-inflammatory agent is known, research about these properties and the production and market of lutein have been increased exponentially during the last decade. The most studies are focused on the influence of lutein consumption in human eye-related diseases wherein the optimal beneficial of this carotenoid in the prevention and improvement of these king of health problems has been demonstrated. However, the studies in relation to other diseases are today scanty and with clinic results not clear or controversial, although the antioxidant role of the lutein from the diet is confirmed. In this aspect, microalgae are a potential good candidate to produce lutein to large scale even being marigold the most used today. In relation to this point, an improvement of the lutein production costs from microalgae via a circular economy using wastewater as the culture medium or coupling the use of CO_2 from main pollutant industries and the use of genetic engineering and synthetic biology could be a great way to be considered for the future of lutein as nutraceutical from microalgae.

References

Alam, M. A., & Wang, Z. (Eds.). (2019). *Microalgae biotechnology for development of biofuel and wastewater treatment* (pp. 1–655). New York, NY: Springer.

An, G., & Cho, E. (2003). Preparation of the red yeast, *Xanthophyllomyces dendrorhous*, as feed additive with increased availability of astaxanthin. *Biotechnology Letters, 25*, 767–771.

Bermejo, E., Ruiz-Domínguez, M. C., Cuaresma, M., Vaquero, I., Ramos-Merchante, A., Vega, J. M., Vílchez, C., & Garbayo, I. (2018). Production of lutein, and polyunsaturated fatty acids by the acidophilic eukaryotic microalga *Coccomyxa onubensis* under abiotic stress by salt or ultraviolet light. *Journal of Bioscience and Bioengineering, 125*(6), 669–675.

Bosma, T. L., Dole, J. M., & Maness, N. O. (2003). Optimizing marigold (*Tagetes erecta* L.) petal and pigment yield crop science. Abstract. *Crop Ecology, Management & Quality, 43*(6), 2118–2124.

Breithaupt, D. E., Bamedi, A., & Wirt, U. (2002a). Carotenol fatty acid esters: Easy substrates for digestive enzymes. *Comparative Biochemistry and Physiology. Part B, Biochemistry & Molecular Biology, 132*, 721–728.

Breithaupt, D. E., Wirt, U., & Bamedi, A. (2002b). Differentiation between lutein monoester regioisomers and detection of lutein diesters from marigold flowers (*Tagetes erecta* L.) and several fruits by liquid chromatography-mass spectrometry. *Journal of Agricultural and Food Chemistry, 50*, 66–70.

Buscemi, S., Corleo, D., Di Pace, F., Petroni, M. L., Satriano, A., & Marchesini, G. (2018). The effect of lutein on eye and extra-eye health. *Nutrients, 10*(9), 1321. https://doi.org/10.3390/nu10091321.

Capa-Robles, W., Paniagua-Michel, J., & Soto, J. O. (2009). The biosynthesis and accumulation of beta-carotene in *Dunaliella salina* proceed via the glyceraldehyde 3-phosphate/pyruvate pathway. *Natural Product Research, 23*(11), 1021–1028.

Casal, C., Cuaresma, M., Vega, J. M., & Vilchez, C. (2011). Enhanced productivity of a lutein enriched novel acidophile microalga grown on urea. *Marine Drugs, 9*(1), 29–42.

Cerón, C. M., Campos, I., Sánchez, J. F., Acién, F. G., Molina, E., & Fernández-Sevilla, J. M. (2008). Recovery of lutein from microalgae biomass: Development of a process for *Scenedesmus almeriensis* biomass. *Journal of Agricultural and Food Chemistry, 56*(24), 11761–11766. https://doi.org/10.1021/jf8025875.

Chen, C. Y., Ho, S. H., Liu, C. C., & Chang, J. S. (2017). Enhancing lutein production with *Chlorella sorokiniana* Mb-1 by optimizing acetate and nitrate concentrations under mixotrophic growth. *Journal of the Taiwan Institute of Chemical Engineers, 79*(Suppl. C), 88–96.

Chen, J. H., Chen, C. Y., Hasunuma, T., Kondo, A., Chang, C. H., Ng, I. S., & Chang, J. S. (2019). Enhancing lutein production with mixotrophic cultivation of *Chlorella sorokiniana* MB-1-M12 using different bioprocess operation strategies. *Bioresource Technology, 278*, 17–25.

Chitchumroonchokchai, C., Schwartz, S. J., & Failla, M. L. (2004). Assessment of lutein bioavailability from meals and a supplement using simulated digestion and caco-2 human intestinal cells. *Journal of Nutrition, 134*, 2280–2286.

Chung, R. W. S., Leanderson, P., Lundberg, A. K., & Jonasson, L. (2017). Lutein exerts anti-inflammatory effects in patients with coronary artery disease. *Atherosclerosis, 262*, 87–93. https://doi.org/10.1016/j.atherosclerosis.2017.05.008.

Cordero, B. F., Obraztsova, I., Couso, I., Leon, R., Vargas, M. A., & Rodriguez, H. (2011). Enhancement of lutein production in *Chlorella sorokiniana* (Chorophyta) by improvement of culture conditions and random mutagenesis. *Marine Drugs, 9*(9), 1607–1624.

Del Campo, J. A., Moreno, J., Rodriguez, H., Vargas, M. A., Rivas, J., & Guerrero, M. G. (2000). Carotenoid content of chlorophycean microalgae: Factors determining lutein accumulation in *Muriellopsis* sp. (Chlorophyta). *Journal of Biotechnology, 76*, 51–59.

Díaz-Santos, E. (2019). Towards the genetic manipulation of microalgae to improve the carbon dioxide fixation and the production of biofuels: Present status and future prospect. In M. Alam & Z. Wang (Eds.), *Microalgae biotechnology for development of biofuel and wastewater treatment*. Singapore: Springer. https://doi.org/10.1007/978-981-13-2264-8_7.

Farrow, W. M., & Tabenkin, K. (1966). *Process for the preparation of lutein*. U.S. *Patent No*. 3,280,502.

Fernandez, R. X. E., Shier, N. W., & Watkins, B. A. (2000). Effect of alkali saponification, enzymatic hydrolysis and storage time on the total carotenoid concentration of Costa Rica crude palm oil. *Journal of Food Composition and Analysis, 13*, 179–187.

Fernandez-Sevilla, J. M., Fernandez, F. G. A., & Grima, E. M. (2010). Biotechnological production of lutein and its applications. *Applied Microbiology and Biotechnology, 86*, 27–40.

Fernández-Sevilla, J. M., Acién Fernández, F. G., & Molina Grima, E. (2010). Biotechnological production of lutein and its applications. *Applied Microbiology and Biotechnology, 86*, 27–40. https://doi.org/10.1007/s00253-009-2420-y.

Francisco, D. V., & Octavio, P. L. (1996). Correlation of HPLC and AOAC methods to assess the all-trans-lutein content in marigold flowers. *Journal of the Science of Food and Agriculture, 72*, 283–290.

Gao, C., Xiong, W., Zhang, Y., Yuan, W., & Wu, Q. (2008). Rapid quantitation of lipid in microalgae by time-domain nuclear magnetic resonance. *Journal of Microbiological Methods, 75*(3), 437–440.

Gong, M., Wang, Y., & Bassi, A. (2017). Process analysis and modelling of a single-step lutein extraction method for wet microalgae. *Applied Microbiology and Biotechnology, 101*, 80–89. https://doi.org/10.1007/s00253-017-8496-x.

Gong, X., Smith, J. R., Swanson, H. M., & Rubin, L. P. (2018). Carotenoid lutein selectively inhibits breast cancer cell growth and potentiates the effect of chemotherapeutic agents through ROS-mediated mechanisms. *Molecules, 23*, 905.

Granado, F., Olmedilla, B., Gil-Martinez, E., & Blanco, I. A. (2001). Fast, reliable and low-cost saponification protocol for analysis of carotenoids in vegetables. *Journal of Food Composition and Analysis, 14*, 479–489.

Granado, F., Olmedilla, B., & Blanco, I. (2002). Serum depletion and bioavailability of lutein in type I diabetic patients. *European Journal of Nutrition, 41*, 47–53.

Granado, F., Olmedilla, B., & Blanco, I. (2003). Nutritional and clinical relevance of lutein in human health. *The British Journal of Nutrition, 90*(3), 487–502. https://doi.org/10.1079/BJN2003927.

Grether-Beck, S., Marini, A., Jaenicke, T., Stahl, W., & Krutmann, J. (2017). Molecular evidence that oral supplementation with lycopene or lutein protects human skin against ultraviolet radiation: Results from a double-blinded, placebo-controlled, crossover study. *British Journal of Dermatology, 176*, 1231–1240. https://doi.org/10.1111/bjd.15080.

Hejazi, M. A., De Lamarliere, C., Rocha, J. M. S., Vermuë, M., Tramper, J., & Wijffels, R. H. (2002). Selective extraction of carotenoids from the microalga *Dunaliella salina* with retention of viability. *Biotechnology and Bioengineering, 79*, 29–36.

Ho, S. H., Chan, M. C., Liu, C. C., Chen, C. Y., Lee, W. L., Lee, D. J., & Chang, J. S. (2014). Enhancing lutein productivity of an indigenous microalga *Scenedesmus obliquus* FSP-3 using light-related strategies. *Bioresource Technology, 152*, 275–282.

Huang, W., Lin, Y., He, M., Gong, Y., & Huang, J. (2018). Induced high-yield production of zeaxanthin, lutein, and β-carotene by a mutant of *Chlorella zofingiensis*. *Journal of Agricultural and Food Chemistry, 66*(4), 891–897.

Jian-Hao, L., Duu-Jong, L., & Jo-Shu, C. (2015). Lutein production from biomass: *Marigold* flowers versus microalgae. *Bioresource Technology, 184*, 421–428.

Khachik, F. (1999). *Process for extraction and purification of lutein, zeaxanthin and rare carotenoids from marigold flowers and plants*. Publication number: WO1999020587A1. U.S. *Patent No.* 7,173,145.

Kimura, M., Rodriguez-Amaya, D. B., & Godoy, H. T. (1990). Assessment of the saponification step in the quantitative determination of carotenoids and provitamins A. *Food Chemistry, 35*, 187–195.

Koh, H.-H., Murray, I. J., Nolan, D., Carden, D., Feather, J., & Beatty, S. (2004). Plasma and macular responses to lutein supplement in subjects with and without age-related maculopathy: a pilot study. *Experimental Eye Research, 79*(1), 21–27.

Koushan, K., Rusovici, R., Li, W., Ferguson, L. R., & Chalam, K. V. (2013). The role of lutein in eye-related disease. *Nutrients, 5*(5), 1823–1839. https://doi.org/10.3390/nu5051823.

Kull, D. R., & Pfander, H. (1997). Isolation and structure elucidation of two (Z)-isomers of lutein from the petals of rape (*Brassica napus*). *Journal of Agricultural and Food Chemistry, 45*, 4201–4203.

Landrum, J. T., Bone, R. A., Joa, H., Kilburn, M. D., Moore, L. L., & Sprague, K. E. A. (1997). One year study of the macular pigment: The effect of 140 days of a lutein supplement. *Experimental Eye Research, 65*, 57–62.

Larsen, E., & Christensen, L. P. (2005). Simple saponification method for the quantitative determination of carotenoids in green vegetables. *Journal of Agricultural and Food Chemistry, 53*, 6598–6602.

Li, H., Jiang, Y., & Chen, F. (2002). Isolation and purification of lutein from the microalga *Chlorella vulgaris* by extraction after saponification. *Journal of Agricultural and Food Chemistry, 50*, 1070–1072.

Liang, S. X., Tang, D. C., & Yang, Y. Z. (2007). Comparative studies of fresh flower yield and lutein content of marigold. *Northern Horticulture, 6*, 124–125.

Liao, S. M., Zheng, W., Zhu, J., et al. (2017). Specific correlation between the major chromosome 10q26 haplotype conferring risk for age-related macular degeneration and the expression of HTRA1. *Molecular Vision, 23*, 318–333.

Lin, J. H., Lee, D. J., & Chang, J. S. (2015). Lutein production from biomass: Marigold flowers versus microalgae. *Bioresource Technology, 184*, 421–428.

Ma, R., Zhao, X., Xie, Y., Ho, S. H., & Chen, J. (2019). Enhancing lutein productivity of *Chlamydomonas* sp. via high-intensity light exposure with corresponding carotenogenic genes expression profiles. *Bioresource Technology, 275*, 416–420.
Maci, S., & Santos, R. (2015). The beneficial role of lutein and zeaxanthin in cataracts. *Nutrafoods, 14*, 63. https://doi.org/10.1007/s13749-015-0014-0.
Manayi, A., Abdollahi, M., Raman, T., Nabayi, S. F., Habtemariam, S., Daglia, M., & NAbayi, S. M. (2016). Lutein and cataract: From bench to bedside. *Critical Reviews in Biotechnology, 36*(5), 829–839.
Maria, A. G., Graziano, R., & D'Orazio, N. (2015). Carotenoids: Potential allies of cardiovascular health. *Food & Nutrition Research, 59*, 1. https://doi.org/10.3402/fnr.v59.26762.
März, U. (2015). FOD025E-The Global Market for Carotenoids. In: BCC Research.
McWilliams, A. (2018). *The global market for carotenoids. FOD025F. BCC research report overview*. Wellesley, MA: BCC Publishing.
Mendes-Pinto, M. M., Raposo, M. F. J., Bowen, J., Young, A. J., & Morais, R. (2001). Evaluation of different cell disruption processes on encysted cells of *Haematococcus pluvialis*: Effects on astaxanthin recovery and implications for bio-availability. *Journal of Applied Phycology, 13*, 19–24.
Natchigal, A. M., Stringheta, A. C. O., Bertoldi, M. C., & Stringheta, P. C. (2012). Quantification and characterization of lutein from Tagetes (*Tagetes patula* L.) and Calendula (*Calendula officinalis* L.) flowers. *Acta Horticulturae, 939*, 309–314.
Nonomura, A. M. (1987). *Process for producing a naturally-derived carotene/oil composition by direct extraction from algae*. U.S. *Patent No*. 4,680,314.
Park, P. K., Kim, E. Y., & Chu, K. H. (2007). Chemical disruption of yeast cells for the isolation of carotenoid pigments. *Separation and Purification Technology, 53*, 148–152.
Pennington, K. L., & DeAngelis, M. M. (2016). Epidemiology of age-related macular degeneration (AMD): Associations with cardiovascular disease phenotypes and lipid factors. *Eye and Vision, 3*, 34. https://doi.org/10.1186/s40662-016-0063-5.
Perrone, S., Tei, M., Longini, M., & Buonocore, G. (2016). The multiple facets of lutein: A call for further investigation in the perinatal period. *Oxidative Medicine and Cellular Longevity, 2016*, 1–8. https://doi.org/10.1155/2016/5381540.
Pintea, A., Bele, C., Andrei, S., & Socaciu, C. (2003). HPLC analysis of carotenoids in four varieties of *Calendula officinalis* L. flowers. *Acta Biologica Szegediensis, 47*, 37–40.
Rodolfi, L., Chini Zittelli, G., Bassi, N., Padovani, G., Biondi, N., Bonini, G., & Tredici, M. R. (2009). Microalgae for oil: Strain selection, induction of lipid synthesis and outdoor mass cultivation in a low-cost photobioreactor. *Biotechnology and Bioengineering, 102*, 100–112.
Rodrigues, D. B., Mercadantea, Z. A., & Mariutti, B. L. R. (2019). Marigold carotenoids: Much more than lutein esters. *Food Research International, 119*, 653–664.
Roukas, T., & Mantzouridou, F. (2001). An improved method for extraction of β-carotene from *Blakeslea trispora*. *Applied Biochemistry and Biotechnology, 90*, 37–45.
Ruane M. (1977). *Extraction of caroteniferous materials from algae*. Australia *Patent No*. 72,395,74.
Sathasivam, R., & Ki, J. S. (2018). A review of the biological activities of microalgal carotenoids and their potential use in healthcare and cosmetic industries. *Marine Drugs, 16*(1), 26. https://doi.org/10.3390/md16010026.
Schwender, J., Gemünden, C., & Lichtenthaler, H. K. (2001). Chlorophyta exclusively use the 1-deoxyxylulose 5-phosphate/2-C-methylerythritol 4-phosphate pathway for the biosynthesis of isoprenoids. *Planta, 212*(3), 416–423.
Seddon, J. M., Ajani, U. A., Sperduto, R. D., Hiller, R., Blair, N., Burton, T. C., Farber, M. D., Gragoudas, E. S., Haller, J., et al. (1994). Dietary carotenoids, vitamins A, C, and E, and advanced age-related macular degeneration. Eye Disease Case-Control Study Group. *JAMA, 272*, 1413–1420.

Shi, X. M., Jiang, Y., & Chen, F. (2002). High-yield production of lutein by the green microalga *Chlorella protothecoides* in heterotrophic fed-batch culture. *Biotechnology Progress, 18*, 723–772.

Sommerburg, O., Keunen, J. E., Bird, A. C., & van Kuijk, F. J. (1998). Fruits and vegetables that are sources for lutein and zeaxanthin: The macular pigment in human eyes. *The British Journal of Ophthalmology, 82*(8), 907–910.

Subagio, A., Sari, P., & Morita, N. (2001). Simultaneous determination of (+)-catechin and (-)-epicatechin in cacao and its products by high-performance liquid chromatography with electrochemical detection. *Phytochemical Analysis, 12*(4), 271–276.

Sun, Z., Li, T., Zhou, Z., & Jiang, Y. (2015). Microalgae as a source of lutein: Chemistry, biosynthesis and carotenogenesis. In C. Posten & S. Feng Chen (Eds.), *Microalgae biotechnology. Advances in biochemical engineering/biotechnology* (p. 153). New York, NY: Springer.

Trumbo, P. R., & Ellwood, K. C. (2006). Lutein and zeaxanthin intakes and risk of age-related macular degeneration and cataracts: An evaluation using the Food and Drug Administration's evidence-based review system for health claims. *The American Journal of Clinical Nutrition, 84*(5), 971–974. https://doi.org/10.1093/ajcn/84.5.971.

Varela, J. C., Pereira, H., Vila, M., & León, R. (2015). Production of carotenoids by microalgae: Achievements and challenges. *Photosynthesis Research, 125*(3), 423–436. Review. Erratum in: Photosynth Res. 2016;127(2):285–286.

Wei, D., Chen, F., Chen, G., Zhang, X., Liu, L., & Zhang, H. (2008). Enhanced production of lutein in heterotrophic *Chlorella protothecoides* by oxidative stress. *Science in China. Series C, Life Sciences, 51*(12), 1088–1093.

Woodall, A. A., Lee, S. W., Weesie, R. J., Jackson, M. J., & Britton, G. (1997). Oxidation of carotenoids by free radicals: Relationship between structure and reactivity. *Biochimica et Biophysica Acta, General Subjects, 1336*, 33–42.

Xiao, Y., He, X., Ma, Q., Lu, Y., Bai, F., Dai, J., & Wu, Q. (2018). Photosynthetic accumulation of lutein in *Auxenochlorella protothecoides* after heterotrophic growth. *Marine Drugs, 16*(8), E283.

Xie, Y., Ho, S. H., Chen, C. N. N., Chen, C. Y., Ng, I. S., Jing, K. J., Chang, J. S., & Lu, Y. (2013). Phototrophic cultivation of a thermo-tolerant *Desmodesmus* sp. for lutein production: Effects of nitrate concentration, light intensity and fed-batch operation. *Bioresource Technology, 144*, 435–444.

Xie, Y., Zhao, X., Chen, J., Yang, X., Ho, S. H., Wang, B., Chang, J. S., & Shen, Y. (2017). Enhancing cell growth and lutein productivity of *Desmodesmus* sp. F51 by optimal utilization of inorganic carbon sources and ammonium salt. *Bioresource Technology, 244*(Pt 1), 664–671. https://doi.org/10.1016/j.biortech.2017.08.022.

Xie, Y., Lu, K., Zhao, X., Ma, R., Chen, J., & Ho, S. H. (2019). Manipulating nutritional conditions and salinity-gradient stress for enhanced lutein production in marine microalga *Chlamydomonas* sp. *Biotechnology Journal, 14*(4), e1800380.

Yen, H. W., Hu, I. C., Chen, C. Y., Ho, S. H., Lee, D. J., & Chang, J. S. (2013). Microalgae-based biorefinery-from biofuels to natural products. *Bioresource Technology, 135*, 166–174.

Zhao, X., Ma, R., Liu, X., Ho, S.-H., Xie, Y., & Chen, J. (2019). Strategies related to light quality and temperature to improve lutein production of marine microalga *Chlamydomonas* sp. *Bioprocess and Biosystems Engineering, 42*(3), 435–443.

Zielińska, M. A., Wesołowska, A., Pawlus, B., & Hamułka, J. (2017). Health effects of carotenoids during pregnancy and lactation. *Nutrients, 9*(8), 838. https://doi.org/10.3390/nu9080838.

Part III
Microalgae for Cosmetic Formulations

Chapter 8
Algae and Ageing

Sakshi Guleri and Archana Tiwari

Abstract Algae are photosynthetic organisms inhabiting diverse habitats and performing an array of pioneer ecological roles. They are extensively researched for a wide range of applications ranging from wastewater remediation, biofuels, therapeutics and cosmeceuticals to plant growth stimulators. The abundance of bioactive compounds from algae imparts unique biochemical and physiological potentials, making them unique organisms with unmatched potentials. Algae serve as an eminent source of cosmeceuticals with anti-ageing benefits as they are rich in carotene and vitamin E concentration that aids in the prevention of cancer and premature skin ageing along with the immune system stimulation. The reactive oxygen species (ROS) are aroused by virtue of oxidative stress that causes the rapid degradation of skin collagen contributing towards the skin wrinkles, a prominent sign of ageing. Marine algae have pronounced effect in combating the skin wrinkles by virtue of their phenolic content. In addition, the algal extracts can also act as natural tyrosinase inhibitor agents leading to skin whitening and fairer look. The marine algae are investigated for their impact on the skin pigmentation with much emphasis on tyrosinase inhibitors. The marine algal extracts contain phlorotannins, which possess great efficacy as anti-inflammatory and hyaluronidase inhibitors. The cosmetic products using algae *Spirulina platensis* extracts are marketed in parts of Europe and South America. The prominent causes of ageing can be well addressed by algal products as they are rich sources of antioxidative enzymes which can effectively encounter the negative effects of ROS. More algal strains with potential secondary metabolites need to be explored for wide application of algae as anti-ageing agents and can be envisaged as novel sources of cosmeceuticals.

Keywords Algae · Ageing · Bioactive compounds · ROS · Cosmeceuticals · Tyrosinase inhibitor

---

S. Guleri · A. Tiwari (✉)
Diatom Research Laboratory, Amity Institute of Biotechnology, Amity University, Noida, India

M. A. Alam et al. (eds.), *Microalgae Biotechnology for Food, Health and High Value Products*, https://doi.org/10.1007/978-981-15-0169-2_8

1 Algae and Its Unique Potential

Algae are explicit class of photosynthetic organisms with enormous potentials and wide uses. The metabolic potential algae are conferred by the magnificent photosynthetic machinery with myriad of photosynthetic pigments. Their success story is validated with their ubiquitous nature, and advent of new technological interfaces has stimulated the attention in algal bioactive compounds and their tremendous applications. The plethora of products derived from algae is quite significant, and their applications are quite vivid and multifaceted in nature ranging from the biofuels like hydrogen, ethanol, butanol, syngas, hydrogen, nutraceuticals, antioxidative enzymes, cosmeceuticals, food and feed. The antioxidative system of algae comprises of supreme enzymes like catalase, superoxide dismutase, peroxidases and many more, which are equipped with the strategy to flourish in extreme circumstances and have evolved many substances for their survival concomitant with their protection. Algae hold immense potentials as source of different products with cosmeceutical value which can open new avenues in dealing with ageing. This chapter elaborates the application of algae in prevention of ageing.

1.1 Bioactive Compounds from Algae

Cosmeceuticals contain dynamic ingredients, for example, vitamins, phytochemicals, chemicals, antioxidants and basic essential oils that are consolidated into the products like creams, moisturizers and treatments. There is a wide diversity in the bioactive compound found in marine algae which envisage their further utilization in cosmeceuticals, more pertinently healthy products for the skin (Kim et al. 2008a, b). Various bioactive compounds are illustrated in Fig. 8.1.

1.1.1 Terpenoids

Terpenoids (isoprenoids) are the biggest and majorly abundant across the broad category of secondary metabolites, in advanced classes of plants inclusive of algae in marine waters, bugs and few microbes. Novel marine terpenoids exhibit extraordinary guarantee as a hotspot for antioxidant compounds in cosmetic product preparations (Paduch et al. 2007; Kang et al. 2004), because of their great penetration enhancing capacities, low fundamental poisonous quality. A steroidal terpenoid extracted from *Phaeophyta* is called fucosterol (Jung et al. 2006; Tang et al. 2002). Fucosterol is generally procured as an essential substance from the extract fractions of algae (Jung et al. 2006). This compound shows increased antioxidative activities by expanding the concentrations of antioxidative enzyme, which regulates the hydrogen peroxide, and examples of these enzymes are superoxide dismutase, catalase, etc. Fucosterol helps the components by counteracting the oxidation of cell

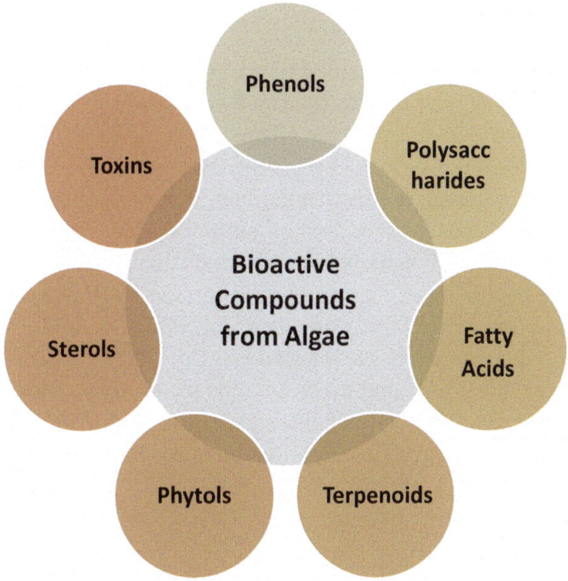

Fig. 8.1 Types of bioactive compounds

membrane because of its imperative function in searching and collecting hydrogen peroxide and re-establishing enzymatic action. Essentially, an expansion of such enzymes shows that this terpenoid likewise aids in the rebuilding of indispensable endogenic antioxidative agents, for example, glutathione (Lee et al. 2003; Pillai et al. 2005).

1.1.2 Carotenoids

Carotenoids are a diversified category of naturally existing tetra terpenoid particles that are incorporated by plants, microscopic organisms and algae. It has been discovered that they protect the cell from oxidative stress (Stahl and Sies 2003). A few carotenoids work as immediate quenchers of reactive oxygen species (Edge et al. 1997). Allenic and acetylenic carotenoids, for example, fucoxanthin and neoxanthin, separately (Bjornland and Aguilarmartinez 1976), are found in red algae, and somewhere around 30 unique carotenoids are distinguished in this class (Schubert et al. 2006). The carotenoids are allotted on the basis of their structures and biosynthetic pathways; they follow the algal classification (Mimuro and Akimoto 2003).

1.1.3 Phenolic Compounds

Marine algae consist of highly potent antioxidative enzymes which are useful in therapeutics. Algae from marine waterbodies constitute a substantial source of different phenolic antioxidative substances (Koivikko et al. 2007). These compounds display a wide scope of physiological properties, for example, hostile to allergenic, against atherogenic, calming, hostile to microbial, against thrombotic, cardioprotective, vasodilatory, and cancer prevention or antibiotic agent impacts (Balasundram et al. 2006). The association of phenolic compounds that are called phlorotannins is observed in some brown algae, for example, *Sargassaceae* (Pavia and Brock 2000; Jormalainen and Honkanen 2004). Also their dry weight is up to 20% and is the concentrates of cell layers and tissues (Koivikko et al. 2005). These are purged from a few brown algae having around eight interconnected rings. Their antioxidative action, which is the result of polyphenols, is exceedingly identified with the phenol rings present that traps electron to search free radicals (Balasundram et al. 2006).

1.1.4 Polysaccharides

Fucoidans

A wide range of algal fusions are continuously utilized in cosmeceutical product formation for a long time as they control consistency, biological active, emollient and skin modifying properties (Kim et al. 2005). For example fucoidans, also called as 'fucan', are a kind of profoundly extended polysaccharide which has a significant percentage of L-fucose, for the most part sulfated and acetylated (Holtkamp et al. 2009), and are found in different types of brown algae (*Phaeophyta*, *Hizikia fusiforme*).

Alginates

Alginates are polysaccharides which are the constituents of the cell structure of the brown algae (Phaeophyceae), namely, *Laminaria* species (*Laminaria digitata*) and also *Durvillaea antarctica*, and are inhabitants of coasts worldwide (Fujimura et al. 2002; Refilda et al. 2009). It is a carboxylic polymer that consists up to 40% w/w of brown algae. The mechanical strength and flexibility are provided by these alginates and help them to adjust to water movements in which they grow. Alginates also help algae to preserve hydration (Rinaudo 2008). For decades, food industry uses alginate as a stabilizer, emulsifier as well as a gelling component. Alginate is likewise utilized as a shear-thinning agent in paper industry. In medical field it is used as a dental impression material and wound dressing (Augst et al. 2006). Alginate has been utilized for ligament recovery and for development of hairlike veins in cell culture and tissue designing (Rinaudo 2008).

1.1.5 Polyunsaturated Fatty Acids (PUFA)

Ocean algae are sources of polyunsaturated fats (Wood 1988), particularly in the lipophilic extracts (Li et al. 2002a). For instance, PUFAs represent huge portion of the total lipids in marine algae, for example, *Tetraselmis suecica*, *Porphyridium cruentum* and *Isochrysis galbana*, containing 20.9%, 17.1% and 17%, respectively. Because marine algae are an enormous and inexhaustible asset, consisting of countless unsaturated fats, it gives promising renewable resources of crude PUFAs for beauty care products (Khotimchenko et al. 2002). Red, brown and green algae have recognizing unsaturated fatty acid profiles which don't rely upon the land area of the algae. In any case, conditions of algae habitat influence quantitative attributes of the unsaturated fats. Each category of marine microalgae consists of a trademark characteristic fatty acid pattern. It is recommended that exceptional acids, some typical acids and the proportion of acids present are helpful chemotaxonomic markers (Zhukova and Aizdaicher 1995; Sahu et al. 2013; Temina et al. 2007).

1.1.6 Mycosporine-Like Amino Acids

The MAA or mycosporine-like amino acids (mycosporine-glycine) are a cluster of more than 20 compounds that absorb UV and are available in a different scope of oceanic organisms, and their application as sunscreens diminishes UV damage. The primary function is protection against the vigorous radiation generated by the ultraviolet light. In any case, MAAs additionally assume a role as protection against harmful effects brought by sunlight through its antioxidant function which converts the dangerous free radicals into less dangerous substance (Dunlap and Yamamoto 1995). Aside from that, they can be good cell protecting options from various stresses (Oren and Gunde-Cimerman 2007).

1.1.7 Toxins

A harmful algal bloom (HAB) is a consequence of eutrophication that has negative effects on different living beings due to its toxins that can poison different animals or humans. Substantial scale marine mortality occasions are frequently connected with HABs. Instances of basic harmful impacts of HABs consist of the production of neurotoxins that can poison and cause death of fish, tortoises, seabirds marine mammals and humans. It causes significant harm to different organisms, for example, intruding the fish epithelial gill tissues that results to unconsciousness and removal of oxygen from the waterbodies (hypoxia or anoxia) which disables the cell from respiration and causes accumulation of harmful bacteria (Tiwari 2010).

Cyanotoxins

Cyanotoxins are toxins produced by cyanobacteria. Blooming cyanobacteria can release cyanotoxins that can toxify and even assassinate animals and people. Cyanotoxins can likewise aggregate in different animals, for example, fish and shellfish, and cause poisonings, for example, shellfish poisoning. The first published report that blue green algae or cyanobacteria could have deadly influences showed up in *Nature* in 1878. The toxins include:

Hepatotoxins (Gr., hepato = liver)

Hepatotoxin is a toxic chemical substance produced within the cell and released in the surrounding water after cell death. This toxic chemical substance damages the liver, e.g. microcystin, produced by cyanobacteria (*Microcystis aeruginosa*) (Tiwari 2014).

Neurotoxins

A neurotoxin is a chemical which inhibits the functions of neurons. In some cases, neurotoxins simply damage the neurons so that they cannot function properly. Others attack the signalling capability of neurons, through the release of various chemicals, e.g. a neurotoxin, that is, saxitoxin (STX), is produced naturally by certain marine species of cyanobacteria (*Anabaena* sp.). It prevents normal cellular function by acting on nerve cells which have voltage-gated sodium channels, leading to the cause of paralysis (Pandey and Tiwari 2014).

Cytotoxins (Gr., cyto = cell)

A cytotoxin is a chemical compound which has adverse poisonous effect on cells and targets only on a cell or organ of specific type instead of whole body. They can destroy a cell by several ways: one is necrosis in which there is loss of cell integrity from their membrane and get collapsed, while another is apoptosis in which the cell death occurs prematurely (Tiwari 2012).

Endotoxins

Endotoxins are the chemicals produced within the Gram-negative bacteria and released after the destruction of its cell wall, causing many health-related problems like illness, nausea, vomiting, diarrhoea, white-blood cell count fluctuations and hypertension or high blood pressure. Now the term ‘endotoxins’ is used synonymously to lipopolysaccharide. Most of the researchers reported the incidents of toxicity by microalgal toxic compounds and their occurrence in freshwater environments, and they are becoming more frequent and extended (Tiwari 2010).

2 The Antioxidative System and Ageing

An antioxidant is a particle or molecule which has potential to moderate or avert the oxidation of different atoms. An exchange of electrons in oxidation results into free radicals which are fatal for the cell. Antioxidants oxidize themselves in order to expel these compounds, for example, ascorbic acid or thiols. Additionally, these substances have numerous medicine and industrial uses, for example, additives in food and cosmetic products, counteracting the decay of elasticity and fuel. At first these are considered as the compounds that help in the oxygen consumption.

Antioxidants can be classified into three general types:

(a) Lipid solvent and bilayer-related tocopherols,
(b) Water solvent reductants, for example, ascorbic acid and glutathione
(c) Enzymes, for example, superoxide dismutase, catalase, peroxidase, ascorbate peroxidase and glutathione reductase

During ordinary metabolic functions, free radicals which are highly reactive compounds or we can say the generation of reactive oxygen species lead to cellular damage as illustrated in Fig. 8.2. Notwithstanding, free radicals may likewise be presented from nature. These compounds are unstable since they have an odd number of electrons. To compensate for their lack of electrons, these free radicals will respond with specific synthetic substances in the body, and in this manner, they interfere with the cell's capacity to function properly. However, similarly as the body normally creates free radicals, it has a way to protect against its harmful impacts. There is an available framework known as antioxidant system which contains antioxidative enzymes or catalysts. Antioxidant compounds are chemical

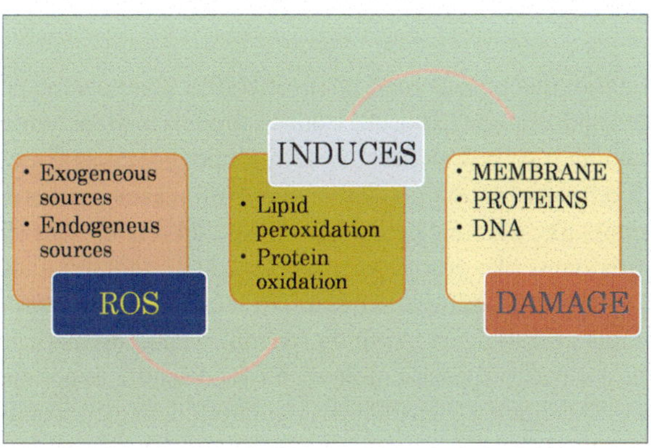

Fig. 8.2 Body cell without antioxidants

substances found in plants that follow up on free radicals. Antioxidant enzymes work in a few different ways. For one, they may lessen the energy of the free radical or surrender a portion of their electrons for its utilization and, hence, become stable. Antioxidants may likewise prevent the free radical from shaping in any case and may intrude an oxidizing bind response to limit the harm brought about by free radicals. In total, the principal capacity of cell antioxidants is killing or neutralizing radicals.

A few antioxidant enzymes act together to keep up redox homeostasis which appeared in the presence of five, two and three isoforms of SOD, ascorbate peroxidase and catalase proteins separately in the crude extracts of naturally collected mats of desiccation-tolerant cyanobacterium *Lyngbya arboricola* on native PAGE (Tripathi and Srivastava 2001). The presence of two and four bands of catalase and SOD as distinguished by native PAGE has additionally been accounted for in *Tolypothrix* (Rajendran et al. 2007), for example, single isoform can be seen in *Nostoc* sp. while staining for catalase activity with the virtue of PAGE, whereas *Aulosira* sp. shows three isoforms of catalase. Haem-containing catalysts are divided in three families, (1) monofunctional catalases (Zamocky et al. 2008a), (2) the haem peroxidase family present in plants or protists (Welinder 1992; Passardi et al. 2007) and (3) the peroxidase-cyclooxygenase (Zamocky et al. 2008b). Additionally, there are small groups of haem-containing peroxidases which can be found in Archaea: non-haem peroxide present in bacteria and parasites (Zubieta et al. 2007) or bacterial di-haem peroxidases (Echalier et al. 2006) which are manganese catalases (Zamocky et al. 2008a), vanadium peroxidases and universal thiol peroxidases which catalyse peroxide reduction by reactant cysteine residues and thiol-containing proteins as reductants (Rouhier and Jacquot 2005).

Another fascinating potential of cyanobacteria is adapting to UVR through UV engrossing/screening compounds, for example, mycosporine-like amino acids (MAAs) and scytonemin. These compounds are characteristic photoprotectants.

2.1 Role of AOS

Another use of antioxidants is as food additive so that food can be prevented from spoilage. Oxygen and sunlight exposure leads to food oxidation; hence, food can be preserved by keeping it in dark and sealing it in proper containers or even wax coating to prevent food deterioration (as with cucumbers). These preservatives include natural antioxidants like ascorbic acid (AA, E300) and tocopherols (E306), as well as synthetic antioxidants like propyl gallate (PG, E310). Industries are utilizing the antioxidants by frequently adding them into products. They are commonly used in fuels and lubricants as stabilizers to inhibit oxidation and polymerization. They are used in gasoline so that polymerization won't lead to the generation of engine-fouling residues. The brain is sensitive to oxidative damage, because of its high metabolic rate and increased levels of polyunsaturated lipids, the target of lipid peroxidation. Thus, antioxidants are generally utilized like drugs to treat different types of brain damage. Here, superoxide dismutase mimetics, sodium thiopental and propofol are utilized to treat reperfusion damage and awful brain damage; antioxidants are likewise being researched as conceivable medications for neurodegenerative illnesses and diseases, like Alzheimer's and Parkinson's malady.

2.2 *Reactive Oxygen Species (ROS)*

During the normal cell functioning linked with the processes like reduction and oxidation, reactive oxygen species are generated specially at the time of photosynthesis that is in chloroplast. Under unstressed conditions, generation and termination of ROS in these life forms are in equilibrium. Diverse ecological factors of stress such as contamination, temperature, excessive light forces and nutritional constraint can generate ROS. Oxidative pressure is intently related to these temperamental however reactive radicals (Borowitzka 1995). At the point when oxygen interacts with metabolic systems, it very well may be transformed into progressively reactive and lethal type of superoxide particle, hydrogen peroxide, hydroxyl radical and singlet oxygen. Arrangement of singlet oxygen (O^2) along these lines stimulates the generation of other reactive oxygen species like hydrogen peroxide (H_2O_2), superoxide anion (O^{2-}) hydroxyl (OH^-) and perhydroxyl radicals (O^2H).

2.2.1 Role of Reactive Oxygen Species (ROS)

Reactive oxygen species produced due to UV-B radiation easily destroys proteins, DNA and other biological molecules (Douglas 1994). Activated oxygen and agents that produce oxygen-free radicals, for example, ionizing radiation, are in charge of inducing various injuries in DNA because of which deletions, transformations or mutations and other deadly genetic impacts are caused. Characterization of such DNA injury shows that sugar and the base moieties are powerless to oxidation leading to degradation of base, breakage of single strand and protein cross-linking (Echalier et al. 2006). Oxidative assault of proteins results in amino acid alterations which are site specific, peptide chain discontinuity, accumulation of reaction products which are cross-linked, changed electrical charge and expanded proteolysis susceptibility. ROS formed in the natural framework during the ordinary course of metabolism are potentially destructive since they attack membranes, proteins and DNA molecules. Also the ROS are the leading cause of skin ageing which is demonstrated in Fig. 8.3. ROS formation is additionally increased in the presence of the measures of iron or other transition metal ions and metal chelators (Chen et al. 2010).

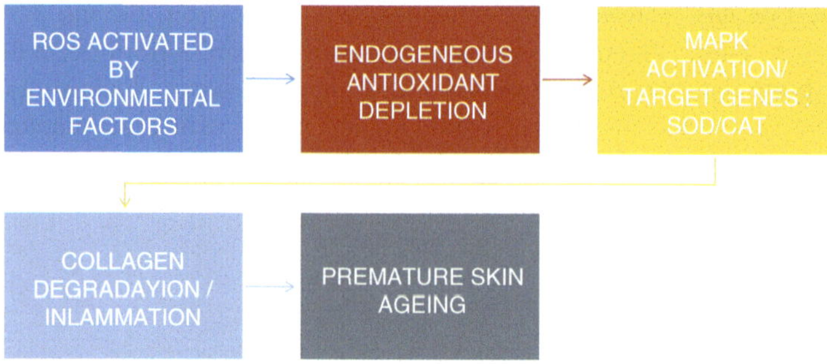

Fig. 8.3 The mechanism of skin ageing and ROS

2.2.2 Protection Against Damages Caused by ROS

The algae can tolerate increased levels of oxygen by means of endogenous mechanisms that scavenge and eliminate the poisonous products effectively before any damage occurs in the cell by the help of antioxidant cascade, where the superoxide dismutase (SOD) acts before the catalase and peroxidizes as shown in Fig. 8.4.

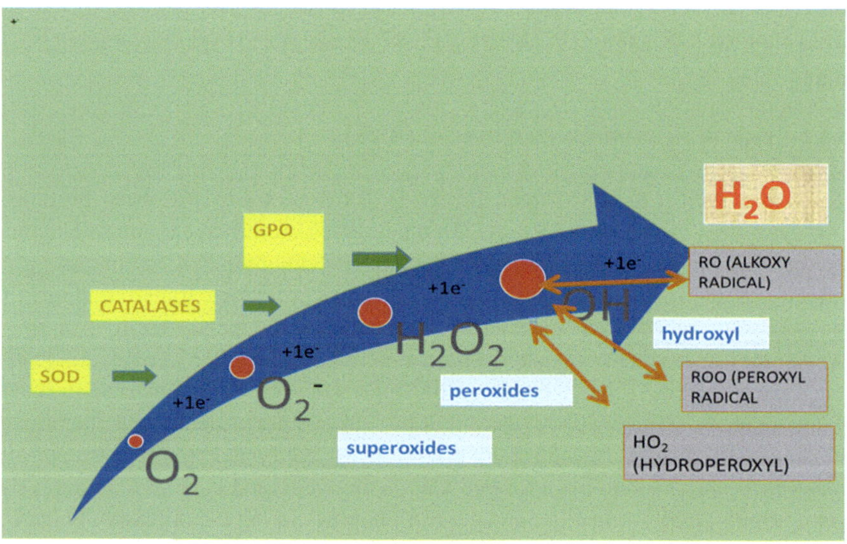

Fig. 8.4 Enzymatic antioxidants working in a cascade fashion

2.3 Antioxidative Enzymes

2.3.1 Catalase

They stand out amongst the most potent catalyst known. The response it catalyses is significant to life. It is a typical protein found in about every single living life form that is oxygen exposed. Catalase (EC 1.11.1.6) catalyses change of hydrogen peroxide, a potential and conceivably unsafe oxidizing agent, to water and molecular oxygen. The ideal pH for human catalase is approximately 7. One molecule of catalase can decay right around a hundred thousand particles of hydrogen peroxide each second. In 1818, it is found that H_2O_2 (hydrogen peroxide), breakdown was possible. In 1900, it was the first named as catalase and discovered its presence in numerous plants and creatures. In 1937 beef liver catalase was crystallized and the atomic weight was found in 1938. In 1969, bovine catalase's amino acid sequence was worked upon. At that point in 1981, the 3D structure of the protein was opened up. Catalase is a tetramer of four polypeptide chains, each more than 500 amino acids long. It contains four porphyrin haeme (iron) bunches which enable the compound to respond with the hydrogen peroxide (Regelsberger et al. 2002). This is a model of catalase, appearing globular structure, somewhat like a tangled mass of string.

Active sites are available as splits or hollows on the surface of the catalyst brought about through the manner in which the protein takes the tertiary structure by folding itself. Molecules having the perfect shape and size can be easily fitted into such active sites. Basically separation of catalases can be done in proteins containing monofunctional haeme and those with a manganese-containing response focus at the active site. The manganese catalases structure moderately small groups (Allgood and Perry 1986). In cyanobacteria, so far KatG was recognized in the unicellular species *Synechococcus* PCC 7942, *Synechococcus* PCC 6301 (*Anacystis nidulans*) and *Synechocystis* PCC 6803. Biochemical as well as genetic examinations showed presence of only a single isoform of an ordinary multifunctional catalase-peroxidase (KatG) as a cytosolic hydrogen peroxide scavenging haemoprotein (Obinger et al. 1997). In the genomes of *Anabaena* PCC 7120, *Nostoc punctiforme*, *Synechococcus* WH8102, *Prochlorococcus marinus* MED4 and *Prochlorococcus* MIT 9313, no KatG quality has been found.

Antioxidant enzymes, which also include catalase, make the primary line of protection against free radicals; in this way their guidelines rely fundamentally upon the oxidant status of the cell. Be that as it may, there are different components associated with their regulation, which also includes the enzyme modulating action of different hormones, for example, growth hormone, prolactin and melatonin. Melatonin is amino acid tryptophan's derivative which acts as a neurohormone in mammals, but at the same time is incorporated by many different species, including plants, green growth and microorganisms. Melatonin appears to extraordinarily secure both lipid membranes and nuclear DNA from oxidative harm. Melatonin can straightforwardly neutralize ROS, including hydrogen peroxide. It can likewise prompt different antioxidant enzymes, including catalase, either by expanding their action or by promoting gene expression for these catalysts. Abatement in melatonin levels observed with age relates to an expansion in neurogenerative clutters, for example, Parkinson's malady, Alzheimer's ailment, Huntington's illness and stroke, all of which may include oxidative pressure. Generally, the formation of ROS increments with maturing and is related with DNA harm to the tissues. On the other hand, development hormone, potentially prolactin, can diminish catalase and other antioxidants in different tissues in mice, proposing that this hormone acts as a silencer of key antioxidant components.

Oxygen has an extraordinary task to carry out in our cells as it offers vitality to our cells; however oxygen is a reactive molecule which causes major issues if its level increased and is not controlled deliberately. And its consequence is that it leads to the generation of reactive oxygen species transported across various sites with the help of carriers, for example, riboflavin and niacin derived carriers; when oxygen exchange the electrons, it will lead to the generation of free radicals like superoxide radicals and hydrogen peroxide, which can degrade the proteins by attacking at their ions, free iron particles in the cell occasionally convert hydrogen peroxide into hydroxyl radicals. All these events become the cause of transformation of DNA. There is still a controversy regarding this hypothesis that oxidative damage contributes to aging (Tiwari 2014).

2.3.2 Superoxide Dismutase (SOD)

Superoxide dismutase (SOD, E.C. 1.15.1.1) plays a significant function in the protection of cell from oxidative damages in aerobes. Like all other aerobes, the algae utilize SODs to separate superoxide anions produced by reactive oxygen species (Herbert et al. 1992).

Within a cell, SOD constitutes the first line of defence against ROS and is responsible for removal of superoxide radicals. SOD enzymes are metalloproteins and are related to a class of oxidoreductase enzyme (Li et al. 2002b).

2.3.3 Types of Superoxide Dismutase

Conclusively, phospholipid bilayers are impermeable to charged O^{2-} particles. Subsequently, it is critical that SOD is available for the expulsion of superoxide radicals in the compartments where they are formed. Algae have three kinds of SOD with various prosthetic metal groups:

Fe-SODs

The site of Fe-SODs is chloroplast. Fe-SOD has two different groups. First group consists of homodimer (Thomas et al. 1999). The second Fe-SOD group, present in advanced plants, is a four-subunit tetramer (Shirkey et al. 2000). It is the most ancient group of SOD group. Hydrogen peroxide is responsible for the inactivation of Fe-SOD and KCN inhibition is resisted by it.

Mn-SODs

Mitochondria and peroxisome are the location of Mn-SOD. They carry only single metal atom for each subunit. They are homodimeric or homotetrameric protein with only Mn (III) particle per subunit. Mn-SOD more often than not has an atomic mass of 40–46 kDa; however higher subatomic mass Mn-SODs have been recognized in a few types of microscopic organisms with subatomic masses of 110–140 kDa. Potassium cyanide (KCN) cannot hinder the compound and H_2O_2 cannot inactivate it.

Cu-SODs

Cu/Zn SODs have different electrical properties than others. Cu/Zn SOD is found all through the algae or plants. They comprise two types of Cu/Zn SOD. The first one constitutes homodimeric and second involves the homotetrameric (Bradford 1976). Cu/Zn SODs are dissolvable enzyme, with atomic mass of 32 kDa, and comprises two indistinguishable subunits.

2.3.4 SOD Genes

Synechocystis sp., a cyanobacterial strain, consists of SODB gene only (Beauchamp and Fridovich 1971), while most of the algae consist of two types of SOD genes: SODA which encodes Mn-SOD and SODB that encodes Fe-SOD. The nitrogen-fixing heterocystous cyanobacterium *Anabaena* sp. strain PCC 7120 has two SOD encoding qualities: a SODB and a SODA. The number of SODA qualities encoding Mn-SOD differs amongst cyanobacteria. *Anabaena* sp. strain PCC 7120, as the majority of the cyanobacteria, consists of membrane systems: plasma layers and thylakoid layers. To think about the area of SODA in *Anabaena* sp. strain PCC 7120, plasma films and thylakoid layers can be isolated by utilizing a two-stage framework (Sambrook and Russell 2001; Chen and Pan 1996). Apparently, in the majority of the cyanobacteria, there are two types of SOD present which are beneficial for the adaptation in fluctuation of metal concentration in their environments. In *Spirulina* the phycocyanobilin has been shown to have antioxidizing activity and therefore may act as an effective antioxidant in humans. It has therapeutic value as it was shown to have immunomodulating activity and anticancer activity. C-Phycocyanins (a blue pigment protein) found in cyanobacterium *Spirulina platensis* has been reported to be a potent peroxy radical scavenger in vitro and in vivo. Phycoerythrin isolated from *Nostoc* species was used as a protein marker for electrophoretic techniques. C-Phycoerythrin (red pigment) was shown to protect rats from CCl_4-induced toxicity and from diabetic complications. It also provides protection against permanganate-mediated DNA damage, $HgCl^{2-}$ caused oxidative stress and cellular damage in the kidney. The colouring agent in food consists of phycobiliproteins (chewing gums, dairy products, gellies, etc.) and cosmetics including lipstick and eyeliners in Japan, Thailand and China.

The presence of SOD appears to help in shielding numerous kinds of cells from the free extreme harm that is essential in maturing, senescence and tissue damage. SOD additionally protects cells from DNA harm, lipid peroxidation, damage from ionizing radiations, denaturation of protein and numerous different types of dynamic cell debasement. SOD is utilized in restorative products to diminish extreme harm to skin, for instance, to lessen fibrosis following radiation for malignancy (Obinger et al. 1998).

2.4 *Tyrosinase Inhibitors and Algae*

Marine algae, ordinarily utilized as foodstuffs in East Asian nations, are likewise potential sources of bioactive metabolites, for example, terpenoids, polyphenols and halogenated compounds (Mikami et al. 2016). Free radicals are persistently produced in human body and cause to fall apart lipids, proteins, compounds, DNA and RNA. They may prompt numerous infections, for example, atherosclerosis, cardiovascular diseases and carcinogenesis (Li et al. 2012). Although manufactured antioxidants are generally used to anticipate these issues, their utilizations are necessary for such toxic and cancer-causing impact (Javan et al. 2013). Characteristic antioxidants with no side reactions are needed for free radical scavenging action and security against oxidative stress (Reyes et al. 2013). The activity of tyrosinase inhibitors is one of the parameters on skin lightening agents. The process of tyrosinase inhibitors is to diminish skin pigmentation by inhibiting the catalysis of the pigmentation related with melanin generation in the pathway of melanogenesis. Skin protecting agents are related with antioxidative movement that can ensure the skin against oxidative compounds due to UVR (Park et al. 2012). Skin exposed to UVR both intensely and ceaselessly can cause different cell and natural changes, for example, DNA harm, loss of homeostasis and cell work, irregular pigmentation because of changes in melanin pigment, immune suppression and aggravation that can cause skin issues, for instance, photoageing and non-obtrusive skin tumours, either non-melanoma or melanoma. Red algae of the family Rhodomelaceae are enhanced with bromophenols which leads to radical formation (Li et al. 2012), protein hindrance (Kurihara et al. 1999), nourishing obstacle, mitigation, cell security, antimicrobial, anticancer, hostile to diabetic (Kim et al. 2008a, b) and antiviral action (Park et al. 2012). Tyrosinase (Wang et al. 2005) is valuable for the synthesis of melanin and a key compound for melanin biosynthesis and natural product sautéing (Chang 2009). Nonetheless, it is directed to hyperpigmentation (Tan et al. 2016). More radical or reactive oxygen species (ROS) may cause secretion of α-melanocyte-stimulating hormone (α-MSH) and that leads to pigmentation, for example, age spot, spots, skin maturing, melasma, etc. (Lan et al. 2013). Numerous regular tyrosinase inhibitors have been identified as polyphenols and lipids.

3 Cosmeceuticals from Algae

The products that follow up on the epidermis (shallow layer of the skin), without changing the physiological states of the skin, are beauty care products or we call them cosmetics. In fact they do not require logical investigations as a proof of their viability and are before long promoted, for instance, lotions. Cosmeceuticals comprise naturally dynamic fixings, for example, phytochemicals, fundamental oils, nutrients, antioxidants and so on (Cannell 2006). The result of cosmeceuticals can be in any way similar to medication, cosmetic or both.

Beautifiers or cosmeceuticals are intended to improve, elevate attractiveness, and cleanse the skin, though they are not endorsed by FDA and available to be purchased. There is wide scope of items like from fat cutter lotions to aromas, hair care products, detanning packs and shower gels (Pereira 2018). These days everybody is showing interest for everything that can enhance their lifestyle, with exceptional belief that a large number of answers of the issues are found in nature itself, so the interest for products obtained using natural sources has expanded in which algae have incredible consideration. For example, consumption of green algae as food, in Europe, mostly of *Laminaria* (brown algae) and *Chondrus* (red algae), is around 70,000 tons/year (Mafinowska 2011).

Around 221 algal species are commercially utilized in food and pharma- and cosmeceutical industries (Table 8.1), and around ten species have been cultivated strongly, for example, brown algae (*Saccharina japonica*), red algae (*Gracilaria* sp.) and green algae (*Monostroma nitidum*). Microalgae (*Spirulina/Arthrospira* spp.) are additionally developed, and the significant makers are Australia, India, Israel, Japan, Malaysia and Myanmar. Today, the worldwide worth of algal industry

Table 8.1 Some cosmetic algal products available in the market

Company	Product name	Reference
Aquarev industries	Red algae-based products	www.aquarev.in
Algatech	Green algae supplies astaxanthin	https://www.algatech.com/
Nykaa	Body serum	http://www.nykaa.com/
L'Oreal Paris	Face mask	https://www.lorealparis.co.in/
La Prairie	Ice crystal dry oil	http://www.laprairieswitzerland.com/
Shiseido	Stemlan-173 (WO2018074606A1)	https://www.shiseidogroup.com
AlgEternal Technologies	PhtcoDerm	https://www.patent-art.com/algae-cosmetics-18
Algenist	Alguronic acid (patented)	
Repêchage Professional	Seaweed wax	
Greenaltech	ALGAKTIV® LightSKN™	
Innisfree	Amorepacific	
Revlon Professional	Eksperience™	

is more than USD 6 billion per year, out of which 85% contains nourishment items for human utilization. The algal extracts like carrageenan and alginates are essential in the cosmeceuticals, and together make up to 40% of the hydrocolloidal market of the world.

Recent Patent Application

(a) The patent US9717932B2 by BASF, entitled "Marine extracts and biofermentations for use in cosmetics" reveals about the marine algae like *Sarcodiotheca gaudichaudii* and its biofermentation use as a skin care active ingredient for anti-ageing cosmetic applications. It is also used with the red algae extracts (*Chondrus crispus*).
(b) The two recent patents (KR1972071B1, KR1901859B1), by Amorepacific Corp., have opened up about incorporation of marine algae extracts. For instance green algae, red algae, brown algae and other algal extracts in cosmeceuticals for improving skin quality, inhibiting skin ageing and enhancing skin elasticity, as well as for masking skin wrinkles and protecting skin from UV rays.
(c) An independent inventor OH, Chung Heon, has developed fucoidan-based mask (KR1841099B1). It is extracted from brown algae, followed by fermenting, drying and pulverizing (https://openaccesspub.org/japb/article/530).

3.1 *Sources of Cosmeceuticals*

Algae are those microbes which are very peculiar and found almost everywhere on this planet. They provide the foundation for aquatic life chains and produce 70% of the air we use to inhale, and in addition they are the major source of antioxidants. Due to these antioxidative properties, they have been utilized in the field of pharmaceuticals, agriculture, medicine and cosmeceuticals. The natural antioxidants present in algae consist of bioactive compounds which have the potential to fight against the different diseases and ageing. Despite of formation of ROS and free radicals by the ultraviolet rays through sunlight, there is the absence of oxidative stress in the cellular components of algae which shows that there is defence system present or we can say that antioxidant protection system is present in their cells.

Algae have chemical diversity and also have very unique properties; algae has been implemented for many studies and has broad range of applications in the cosmeceuticals (Table 8.2). It consists of different biochemical or bioactive compounds like pigments, proteins, lipids, vitamins, phenolic compounds, polysaccharides and others (Paul and Pohnert 2011), including macro- and micro-elements (Majmudar 2012). Algae produce both primary metabolites and secondary metabolites. The primary metabolites are responsible for the adequate growth or reproduction conditions so that they can perform various functions properly, and secondary metabolites are the substances or compounds which play their role under various adverse conditions such as stress generated by UV radiations, alkalinity, variations in temperature or pH or environmental toxins (Thalgo La 2008).

Table 8.2 Algal species and their bioactive compounds in cosmetics

Algae species	Bioactive compound	Cosmetic property	References
Cladophora glomerata	Chlorophylls (a, b, c, d)	Antibacterial, antioxidant	Lanfer-Marquez et al. (2005)
Caulerpa sp.	Extracts: steroids, flavonoids, phenols hydroquinone and saponin	Tyrosinase inhibitor	Demais et al. (2007)
D. tertiolecta	Phenolic compound	Anti-ageing	Norzagaray-Valenzuela et al. (2017)
H. lacustris	Astaxanthin	Anti-ageing	Huangfu et al. (2013)
U. lactuca	Carotenoids (fucoxanthin, lutein)	Anti-inflammatory, antioxidant, tyrosinase inhibitors, anti-ageing	Christaki et al. (2013)
Cladosiphon okamuranus	Extract	Moisturizer	Wang et al. (2013a, b)
E. cava	Phlorotannins: dieckol	Hair growth	Kang et al. (2012)

The secondary metabolites produced are pigments, sterols and other bioactive agents. Algae are generally classified as:

(a) Rhodophyta phylum includes red algae and consists of pigments like chlorophyll a, phycobilins and carotenoids.
(b) Ochrophyta phylum consists brown algae comprised of chlorophylls a and c and carotenoids (fucoxanthin).
(c) Chlorophyta phylum includes green algae with pigments that include chlorophylls a and b and carotenoids.

Due to variety of components, there are numerous applications of algae as it is utilized for cosmetics, pharmaceuticals and as food supplement too. They are sources of many bioactive compounds of incredible quality, since they comprise basic amino acids, omega-3 family and other fatty acids and nutrients.

3.2 Algae as Source of Cosmeceuticals

The benefits of algae, especially its use in skin treatment or cosmeceuticals, are of high significance due to their ability to fight against acne and reactive oxygen species as they are antioxidants. They are used as anti-ageing and anti-inflammatory agent (Berthon et al. 2017) and inhibit melanogenesis as well. It has UV photoprotective and anti-melanoma effects (C. Corinaldesi et al. 2017). Lots of investigation have been done to show the pathological effects of algal bioactive compounds in proper downregulation of MMPs and tyrosinase inhibitor activity.

The compounds obtained from algae have various beneficial effects on health and care of skin (Table 8.3). Brown algae-derived phlorotannins and various polysaccharides have many applications in cosmeceuticals (Choi et al. 2018). There are many other applications of algae. The use of algae for various purposes like biotechnological applications or in cosmetics is possibly because of its simple harvesting and fast growth rate.

Some of the applications of algae in cosmetics in details are as follows.

3.2.1 Algae as Anti-Tanning and Moisturizing Agent

Melanin is the pigment present in the skin which absorbs the UV radiations when there is direct exposure with skin (Gong et al. 2007). It is a complex polymer, which gives colour to the human skin, and also it is the protection for human skin cells. Therefore the increased and constant exposure to sunlight increases the level of melanin pigment which results in tanning (Park 2006). Tyrosinase is synthesized by the radiations from sunlight which catalyses the reaction and leads to the formation of melanosomes which then get matured and become melanin. Further the differentiation takes place and keratinocytes are formed to raise the deterioration of skin (Silab 2018). Here we conclude that excess production of melanin causes hyperpigmentation, so tyrosinase inhibitors are implemented in pigmentation (Dae et al. 2007). Pigments like fucoxanthin from brown alga like *Macrocystis* inhibit melanogenesis. Melanin is delivered by the cells called melanocytes situated in the basal epidermal layer (Solano et al. 2006). The roles of tyrosinase inhibitors are significant in the process of depigmenting and brightening (Liang et al. 2012).

Table 8.3 Algal species and their applications in cosmetics

Algal species	Type	Pigment	Metabolites	Application
Ulva lactuca	Green algae	Chlorophyll (a, b), β-carotene	Oleic and linolenic acid	Antioxidant, anti-inflammatory, anti-wrinkle, moisturizing agent (2017).
Postelsia palmaeformis	Brown algae	Chlorophyll c, fucoxanthin	–	Skin softening, anti-wrinkle, anti-inflammatory (2017)
Porphyra umbilicalis	Red algae	Phycoerythrin	Linolenic acid	Skin conditioner (2017)
Spirulina	Cyanobacteria	Phycocyanin	Linolenic acid, phycocyanobilin, phycoerythrobilin	Anti-ageing, collagen synthesis, anti-inflammatory, antioxidant (2013)
Dunaliella salina	Green algae	Chlorophylls (a, b), β-carotene	Palmitic and linolenic acid, β-cryptoxanthin	Antioxidants, smoothing agent (2017)

The side effects like pigmented contact dermatitis (García-Gavín et al. 2010) due to kojic acid and a conceivable genotoxic impact brought about by arbutin can be caused by chemical tyrosinase inhibitor. To make the epidermis of the skin softer, one needs to apply moisturizer over it, which is made up of such mixture of chemicals. Unmoisturized skin is prone to various skin problems like acne or even eczema. Therefore the moisturizer helps the skin to retain its moisture and prevent it from bruising, wrinkling and drying. There are certain acids like hyaluronic acid and various polysaccharides like agar, alginate, carrageenan and fucoidans from algae that help retain moisture in the skin along with water. The polysaccharides from certain algal species absorb water or moisture and provide soothing effect, for example, *S. japonica* and *Codium tomentosum*, which helps in the proper water movement and distribution. This keeps the skin hydrated and moisturized even in adverse climatic conditions like hot and dry environments (Wang et al. 2013a, b).

3.2.2 Algae as Anti-Ageing Agent

The deprivation in the elasticity of the skin and the occurrence of lines, wrinkles or crest including the discolouration or the loss of glow from the skin are the indication of ageing. With growing age, there is the skin ageing. The factor which is responsible for this condition of the skin not only includes the growing age but also the harsh environmental factors. The harsh environmental conditions lead to the dryness, thinning, sagging, enlarged pores, etc., of the skin, hence causing premature wrinkles and skin deterioration. These factors are continuous vulnerability of heavy metals, nutrient scarcity and moisture deficiency in epidermal layer of the skin. And the most common reason for skin ageing is the reactive oxygen species. Various researchers have concluded about the algal products and their anti-ageing properties. For instance, vitamin E and carotene nourish the skin to rejuvenate and get immune to skin maturation and also eliminate the chances of getting sarcoma or cancer of the skin. Algal species like *Colpomenia*, *Halymenia*, *Padina*, *Polysiphonia*, etc., are used as anti-ageing agents. Mycosporine like amino acid provides ultraviolet ray protection which usually causes damage to the skin and leads to skin prematuration (Christaki et al. 2013).

The debasement of flexible fibre and putrefaction of collagen are caused by elastases and MMP (matrix metalloproteinase). At the point when there is exposure to radiation, the skin starts losing elasticity and lines or, we can say, wrinkles start appearing. In this way, collection of MMPs is unsafe for the skin. Marine organisms, particularly macroalgae, produce a variety of defensive compounds against photoageing (Pallela et al. 2010). Macroalgae having such bioactive compounds can assimilate both types of UV radiations, that is, UV-A and UV-B, and some may inhibit ROS and the progression of matrix metalloproteinase. The most productive naturally occurring UV-A-engrossing compound is the mycosporine-like amino acids (MAAs). One of the characteristics of MAAs is that they are water soluble and present in various living forms, similar to algae, and numerous marine creatures like corals. For instance, the MMAs, porphyra-334, can be obtained from the red algae

Porphyra umbilicalis. Its retention coefficient at 334 nm is 42,300, demonstrating that its channel ability is like those engineered UV-A sunscreens (Ryu et al. 2014), for example, butyl methoxydibenzoylmethane (Daniel et al. 2004).

3.2.3 Algae as Wound Healer and Antibacterial for the Skin

The scar formation can be inhibited by the alginic acid (Borowitzka 2013). Not only this, it also acts as a physical barrier in order to invade the fibroblasts and hence help in wound healing. There are many applications of alginates especially in the field of tissue engineering and clinical trials. Soluble algae extract-based dressings are helpful in wound healing. The ethanolic extract of *Kappaphycus alvarezii* (a red alga) promotes wound healing and hair growth (Lanfer-Marquez et al. 2005).

Algae act against *acne vulgaris*, which causes skin inflammation. Acne vulgaris is a typical skin malady or state influencing numerous youths and youthful grown-ups. It is described by clogged pores or we can say whiteheads, pimples, oily skin and conceivable scarring. Skin breakout can persevere for a considerable length of time and result in changeless scars and deformation and effectively affects physiological advancement (Leyden 1995). The pathogenesis of skin breakout is intricate and multifactorial. For the most part, it is seen as an incendiary ailment with different responsible factors, for example, keratin treatment of hair follicles, sebum emission and microorganisms that cause skin inflammation (Farrar and Ingham 2004). *Staphylococcus epidermidis*, *Pseudomonas aeruginosa* and *S. aureus* are normally engaged with acne development (Yamaguchi et al. 2009). Specifically, anaerobic microorganism (the gram-positive), *P. acnes*, is frequently perceived in acne. *Acne vulgaris*, because of the development of microbes, is treated with anti-toxin treatments, for example, clindamycin and erythromycin. However, broad use of anti-infection compounds has prompted bacterial obstruction. Additionally, antimicrobials may cause skin hypersensitivities and skin aggravation. Thus, the bioactive compounds extracted from marine algae could be the best option. Macroalgal extracts have been found to have antibacterial and antifungal properties (Pérez et al. 2016). Likewise, extracts from some macroalgae show antimicrobial impacts and can regulate the development components of the skin and collagen too, which could improve the skin inflammation and accelerate skin nourishment.

3.2.4 Algae as Hair Growth Promoter

Hair fall or hair loss is the major problem of many millions of people and there are many reasons behind this, which includes ageing. Many studies have been conducted to prevent hair loss and promote hair growth, for instance, dieckol isolated from *E. cava* which is further used for hair problems (Kang et al. 2012).

3.3 Advantages of Algal Products

A few compounds extracted from algae are utilized in cosmetic industry as thickening specialists, water-restricting agents and cancer prevention agents in facial and healthy skin products. Natural carotenoids are favoured than synthetic carotenoids and have great applications in the nourishment and feed industries, just as in beauty care products and pharmaceuticals. It is notable that microalgae contain significant compounds including proteins, β-carotene, lutein, sugar, phenolic and flavonoids that are helpful for different modern applications. Carotenoids from microalgae are a basic class of cancer prevention agents which play important role in extinguishing responsive oxygen species produced amid photosynthesis. The algal products have many advantages which are illustrated in Fig. 8.5, and this is all because of their unique, novel and potential bioactive compounds as described earlier.

4 Future Prospects

The microalgae are promising tools inferable from expanded interest for bioactive compounds like carotenoids, phenolic, flavonoids, proteins and starches. Segregation and identification of novel metabolites from microalgae will help the improvement of new therapeutics, cosmeceuticals, nutraceuticals and food industries. The dynamic products are finding a degree scope of utilizations in the beauty care products, pharmaceuticals, sustenance and feed enterprises. Carotenoids, phenolic and

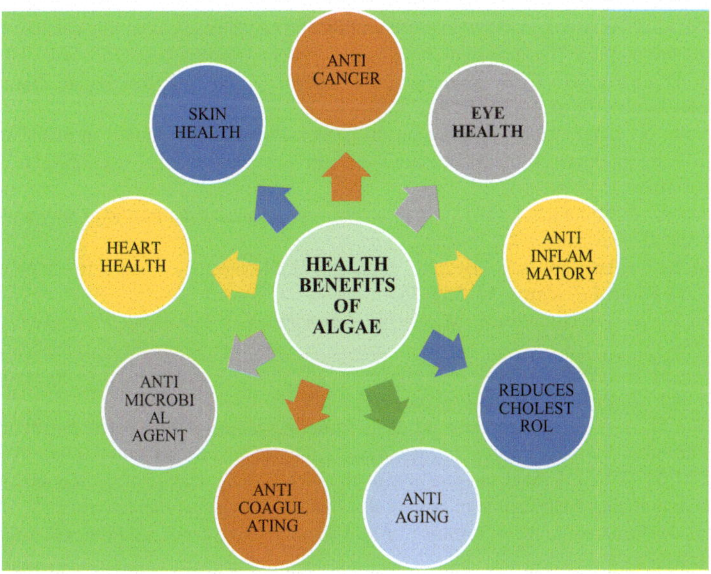

Fig. 8.5 Health benefits from algae

flavonoid are solid cancer prevention agents, fit for responding with searching oxygen species as a result of their hydroxyl gatherings. Algal species are presently effectively being used capably in the various restorative items as a solid natural fixing and furthermore to increase the value of these items. The main component of the algal products is the colours obtained from their pigments. The various bioactive compounds metabolites, for example, polysaccharides, or proteins, and so on constitutes differing capacities and benefits. These metabolites improve the skin quality by acting an agent against skin maturing, cancer, mitigation and wrinkle. These microorganisms are additionally utilized in different applications based on biotechnology, for example, bioenergy, biofertilizers, supplements and so forth. This audit primarily centres on the developing uses of green growth in the restorative business.

Yet there is very less fact known about the algal compounds and a need to explore more to obtain the cosmeceutical benefits; however recent studies demonstrate that many algae-derived products can lead to the development in cosmeceuticals in the coming time. With the increasing curiosity and demand for the improvement and generation of progressively compelling cosmeceuticals that fight the presence of maturing, the properties of a significant number of the marine algae merit further examination.

Acknowledgement The work was supported by the research grant from the Department of Biotechnology, Ministry of Science and Technology, India (BT/PR15650/AAQ/3/815/2016).

References

Allgood, G. S., & Perry, J. J. (1986). Characterization of a manganese-containing catalase from the obligate thermophile Thermoleophilum album. *Journal of Bacteriology, 168*, 563–567.

Augst, A. D., Kong, H. J., & Mooney, D. J. (2006). Alginate hydrogels as biomaterials. *Macromolecular Bioscience, 6*, 623–633.

Balasundram, N., Sundram, K., & Samman, S. (2006). Phenolic compounds in plants and agri-industrial by-products: Antioxidant activity, occurrence, and potential uses. *Food Chemistry, 99*, 191–203.

Beauchamp, C., & Fridovich, I. (1971). Improved assays and assay applicable to acrylamide gels. *Analytical Biochemistry, 44*, 276–287.

Berthon, J. Y., Nachat-Kappes, R., Bey, M., et al. (2017). Marine algae as attractive source to skin care. *Free Radical Research, 51*, 555–567.

Bjornland, T., & Aguilarmartinez, M. (1976). Carotenoids in red algae. *Phytochemistry, 15*, 291–296.

Borowitzka, M. A. (1995). Microalgae as sources of pharmaceuticals and other biologically active compounds. *Journal of Applied Phycology, 7*, 3–15.

Borowitzka, M. A. (2013). High-value products from microalgae—Their development and commercialization. *Journal of Applied Phycology, 25*, 743–756.

Bradford, M. M. (1976). A rapid and sensitive method for the quantization of microgram quantities of protein utilizing the principle of protein-dye binding. *Analytical Biochemistry, 72*, 248–254.

Cannell, R. J. P. (2006). Algae as a source of biologically active products. *Pesticide Science, 39*, 147–153.

Chang, T. S. (2009). An updated review of tyrosine inhibitors. *International Journal of Molecular Sciences, 10*, 2440–2475.

Chen, C.-N., & Pan, S.-M. (1996). Assay of superoxide dismutase activity. *Botanical Bulletin of Academia Sinica, 37*, 107–111.

Chen, Y. L., Juang, Y. C., Liao, G. Y., Ho, S. H., Yeh, K. L., Chen, C. Y., Chang, J. S., Liu, J. C., & Lee, D. J. (2010). Dispersed ozone flotation of *Chlorella vulgaris*. *Bioresource Technology, 101*, 9092–9096.

Choi, J. S., Bae, H. J., Kim, S. J., et al. (2018). *In vitro* antibacterial and anti-inflammatory properties of seaweed extracts against acne inducing bacteria, *Propionibacterium acnes*. *Journal of Environmental Biology, 32*, 313–318.

Christaki, E., Bonos, E., Giannenas, I., & Florou-Paner, P. (2013). Functional properties of carotenoids originating from algae. *Journal of the Science of Food and Agriculture, 93*, 5–11.

Corinaldesi, C., Barone, G., Marcellini, F., et al. (2017). Marine microbial derived molecules and their potential use in cosmeceutical and cosmetic products. *Marine Drugs, 15*, 118.

Dae, H. P., Won, S. C., Sean, H. Y., Jung, S. S., & Chul, H. S. (2007). A developmental study of artificial skin using the alginate dermal substrate. *Key Engineering Materials, 342*, 125–128.

Daniel, S., Cornelia, S., & Fred, Z. (2004). UV-A sunscreen from red algae for protection against 22 premature skin aging. *Cosmetics and Toiletries Manufacture Worldwide, 2004*, 139–143.

Demais, H., Brendle, J., Le Deit, H., Laza Anca, L., Lurton, L., & Brault, D. (2007). *Argiles Intercalés*. European *Patent No*. EP1786862 A1. Retrieved September 27, 2018, from https://patents.google.com/patent/EP1786862A1.

Douglas, S. E. (1994). Chloroplast origins and evolution. In D. A. Bryant (Ed.), *The molecular biology of Cyanobacteria* (pp. 91–118). Dordrecht: Kluwer Academic Publishers.

Dunlap, W. C., & Yamamoto, Y. (1995). Small-molecule antioxidants in marine organisms: Antioxidant activity of mycosporine-glycine. *Comparative Biochemistry and Physiology B, 112*, 105–114.

Echalier, A., Goodhew, C. F., Pettigrew, G. W., & Fülöp, V. (2006). Activation and catalysis of the di-heme cytochrome c peroxidase from *Paracoccus pantotrophus*. *Structure, 14*, 107–117.

Edge, R., McGarvey, D. J., & Truscott, T. G. (1997). The carotenoids as anti-oxidants-a review. *Journal of Photochemistry and Photobiology B, 41*, 189–200.

Farrar, M. D., & Ingham, E. (2004). Acne: Inflammation. *Clinical Dermatology, 22*(5), 380–384.

Fujimura, T., Tsukahara, K., Moriwaki, S., Kitahara, T., Sano, T., et al. (2002). Treatment of human skin with an extract of *Fucus vesiculosus* changes its thickness and mechanical properties. *Journal of Cosmetic Science, 53*, 1–9.

García-Gavín, J., González-Vilas, D., Fernández-Redondo, V., & Toribio, J. (2010). Pigmented contact dermatitis due to kojic acid. A paradoxical side effect of a skin lightener. *Contact Dermatitis, 62*(1), 63–64.

Gong, T. F., Fang, C. Y., Chen, W., & Wang, P. N. (2007). Physiochemical properties of alginate in tissue engineering research and its clinical application. *Journal of Clinical Rehabilitative Tissue Engineering Research, 11*, 3613–3616. Retrieved October 5, 2018, from http://www.fao.org/3/CA1121EN/ca1121en.pdf.

Herbert, S. K., Samson, G., Fork, D. C., & Laudenbach, D. E. (1992). Characterization of damage to photosystem I and II in a cyanobacterium lacking detectable iron superoxide dismutase activity. *Proceedings of National Academy of Sciences U S A, 89*, 8716–8720.

Holtkamp, A. D., Kelly, S., Ulber, R., & Lang, S. (2009). Fucoidans and fucoidanases focus on techniques for molecular structure elucidation and modification of marine polysaccharides. *Applied Microbiology and Biotechnology, 82*, 1–11.

Huangfu, J., Liu, J., Sun, Z., Wang, M., Jiang, Y., Chen, Z. Y., & Chen, F. (2013). Antiaging effects of astaxanthin-rich alga *Haematococcus pluvialis* on fruit flies under oxidative stress. *Journal of Agricultural and Food Chemistry, 61*, 7800–7804.

irishseaweeds.com. (2017, June 28). *Sea Lettuce (Ulva lactuca)*. Retrieved from http://www.irishseaweeds.com/sea-lettuce-ulva-lactuca.

Javan, A. J., Javan, M. J., & Tehrani, Z. A. (2013). Theoretical investigation on antioxidant activity of bromophenols from the marine red alga *Rhodomela confervoides*: H-atom vs electron transfer mechanism. *Journal of Agricultural and Food Chemistry, 61*, 1534–1541.

Jormalainen, V., & Honkanen, T. (2004). Variation in natural selection for growth and phlorotannins in the brown alga *Fucus vesiculosus*. *Journal of Evolutionary Biology, 17*, 807–820.
Jung, H. A., Hyun, S. K., Kim, H. R., & Choi, J. S. (2006). Angiotensin-converting enzyme inhibitory activity of phlorotannins from *Ecklonia stolonifera*. *Fisheries Science, 72*, 1292–1299.
Kang, H. S., Chung, H. Y., Kim, J. Y., Son, B. W., Jung, H. A., et al. (2004). Inhibitory phlorotannins from the edible brown alga *Ecklonia stolonifera* on total reactive oxygen species (ROS) generation. *Archives of Pharmacal Research, 27*, 194–198.
Kang, J. I., Kim, S. C., Kim, M. K., Boo, H. J., Jeon, Y. J., Koh, Y. S., Yoo, E. S., Kang, S. M., & Kang, H. K. (2012). Effect of dieckol, a component of *Ecklonia cava*, on the promotion of hair growth. *International Journal of Molecular Sciences, 13*, 6407–6423.
Khotimchenko, S., Vaskovsky, V., & Titlyanova, T. (2002). Fatty acids of marine algae from the Pacific coast of North California. *Botanica Marina, 45*, 17–22.
Kim, H. H., Shin, C. M., Park, C. H., Kim, K. H., Cho, K. H., et al. (2005). Eicosapentaenoic acid inhibits UV-induced MMP-1 expression in human dermal fibroblasts. *Journal of Lipid Research, 46*, 1712–1720.
Kim, K. Y., Choi, K. S., Kurihara, H., et al. (2008a). β-glucuronidase inhibitory activity purified from *Grateloupia elliptica*. *Food Science and Biotechnology, 17*, 1110–1114.
Kim, S. K., Ravichandran, D., Khan, S. B., & Kim, Y. T. (2008b). Prospective of the cosmeceuticals derived from marine organisms. *Biotechnology and Bioprocess Engineering, 13*, 511–523.
Koivikko, R., Loponen, J., Honkanen, T., & Jormalainen, V. (2005). Contents of soluble, cell-wall-bound and exuded phlorotannins in the brown alga *Fucus vesiculosus*, with implications on their ecological functions. *Journal of Chemical Ecology, 31*, 195–212.
Koivikko, R., Loponen, J., Pihlaja, K., & Jormalainen, V. (2007). High-performance liquid chromatographic analysis of phlorotannins from the brown alga *Fucus vesiculosus*. *Phytochemical Analysis, 18*, 326–332.
Kurihara, H., Mitani, T., Kawabata, J., et al. (1999). Two new bromophenols from the red alga *Odonthalia corymbifera*. *Journal of Natural Products, 62*, 882–884.
Thalgo La. (2008). *Beaute marine*. Retrieved from http://www.thalgo.com.
Lan, W. C., Tzeng, C. W., Lin, C. C., et al. (2013). Prenylated flavonoids from *Artocarpus altilis*: Antioxidant activities and inhibitory effects on melanin production. *Phytochemistry, 89*, 78–88.
Lanfer-Marquez, U. M., Barros, R. M. C., & Sinnecker, P. (2005). Antioxidant activity of chlorophylls and their derivatives. *Foodservice Research International, 38*, 885–891.
Lee, S., Lee, Y. S., Jung, S. H., Kang, S. S., & Shin, K. H. (2003). Anti-oxidant activities of fucosterol from the marine algae *Pelvetia siliquosa*. *Archives of Pharmacal Research, 26*, 719–722.
Leyden, J. J. (1995). New understandings of the pathogenesis of acne. *Journal of the American Academy of Dermatology, 32*(5), S15–S25.
Li, X. C., Fan, X., Han, L. J., Yan, X. J., & Lou, Q. X. (2002a). Fatty acids of common marine macrophytes from the yellow and bohai seas. *Oceanologia et Limnologia Sinica, 33*, 215–223.
Li, T., Xu, H., Zhou, R., Liu, Y., Li, B., Nomura, C., & Zhao, J. (2002b). Differential expression and localization of Mn and Fe superoxide dismutases in the Heterocystous Cyanobacterium *Anabaena* sp. strain PCC 7120. *Journal of Bacteriology, 184*(18), 5096–5103.
Li, K., Li, X. M., Gloer, J. B., et al. (2012). New nitrogen-containing bromophenols from the marine red alga *Rhodomela confervoides* and their radical scavenging activity. *Food Chemistry, 135*, 868–872.
Liang, C., Lim, J. H., Kim, S. H., & Kim, D. S. (2012). Dioscin: A synergistic tyrosinase inhibitor from the roots of *Smilax china*. *Food Chemistry, 134*(2), 1146–1148.
Mafinowska, P. (2011). Algae extracts as active cosmetic ingredients. *Zeszyt Naukowy, 212*, 123–129.
Majmudar, G. (2012). *Compositions of marine botanicals to provide nutrition to aging and environmentally damaged skin*. US *Patent No*. 8318178 B2. Retrieved from https://patentimages.storage.googleapis.
Mikami, D., Kurihara, H., Ono, M., et al. (2016). Inhibition of algal bromophenols and their related phenols against glucose 6-phosphate dehydrogenase. *Fitoterapia, 108*, 20–25.

Mimuro, M., & Akimoto, S. (2003). Carotenoids of light harvesting systems: Energy transfer processes from fucoxanthin and peridinin to chlorophyll. *Advances in Photosynthesis and Respiration, 14*, 335–349.

Norzagaray-Valenzuela, C. D., Valdez-Ortiz, A., Shelton, L. M., Jiménez-Edeza, M., Rivera-López, J., Valdez-Flores, M. A., & Germán-Báez, J. (2017). Residual biomasses and protein hydrolysates of three green microalgae species exhibit antioxidant and anti-aging activity. *Journal of Applied Phycology, 29*, 189–198.

Obinger, C., Regelsberger, G., Strasser, G., Burner, U., & Peschek, G. A. (1997). Purification and characterization of a homodimeric catalase-peroxidase from the cyanobacterium *Anacystis nidulans*. *Biochemical and Biophysical Research Communications, 235*, 545–552.

Obinger, C., Regelsberger, G., Pircher, A., Strasser, G., & Peschek, G. A. (1998). *Physiologia Plantarum, 140*, 693–698.

Oren, A., & Gunde-Cimerman, N. (2007). Mycosporines and mycosporine-like amino acids: UV protectants or multipurpose secondary metabolites? *FEMS Microbiology Letters, 269*, 1–10.

Paduch, R., Kandefer-Szersze, Å. M., Trytek, M., & Fiedurek, J. (2007). Terpenes: Substances useful in human healthcare. *Archivum Immunolgiae et Therapiae Experimentalis (Warsz), 55*, 315–327.

Pallela, R., Na-Young, Y., & Kim, S. K. (2010). Anti-photoaging and photoprotective compounds derived from marine organisms. *Marine Drugs, 8*(4), 1189–1202.

Pandey, A., & Tiwari, A. (2014). *Toxicogenomics of microalgal microcystins*. Atlanta, GA: Scholars' Press. ISBN: 978-3639713640.

Park, Y. I. (2006). Skin whitening effect. In Y. I. Park & S. K. Lee (Eds.), *New perspectives on aloe* (pp. 127–135). New York, NY: Springer. ISBN: 978-0-387-31799-1.

Park, S. H., Song, J. H., Kim, T., et al. (2012). Anti-human rhinoviral activity of polybromocatechol compounds isolated from the rhodophyta, *Neorhodomela aculeate*. *Marine Drugs, 10*, 2222–2233.

Passardi, F., Zamocky, M., Favet, J., Jakopitsch, C., Penel, C., Obinger, C., & Dunand, C. (2007). Phylogenetic distribution of catalase peroxidases: Are there patches of order in chaos? *Gene, 397*, 101–113.

Paul, C., & Pohnert, G. (2011). Production and role of volatile halogenated compounds from marine algae. *Natural Product Reports, 28*, 186–195.

Pavia, H., & Brock, E. (2000). Extrinsic factors influencing phlorotannin production in the brown alga *Ascophyllum nodosum*. *Marine Ecology Progress Series, 193*, 285–294.

Pereira, L. (2018). *Therapeutic and nutritional uses of algae* (1st ed., pp. 2–64). Boca Raton, FL: CRC Press. ISBN: 9781498755382.

Pérez, M. J., Falqué, E., & Domínguez, H. (2016). Antimicrobial action of compounds from marine seaweed. *Marine Drugs, 14*(3), 52.

Pillai, S., Oresajo, C., & Hayward, J. (2005). Ultraviolet radiation and skin aging: Roles of reactive oxygen species, inflammation and protease activation, and strategies for prevention of inflammation-induced matrix degradation–A review. *International Journal of Cosmetic Science, 27*, 17–34.

Rajendran, U. M., Kathirvel, E., & Anand, N. (2007). Desiccation-induced changes in antioxidant enzymes, fatty acids, and amino acids in the cyanobacterium *Tolypothrix scytonemoides*. *World Journal of microbiology Biotechnology, 23*, 251–257.

Refilda, R., Munaf, E., Zein, R., Dharma, A., Indrawati, I., et al. (2009). Optimation study of carrageenan extraction from red algae (*Eucheuma cottonii*). *Jurnal Riset Kimia, 2*, 120–126.

Regelsberger, G., et al. (2002). Occurrence and biochemistry of hydroperoxidase in oxygenic phototrophic prokaryotes (cyanobacteria). *Plant Physiology and Biochemistry, 40*, 479–490.

Reyes, C. D. L., Zbakh, H., Motilva, V., et al. (2013). Antioxidant and anti-inflammatory meroterpenoids from the brown alga *Cystoseira usneoides*. *Journal of Natural Products, 76*, 621–629.

Rinaudo, M. (2008). Main properties and current applications of some polysaccharides as biomaterials. *Polymer International, 57*, 397–430.

Rouhier, N., & Jacquot, J. P. (2005). The plant multigenic family of thiol peroxidase. *Free Radical Biology & Medicine, 38*, 1413–1142.

Ryu, J., Park, S. J., Kim, I. H., Choi, Y. H., & Nam, T. J. (2014). Protective effect of porphyra-334 on UVA-induced photoaging in human skin fibroblasts. *International Journal of Molecular Medicine, 34*(3), 796–803.

Sahu, A., Pancha, I., Jain, D., Paliwal, C., Ghosh, T., et al. (2013). Fatty acids as biomarkers of microalgae. *Phytochemistry, 89*, 53–58.

Sambrook, J., & Russell, D. W. (2001). *Molecular cloning: A laboratory manual* (Vol. 3, 3rd ed., pp. A8.40–A8.50). New York, NY: Cold Spring Harbor Laboratory Press.

Schubert, N., Garcia-Mendoza, E., & Pacheco-Ruiz, I. (2006). Carotenoid composition of marine red algae. *Journal of Phycology, 42*, 1208–1216.

Shirkey, B., Kovarcik, D. P., Wright, D. J., Wilmoth, G., Prickett, T. F., Helm, R. F., Gregory, E. M., & Potts, M. (2000). Active Fe-containing superoxide dismutase and abundant sodF mRNA in *Nostoc commune* (cyanobacteria) after years of desiccation. *Journal of Bacteriology, 182*, 189–197.

Silab. (2018). *WHITONYL® and the complexion becomes porcelain*. Retrieved October 20, 2018, from https://www.silab.fr/produit-55-whitonyl_usa.html.

Solano, F., Briganti, S., Picardo, M., & Ghanem, G. (2006). Hypopigmenting agents: An updated review on biological, chemical and clinical aspects. *Pigment Cell Research, 19*(6), 550–571.

SpecialChem. (2017, June 28). *Porphyra umbilicalis extract*. Retrieved from https://cosmetics.specialchem.com/inci/porphyra-umbilicalis-extract.

Stahl, W., & Sies, H. (2003). Antioxidant activity of carotenoids. *Molecular Aspects of Medicine, 24*, 345–351.

Tan, X., Song, Y. H., Park, C., et al. (2016). Highly potent tyrosinase inhibitor, neorauflavane from *Campylotropis hirtella* and inhibitory mechanism with molecular docking. *Bioorganic & Medicinal Chemistry, 24*, 153–159.

Tang, H. F., Yang-Hua, Y., Yao, X. S., Xu, Q. Z., Zhang, S. Y., et al. (2002). Bioactive steroids from the brown alga *Sargassum carpophyllum*. *Journal of Asian Natural Products, 4*, 95–101.

Temina, M., Rezankova, H., Rezanka, T., & Dembitsky, V. M. (2007). Diversity of the fatty acids of the *Nostoc* species and their statistical analysis. *Microbiological Research, 162*, 308–321.

Thomas, D. J., Thomas, J. B., Prier, S. D., Nasso, N. E., & Herbert, S. K. (1999). Iron superoxide dismutase protects against chilling damage in the cyanobacterium *Synechococcus* species PCC7942. *Plant Physiology, 120*, 275–282.

Tiwari, A. (2010). *Toxins of cyanobacteria: The nature and impact of cyanobacterial toxins*. Saarbrücken: LAP LAMBERT Academic Publishing. ISBN: 978-3843368780.

Tiwari, A. (2012). *Molecular and biochemical analysis of bloom forming cyanobacteria*. Saarbrücken: LAP LAMBERT Academic Publishing. ISBN: 978-3659162107.

Tiwari, A. (2014). *Cyanobacteria nature, potentials and applications*. Delhi India: Daya Publishing House. ISBN: 9789351301332.

Tripathi, S. N., & Srivastava, P. (2001). Presence of stable active oxygen scavenging enzymes superoxide dismutase, ascorbate peroxidase and catalase in a desiccation-tolerant cyanobacterium Lyngbya arboricola under dry state. *Current Science, 81*(2), 25.

Wang, W., Okada, Y., Shi, H., et al. (2005). Structure and aldose reductase inhibitory effects of bromophenols from the red alga *Symphyocladia latiuscula*. *Journal of Natural Products, 68*, 620–622.

Wang, J., Jin, W., Hou, Y., Niu, X., Zhang, H., & Zhang, Q. (2013a). Chemical composition and moisture-absorption/retention ability of polysaccharides extracted from five algae. *International Journal of Biological Macromolecules, 57*, 26–29.

Wang, R., Paul, V. J., & Luesch, H. (2013b). Seaweed extracts and unsaturated fatty acid constituents from the green alga *Ulva lactuca* as activators of the cytoprotective Nrf2-ARE pathway. *Free Radical Biology & Medicine, 57*, 141–153.

Welinder, K. G. (1992). Superfamily of plant, fungal and bacterial peroxidases. *Current Opinion in Structural Biology, 2*, 388–393.

Wood, B. J. B. (1988). *Lipids of algae and protozoa. Microbial lipids*. London: Academic Press.
Yamaguchi, N., Satoh-Yamaguchi, K., & Ono, M. (2009). In vitro evaluation of antibacterial, anti-collagenase, and antioxidant activities of hop components (*Humulus lupulus*) addressing acne vulgaris. *Phytomedicine, 16*(4), 369–376.
Zamocky, M., Furtmüller, P. G., & Obinger, C. (2008a). Evolution of catalases from bacteria to humans. *Antioxidants and Redox Signaling, 10*, 1527–1547.
Zamocky, M., Jakopitsch, C., Furtmüller, P. G., Dunand, C., & Obinger, C. (2008b). The peroxidase-cylooxygenase superfamily: Reconstructed evolution of critical enzymes of the innate immune system. *Proteins, 71*, 589–605.
Zhukova, N. V., & Aizdaicher, N. A. (1995). Fatty acid composition of 15 species of marine micro-algae. *Phytochemistry, 39*, 351–356.
Zubieta, C., Krishna, S. S., Kapoor, M., et al. (2007). Crystal structures of two novel dye-decolorizing peroxidases reveal a b-barrel fold with a conserved heme binding motif. *Proteins, 70*, 223–233.

Further Reading

http://www.laprairieswitzerland.com/.
http://www.nykaa.com/.
https://openaccesspub.org/japb/article/530.
https://www.algatech.com/.
https://www.lorealparis.co.in/.
https://www.patent-art.com/algae-cosmetics-18.
https://www.shiseidogroup.com/.
www.aquarev.in.

Chapter 9
Extracts and Bioactives from Microalgae (Sensu Stricto): Opportunities and Challenges for a New Generation of Cosmetics

Lorenzo Zanella and Md. Asraful Alam

Abstract The cosmetic industry, which increasingly aims to develop products that affect body appearance, prevent aging and promote skin and hair well-being, has changed over the last decades. The increased sensibility of consumers to the ethics of green economy drew the attention of this industry to microalgae as novel source of active ingredients. Microalgae, which are often improperly considered as inclusive of prokaryotic microorganisms, i.e. cyanobacteria, are eukaryotic microorganisms capable of synthesising biologically active molecules that affect human metabolism.

Many classes of beneficial compounds, including carotenoids, polyphenols, vitamins and polysaccharides, can be obtained from microalgae cultivated with sustainable and environment-friendly techniques. Microalgal extracts are already commercialised in products that claim several biological activities, such as hair growth stimulation, prevention of solar radiation damages, modulation of skin pigmentation, skin tightening and anti-aging. However, their mechanisms of action and metabolic effects are not fully understood, and the related beneficial effects are probably underestimated. This contribution aims to review the state-of-the-art cosmetic applications of microalgae with a critical discussion of the experimental methods adopted and potential perspectives.

Keywords Microalgae · Natural extracts · Cosmetics · Skin · Pigmentation · Dermis · Keratinocytes · Melanogenesis · Sebogenesis · Hair follicle · Biological activity · Oxidative stress · Transdermal delivery

---

L. Zanella (✉)
Venice, Italy
e-mail: lorenzo.zanella@libero.it

M. A. Alam
School of Chemical Engineering, Zhengzhou University, Zhengzhou, China
e-mail: alam@zzu.edu.cn

M. A. Alam et al. (eds.), *Microalgae Biotechnology for Food, Health and High Value Products*, https://doi.org/10.1007/978-981-15-0169-2_9

Abbreviations

ACTH	adrenocorticotropic hormone
Akt	Protein kinase B
AP-1	Activator protein 1
ARE	Antioxidant response elements
ATX	Astaxanthin
BAD	Bcl-2-associated death promoter
Bax	Bcl-2-associated X
Bcl-2	B-cell lymphoma 2
βC	β-Carotene
Casp	Caspase
CE	Cornified envelope
COX-2	Cyclooxygenase-2
CRH	corticotropin-releasing hormone
CT	Carotenoid
CTX	Canthaxanthin
cyt-c	Cytochrome-c
DGDG	Digalactosyl diacylglycerol
DP	Dermal papilla
DW	Dry weight
ECM	Extracellular matrix
EPA	Eicosapentaenoic acid
SEP	Sulphated exopolysaccharide
ERK	Extracellular signal-regulated kinase
FB	Fibroblast
FoxOs	Forkhead box, class O family member proteins
FT	Ferritin
FXT	Fucoxanthin
GABA	γ-Aminobutyric acid
GAG	Glycosaminoglycan
GCL	Glutamate-cysteine ligase
GM-CSF	Granulocyte-macrophage colony-stimulating factor
GSR	Glutathione-disulphide reductase
h	Human
HA	Hyaluronic acid
4HNE	4-Hydroxy-2-nonenal
HO-1	Haeme oxygenase (heat shock protein fam.)
hSE	3D human skin equivalents
hFTS	Human full-thickness skin
HF	Hair follicle
IL	Interleukin
iNOS	Inducible nitric oxide synthase
IR	Infrared rays
JNK	c-Jun N-terminal kinase

KC	Keratinocyte
Keap1	Kelch-like ECH-associated protein 1
L-DOPA	3,4-Dihydroxy-L-phenylalanine
LF	Lipofuscin
MA	Microalga
MAs	Microalgae
MAA	Mycosporine-like amino acid
MAPK	Mitogen-activated protein kinase
Mcl-1	Induced myeloid leukemia cell differentiation protein
MGDG	Monogalactosyl diacylglycerol
MITS	Microphthalmia-associated transcription factor
MMP	Matrix metalloproteinase
MW	Molecular weight
NEP	Neprilysin or neutral endopeptidase
NF-kB	Nuclear factor kappa-light-chain-enhancer of activated B cells
NMF	Natural moisturising factors
Nrf2	NF-E2 p45-related factor 2
NQO-1	NAD(P)H dehydrogenase (quinone) 1
NT	Neurotrophin
PKCδ	Protein kinase C-δ type
PMA	Phorbol myristate acetate
PUFA	Polyunsaturated fatty acid
ROS	Reactive oxygen species
RNS	Reactive nitrogen species
5α-R1	5α-reductase type-1
SAGE	Semisynthetic glycosaminoglycan ether
SC	Stratum corneum
SEP	Sulphated exopolysaccharide
SOD	Superoxide dismutase
SPR	Small proline-rich protein
STAT3	Signal transducer and activator of transcription 3
TGF-β	Transforming growth factor-β
T-Iso	Tahitian *Isochrysis*
TNFα	Tumour necrosis factor-α
UCA	Urocanic acid
UV	Ultra violet (UVR: rays; UVA: type A; UVB: type B; UVC: type C)

1 Introduction

In the biological world, all organisms are endowed with enzymes necessary for essential metabolic cell processes, but some have also developed enzymes that produce special secondary metabolites, especially amongst prokaryotes, plants and fungi. These metabolites, which include antioxidants or some toxins against preda-

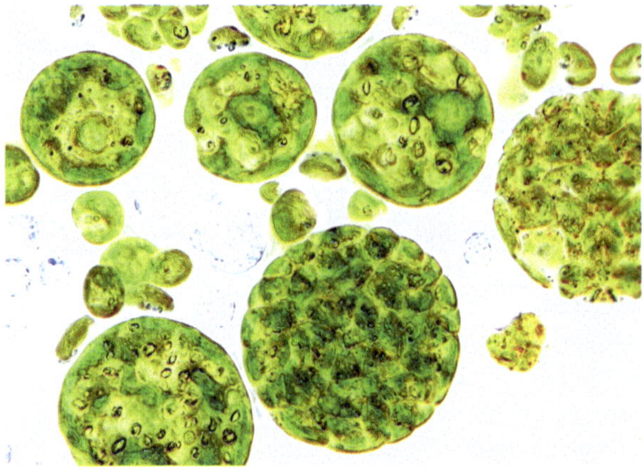

Fig. 9.1 *Chlorococcum minutum* (Source: courtesy of Cutech Srl)

tors, generally protect against potential damages from the external environment. Amongst autotrophic organisms, they comprise vitamins, special macromolecules (e.g. long-chain unsaturated fatty acids) and some accessory pigments for photosynthesis, e.g. carotenoids (CTs), which are able to work as free radical scavengers in humans (Zhang et al. 2014). Some of these compounds are irreplaceable dietary vitamins and micronutrients for animals and humans. The cosmetic industry is increasingly getting aware that the microorganism biochemistry offers numerous compounds to preserve youth and beauty. Amongst the richest sources of active ingredients, microalgae (MAs) have become the object of special attention due to their extraordinary capability to synthesise and store bioactives that are still relatively unknown (Plaza et al. 2009). In this regard, some clarifications are appropriate, because with the term MAs (sensu lato), very heterogeneous microorganisms are generally indicated (see Andersen 2013 for a concise systematic overview). For instance, prokaryotes belonging to the vast group of cyanobacteria (also called blue-green algae) are generally included. Cyanobacteria have biological traits that are typical of bacteria, including prokaryotic polysaccharides in their cell walls and special pigments with high biological activity (phycocyanins). The MAs sensu stricto, however, have a nucleus; therefore, they belong to the domain Eukaryota and have a biochemical composition of their own, which can sometimes include molecules in common with macroalgae or even higher plants depending on the case. This contribution will refer to MAs sensu stricto (Fig. 9.1 shows an eukaryotic microalga of cosmetic interest).

1.1 The Exploitation of Microalgae in Cosmetics: A Recent Story

The exploitation of MAs for industrial processes has become significant due to marine aquaculture, from which most of the knowledge that has inspired recent studies was derived. The need to cultivate MAs in monoculture was born at the beginning of the last century for nutrition of microinvertebrates (Allen and Nelson 1910) and assumed commercial applications when microalgal cultures were used for the rearing of bivalve molluscs intended for human consumption (Bruce et al. 1940; Rhyter and Goldman 1975).

Many steps forward, such as the selection of monospecific strains for the nutrition of bivalves and phytophagous larval phases of prawns, were made in the following decades (De Pauw et al. 1984). However, in these pioneering experiences, MAs were intended for animal species that physiologically feed on MAs. The use of MAs as live food was an obvious choice, whereas the biological value of their biochemical composition for other applications was established only later. In the 1970s, the first attempts to reproduce and breed valuable marine fish had little success due to the low survival of early larval stages, which fed on zooplankton (rotifers), and the high rate of malformation suffered by fingerlings that reached weaning. These difficulties were overcome only when some researchers realised that the high nutritional requirements of many marine fish larvae could be satisfied by administering 'enriched' rotifers, i.e. fed with some MAs. The high content of polyunsaturated fatty acids (PUFAs) and other essential nutrients in MAs was recognised as the key factor. Japanese aquaculture gave a significant contribution (Watanabe et al. 1983) to the achievement of the first successes. Since the 1970s, fish farmers have selected many microalgal strains endowed with some fundamental properties, such as the presence of secondary metabolites with high biological value, absence of toxicity and adaptability to intensive culture conditions, which are useful for exploitation in cosmetics. Indeed, these three elements made phytoplankton cultivation an ideal source of natural ingredients for the cosmetic industry.

Another pivotal element of innovation was introduced by the development of intensive phytoplankton cultures in photobioreactors, which allow very intensive production without using pesticides, are eco-sustainable and can be upscaled to industrial production with fully traceable and certifiable processes. However, although the use of modern photobioreactors has reduced production costs (Molina Grima et al. 2003; Tredici et al. 2016), the final price of MA biomasses remains relatively high (Barsanti and Gualtiero 2018).

The cosmetic industry is amongst the sectors that can exploit the biochemical characteristics of MAs, because value-added products can be developed using a relatively small quantity of biomass.

1.2 A New Concept of Cosmetic Treatment Based on Bioactives

Cosmetics are traditionally classified into skin care, makeup, body and hair care, oral cosmetics and fragrances (Mitsui 1997). Before the 1980s, they were principally used to beautify or cover minor, visible imperfections or, at the most, improve the structure of the skin and its annexes. The main biological activities were due to physicochemical properties of some ingredients, e.g. emollients and moisturisers. The use of beneficial natural ingredients has always been present in cosmetics, but this aspect has become predominant only recently along with environmental sensibility. A new range of cosmetic products, which are designed as a vehicle of natural principles endowed with biological activity, have been developed (Kumar 2005; Paye et al. 2009). The concept of cosmeceutics, a neologism obtained from the fusion of the word 'cosmetics' and 'pharmaceutical' introduced by dermatologist Albert Kligman (Tsai and Hantash 2008), was born. Cosmeceutics is defined as '… *topical formulations which were neither pure cosmetics, like lipstick or rouge, nor pure drugs, like corticosteroids. They lay between these poles, constituting a broad-spectrum intermediate group*' (Kligman 2005).

The search for active ingredients suitable for the cosmetic sector received great impetus, and natural extracts are the main sources of inspiration due to their wealth of molecules that are already known as ingredients of traditional herbal medicines. The cosmetic exploitation of MAs is part of this sociocultural and industrial trend due to their richness of active compounds with a strong commercial appeal.

2 Bioactives from Microalgae and Their Applications in Cosmetics

2.1 Microalgae as Novel Sources of Active Compounds

Traditional medicine and, to a lesser extent, cosmetics have exploited the biological properties of plants since time immemorial to obtain natural extracts containing molecules that are active on human wellness. In terrestrial plants, many metabolites are often concentrated or stored in different organs (e.g. roots, leaves, flowers and fruits), depending on their specific functions, so that only that part is used for the preparation of cosmetic ingredients. Meanwhile, MAs are single-celled organisms with a whole library of enzymes and metabolites concentrated in the same cell. Sometimes cells are organised in colonial forms, such as in many Bacillariophyceae; however, metabolic self-sufficiency is generally maintained in each cell. Therefore, the microalgal biomass is homogeneous and entirely used for the extraction of active ingredients, whose composition depends on cell characteristics, solvent used and extraction procedure (Chojnacka and Kim 2015).

Some strains were identified as sources of specific compounds that were accumulated in large amounts, especially if cultivated under appropriate environmental conditions. Indeed, the prevailing approach in the exploitation of MAs was to use them as biofactories for the production of compounds known for their beneficial properties (Barclay et al. 1994; Jin et al. 2003; Spolaore et al. 2006; Catalina Adarme-Vega et al. 2012; Priyadarshani and Rath 2012; Guarnieri and Pienkos 2015; Guedes et al. 2011; Koller et al. 2014; Wobbe and Remacle 2015; Singh et al. 2017; Islam et al. 2017). Representative examples are the extraction of astaxanthin (ATX) from cysts of *Haematococcus pluvialis* (Guerin et al. 2003), CTs from *Dunaliella salina* (Jin and Melis 2003; Pisal and Lele 2005; Del Campo et al. 2007) and omega-3 fatty acids from *Phaeodactylum tricornutum* (Reis et al. 1996), *Porphyridium cruentum* (Asgharpour et al. 2015), *Crypthecodinium cohnii* (Mendes et al. 2009) and *Nannochloropsis* spp. (Forján Lozano et al. 2007; Chini Zittelli et al. 1999).

Unfortunately, this approach is affected by competition with traditional sources of the same molecules, such as fish oil and chemical industries. However, an interesting alternative way to exploit the richness of MAs is to develop multifunctional extracts, which allow to achieve beneficial effects by acting simultaneously on various metabolic processes of the treated tissue. This topic will be discussed in more detail below, but it is worth noting here that the extraction of specific bioactives from the biomass involves relevant purification costs, whereas the multifunctional approach takes advantage of a simplified extraction process. Extracts with a broad spectrum of action require more research effort to characterise their effects on the tissues or organs, but their industrial production is less expensive and their composition cannot be synthetically reproduced by commercial competitors.

2.2 *Effects on the Skin and Its Accessory Structures*

Historically, the most investigated application is probably in the prevention of aging and reduction of wrinkles. Aging causes the overall loss of structural organisation of the skin, with particularly evident consequences in the dermis, whose biomechanical properties are attributable to the fibrous and amorphous connective tissue (extracellular matrix, ECM), which is composed of proteins, proteoglycans and glycosaminoglycans (GAGs). Many cosmetic products therefore claim the ability of stimulating the synthesis of the dermis ECM and protecting it from degradation processes.

To delay aging, however, modern cosmetic science has also developed a wide range of active products aimed at improving other aspects of skin metabolism, e.g. state of hydration, smoothness of the stratum corneum (SC), modulation of sebum production and melanogenesis. Many extend to the treatment of problems that are borderline with pathological disorders, such as melasma, acne, seborrheic dermatitis, various forms of dermatitis or psoriasis and solar erythema. To date, preparations obtained from MAs showed various activities on the skin and its annexes, confirming that they are a valuable source of active compounds of high cosmetic interest. However, an appropriate exploitation of them requires a deep comprehension of skin biology and molecular signals that regulate its metabolism.

2.2.1 Skin Anatomy

The skin is the largest organ of the body (1.5–2 m^2) and has an average thickness of 1–2 mm, but with variation from 0.5 mm of eyelids to over 6 mm between the shoulder blades (Saladin 2007) (Fig. 9.2).

The skin is composed of epidermis and dermis, whereas the underlying adipose panniculus (called hypodermis or subcutis) is generally considered distinct even though it is closely connected to the skin both anatomically, because some fat cells

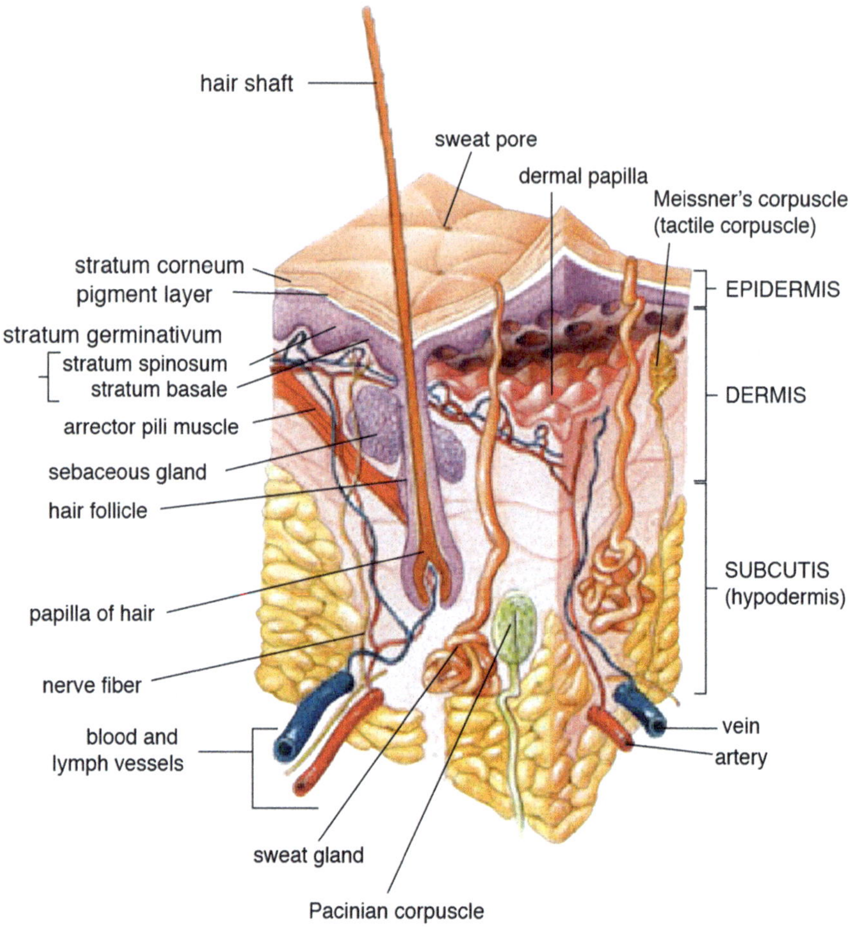

Fig. 9.2 A cross section of the skin and its underlying structures (image contributed by Wikimedia Commons, USGOV (Public Domain), from: Anatomy, Skin (Integument), Epidermis (Yousef and Sharma 2019), book distributed under the terms of the Creative Commons Attribution 4.0 International License (http://creativecommons.org/licenses/by/4.0/))

can be dispersed in the deep dermis, and functionally through a consistent exchange of cytokines. The epidermis is the superficial layer and is divided from the underlying dermis by the basal lamina, which is a dense planar-reticular structure mainly consisting of glycoproteins (e.g. fibronectin and laminin) and collagen (COL) type IV (COL-IV). Epidermis is a densely cellularised epithelial tissue composed of keratinocytes (KCs), which account for about 95% (McGrath et al. 2010) and proliferate starting from the primary basal layer, called *stratum basale*. Amongst KCs of the stratum basale occur the melanocytes, which are specialised pigmentary cells of neuroectodermal origin that synthesise melanin in special subcellular organelles called melanosomes. Melanocytes can assume a dendritic shape and develop temporary cell projections, known as pseudopodia, which carry melanosomes away from the centre of the cell. KCs can engulf the tips of the melanocyte pseudopodia through phagocytosis, receiving a certain quantity of melanin (Nordlund et al. 1989). This process modulates the skin pigmentation and is stimulated by exposure to solar radiation (skin tanning).

Following these important cell interactions in the basal layer, the KCs move towards the epidermal surface undergoing a process of differentiation that leads to the formation of the SC.

Keratinocyte Differentiation

KCs proliferate from the basal layer and undergo differentiation, thereby forming the following layers (Fig. 9.3): *stratum basale* or *stratum germinativum*, *stratum spinosum*, *stratum granulosum* and *stratum corneum*. In palmoplantar skin is observed an additional electrolucent stratum, called stratum lucidum, interposed between the granulosum and corneum (McGrath et al. 2010). The differentiation process is called cornification and is regulated by the concentration of Ca^{2+}, which increases from the stratum basale to the SC (Eckhart et al. 2013). KC differentiation

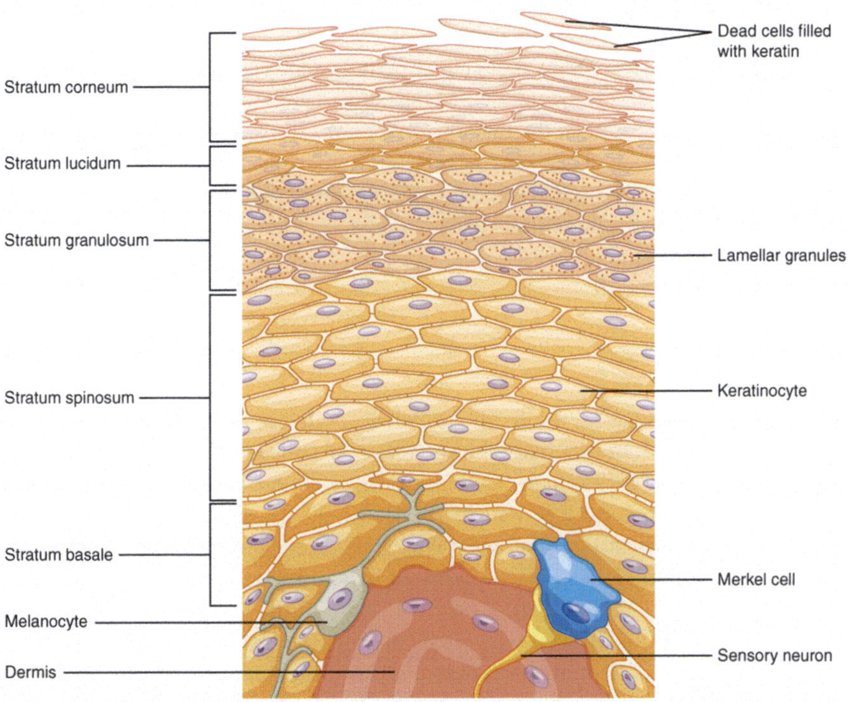

Fig. 9.3 Layers of the Epidermis. The epidermis of thick skin has five layers: stratum basale, stratum spinosum, stratum granulosum, stratum lucidum and stratum corneum (Access for free at https://openstax.org/books/anatomy-and-physiology/pages/1-introduction) (license conditions at https://creativecommons.org/licenses/by/4.0/)

occurs with synthesis of cytoskeletal scleroproteins, organised in the cornified envelope (CE) and lipids, which are confined in lamellar bodies (for review see Candi et al. 2005 and Eckhart et al. 2013). The main CE proteins are loricrin (the most abundant), involucrin, filaggrin (which aggregates keratin filaments into tight bundles), elafin (a serine proteinase inhibitor) and small proline-rich repeat proteins (SPRs) that have antioxidant properties (Steinert and Marekov 1995; Rinnerthaler et al. 2015). These proteins constitute about 7–10% of the mass of the epidermis (Candi et al. 2005) and are synthesised at different phases of KC differentiation to be cross-linked by transglutaminases, especially transglutaminase-1 and transglutaminase-3, which are Ca^{2+}-dependent enzymes that catalyse ε-(ϒ-glutamyl)lysine cross-linking reactions (Terazawa et al. 2015).

In the stratum granulosum, KC develops lamellar bodies, which are derived from the Golgi apparatus and filled with phospholipids, glucosylceramides, sphingomyelin and cholesterol (Feingold 2007; Rinnerthaler et al. 2015). In the final phase of differentiation, KCs collapse into dead cells called corneocytes that are connected by keratin bridges, whereas the lamellar bodies are secreted in the extracellular space, where some enzymes complete the process of maturation of the corneal matrix. Part of filaggrin is degraded by caspase-14 (Casp-14) into amino acids, some of which act as natural moisturising factors (NMF). Filaggrin is also a major source of histidine, which is further metabolised into the potent UVB scavenger urocanic acid (UCA) in the cornifying layers (Eckhart et al. 2013).

The mature SC is a complex model of structure called 'brick and mortar barrier', wherein the lipid matrix is the mortar and the corneocytes are the bricks (Nemes and Steinert 1999). Interestingly, the pyknotic cytoplasm of the corneocyte is occupied by NMFs, i.e. amino acids and their derivatives and salts, which contribute to the hydration and elasticity of SC. The epidermis is superficially lubricated by sebum, which contributes to the proteolipid barrier, interacts with the microbiome and regulates the SC's exfoliation process.

The Dermis

Under the basal lamina lies the dermis, which is composed of proteins and polysaccharides of the ECM, in which fibroblasts (FBs) and cells of the immune system are dispersed. Amongst the proteins, COL contributes 70–80% to the dry weight and confers tensile properties, followed by elastin (2–4% of the dermis per volume), which provides resilience and softness (Waller and Maibach 2006). The most abundant GAG is hyaluronic acid (HA), followed by several derivatives of chondroitin sulphate. Although GAGs represent only 0.1–0.3% of the dry weight of skin, they can bind up to 1000 times their own volume in water (Bernstein et al. 1996), thereby regulating the state of hydration and plumpness of the organ. The intrinsic and photoinduced aging processes determine the alterations of all these structural molecules, thereby compromising the mechanical properties of the skin and significantly reducing its ability to maintain water in the bound state (Waller and Maibach 2006). FBs are responsible for the synthesis of ECM, but together with immune cells, they also participate in its degradation, releasing matrix metalloproteases (MMPs) and other proteases and hyaluronidases (Pittayapruek et al. 2016).

Skin Appendages

The main skin appendages are the sebaceous glands (SGs), sweat glands, hair follicles (HFs) and nails. These organs are strongly integrated with the surrounding skin environment but have their own metabolism. SGs are holocrine glands comprising sebocytes that change into lipid-producing cells from the undifferentiated basal layer, which finally die to secrete the oily and waxy substance called *sebum* (Mitsui 1997). Sebum consists of squalene, esters of glycerol (glycerides) and wax, free fatty acids and free and esterified cholesterol (Picardo et al. 2009; Wertz 2009). It is excreted through SG ducts to the skin surface, almost always by way of the HF infundibulum, or HF canal, because HFs and SGs are anatomically associated in the so-called pilosebaceous unit.

HFs of scalp and body have an enormous impact on the appearance and related psychological, social and cultural implications. For this reason, the hair care market has huge commercial value. HF is a complex organ characterised by continual and cyclical transition amongst growth stage (anagen), in which the development of the hair is observed; a subsequent regression stage (catagen), in which the apoptosis of a considerable part of HF cells takes place; and a stage of quiescence at the end (telogen), following which the HF returns to the anagen stage with the formation of a new hair shaft. This life cycle is repeated over time with different rhythms depending on the region of the body (for more information on hair biology, see Paus and Peker 2003) and is controlled by the dermal papilla (DP), an inner region of the basal bulb comprising specialised FBs. The DP is in close contact with the *matrix*, a population of special KCs with high proliferative activity that occupies the upper part of the basal bulb and from which the hair develops.

During the telogen phase, the DP enters a resting phase, the basal bulb degenerates and the hair shaft remains in the scalp until it is pushed out by the growth of a new anagen hair (*exogen*). The telogen ends when DP releases signals that activate follicle regeneration, a process that starts from stem cells stored in a specialised follicular region called *bulge region*.

2.2.2 Effects of Microalgae Extracts on Epidermis

The effects of MA extracts on KC differentiation have been often studied using some protein markers indicative of CE development, but the findings obtained can be considered representative of the whole differentiation process, including the formation of the lamellar bodies. KC differentiation is of relevant cosmetic interest because its anomalies affect the proteolipid barrier with consequences on softness and smoothness of the epidermis.

Tests performed on human skin cultivated ex vivo showed that some extracts of *Tetraselmis suecica* can stimulate the synthesis of involucrin and filaggrin in KCs (Pertile et al. 2010). Involucrin modulation was obtained on the same experimental model by treatment with extracts of *Monodus subterraneus* and *Chlorococcum* sp., but with different results depending on the solvent used for the preparation of the extracts (Zanella et al. 2012). An aqueous extract of *Chlorella vulgaris*

(Dermochlorella, Codif) stimulated CE proteins, SPRs and elafin (Morvan and Vallee 2007). Thus far, little is known about the effects of MAs on the lipid composition of SC.

The cosmetic industry is also very interested in preventing or repairing the damages induced by ultraviolet radiation (UV), which causes photoaging. Nizard et al. (2004) showed in vivo that extracts of *P. tricornutum* stimulate the protective activity of 20S proteasome in KCs, preventing the increase of oxidised proteins and improving the protection of cell from UVB damages. Other molecules of interest for the same application are mycosporine-like amino acids (MAAs), which are secondary metabolites characterised by a cyclohexenone or cyclohexenimine chromophore conjugated with one or two amino acids (Cardozo et al. 2007). These compounds do not regulate epidermis metabolism but protect from UV due to their screening action in absorptions from 309 to 360 nm (Hartmann et al. 2015). In Table 9.1, some potential microalgal sources of MAAs are listed, but researchers need to beware, because this list also includes some toxic species not suitable for cosmetics (e.g. *Alexandrium tamarense*).

2.2.3 Activity of Microalgae Extracts on Dermis

The dermis is an object of special attention in the cosmetic field because its structure plays a primary role in determining the tensile properties and plumpness of the skin. Alterations connected to aging determine flaccidity and the formation of wrinkles. Intrinsic aging occurs physiologically, but it is accelerated by oxidative stress induced by solar radiation (photoaging) and by other stressful factors related to lifestyle and environment.

Chung et al. (2001) showed by in vivo analysis that intrinsic aging leads to a reduction of COL synthesis by FBs, whereas chronic photoaging leads to an increase, which does not compensate for the increased degradation due to the secretion of collagenases (MMP1 and MMP2) by the same cells. In both processes, the dermis COL undergoes qualitative and quantitative decrease with aging (Waller and Maibach 2006). This has decidedly oriented cosmetics towards the search for natural preparations suitable for promoting the production of COL, especially the type I (85–90% of this protein) and type III (10–15%) (Cheng et al. 2011). Amongst the preparations obtained from MAs, an aqueous extract of *Nannochloropsis oculata* exerted strong protection from oxidative stress and stimulated the production of COL in FB cultures (Stolz and Obermayer 2005). A similar preparation obtained from *D. salina* also stimulated COL production and cell proliferation (Stolz and Obermayer 2005). The already mentioned extract of *C. vulgaris* marketed under the name *Dermochlorella* was tested on cultures of primary FBs and KCs, as well as in clinical trials, and produced the following anti-aging and anti-inflammatory effects (Morvan and Vallee 2007):

- Stimulation of the synthesis of COL-I, COL-III, elastin, collagenase inhibitors and *plasminogen activator inhibitor-2*

Table 9.1 Mycosporine-like amino acids with UV shielding properties identified as compounds of microalgal species. Nomenclature according to www.algaebase.org (re-elaborated from Llewellyn and Airs 2010, Priyadarshani and Rath 2012, Flaim et al. 2014 and Suh et al. 2014)

Compound	Microalgae source
Sporopollenin	*Characium terrestre*, *Coelastrum microporum*, *Enallax coelastroides*, *Scenedesmus* sp., *Scotiella chlorelloidea*, *Scotiellopsis rubescens*, *Spongiochloris spongiosa*, *Dunaliella salina*, *Chlorella fusca*
Mycosporine glycine	*Chlamydomonas hedleyi*, *Alexandrium tamarense*, *Karlodinium venificum*, *Gymnodinium galatheanum*, *Prorocentrum lima*, *P. micans*, *P. cordatum*, *Scrippsiella trochoidea* and *Oxyrrhis marina*
Shinorine	*Chlamydomonas hedleyi*, *Peridinium aciculiferum*, *Chlorarachnion reptans*, *Alexandrium tamarense*, *Karlodinium venificum*, *Gymnodinium galatheanum*, *Kryptoperidinium foliaceum*, *Prorocentrum lima*, *P. micans*, *P. cordatum*, *Scrippsiella trochoidea* and *Oxyrrhis marina*
Porphyra-334	*Chlamydomonas hedleyi*, *Peridinium aciculiferum*, *Thalassiosira weissflogii*, *Alexandrium tamarense*, *Karlodinium venificum*, *Gymnodinium galatheanum*, *Prorocentrum micans*, *P. cordatum* and *Scrippsiella trochoidea*
Palythene	*Peridinium aciculiferum*, *Alexandrium tamarense*, *Karlodinium venificum*, *Prorocentrum lima*, *P. micans*, *P. cordatum* and *Scrippsiella trochoidea*
Palythine	*Peridinium aciculiferum*, *Alexandrium tamarense*, *Karlodinium venificum*, *Gymnodinium galatheanum*, *Prorocentrum lima*, *P. cordatum* and *Scrippsiella trochoidea*
Palythic acid	*Karlodinium venificum*, *Gymnodinium galatheanum*, *Kryptoperidinium foliaceum*, *Prorocentrum lima*, *P. micans*, *P. cordatum* and *Scrippsiella trochoidea*
Asterina-330	*Peridinium aciculiferum* and *Heterosigma akashiwo*
Undetermined mycosporine-like amino acids	*Ankistrodesmus spiralis*, *Chlorella minutissima*, *Chlorella sorokiniana*, *Dunaliella tertiolecta*, *Pelagococcus subviridis*, *Porphyridium purpureum*, *Rhodomonas maculata*, *R. salina*, *Cryptomonas reticulata*, *Cryptomonas baltica*, *Scotiella chlorelloidea*, *Isochrysis* sp., *I. galbana*, *Pavlova gyrans*, *Emiliania huxleyi*, *Corethron criophilum*, *Thalassiosira tumida*, *T. weissflogii*, *Porosira pseudodenticulata*, *Stellarima microtrias*, *Alexandrium catenella*, *Euglena gracilis*, *Nannochloropsis oculata and Nephroselmis rotunda*

- Inhibition of the expression of collagenase activators: *tissue plasminogen activator* and *urokinase plasminogen activator*
- Increased synthesis of the antioxidant enzyme *thioredoxin-2*

The extracts of *Chlorococcum* sp., *Chaetoceros* sp. and *M. subterraneus* stimulated COL-I production in primary FB cultures (Zanella et al. 2012). The extracts of *Porphyridium purpureum*, *Rhodosorus marinus*, *Chlorella pyrenoidosa* and *D. salina* showed high inhibitory effect on hyaluronidase, the enzyme that degrades the polysaccharide fraction of ECM (Fujitani et al. 2001).

2.2.4 Effects of Microalgae Extracts on Skin Pigmentation

MAs offer opportunities in the development of novel cosmetics for skin pigmentation. Products that inhibit melanogenesis (skin lighteners) and stimulate it (skin darkeners or tanners) are both appreciated. Skin lighteners are used to obtain a lighter complexion or to treat unwanted hyperpigmentation (e.g. lentigo solaris and melasma), whereas skin darkeners promote a tan without exposure to solar radiation or prepare the skin for sun exposure, thereby preventing erythema or burns.

Many microalgal compounds exert an activity on tyrosinase, the key enzyme that controls melanin synthesis (Nordlund et al. 1989). Tyrosinase catalyses melanin synthesis by hydroxylation of L-tyrosine to *3,4-dihydroxy-L-phenylalanine* (L-DOPA) and by oxidation of L-DOPA to dopaquinone followed by further conversion to melanin (Godin and Touitou 2007). Some microalgal compounds, particularly fatty acids and CTs, have been shown to exert an activity on tyrosinase. Interestingly, although saturated fatty acids often stimulate melanogenesis by delaying the degradation of tyrosinase, PUFAs have a predominantly inhibitory effect (Table 9.2) by downregulating the activity of the enzyme and by accelerating its degradation (Ando et al. 1998; Chiang et al. 2011). Since MAs are major producers of PUFAs, most preparations obtained from their extracts are expected to act as skin lighteners. In fact, extracts of *Nannochloropsis gaditana* showed an inhibition of tyrosinase (Letsiou et al. 2017), and extracts of *T. suecica* (Pertile et al. 2010), *Chaetoceros calcitrans* f. *pumilus*, *M. subterraneus*, *Chlorococcum minutum*, *Thalassiosira pseudonana* (Zanella et al. 2012) and *Nannochloropsis* sp. (Zanella and Pertile 2016) acted as skin lighteners in ex vivo skin cultures. Kurfurst et al. (2010) showed that *T. pseudonana* reduces melanin synthesis and inhibits its delivery to KC by downregulating the expression of *Myosin-X protein*, a protein involved in this process. Finally, a hydro-alcoholic extract of *Chlamydomonas reinhardtii* inhibited melanogenesis in melanoma cell cultures and in 3D human skin equivalent (hSE) (Lee et al. 2018).

However, it is wrong to assume that the high concentration of PUFAs is a guarantee of skin whitening activity, because microalgal extracts are complex mixtures of bioactives, whose final effects depend on the overall balance of their combined effects. For example, the ethyl acetate extract of Tahitian *Isochrysis* (T-Iso), a cos-

Table 9.2 Activity of some fatty acids on tyrosinase

Compound	Activity	Reference
Palmitic acid (C16:0)	stimulating	Ando et al. (1998, 1999)
Stearic acid (C18:0)	stimulating	Ando et al. (1998)
Oleic acid (C18:1n – 9)	Inhibiting	Ando et al. (1998)
Linoleic acid (C18:2n – 6)	Inhibiting	Ando et al. (1998, 1999)
α-Linolenic acid (C18:3n – 3)	Inhibiting	Ando et al. (1998)
Arachidonic acid (C20:4)	Inhibiting	Mishima et al. (1993)
Eicosapentaenoic acid (C20:5n – 3)	Inhibiting	Mishima et al. (1993)
Docosahexaenoic acid (C22:6n – 3)	Inhibiting	Balcos et al. (2014)

metic preparation marketed with the name *BIO1659* (Symrise AG), can stimulate melanogenesis in the skin and hair (Herrmann et al. 2012a, 2013), even though this Haptophyta is an excellent producer of PUFAs (Mishra and Mishra 2018).

Amongst the microalgal compounds that show activity on skin melanogenesis are CTs. Fucoxanthin (FXT) has been reported to decrease tyrosinase activity in UVB-irradiated guinea pigs, melanogenesis in UVB-irradiated mice and the mRNA levels of proteins linked to melanogenesis in skin cells (Sathasivam and Ki 2018). Also, orally administered lutein and zeaxanthin promoted skin lightening in a clinical trial involving 46 healthy subjects (Juturu et al. 2016). Zeaxanthin purified from *N. oculata* showed anti-tyrosinase action (Shen et al. 2011), thereby confirming that contributes to the skin whitening properties of this MA. Finally, ATX can inhibit skin pigmentation by interfering with the signaling of the *stem cell factor* released by KCs, which regulates different aspects of the melanocyte activity, including proliferation, differentiation and melanogenesis (Pillaiyar et al. 2017)

2.2.5 Preliminary Evidence of Microalgae Activities on Signals Released by the Peripheral Nervous System

A further aspect that should be considered for skin homeostasis and wellness is the effect of neurogenic inflammation, which is a challenging issue of great cosmetic interest due to its implications on irritation and itching. In vivo, skin responds to stress-induced brain nervous stimuli producing numerous local signals. KCs and melanocytes secrete *corticotropin-releasing hormone* (CRH), *adrenocorticotropic hormone* (ACTH) and catecholamines. Dermal FBs secrete ACTH, cortisol and prolactin. Skin nerve endings secrete adrenaline, noradrenaline and substance P. SGs secrete CRH and prolactin (Zmijewski and Slominski 2011; Alexopoulos and Chrousos 2016). In addition, cutaneous nerve endings and almost all skin cells share the ability to produce and respond to special cytokines called *neurotrophins* (NTs) in a paracrine and autocrine way. These affect many metabolic processes of the skin (e.g. FB migration, melanocyte response to UV stress and epidermal differentiation) and stimulate the development of nerve endings (Borroni et al. 2009; Truzzi et al. 2011). The NT signalling network is important in some inflammatory processes, such as atopic dermatitis and psoriasis, in which nervous stress plays a role. NTs can induce proliferation of cutaneous nerve endings with important consequences on symptoms, such as itching and pain (Grewe et al. 2000; Pavlovic et al. 2008). Studies on the activity of microalgal extracts on skin disorders due to NTs are limited, but some important evidences concerning compounds from MAs suggest that they might be very effective. Horváth et al. (2015) verified that βC and lutein are effective in the treatment of neurogenic inflammation induced on mouse ear by stimulation with mustard oil. These two CTs (but not lycopene) negatively modulated the expression of *transient receptor mustard oil potential ankyrin 1* in peptidergic nerve terminals. Sharma et al. (2018) showed that ATX inhibited neuropathic pain in rats subjected to thermal and mechanical trauma. This is consistent with the efficacy of ATX in the inhibition of *N*-methyl-D-aspartate receptors, which are also implicated in the mechanism of action of pain from neurogenic inflammation (Kinkelin et al. 2000).

Scandolera et al. (2018) recently tested a *Rhodosorus marinus* extract (Rhodophyta) on in vitro cultures of human (h) KC, astrocytes and hSE, thereby demonstrating its effectiveness in reducing the secretion of *interleukin-1α* (IL-1α) and *nerve growth factor* following an inflammatory stimulus with phorbol myristate acetate (PMA). The same preparation inhibited PMA-induced overexpression of *transient receptor mustard oil potential vanilloid 1*, another receptor implicated in the inflammation induced by mustard oil. These activities were attributed to the γ-aminobutyric acid (GABA) and GABA-alanine derivatives contained in the tested MA.

2.2.6 Activity of Microalgae on Skin Appendages

The cosmetic industry is strongly interested in achieving novel natural principles suitable to modulate sebogenesis, because sebum overproduction affects the appearance of the skin and hair, making them shiny and oily. Sebum is important in skin wellness, because the hydrolipidic film derived from secretions of the sebaceous and sweat glands contributes in regulating water loss and in protecting the skin against mechanical damage and UV. Its composition is relevant for both UV-induced photo-oxidation processes and the effects on the skin inflammasome (Oyewole and Birch-Machin 2015), since the presence of oleic acid and other unsaturated fatty acids can irritate sensitive subjects (DeAngelis et al. 2005; Schwartz et al. 2012). In addition, it is the main source of tocopherol for the skin (Mackenna et al. 1950; Thiele et al. 1999) and one of the main sources of CTs (Darvin et al. 2011a).

Excessive sebum production can also lead to skin disorders, such as seborrheic dermatitis, acne and dandruff. Although these disorders have multifactorial causes that often involve the skin microbiome, excessive sebum is an important condition for their onset (DeAngelis et al. 2005; Schwartz et al. 2012). Few studies have addressed the exploitation of MAs for the treatment of skin appendages, but early findings are promising. A hydrophilic extract of *Galdieria sulphuraria* reduced the expression of *5α-reductase type-1* (5α-R1), an enzyme involved in testosterone metabolism, in immortalised hFBs and hKCs (Bimonte et al. 2016). The reduction of 5α-R1 was considered responsible for the downregulation of sebogenesis, which has been documented also in vivo. In fact, SG activity is largely affected by male hormones (Mitsui 1997, p. 18).

Extracts of *C. calcitrans* f. *pumilus*, *T. pseudonana*, *M. subterraneus*, *C. minutum* and *Nannochloropsis* sp. decreased sebum production in human SGs cultivated ex vivo and were found comparable with or superior to treatments with reference compounds, e.g. capsaicin (Zanella and Pertile 2016; Zanella et al. 2016). MA extracts can regulate sebum quantitative production, but to date, no information is available on their effects on sebum composition, despite both have relevant effects on the skin microbiome (DeAngelis et al. 2005; Byrd et al. 2018). For instance, two yeasts considered amongst those responsible for dandruff, *Malassezia globosa* and

M. restricta, grow only in areas of the scalp where sebum is overabundant (Schwartz et al. 2012). Concerning sebum composition, *Propionibacterium acnes*, a bacterium predominant in sebaceous follicles, metabolises some lipids to short-chain fatty acids that act as antimicrobials (Christensen and Brüggemann 2013). Since MAs regulate the quantitative production of sebum by SGs, it is likely that they may also influence its composition, but it has not been possible to find studies in this regard.

HF is another appendage of great interest in the cosmetic industry, but surprisingly, the disclosure of active ingredients from MAs still has few well-documented case studies. Amongst these, the methanolic extract of T-Iso marketed with the name *BIO1631* (Symrise AG), showed anti-hair loss effects due to prolonged HF anagen phase and a reduced ratio between apoptotic and proliferating KCs of the matrix (Herrmann et al. 2012b, 2013). Similar results were obtained in ex vivo HF cultures with ethanol extracts of *Chaetoceros* sp., *Chlorococcum* sp. and *M. subterraneus* (Zanella et al. 2012), whereas some extracts of *T. suecica* had been shown to reduce hair growth (Pertile et al. 2010). A number of patent applications have been filed for MA embodiments aimed at treating hair, protecting against environmental agents (e.g. UV and pollution) and increasing mechanical resistance (Table 9.3).

2.3 *Role of Oxidative Stress in Skin Photoaging and Inflammation*

Aging processes are closely linked to oxidative stress produced by highly reactive compounds, which are free radicals or, more correctly, reactive oxygen species (ROS) and reactive nitrogen species (RNS). These highly reactive compounds include a heterogeneous group of molecules, some electrically charged and others neutral, characterised by the presence of atoms with an unpaired electron in the outermost atomic orbital (Halliwell 2006). This condition makes them extremely unstable, since they attempt to lose or add an electron to restore the equilibrium of the orbital, reacting and modifying different molecules, such as glucides, proteins, lipids and DNA, with which they come into contact in the cellular environment. Some ROS are produced physiologically through cell metabolism in cytosol organelles, such as mitochondria, endoplasmic reticulum and peroxisomes (Rinnerthaler et al. 2015). For example, hydrogen peroxide is normally produced by some reactions of the mitochondrial respiratory chain or by lymphocytes in immune defence processes. For this reason, cells also have enzymes that can neutralise ROS, thereby protecting themselves from damage.

Unfavourable external factors, such as exposure to UV, aggressive agents of chemical or biological origin, inflammatory agents and atmospheric pollution, can increase ROS production up to a level that that overcomes cell defence and causes cell damage. These events are strongly connected with skin aging. Therefore, ROS toxicology has become a core issue for the cosmetic industry. Although many oxidised molecules can be repaired or catabolised and replaced, some instances of

Table 9.3 Some granted and pending patents concerning hair treatments with preparations or extracts obtained from MAs

MA species	Claim	Applicant		Reference
Porphyridium, Rhodosorus, Chlorella, Dunaliella, Closterium, Phaeocystis, Pleurochrysis, Primnesium, Euglena	5alpha-reductase inhibitor and hair-growing agent containing the same	Microalgae Corp	JP2002068943 (A)	Fujitani et al. (2002)
Chaetoceros gracilis	Scalp hair loss prevention and improvement	Park SH	KR20140062249 (A)	Park et al. (2014)
Chlorella	Grey hair prevention agent	Shiseido Co Ltd	JP2002212039 (A)	Suzuki et al. (2002)
Chlorella vulgaris	Hair papilla cell growth agent and vascular endothelial growth factor production promoter	Naris Cosmetics Co. Ltd	JP2006282597 (A)	Megata (2006)
Tetraselmis tetrathele	Improves scalp and prevents hair loss	Park SH	KR20150100302 (A)	Park et al. (2015)
Prototheca moriformis, Chlorella protothecoides	Increases hair shine, combability, strength, prevents UV and pollution damages, moisture loss or spit ends	Solazyme Inc.	US20150352034 (A1)	Schiff-Deb and Sharma (2015)
Chlorella	Hair growth and care	Nakano Seiyaku KK, Chlorella Ind	JPS63135315 (A)	Katsuyama and Obata (1988)
Porphyridium cruentum	Enhances beta-catenin activity and cell differentiation in dermal papilla cells	Radiant Co Ltd. (KR), Seoul Cosmetics Ltd. (KR)	KR101856480 (B1)	Kee et al. (2018)
Haematococcus pluvialis	Source of ATX for HF protection from oxidative stress and anti-hair loss	Cognis Deutschland GmbH & Co. KG.	WO03105791 (A1)	Eisfeld and Mehling (2003)
Isochrysis	Improves combability, strength, volume and stress resistance and reduces frizz and breakage	Symrise AG	WO2019037843 (A1)	Nakano et al. (2019)

damage are permanent and accumulate over time, thereby leading to many of the effects we observe in aged tissues. There are several environmental factors that produce chronic oxidative stress (e.g. air pollution and aggressive detergents), but solar radiation is the most relevant and studied. UVR associated with solar radiation comprises UVC (100–280 nm), UVB (280–315 nm) and UVA (315–400 nm). UVC is blocked by the atmospheric ozone layer, whereas UVB (<5% of the UVR), which does not penetrate far beyond the epidermis, produces DNA damage, burns and erythema (Svobodova et al. 2006). Oxidative damage at the epidermis, which as mentioned is densely cellularised, is mainly caused by UVB (Van Laethem et al. 2005).

Most of UVR is composed of UVA, which has lower energy content than UVB and requires doses of 600–800 times greater to produce erythema, but they are able to penetrate deeply into the dermis (Gilchrest 1996). ROS toxicology is extremely complex, and possible consequences in the cell environment strongly depend on several conditions, which include overall energy dose of the radiant spectrum, individual characteristics of the skin (e.g. pigmentation and epidermis thickness), diet and lifestyle of the subject.

Importantly, at low doses, solar irradiation stimulates autogenous defences against ROS by activating the transcription factors of *forkhead box, class O family member proteins* (FoxOs), which promote the transcription of antioxidant enzyme genes, e.g. *superoxide dismutase-2* (SOD2), *peroxiredoxins 3* and *5* and *catalase* (CAT) (Klotz et al. 2015).

2.3.1 Skin Inflammation by Oxidative Stress

Figure 9.4 summarises and simplifies some biochemical pathways through which ROS can induce oxidative damage. UVB activity is expressed at the level of epidermis KCs, whereas it affects the dermis only marginally. The absorption of energy can denature DNA by the formation of pyrimidine dimers, leading to apoptosis by p53 activation (Van Laethem et al. 2005). This signal activates the cytosolic protein *Bcl-2-associated-X-protein* (Bax), which is stimulated by ROS also through the activation of *mitogen-activated protein kinases* (MAPKs), particularly via the p38MAPK. In the activated form, Bax moves to the outer mitochondrial membrane, where it produces two effects: (1) inhibition of *B-cell lymphoma-2* (Bcl-2), an antagonistic signal that promotes cell survival by activating mechanisms of protection of mitochondrial integrity (Dewson and Kluck 2010), and (2) release of *cytochrome-c* (cyt-c), which is the main signal of apoptosis activation (Van Laethem et al. 2005).

Cyt-c activates Casp-9, which in turn activates Casp-3, an effector protease that is the primary performer of apoptotic death (Brentnall et al. 2013). Casp-3 cleaves and activates *protein kinase C delta type* (PKC-δ), a pro-apoptotic factor that cooperates in the activation of Bax by downregulating *induced myeloid leukaemia cell differentiation protein* (Mcl-1), a potent anti-apoptotic factor (D'Costa and Denning 2005). This cascade triggers a self-amplifying cycle that results in the expression of pro-inflammatory signals IL-1α and IL-6 (Fig. 9.4). IL-1α acts as an autocrine signal stimulating the same KC to release *granulocyte-macrophage colony stimulatory factor* (GM-CSF) (Yano et al. 2008; Imokawa et al. 2015). GM-CSF and IL-1α are released into the surrounding tissue and penetrate the dermal layer, stimulating FBs to release neprilysin (also called *neutral endopeptidase*, NEP), a protein that degrades elastin, thereby favouring the formation of wrinkles (Imokawa et al. 2015).

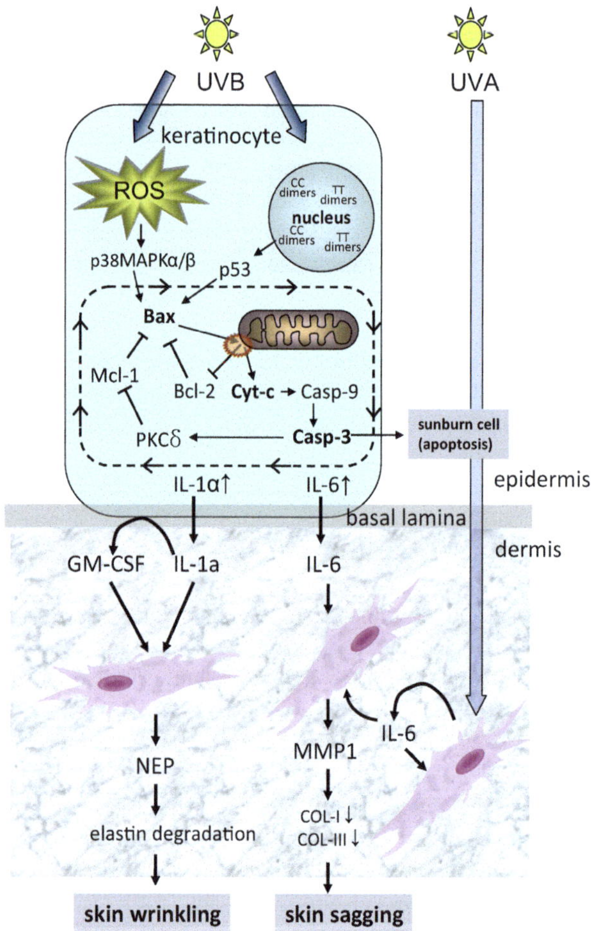

Fig. 9.4 Some major inflammatory and apoptotic routes in the skin epidermis. UVB induces DNA denaturation and release of p53, as well as the activation of p38MAPKs. These effects activate Bax, which in turn promotes a self-amplifying pro-apoptotic cascade of signals and effector proteins (schematised within the dashed line). Besides, Bax is sustained and amplified by the inhibition of some anti-apoptotic proteins (Bcl-2 and Mcl-1). This scenario leads to the KC death and release of inflammatory signals that induce dermis ECM degradation, especially by lysis of elastin. UVA reaches the dermis FBs and stimulate the release of inflammatory signals and MMPs, promoting mainly the degradation of COLs (Re-elaborated from Van Laethem et al. 2005 and Imokawa et al. 2015)

At the same time, IL-6 promotes the release of collagenases by FBs, particularly MMP1, which acts on COL-I and COL-III by promoting skin flaccidity (for more information on MMPs, see also Pittayapruek et al. 2016).

UVA radiation acts mainly on dermal FBs inducing the secretion of IL6, which in turn promotes in autocrine and paracrine manner the production and release of MMP1, whereas NEP to a lesser extent (Imokawa et al. 2015; Wlaschek et al. 1993).

Imokawa et al. (2015) showed that UVB acts mainly on KCs favouring the signal cascade that promotes FB release of NEP and degradation of elastin, whereas UVA is less active on KCs and causes dermal FBs to secrete IL-6 and MMP1, with more intense degradation of COLs. According to these findings, UVB mainly favours the formation of wrinkles, whereas UVA is the main factor responsible for skin sagginess.

2.3.2 Activation of the MAPKs/AP-1 and PI3K/Akt Pathways by Oxidative Stress

Figure 9.5 shows two biochemical paths promoted by ROS as a result of alternative combinations of signals and transcription factors outlined below.

Activator Protein 1 (AP-1) pathway

ROS activates several MAPKs, which include p38, *extracellular signal-regulated kinase* (ERK) and *c-Jun N-terminal kinase* (JNK). These MAPKs cooperate by activating the nuclear transcription factor AP-1 (Berthon et al. 2017). The latter inhibits the *transforming growth factor-β* (TGF-β) that stimulates procollagen production (Pittayapruek et al. 2016) and promotes the expression of pro-apoptotic factors, such as *BCL2-antagonist of cell death* (BAD) and *signal transducer and activator of transcription 3* (STAT3). Besides, AP-1 induces ECM degradation by stimulating the secretion of MMPs (Akhalaya et al. 2014; Berthon et al. 2017).

Nuclear Factor Kappa-Light-Chain-Enhancer of Activated B Cells (NF-kB) pathway

ROS can activate *phosphoinositide-3-kinase* (PI3K) by triggering the sequential phosphorylation of *protein kinase* B (Akt), *IkB kinase* (IKK) and *inhibitor of kB* (I-kB). Akt also inhibits transcriptional functions of FoxOs reducing the expression of antioxidant factors, such as CAT and SOD (Berthon et al. 2017). I-kB loses its inhibitory function on *nuclear factor kappa-light-chain-enhancer of activated B cells* (NF-kB), which consist of two proteins, namely, p50 and p65. In the absence of inhibition, NF-kB moves from cytosol into the nucleus and stimulates the expression of various pro-inflammatory molecules, i.e. IL-1β, IL-6, IL-8, tumour necrosis factor-α (TNFα), *inducible nitric oxide synthase* (iNOS) and *cyclooxygenase-2*

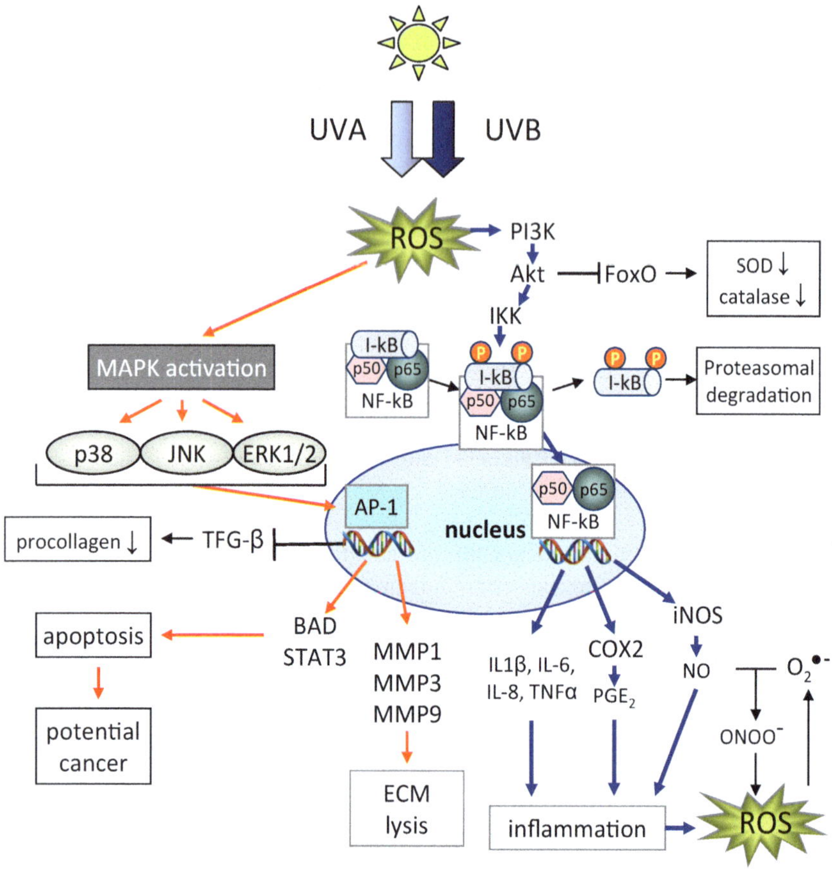

Fig. 9.5 Biochemical pathways governed by the key nuclear transcription factors AP-1 and NF-kB (re-elaborated from Chiou et al. 2011, Zhang et al. 2014 and Berthon et al. 2017)

(COX-2) (Zhang et al. 2014; Berthon et al. 2017). These signals and enzymes are involved in promoting inflammation and RNS production (Fig. 9.5), resulting in the enhancement of oxidative stress to which contribute also the immune cells stimulated by ILs and TNFα.

2.3.3 Cytoprotective Response Involving the Keap1/Nrf2 Pathway

The cell reacts to the formation of ROS by activating signals that enhance antioxidant defences and repair or degrade dysfunctional molecules. Emphasis is attributed to the regulatory factor *NF-E2 p45-related factor 2* (Nrf2), which controls the gene expression of *antioxidant response elements* (AREs) and a large range of other proteins related to cytoprotective function (Baird and Dinkova-Kostova 2011).

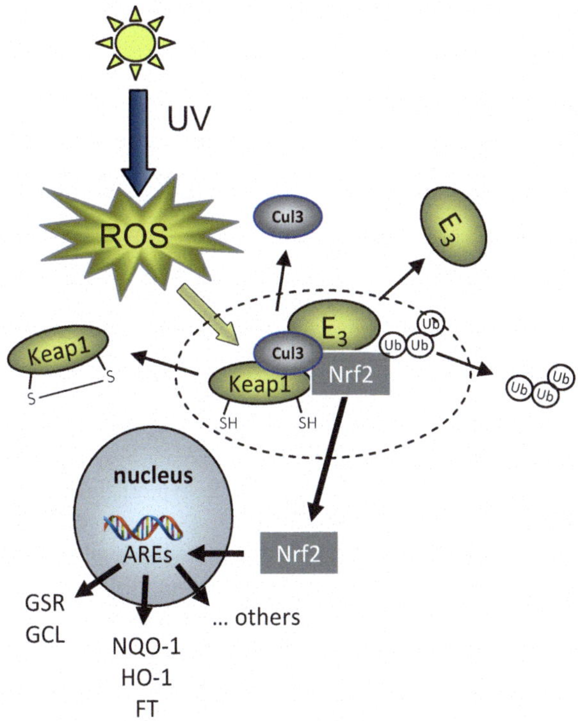

Fig. 9.6 Signalling pathway controlled by Nrf2/Keap1 complex. The basal repressed condition of Nrf2 depends on the molecular complex within the dashed line. The redox perturbation of the citosol triggers the detachment of Keap1 and consequently of Cul3 and ligase E3 (re-elaborated from Kobayashi et al. 2004, Baird and Dinkova-Kostova 2011 and Kansanen et al. 2013)

Under homeostatic conditions, Nrf2 is found at the inhibited state in the cytosol and bound to *Kelch-like ECH-associated protein 1* (Keap1), a cysteine-rich dimeric protein. It seems that Keap1 acts as an adapter for *cullin 3* (Cul3), a protein that interacts with *E3 ligase*, thereby resulting in the proteasomal degradation of Nrf2 by polyubiquitination (Kobayashi et al. 2004). Therefore, Nrf2 is characterised by a rapid turnover and is continuously synthesised, inhibited by Keap1 and degraded via proteasome. Due to its high cysteine content, Keap acts as a sensor for the alterations of the cellular redox environment (Baird and Dinkova-Kostova 2011). In the presence of ROS, Keap1 cysteine groups are oxidised to cystines, and through conformational changes and interactions that are not yet fully elucidated, Nrf2 is rapidly released and deubiquinated, thereby escaping degradation and moving into the nucleus (Fig. 9.6). Moreover, the newly synthesised Nrf2 continues to move into the nucleus as long as the cellular environment maintains Keap1 in a reduced form (Baird and Dinkova-Kostova 2011; Kansanen et al. 2013). In the nucleus, Nrf2 cooperates with *small Maf protein* (sMaf), thereby promoting the expression of over 600 target genes (Baird and Dinkova-Kostova 2011). Amongst these genes, the expression of AREs produces proteins of relevant protective impact, e.g. *NAD(P)H quinone dehydrogenase 1* (NQO-1), *haeme oxygenase 1* (HO-1), *glutamate-cysteine ligase* (GCL), *glutathione-disulphide reductase* (GSR), *leukotriene B4 12-hydroxydehydrogenase* (LB4DH) and *ferritin* (FT) (Baird and Dinkova-Kostova 2011; Kansanen et al. 2013). In the nucleus, Nrf2 undergoes a slow turnover due to non-proteasomal degeneration (Kobayashi et al. 2004).

2.3.4 Permanent Damages by Oxidative Stress in Human Skin

In the previous paragraphs, attention was on the signals that govern reactions to oxidative stress. Here, we focus on some oxidative damages to the enzymatic and structural components of the cell. Cells can survive oxidative damage by repairing or recycling processes, but some damaged molecules persist and tend to accumulate over time.

Oxidative stress frequently causes protein carbonylation and peroxidation of membrane lipids with formation of malonaldehyde and 4-hydroxy-2-nonenal (4HNE). The 4HNE groups of peroxidised lipids can give rise to non-enzymatic reactions and cross-link with carbonylated proteins, which in turn can contain proline and lysine modified into glutamic semialdehyde and aminoadipic semialdehyde, respectively (Castro et al. 2017). Such reactions, which establish covalent bonds between proteins and lipids, can prevent molecule unfolding, which is necessary for the enzymatic degradation and can favour aggregation in clusters of progressively increasing dimensions. Thus, heteropolymeric macroaggregates are formed; besides being unaffected by cytosolic proteases, they interfere with lysosomal functionality (Terman and Brunk 2004) and irreversibly bind to proteasomes, thereby blocking the activity of these organelles (Höhn et al. 2011). The inactivation of lysosomes and proteasome deprives the cell of its main tools for recycling dysfunctional molecules, thereby triggering a vicious cycle that favours the incorporation of other oxidised molecules into these 'ceroid' macroaggregates termed 'lipofuscin' (LF). LF has variable composition that includes proteins (30–58%), lipids (19–51%), carbohydrates (4–7%), metal ions and mineral elements, such as Fe, Cu, Al, Zn, Mn and Ca, which account for <2% (Terman and Brunk 2004; Jung et al. 2007). Being nondegradable, it accumulates indefinitely in postmitotic cells, occupying up to 75% of cell volume in motor neurons (Rinnerthaler et al. 2015). In proliferative cells, such as epidermal KCs, it ends up accumulating in the intercellular space upon cell death. The formation of age spots on the back of the hand is a common consequence of its accumulation. Age spots of the skin may have different origins, e.g. melanin overproduction, but the so-called senile lentigo or liver spot is due to LF (Skoczyńska et al. 2017), whose brownish colour is due to oxidation of lipids and metal ions. Wang-Michelitsch and Michelitsch (2015) reported interesting observations and hypothetical models to explain the isolation of LF in extracellular fibrotic capsules that increase in size and change shape over time.

2.3.5 Microalgal Products for Preventing and Treating the Oxidative Stress in Human Skin

The biochemical scenarios described above do not exhaust the possible reactions triggered by ROS but provide an idea of their complexity and highlight how these events are closely interconnected with inflammatory processes and skin aging. The comprehension of these pathways is important for an appropriate interpretation of the antioxidant activity of many MAs. Herein, apart from their chemical activity of ROS scavenging, some MA metabolites regulate key factors that affect the inflammasome of skin cells.

On this regard, interesting reviews have recently reported extensive tables that summarise the cosmetic applications from microalgal compounds (Ariede et al. 2017; Mourelle et al. 2017) and list microalgal sources of active compounds of cosmetic interest (Berthon et al. 2017; Brunt and Burgess 2018; García et al. 2017). However, case studies based on extracts or preparation from MAs are limited, even if some of them have a relevant significance.

In addition to the aforementioned anti-oxidative activity of *N. oculata* (see Sect. 2.2.3), a hydro-alcoholic extract of *T. suecica* showed anti-stress properties with modulation of genes involved in the protection against oxidative damages (Sansone et al. 2017). An aqueous extract of *Scenedesmus rubescens* tested in vitro both on primary skin cells and ex vivo full-thickness skin (hFTS) exerted protective effects against UVR damages, stimulated COL, reduced DNA impairment (sunburns cells) and increased both mitochondrial efficiency and cell proliferation (Campiche et al. 2018).

An aqueous extract of *C. pyrenoidosa* showed intense protective activity against UVC damage in cultured FBs, reducing the expression of pro-apoptotic proteins, in particular *Fas-associated death domain-containing protein* and the activated Casp-3 (Shih and Cherng 2012). A similar preparation obtained from commercial *Chlorella* inhibited the expression of MMP1 and the pro-inflammatory signal *cysteine-rich angiogenic inducer 61* in cultures of FBs treated with UVB, thereby preventing the reduction of pro-COL (Chen et al. 2011). Finally, concerning the inhibition of inflammatory signals, the release of IL-1α by ex vivo hFTS stimulated with an irritant (SDS) was inhibited by treatments with lipophilic and hydrophilic extracts of *Nannochloropsis* to a comparable or greater extent than dexamethasone (Zanella and Pertile 2016).

However, most information concerning the potential anti-oxidative power of MAs is inferred from studies on the isolated compounds of which they are important sources. Subsequently, some relevant case studies are discussed below.

Carotenoids

CTs are yellow to red-brown pigments rich with unsaturated double bonds or phenolic rings, which are easily oxidised by ROS, thereby protecting cells from oxidative damage. They are polymeric molecules with isoprenic derivation and are divided into carotenes and xanthophylls. Carotenes, such as β-carotene (βC) and lycopene, lack bonds with oxygen. Xanthophylls, such as lutein, zeaxanthin, violaxanthin, canthaxanthin (CTX) and ATX, contain oxygen atoms. Ketocarotenoids, such as CTX and ATX, are synthesised in MAs and other microorganisms, but they are generally lacking in higher plants (Zhang et al. 2014; Safafar et al. 2015). Many green MAs synthesise mixtures of CTs, which are intermediate or final compounds of complex biosynthetic pathways and are differently arranged amongst species depending on their enzymatic machinery (Jin and Melis 2003; Sathasivam and Ki 2018). However, each species is generally characterised by few prevailing CTs. For example, *Dunaliella bardawil* and *D. salina* produce mainly βC (Jin and Melis 2003), and *H. fluviatilis* is the richest source of ATX (Guerin et al. 2003). *Nannochloropsis* spp. synthesise several xanthophylls, amongst which violaxanthin and vaucheriaxanthin are prevalent, accompanied by antheraxanthin, zeaxanthin and other less abundant CTs (Antia and Cheng 1982; Lubián et al. 2000; Faé Neto et al. 2018).

CTs are chemically lipophilic and suitable to protect the integrity of cell membranes, as occurs with tocopherols. Aboul-Enein et al. (2003) quantified CTs, vitamin E and vitamin C of seven strains belonging to the genera *Dunaliella*, *Chlorella* and *Scenedesmus* and tested their extracts for the efficacy against lipid peroxidation of mice liver microsomes. In that case, the antioxidant efficacy was proportional to the microalgal concentration of active molecules. However, the efficacy of an antioxidant depends on the chemical structure of the ROS with which it reacts; hence, the ranking of extract strength from different MAs can change depending on the antioxidant test (Safafar et al. 2015). In Table 9.4, the scavenging efficiency of some CTs is shown, estimating their relative strength compared with some benchmark antioxidants (trolox, ascorbic acid and cysteine) (Rodrigues et al. 2012). The efficiency of each CT depends on the test considered (i.e. from the ROS involved in the

Table 9.4 Peroxyl radical hydroxyl radical (HO•), hypochlorous acid (HOCl) and peroxynitrite anion ($ONOO^-$) scavenging capacity of seven carotenoids and other compounds incorporated into liposomes

	Scavenging capacity[a]			
Compound	ROO•	HO•	HOCl	$ONOO^-$
β-Carotene	0.14	0.71	NA[b]	1.02
Zeaxanthin	0.56	1.41	3.87	0.77
Lutein	0.6	0.97	4.81	0.78
Lycopene	0.08	0.35	0.4	0.31
Fucoxanthin	0.43	1.18	6.26	NA
Canthaxanthin	0.04	0.28	0.1	NA
Astaxanthin	0.64	1.66	9.4	0.73
α-Tocopherol	0.48	1.77	NA	0.37
Quercetin	0.84	1.42	5.63	0.97
Trolox	**1.00**	**1.00**	NA	NA
Ascorbic acid	NA	NA	0.41	**1.00**
Cysteine	0.04	NA	**1.00**	0.02

The values are the mean of two independent experiments (from Rodrigues et al. 2012, a column of the original table was omitted; license conditions available at http://creativecommons.org/licenses/by/3.0/)

[a]The scavenging capacity was calculated by considering the following as references (in bold): trolox for ROO• and HO•, cysteine for HOCl and ascorbic acid for $ONOO^-$

[b]NA: no activity was found for the tested concentrations

reaction), which explains the importance of introducing mixtures of different antioxidant molecules, rather than high quantities of a single compound. In addition to their specific chemical protection from ROS, many CTs are natural inhibitors of NF-kB (see Sect. 2.3.2) that governs most inflammatory reactions due to oxidative stress (Zhang et al. 2014).

Studies in vivo showed that the skin content in CTs is directly proportional to the consumption of fruit and vegetables and inversely proportional to stress factors, which is reflected in the condition of skin aging (Darvin et al. 2011a). βC and lycopene (carotenes) are the most abundant CTs in humans and constitute approximately 70% of the CTs ordinarily present in the skin (Choi et al. 2018), whereas xanthophylls are less common in the human diet. Research attention has been dedicated to ATX, which is one of the few microalgal molecules produced at the industrial scale (Spolaore et al. 2006). This powerful superoxide anion scavenger inhibits the release of MMP1 and NEP following UVA radiation upon treatment at low concentrations (Imokawa 2019). Its anti-inflammatory activity is also effective for stimuli other than photo-oxidation; for example, ATX inhibited the NF-kB activity, the expression of iNOS and COX-2 and release of TNF-α, IL-1β, IL-6 and IgE in a phthalic anhydride-induced atopic dermatitis animal model (Park et al. 2018). Camera et al. (2009) compared the protective activity of βC, CTX and ATX in FBs treated with UVA, disclosing that although βC is a strong $^1O_2$ quencher, it showed limited protective effects and resulted in phototoxicity at concentrations >2 μM. CTX did not prevent oxidative damage but increased the antioxidant enzyme HO-1, whereas ATX showed the greatest protective activity, such as reduction of Casp-3 and preservation of both the membrane integrity and the antioxidant enzymes (catalase and SOD). Nevertheless, βC showed an effective protection against damages from IR irradiation in clinical tests (Darvin et al. 2011b), thereby showing the complexity of the biological interactions caused by phytochemicals. FXT, another CT occurring in several MAs, promotes the expression of ARE genes (Fig. 9.6) via the stimulation of Nrf2 transcription factor (Berthon et al. 2017).

Overall, these data show that CTs exhibit metabolic interactions that cannot be explained with the mere ROS scavenging. Direct or indirect modulation of gene expression is often performed. More importantly, the chemical reactivity of an antioxidant compound is supposed to be independent from its isomeric conformation, but the modulation of gene expression may require molecular interactions that are isomer dependent. In this case, synthetic isomers could produce effects different from the natural blends. Studies on this topic are still limited. However, some interesting findings have been reported. For instance, Sun et al. (2016) showed that the isomer (3S,3′S)-trans-ATX, the form prevalent in *H. pluvialis*, is much more effective as a stimulant of mouse immune cells than the two other stereoisomers, i.e. (3R,3′R)-trans- and meso-trans-ATX, that contribute up to 75% of synthetic ATX. Analogously, the biological activity of the natural isomer of βC is superior to the synthetic all-trans forms (Spolaore et al. 2006).

Tocopherols and Polyphenols

Tocopherols and polyphenols are important for the protection of the skin. Polyphenols include a large family of molecules, comprising flavonoids, flavones, anthocyanidins, tannins and phlorotannins. Safafar et al. (2015) showed that both tocopherols and polyphenols are abundant in *Phaeodactylum* sp., *Nannochloropsis* sp., *Chlorella* sp., *Dunaliella* sp. and *Desmodesmus* sp. and can be efficiently extracted with methanol.

Tocopherols are a family of antioxidant molecules of which the most biologically active is α-tocopherol, namely, vitamin E, especially effective in preventing cell membrane oxidation (Marquardt et al. 2013).

Polyphenols, which are widespread even in the composition of higher plants, comprise a diversified class of hydrosoluble compounds that can perform an action in some way complementary to CTs and tocopherols. Goiris et al. (2014) studied six MAs from different classes and showed that they synthesise several polyphenols, which include phloroglucinol (39–81 μg/g dry weight (DW)), *p*-coumaric acid (540–7000 ng/g DW) and apigenin (7.3–13.6 ng/g DW). However, these values of concentrations are low in comparison with contents detected in many superior plants. Goiris et al. (2012) analysed CTs and the polyphenol content of hydro-alcoholic extracts of some MAs and then measured their respective antioxidant capacity via three different assays: trolox equivalent antioxidant capacity, ferric reducing antioxidant potential and AAPH-induced oxidation of linoleic acid. Analysis of their findings shows that the antioxidant strength of the extracts is not always proportional to the total content of CTs and/or polyphenols and can vary with the test performed. Hence, the quantitative content in antioxidant compounds is not sufficient to establish the efficacy of microalgal preparations, because each species produces a combination of compounds with its own properties. Biological tests in ex vivo organ culture or in clinical trials are necessary to provide a reliable estimation of the efficacy of natural extracts.

Polysaccharides, Galactolipids and Lipids

Microalgal polysaccharides offer some interesting examples of antioxidant activity and other beneficial effects (see Raposo et al. 2013 for a review of their properties and applications). *P. cruentum* (Rhodophyta) produces sulphoglycolipids with important anticoagulant and antiviral properties, as well as antioxidant and anti-inflammatory activities (Plaza et al. 2009). The sulphated exopolysaccharides (SEP) produced by this MA inhibit NF-kB activity and the release of pro-inflammatory cytokines (Berthon et al. 2017). Biochemical techniques were also proposed for increasing the sulphation of these polysaccharides and their biological activity (Gersh et al. 2002).

Amongst Bacillariophyta species, the antioxidant activity of a β-D- glucan, also called chrysolaminarin or leucosin, was characterised. This glucose polymer contains β-1:3′- and β-1:6′-bonds in the ratio of 11:1 (Beattie et al. 1961). It is accumulated as an energy reserve in *Odontella aurita* but also shows strong activity as a scavenger of hydroxyl radical (Xia et al. 2014). This polysaccharide was also isolated and quantified in *Cyclotella cryptica* (Roessler 1987), *P. tricornutum* (Caballero et al. 2016) and *T. pseudonana* (Hildebrand et al. 2017), and it might be present in all Bacillariophyceae species, as well as in other algal groups.

Interesting cosmetic applications connected with anti-inflammatory effects are also attributable to galactolipids (Fig. 9.7). MA extracts comprising *monogalactosyl diacylglycerol* (MGDG) and *digalactosyl diacylglycerol* (DGDG) demonstrated intense anti-inflammatory activity by reducing ear oedema after croton oil challenge in animal model, especially if the compounds had the esterified two eicosapentaenoic acid (EPA) residues (MGDG-EPA and DGDG-EPA) (Winget 1994). Their anti-inflammatory activity was showed using a *Chlorella minutissima* extract, but several other MAs were indicated as potential sources of this active compound (information worthy of confirmation), including *Chaetoceros*, *Cyclotella*, *Ellipsoidon*, *Isochrysis*, *Nannochloris*, *Nannochloropsis*, *Nitzschia*, *Phaeodactylum*, *Porphyridium*, *Skeletonema*, *Thalassiosira*, *Monochrysis* and *Monoraphidium*.

Bruno et al. (2005) showed that MGDG exerts a dose-dependent activity, which is higher than that of DGDG, and is optimised by the presence of EPA in its composition with anti-inflammatory efficacy at 20 mg/kg higher than the indomethacin control treatment (10 mg/kg). MGDGs have also been isolated from *Tetraselmis chui* and showed a strong inhibition of the release of NO by RAW264.7 macrophage cells (Banskota et al. 2013).

Intriguingly, PUFAs, of which MAs are elective sources, can exhibit anti-inflammatory effects in the skin via metabolisation to monohydroxy acids (Ziboh et al. 2000). Finally, anti-inflammatory properties were recognised to lipid mediators called resolvins (E- and D-series), which are derived from the cellular metabolism of long-chain PUFAs, such as EPA and docosahexaenoic acid (Calder 2009; Weylandt et al. 2012).

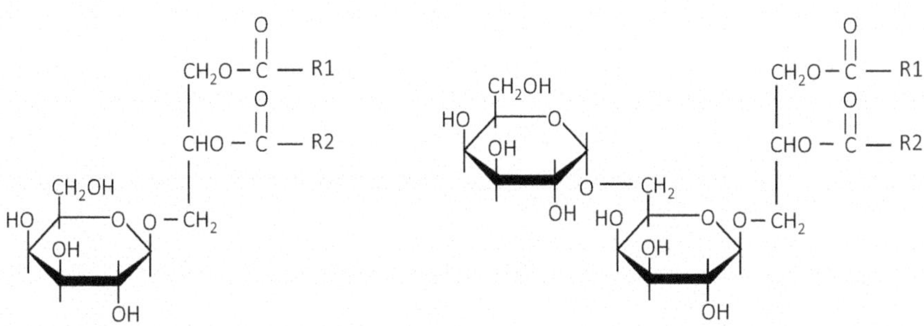

Fig. 9.7 Monogalactosyl diacylglycerol (left) and digalactosyl diacylglycerol (right). R1 and R2 are polyunsaturated fatty acids

2.4 Issues Related with the Multifunctional Bioactivity of the Extracts

Multifunctionality is a typical trait of microalgal extracts that has not been sufficiently appreciated. The great number of active compounds comprised in their composition makes possible to interact simultaneously with different biochemical pathways governing metabolism of cells and tissues. For example, treatments with an ethanol extract of *Chaetoceros* on ex vivo cultures of human organotypic cultures promoted hair follicle growth, modulated pigmentation, ECM composition and cell proliferation in skin, enhanced lipolysis in adipocytes and reduced sebogenesis in SGs (Zanella et al. 2012, 2016). This richness in active compounds is a trait that could be conveniently exploited in cosmetics, especially for treating multifactorial inflammatory processes at the basis of aging and other skin problems. Although MAs and other marine organisms are optimal sources of biologically active compounds (Pulz and Gross 2004; Spolaore et al. 2006; Kim 2014; Balboa et al. 2015), the added value related to the composition of their phytocomplex is still largely undervalued.

2.4.1 Chemical Antioxidant Activity Versus Signal Modulation

The mechanism of action of some MA extracts is still insufficiently elucidated. Many experimental findings cannot be explained only as effect of the chemical antioxidant activity. For example, extracts of *Isochrysis*, *Chaetoceros*, *Monodus* and *Chlorococcum* stimulate growth and prolong the anagen phase in hair follicles under ex vivo culture conditions at very low concentrations (Herrmann et al. 2012b; Zanella et al. 2012), thereby exerting a negligible antioxidant activity. Furthermore, considering that oxidative stress is not present in standard culture conditions, the mentioned extracts should affect the hair metabolism via a different mechanism, perhaps by modulating cytosolic or nuclear signals. Other case studies have shown that compounds present in MAs can modulate the genetic expression in human and animal cells, also in the absence of oxidative stress. FXT topically administered at 1% depressed the mRNA expression of COX-2, *endothelin receptor-A*, *p75 neurotrophin receptor*, *prostaglandin E receptor 1*, *melanocortin 1 receptor* and *tyrosinase-related protein 1* (Muthuirulappan and Francis 2013). An aqueous extract of *C. vulgaris* orally administered to mice modulated some immune cells by regulating the expression of IL-12 and interferon-γ (IFN-γ) with important antiallergic effects of potential cosmetic interest (Hasegawa et al. 1999). A similar preparation promoted the production of IL-1α, TNF-α, IFN-c, IL-10 and IL-6 in mouse natural killer cells following exposure to lead, thereby minimising the immune defects determined by this contaminant (Queiroz et al. 2011). An extract of *C. pyrenoidosa* inhibited the release of IL-5 and GM-CSF in mast mouse cells treated with allergenic stimuli (Kralovec et al. 2005).

These data indicate that MA active compounds can regulate several signals of the immune system. These properties deserve to be studied thoroughly, because they could alleviate problems of sensitisation, irritation and skin contact allergies, but also improve the immune defense.

2.4.2 Modulation of Fat Management

Another topic of cosmetic interest concerning MAs is the modulation of fat metabolism. The subcutis, skin and its appendages harbour three types of cells specialised in lipid metabolism, with different functions: KC (epidermis), sebocytes (SGs) and adipocytes (hypodermis or subcutis). The first two cell types are involved in the synthesis of the skin barrier and sebum, respectively, as discussed in Sect. 2.2.1. Adipocytes not only contribute to skin plumpness and provide a lipid reserve and thermal insulation but are also sources of adipokines that regulate many aspects of skin biology. Adipose tissue is a primary source of paracrine and endocrine signals with a potential secretome estimated at over 600 proteins (Fasshauer and Blüher 2015). The influence of fat secretome on skin well-being and beauty has become an increasingly important issue in cosmetics.

The anatomical contiguity of subcutis with the dermis allows the adipocytes to exert a relevant paracrine action on the dermal and epidermal tissues, thereby affecting the healing processes, hair follicle cycle and thermoregulation (Kruglikov and Scherer 2016). Chronic UV exposure inhibits the release of some adipokines, i.e. adiponectin and leptin, with increasing photo-oxidative skin damage (Kim et al. 2016). Leptin acts on skin cells via the membrane receptor *Janus kinase 2*, which transduces the signal to different secondary messengers, thereby influencing the processes of preservation and regeneration of the skin and other skin appendages (Poeggeler et al. 2010). Considering this background, the preliminary data that indicate MAs as a source of compounds active on adipocytes deserve attention. Preparations obtained from *Chromulina*, *Asterionella* and *Tetraselmis* algal cultures were proposed to inhibit different enzymes involved in fat metabolism, including *acetyl coenzyme A carboxylase*, *phosphodiesterase*, *glyceraldehyde 3-phosphate dehydrogenase*, *fatty acid synthase* and *lipoprotein lipase* (Hugues and Joel 2012). Extracts of *Chaetoceros*, *Chlorococcum*, *Monodus* and *Nannochloropsis* stimulated lipolysis in hFTS with subcutis (Zanella et al. 2012; Zanella and Pertile 2016).

Moreover, some CTs that are often included in the composition of several MA strains affect adipocyte metabolism. FXT is metabolised to fucoxanthinol and amarouciaxanthin-A, which inhibit the differentiation and development of the adipocytes (Muthuirulappan and Francis 2013). Neoxanthin, another CT, shows similar properties, whereas FXT promotes fat loss through higher expression levels of *uncoupling protein 1* and *3-adrenergic receptor* in abdominal fat tissues (Sathasivam and Ki 2018).

3 Relevance of the Adopted Experimental Model in the Development of Multifunctional Cosmetics

The previous topics have highlighted the great complexity and organisation of the skin organ with epithelial and mesenchymal tissues at close contact that exchange signals suitable to modulate the activity of their respective cells. The isolation of a cell type from this context allows the investigation of the responses to a stimulus under conditions very different from in vivo. Furthermore, skin appendages play a relevant role in the dynamic of this signalling interaction. These organs are exposed to the same cosmetic treatment of the skin, can develop specific responses and produce further signals capable of influencing the metabolism of skin cells.

All these elements should be considered when interpreting results obtained using different experimental models to achieve an appropriate characterisation of the active cosmetic ingredients. Cells in culture, 3D human skin equivalents (hSE) and cultures of human tissues ex vivo or live animals have increasing capacity to produce results that reflect interactions between cells and tissues similar to those of the human body. Each model can suitably provide useful information for characterising the biological properties of molecules and preparations, but with different predictivity values concerning the effects in vivo. This issue is of special interest in Europe, which is one of the main markets for the cosmetics industry, since the experimentation on live animals was disallowed (EC regulation No. 1223/2009).

3.1 Cell Cultures Versus Organotypic Cultures and 3D Human Skin Equivalents

In vitro cultures of skin cells are the most widely used tool for screening and characterising active ingredients, as they are easy to manage and relatively inexpensive. Generally, hFBs, hKCs and less commonly melanocytes are used. A clear distinction should be made between primary cells and immortalised line cells. Primary cells are obtained via isolation from explanted human tissues. They can survive for a limited number of generations in culture and retain some features of the donor for some time. For example, hKCs and hFBs show a proliferative capacity in vitro that decreases with the age of the donor, which also affects the maximum number of generations under culture conditions (Martin et al. 1970; Gilchrest 1983). Furthermore, hFBs isolated from elderly subjects show an unbalanced oxidative homeostasis compared with cells isolated from young donors (Boraldi et al. 2010). The primary cells, however, undergo various phenotypic changes as the passages in culture proceed and then completely lose their replicative capacity (Martin et al. 1970; Boraldi et al. 2010). Hence, the use of cells at their first passages in culture is important. Cellular alterations in the cells in culture are sometimes exploited as ‘in vitro aging model’, but important limitations occur, because these changes involve all the cell machinery instead of reflecting the typical damages of in vivo aging (Boraldi et al. 2010).

The immortalised cells (line cells) are derived from the transformation of primary cultures, which can be spontaneous or induced by viral infections, but more frequently, they are obtained via the isolation of tumour cells. Typically, these cells lack contact inhibition and have various modified biological characteristics while maintaining some basic traits of the cell type to which they belong (Jedrzejczak-Silicka 2017). Chromosomal anomalies and the loss of functionality of the p53 pro-apoptotic signal are some of the most significant anomalies (Oh et al. 2007).

The characterisation of active compounds conducted on isolated cells is affected by lack of cytokines or secondary metabolites that, in vivo, could be released from proximal different cell types exposed to the same stimulus. This limitation can be particularly relevant in the case of MA extracts, because a single cell type is unsuitable to disclose the signalling crosstalk triggered by the combined action of several active compounds (for more information on crosstalk in cellular signalling, see Vert and Chory 2011).

To overcome these problems, researchers have developed various co-culture protocols with different cell types and provided significant evidence of the relevant effect produced by cross-talking on their respective metabolism (Maas-Szabowski et al. 1999; Ghahary and Ghaffari 2007; Singh et al. 2008; Hirobe 2014). This technical approach led to the development of various hSEs, consisting of simple epidermis or full-thickness skin, with or without polymeric scaffolds for dermal matrix engineering (Stark et al. 2004, 2006; Griffith and Swartz 2006; Poumay and Coquette 2007; Li et al. 2009; Canton et al. 2010). The development of increasingly advanced hSE has substantially improved the dermatological research opportunities and led to the commercialisation of 3D models designed for different applications, such as those proposed by Episkin SA (de Brugerolle 2007; Alépée et al. 2017), MatTek Corporation (Danilenko et al. 2016) and Henkel AG & Co. KGaA (under the Phenion® brand, Mewes et al. 2017). The use of these hSEs is also important to conduct some safety tests on ingredients and cosmetic products intended for the European market, because they can no longer be performed on live animals (Nakamura et al. 2018).

As an alternative, the active compounds for cosmetics can be tested on ex vivo organotypic cultures, such as the cultures of skin, HFs, SGs and hypodermic fat. Most of these biological materials are waste tissues obtained from cosmetic or reconstructive surgery, which can be kept in culture for a few days (Fig. 9.8). The ex vivo cultures have the advantage of presenting the anatomical organisation of the tissue in vivo, including nerve endings, Langerhans and Merkel cells, and preserving some individual characteristics of the donor (e.g. sex, age and sensitivity). Xu et al. (2012) performed wound-healing studies on human skin samples and verified that they maintained biological performance similar to the skin in vivo for 6 days. Ex vivo cultures are almost irreplaceable for studies on complex annexed organs, such as HFs, because examples of in vitro reconstructed models are limited (e.g. Havlickova et al. 2009). To date, results remain far from the complexity of the human organ.

hSEs are easy to handle, available and suitable for providing replicable data (Danilenko et al. 2016), but the ex vivo skin is more representative of the in vivo

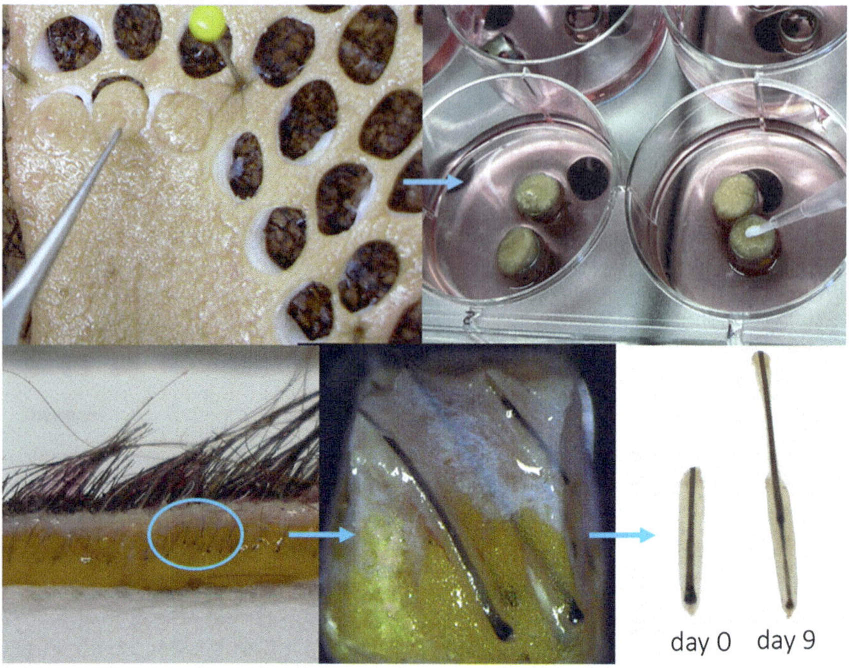

Fig. 9.8 Ex vivo models of human skin and related appendages: dissection of skin samples and cultivation (upper), scalp sample, tissue detail and hair follicles during the dissection, then the isolated hair follicles at day 0 and 10 of culture, respectively (lower) (Source: courtesy of Cutech Srl)

condition in terms of many aspects, as well as with the delivery processes following topical administration (Reus et al. 2012; Andrade et al. 2015; Sidgwick et al. 2016).

Most of the information about the biological properties of algal extracts or single compounds that they contained was obtained via experiments with cells in culture or hSEs. However, experiments conducted with the ethanol extracts of *Nannochloropsis* sp. on ex vivo organs (human skin, subcutis, sebaceous glands and hair follicles) showed that they were active in most skin compartments and appendages (Zanella and Pertile 2016). For example, the topical application of the extract to ex vivo hFTS reduced IL-1α release in response to inflammatory stimulus to a comparable extent to dexamethasone and inhibited melanogenesis to an extent comparable to retinoic acid. Besides, also skin appendages responded to systemic treatments with the same extract; reduced sebogenesis in ex vivo SGs in measure comparable or superior to benchmark compounds (e.g. Asebiol™, 5α-Avocuta® and capsaicin), stimulated growth in ex vivo HF and lipolysis in ex vivo subcutis. This combination of biological activities is not an exception but the consequence of the complex composition of many microalgal extracts, which makes them suitable for affecting the metabolism of different skin compartments.

3.2 Issues Related with Effects In Vivo

Although hESs and ex vivo organotypic cultures are suitable tools for preclinical studies, they still lack some relevant traits of animal models. The blood and lymphatic circulation is completely absent, the microbiome is altered or absent, and the stimuli due to the mechanical solicitation and variation of environmental conditions are lacking (e.g. temperature and solar radiation).

Hence, remarkable differences may occur between preclinical and in vivo findings. The clinical test is a necessary confirmation to validate the results obtained on simplified experimental models.

Skin metabolism is affected by active compounds obtained through diet (Choi et al. 2018) or topical administration, as well as with stress factors derived from lifestyle (Darvin et al. 2011). The cosmetic activity observed in a clinical trial will therefore depend on many variables that interact with the individual sensitivity exhibited by the treated subject. The active compounds applied in vivo can be ineffective in a certain number of subjects who, for unknown reasons, are not responsive. Generally, this condition does not happen with isolated cells, but can sometimes be observed in ex vivo tissues (a personal experience of one of the authors). Intriguingly, some subjects can also be non-responsive to treatments with compounds considered as gold standards (e.g. skin lighteners β-arbutin or kojic acid) as disclose a critical examination of statistical responsiveness in clinical trials (Curto et al. 1999; Solano et al. 2003). On this regard, the use of a mixture of active ingredients with synergistic activities may be advantageous to stabilise the subject's response. This strategy perfectly fits with the use of microalgal extracts. In fact, poor responsiveness or sensitivity to a particular active ingredient could be compensated by other compounds present in the extract.

As evidence of the validity of this approach, dermatologists often treat certain skin disorders, such as melasma or acne, with mixtures of several active ingredients (Kligman and Willis 1975; Lim 1999; Fabbrocini and Saint Aroman 2014; Shankar et al. 2014) to resolve the ineffectiveness of single drugs in a certain number of subjects or exploit their synergistic effects. Skin disorders are often multifactorial and could deal great benefits from multifunctional products.

Finally, even the most advanced in vitro models lack the nervous and vascular components, which affect the release of NTs and the contribution of hormones with great relevance for skin homeostasis and some important forms of inflammations (see Sect. 2.2.5). Ex vivo skin can allow some experimental options in reason of the partial conservation of some parts of the immune system and the nervous system (nerve endings, Merkel and Langerhans cells). However, studying the response of the nervous system on experimental models other than in vivo is difficult.

3.3 Product Formulation and Active Transdermal Delivery

The use of isolated cells or tissues allows researchers to perform treatments at desired concentrations for established times. Culture conditions allow cells to be placed in contact with any compound suitable to be supplemented into the medium, regardless of molecular weight. However, obtaining the same effects in vivo is difficult for at least two reasons: (1) the substance of interest does not necessarily cross the protein-lipid barrier of the SC, and (2) if this event occurs, the concentration of the substance of interest will decrease with both the diffusion through the tissue and the time elapsed since administration. Transdermal absorption is one of the most complex problems affecting the formulation of topical treatments, and only a few experimental models are available as alternatives to in vivo tests. This process consists of multiple steps that include the following: (1) partitioning of the active from the cosmetic vehicle into the SC, (2) molecular diffusion through the SC and partitioning into the epidermal viable cells and (3) diffusion through the epidermis and dermis until it eventually reaches the blood vessels (Pillai et al. 2016) that spread the compounds via a systemic way. The main route of entry is the crossing of the SC via trans- or intercellular pathways; however, some facilitating entry routes are available, such as the hair follicle infundibulum and sweat glands (the latter is preferred for hydrophilic substances) (Abd et al. 2016; Pillai et al. 2016).

In general, the absorption of a compound is inversely proportional to its molecular weight and electric charge. The compounds should not exceed 500 Da (Pillai et al. 2016), although this paradigm does not represent an insurmountable limit, especially in elderly and/or very dry skin (Fields et al. 2009).

Studies conducted on hSEs, although useful, could provide results not replicable under in vivo conditions. In some tests, the absorption of hydrophilic compounds was similar to that in human skin, but lipophilic compounds were absorbed up to 800 times faster (Godin and Touitou 2007). In another hSE, a Raman spectroscopic analysis showed SC anomalies in the continuity and distribution of ceramides, fatty acids and cholesterol with important consequences on permeability (Tfayli et al. 2014). Even studies conducted on animal models showed that the permeability of the SC varies across species and with the body site in the same species consistently with skin thickness, composition of the protein-lipid matrix and density of hair follicles (Godin and Touitou 2007).

The most reliable model is probably the ex vivo human skin (Abd et al. 2016), although this model lacks blood circulation and the lymphatic vessel network, which are essential for evaluating inflammatory responses and clearance of the compounds of interest. Unfortunately, studies on MA preparations conducted on human skin in vivo or on ex vivo are still limited.

3.3.1 Lipophilic Extracts

Considering that the SC is a barrier that is rich in lipids and covered sebum, lipophilic molecules are generally absorbed more easily than hydrophilic ones. Many MAs are known to be able to accumulate CTs and lipids in high amounts, including long-chain PUFAs, so their extracts obtained with non-polar solvents are usually very rich in active compounds of cosmetic interest.

Some studies confirmed that many lipophilic compounds can be effectively absorbed but with important variations in relation to their molecular structure. For example, a good absorption of triglycerides and CTs (βC ≫ α-Carotene) was demonstrated via topical application of crude palm oil to ex vivo human skin (Sri et al. 2013).

Preparations for topical application containing FXT were tested on hairless mice ex vivo and in vivo, which showed that this compound was adsorbed and was active against inflammation and hyperplasia (Rodríguez-Luna et al. 2018).

All the compounds of the given examples are often present in lipophilic MA extracts and it can be presumed that they are being adsorbed with the same efficiency.

3.3.2 Hydrophilic Extracts

Aqueous extracts of MAs contain many small hydrophilic compounds, such as vitamins and polyphenols, as well as significant amounts of polysaccharides and polypeptides with high molecular weight (MW). Large polypeptides could remain unadsorbed, which may be advantageous, by considering the high antigenic power of some vegetable proteins. The active peptides currently designed and synthesised to modulate skin metabolism are generally made up of 3–6 amino acids and are often prepared using lipoamino acids obtained via esterification with palmitic acid in order to improve their adsorption (Fields et al. 2009).

In aqueous microalgal extracts, small peptides may be present, but their occurrence as lipoamino acids and with sequences suitable for acting as signals for human epidermal cells is unlikely. Nevertheless, Chen et al. (2011) showed that an aqueous extract of *Chlorella* containing 430–1350 kDa peptides exhibited protective effects on UVB-induced damages in FB culture.

Amongst hydrophilic compounds, significant activities may be derived from small molecules, such as glutathione, which may have topical activity (Kopal et al. 2007), and modified amino acids such as MAAs, whose characteristics have already been discussed in relation to their property to quench UVR (see Sect. 2.2.2).

The aqueous extracts of MAs also comprise vitamins, polyphenols, flavonoids, single amino acids and small glucides, which are compatible with transcutaneous absorption. Amongst these, special attention is given to the glycolipid MGDGs and DGDGs, which have anti-inflammatory activity. In vivo tests conducted on mice by using ointments of fractionated extracts of *I. galbana* rich in MGDGs confirmed their penetration through the SC and diffusion in the thickness of the skin, with strong anti-inflammatory effect (Rodríguez-Luna et al. 2017). The adsorption of polyphenols was demonstrated ex vivo via the topical application of natural non-microalgal extracts obtained from cocoa, thereby resulting in the stimulation of the GAGs and COL synthesis in skin dermis (Gasser et al. 2008).

More difficult is reconciling the principles of the transdermal delivery with the intense anti-inflammatory activity of high-MW SEPs (5–7 × 10^6 Da) obtained from *Porphyridium*, which were used by Matsui et al. (2003) in a clinical trial to treat the irritation induced with balsam of Peru. Also in this case, however, the SEP activity was remarkably increased following molecular fragmentation obtained by sonication, thereby attesting that the molecular weight and treatment efficacy are directly correlated. In a different experiment, Zhang et al. (2011) studied via a mouse model of Rosacea the anti-inflammatory activity of 5500 Da semisynthetic glycosaminoglycan ethers (SAGEs) obtained by sulphation from fermented 53 kDa HA derivatives. The croton-induced skin inflammation was effectively inhibited by topical treatment with SAGEs, whereas the native 53 kDa HA was ineffective. These data confirm that large dimensions of some polysaccharides hinder absorption. Potent absorption enhancers can be used to improve the absorption of some compounds, but they can be expensive and not always resolutive. Concerning the reported examples, it should be considered that the absorption of large molecules in inflamed skin could be due, at least partially, to the disruption of the SC integrity, which in turn is due to the application of irritants and solvent vehicles. Croton oil, for example, produces important histological alterations of the skin (Moon et al. 2001).

4 Conclusions

Cosmetic applications of extracts and preparations from MAs have achieved development and progress over the past two decades due to the availability of data concerning their activities on skin. Some products have already been marketed but much less compared with their potential as sources of active ingredients (Table 9.5). Some mechanisms of action were documented, especially those related to antioxidant activities and protection from photoaging processes. Much work remains to be done, especially in applications intended for the treatment of skin appendages, modulation of fat's adipokines and effects on the skin microbiota.

The topics addressed in this review highlight how the concentration of many active compounds in a single cell, according to combinations that vary across species, allows the preparation of multifunctional extracts. This trait is not found to the same extent in other natural ingredients and is worthy to be further explored. The exploitation of MAs as a source of isolated active compounds is often economically disadvantageous compared with traditional or synthetic alternatives (see costs for carotenoids in Spolaore et al. 2006, for EPA in Molina Grima et al. 2003 and Koller et al. 2014; for a comprehensive analysis see also Barsanti and Gualtiero 2018). This condition is due to the biomass production costs, which are still relatively high despite the significant progress (Molina Grima et al. 2003; Tredici et al. 2016) and the costs of purification. However, the advisable use of microalgal extracts is to

Table 9.5 Some commercial cosmetic ingredients or products obtained from MAs

Species	Cosmetic application	Active ingredient	Commercial name	Company	Reference
Porphyridium sp.	Skin anti-wrinkle, protection from UV damage	Sulphated polysaccharide	Alguard®	Frutarom	Ryu et al. (2015), Guillerme et al. (2017)
Porphyridium cruentum	Vascular tonicity, improves the skin's aspect and helps decrease rosacea effects and redness	Extract	SILIDINE®	Greentech	Mourelle et al. (2017)
Porphyridium cruentum	Antioxidant, anti-aging and pro-healing	Extract	Cicatrol®	Greensea	Maiz (2007)
Isochrysis T-Iso	Skin tanner	Ethyl acetate extract	BIO1659	Symrise AG	Herrmann et al. (2012a)
Isochrysis T-Iso	Anti-hair loss and hair promoter	Methanol extract	BIO1631	Symrise AG	Herrmann et al. (2012b)
Tetraselmis suecica	Anti-inflammatory and anti-rheumatic	Gold-bearing extract	O+ Gold Microalgae Extract®	Greensea	Maiz (2007)
Scenedesmus rubescens	Skin anti-photoaging and procollagen	Aqueous extract	Pepha®-Age	DSM Nutritional Products	Campiche et al. (2018)
Nannochloropsis oculata	Protection from oxidative stress, COL stimulation and skin tightening	Aqueous extract	Pepha®-Tight	DSM Nutritional Products (Pentapharm)	Stolz and Obermayer (2005)
Dunaliella salina	Stimulation of skin energy metabolism, cell proliferation and turnover, collagen synthesis	Aqueous extract	Pepha®-Ctive	DSM Nutritional Products (Pentapharm)	Stolz and Obermayer (2005)
Dunaliella salina, *Haematococcus pluvialis*	For brighter complexion, it treats uneven and dull skin, dark circles around the eye	extracts	REVEAL Color Correcting Eye Serum Brightener	Algenist	Joshi et al. (2018)

(continued)

Table 9.5 (continued)

Species	Cosmetic application	Active ingredient	Commercial name	Company	Reference
Chlorella vulgaris	Antiaging, anti-wrinkle, anti-cellulite, anti-stretch marks, anti-vascular imperfections, anti-dark circles	Extracted by alkaline hydrolysis	Dermochlorella	CODIF Recherche & Nature	Morvan and Vallee (2007)
Chlorella vulgaris	Neutralises inflammation and improves skin's natural protection	Extract	Phytomer	Phytomer	Ryu et al. (2015)
Chlorella sp.	Hydrates and beautifies skin and hair	Microalgal oil	Golden Chlorella™	Terravia Holdings	Mourelle et al. (2017)
Microalgae	Benefits for skin and hair	Microalgal oil	AlgaPür™	Terravia Holdings	Mourelle et al. (2017)
Microalgae	Skin anti-aging and rejuvenating	Polysaccharides	Alguronic Acid®	Algenist	Martins et al. (2014)

develop multifunctional ingredients, which are entirely natural and with highly sustainable ecological footprint (Fernandez et al. 2017). Their exploitation can be advantageous for treating skin disorders that have multiple causes, such as acne, dermatitis and psoriasis, but also as anti-aging and homeostasis protective agent, whose full effects require long-term application.

Moreover, the costs of biomass can be partly amortised by the high extraction yield obtained using some solvents (Table 9.6), and a great amount of product can be finalised due to their biological activity at a very low concentration (Pertile et al. 2010; Zanella et al. 2012). Consequently, the incidence of these ingredients on the production costs of high value-added products such cosmetics can be sustainable in many cases.

Unfortunately, unlike the synthetic principles, natural extracts have a composition that is partly unknown and subject to variation with the season, cultivation technique and method of extraction (Chojnacka and Kim 2015). This condition does not satisfy the practice of the cosmetic industry in terms of product standardisation and quality control. However, in order not to downplay the advantage of the multifunctional composition of microalgal extracts, a characterisation based on a metabolite fingerprinting obtained by mass spectrometry techniques may be employed, as already proposed for natural preparations intended for nutritional and phytopharmaceutical use (Mattoli et al. 2006, 2011).

The change from a 'one active → one claim' approach to natural phytocomplexes suitable to produce a combination of desirable effects leads to the problem on the functional characterisation of the ingredient. Establishing a quantitative correlation between a biological effect and the responsible active agent is difficult and sometimes impossible, because the ingredient formulation should be standardised. As mentioned for chemical characterisation, to determine the advantages offered by very active complex mixtures, a different standard of product characterisation should be developed. For example, each extract might be evaluated in terms of con-

Table 9.6 Extract yield obtained from some MAs using different solvents

Strain \ solvent	hydrophilic → water	methanol	ethanol	isopropanol	ethyl acetate	→ lipophilic hexane
Tetraselmis suecica	32%	17%	15%	3%	7%	5%
Chaetoceros calcitrans f. pumilus	64%	54%	22%	10%	9-12%	
Thalassiosira pseudonana	22%	39%	27%	15%	18%	
Monodus subterraneus	20%	20%	7-9%	2%	4%	
Chlorococcum minutum	22%	24%	16%	10%	5-8%	

Values expressed as percentage of the dried biomass (data from Pertile et al. 2010 and Zanella et al. 2016)

centration and activity according to an efficacy ratio by comparison with a panel of reference compounds, which are selected as golden standards for each application of interest. Figure 9.9 shows an exemplificative diagram of functional quantification

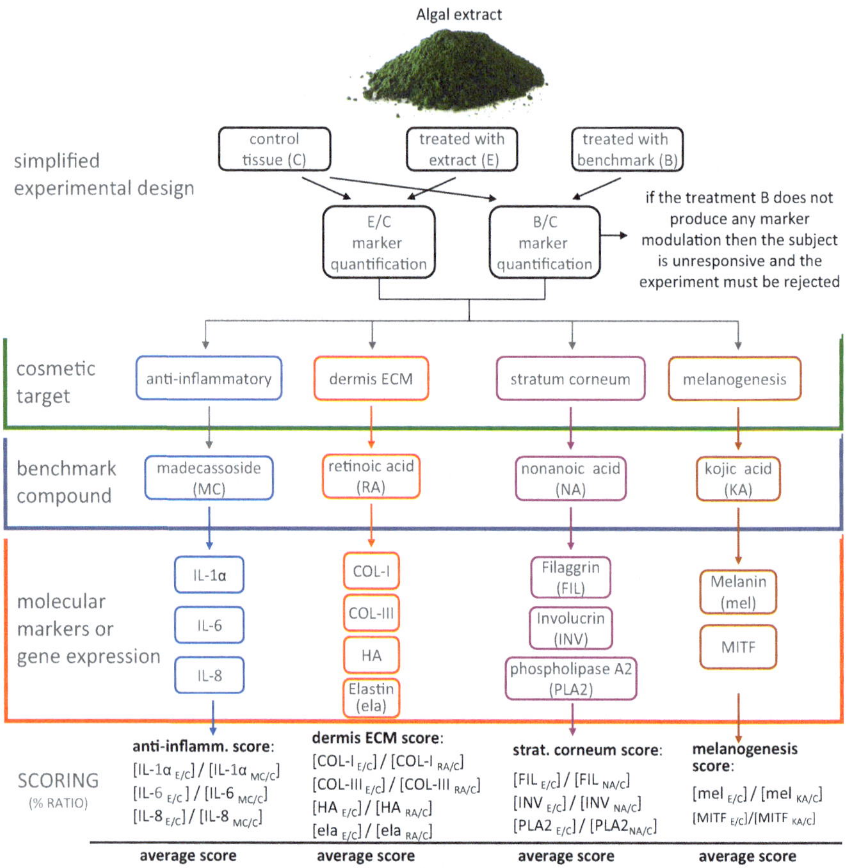

Fig. 9.9 Example of test panel for functional characterisation of a MA extract. The extract (E) is tested on an advanced experimental model (e.g. hSE or hFTS) in comparison with golden standard actives, namely, benchmarks (B), here arbitrarily selected for exemplification. The "simplified" experiment requires the comparison of the effectiveness between E and B treatments. The response to the B treatment in comparison to C defines subject's responsiveness; in case of no marker variation the experiment should be abandoned. Below the experimental design, possible markers and benchmarks (that cosmetic operators should define in a shared way) are schematised. For each application, the variation of the marker in response to treatments E and B is estimated as a ratio on sample C, then the values obtained are used to calculate the % ratio between the marker variation produced by treatment E on that produced by treatment B (i.e. $\text{marker}_{E/C}$ / $\text{marker}_{B/C} \times 100$). This approach would allow the characterisation of multifunctional extracts using the response to B as a normalising factor, thereby making the results comparable between different subjects with different sensitivities (from which depend the range of variation in response to treatments). At least three responsive subjects need to be tested for each application. (*MITF*: microphthalmia-associated transcription factor)

as a result of a panel of comparative tests aimed to specific applications (to be differently arranged with the targets of the cosmetic product category). A similar 'report sheet' could be compiled following tests on ex vivo human skin cultures and averaging the results obtained from no less than three responsive subjects. This approach can allow to determine in which application the extract is more active and the related ranking of effectiveness.

The proposals above, which are intended as representative hypotheses, imply important changes in the current business model used for cosmetic products, which is a significant obstacle. Nevertheless, this issue should be framed in the context of rapid aging experienced by populations in advanced economies. Health costs will grow at unsustainable rate. Hence, both cosmetics and nutrition science can and should play an increasingly important role in the promotion of the well-being and prevention of metabolic disorders. The development of multifunctional cosmeceuticals is in line with this approach, and the related costs should be assessed considering the connected substantial healthcare savings.

Acknowledgements We are very grateful to Dr. Paolo Pertile (Cutech srl, Italy) for kindly granting the permission of use for some images shown in this chapter. We would also like to thank two anonymous reviewers and especially Mr. Balaji Padmanaban (SPi Global) for his contribution to the finalisation of this chapter.

References

Abd, E., Yousuf, S., Pastore, M., Telaprolu, K., Mohammed, Y., Namjoshi, S., Grice, J., & Roberts, M. (2016). Skin models for the testing of transdermal drugs. *Clinical Pharmacology: Advances and Applications, 8*, 163–176.

Aboul-Enein, A. M., El-Baz, F. K., El-Baroty, G. S., Youssef, A. M., & Abd El-Baky, H. H. (2003). Antioxidant activity of algal extracts on lipid peroxidation. *Journal of Medical Sciences, 3*(1), 87–98.

Adarme-Vega, T. C., Lim, D. K., Timmins, M., Vernen, F., Li, Y., & Schenk, P. M. (2012). Microalgal biofactories: A promising approach towards sustainable omega-3 fatty acid production. *Microbial Cell Factories, 11*(1), 96.

Akhalaya, M. Ya., Maksimov, G. V., Rubin, A. B., Lademann, J., & Darvin, M. E. (2014). Molecular action mechanisms of solar infrared radiation and heat on human skin. *Ageing Research Reviews, 16*, 1–11.

Alépée, N., Grandidier, M. H., Tornier, C., & Cotovio, J. (2017). An in vitro skin irritation test using the SkinEthic™ reconstructed human epidermal (RHE) model. In C. Eskes, E. van Vliet, & H. Maibach (Eds.), *Alternatives for dermal toxicity testing* (pp. 59–72). Cham: Springer.

Alexopoulos, A., & Chrousos, G. P. (2016). Stress-related skin disorders. *Reviews in Endocrine & Metabolic Disorders, 17*(3), 295–304.

Allen, E. J., & Nelson, E. W. (1910). On the artificial culture of marine plankton organism. *Journal of the Marine Biological Association of the United Kingdom, 8*(5), 421–474.

Andersen, R. A. (2013). The microalgal cell. In A. Richmond (Ed.), *Handbook of microalgal culture: Applied phycology and biotechnology* (2nd ed., pp. 3–20). Oxford: Wiley.

Ando, H., Ryu, A., Hashimoto, A., Oka, M., & Ichihashi, M. (1998). Linoleic acid and α-linolenic acid lightens ultraviolet-induced hyperpigmentation of the skin. *Archives of Dermatological Research, 290*(7), 375–381.

Ando, H., Funasaka, Y., Oka, M., Ohashi, A., Furumura, M., Matsunaga, J., Matsunaga, N., Hearing, V. J., & Ichihashi, M. (1999). Possible involvement of proteolytic degradation of tyrosinase in the regulatory effect of fatty acids on melanogenesis. *Journal of Lipid Research, 40*, 1312–1316.

Andrade, T. A., Aguiar, A. F., Guedes, F. A., Leite, M. N., Caetano, G. F., Coelho, E. B., et al. (2015). Ex vivo model of human skin (hOSEC) as alternative to animal use for cosmetic tests. *Procedia Engineering, 110*, 67–73.

Antia, N. J., & Cheng, J. Y. (1982). The keto-carotenoids of two marine coccoid members of the Eustigmatophyceae. *British Phycological Journal, 17*, 39–50.

Ariede, M. B., Candido, T. M., Jacome, A. L. M., Velasco, M. V. R., de Carvalho, J. C. M., & Baby, A. R. (2017). Cosmetic attributes of algae-A review. *Algal Research, 25*, 483–487.

Asgharpour, M., Rodgers, B., & Hestekin, J. A. (2015). Eicosapentaenoic acid from *Porphyridium cruentum*: Increasing growth and productivity of microalgae for pharmaceutical products. *Energies, 8*(9), 10487–10503.

Baird, L., & Dinkova-Kostova, A. T. (2011). The cytoprotective role of the Keap1–Nrf2 pathway. *Archives of Toxicology, 85*(4), 241–272.

Balboa, E. M., Conde, E., Soto, M. L., Pérez-Armada, L., & Domínguez, H. (2015). Cosmetics from marine sources. In S. K. Kim (Ed.), *Springer handbook of marine biotechnology. Springer handbooks* (pp. 1015–1042). Berlin: Springer.

Balcos, M. C., Kim, S. Y., Jeong, H. S., Yun, H. Y., Baek, K. J., Kwon, N. S., et al. (2014). Docosahexaenoic acid inhibits melanin synthesis in murine melanoma cells in vitro through increasing tyrosinase degradation. *Acta Pharmacologica Sinica, 35*(4), 489–495.

Banskota, A. H., Gallant, P., Stefanova, R., Melanson, R., & O'Leary, S. J. (2013). Monogalactosyldiacylglycerols, potent nitric oxide inhibitors from the marine microalga *Tetraselmis chui*. *Natural Product Research, 27*(12), 1084–1090.

Barclay, W. R., Meager, K. M., & Abril, J. R. (1994). Heterotrophic production of long chain omega-3 fatty acids utilizing algae and algae-like microorganisms. *Journal of Applied Phycology, 6*(2), 123–129.

Barsanti, L., & Gualtiero, P. (2018). Is exploitation of microalgae economically and energetically sustainable? *Algal Research, 31*, 107–115.

Beattie, A., Hirst, E. L., & Percival, E. (1961). Studies on the metabolism of the Chrysophyceae. Comparative structural investigations on leucosin (chrysolaminarin) separated from diatoms and laminarin from the brown algae. *Biochemical Journal, 79*(3), 531–537.

Bernstein, E., Underhill, C., Hahn, P., Brown, D., & Uitto, J. (1996). Chronic sun exposure alters both the content and distribution of dermal glycosaminoglycans. *British Journal of Dermatology, 135*(2), 255–262.

Berthon, J.-Y., Nachat-Kappes, R., Bey, M., Cadoret, J.-P., Renimel, I., & Filaire, E. (2017). Marine algae as attractive source to skin care. *Free Radical Research, 51*(6), 555–567.

Bimonte, M., De Lucia, A., Carola, A., et al. (2016). *Galdieria sulphuraria* relieves oily and seborrheic skin by inhibiting the 5-α Reductase expression in skin cells and reducing sebum production in vivo. *Journal of Cosmetology and Trichology, 1*(1), 11–18.

Boraldi, F., Annovi, G., Tiozzo, R., Sommer, P., & Quaglino, D. (2010). Comparison of ex vivo and in vitro human fibroblast aging models. *Mechanisms of Aging and Development, 131*(10), 625–635.

Borroni, R. G., Truzzi, F., & Pincelli, C. (2009). The skin neurotrophic network in health and disease. *Actas Dermo-Sifiliográficas, 100*, 70–74.

Brentnall, M., Rodriguez-Menocal, L., De Guevara, R. L., Cepero, E., & Boise, L. H. (2013). Caspase-9, caspase-3 and caspase-7 have distinct roles during intrinsic apoptosis. *BMC Cell Biology, 14*(1), 32.

Bruce, J. R., Knight, M., & Parke, M. W. (1940). The rearing of oyster larvae on an algal diet. *Journal of the Marine Biological Association of the United Kingdom, 24*, 337–374.

Brunt, E. G., & Burgess, J. G. (2018). The promise of marine molecules as cosmetic active ingredients. *International Journal of Cosmetic Science, 40*(1), 1–15.

de Brugerolle, A. (2007). SkinEthic laboratories, a company devoted to develop and produce in vitro alternative methods to the animal use. *ALTEX-Alternatives to Animal Experimentation, 24*(3), 167–171.

Bruno, A., Rossi, C., Marcolongo, G., Di Lena, A., Venzo, A., Berrie, C. P., & Corda, D. (2005). Selective in vivo anti-inflammatory action of the galactolipid monogalactosyldiacylglycerol. *European Journal of Pharmacology, 524*(1-3), 159–168.

Byrd, A. L., Belkaid, Y., & Segre, J. A. (2018). The human skin microbiome. *Nature Reviews Microbiology, 16*(3), 143.

Caballero, M. A., Jallet, D., Shi, L., Rithner, C., Zhang, Y., & Peers, G. (2016). Quantification of chrysolaminarin from the model diatom *Phaeodactylum tricornutum. Algal Research, 20*, 180–188.

Calder, P. C. (2009). Polyunsaturated fatty acids and inflammatory processes: New twists in an old tale. *Biochimie, 91*(6), 791–795.

Camera, E., Mastrofrancesco, A., Fabbri, C., Daubrawa, F., Picardo, M., Sies, H., & Stahl, W. (2009). Astaxanthin, canthaxanthin and β-carotene differently affect UVA-induced oxidative damage and expression of oxidative stress-responsive enzymes. *Experimental Dermatology, 18*(3), 222–231.

Campiche, R., Sandau, P., Kurth, E., Massironi, M., Imfeld, D., & Schuetz, R. (2018). Protective effects of an extract of the freshwater microalga *Scenedesmus rubescens* on UV-irradiated skin cells. *International Journal of Cosmetic Science, 40*, 187–192.

Candi, E., Schmidt, R., & Melino, G. (2005). The cornified envelope: A model of cell death in the skin. *Nature Reviews Molecular Cell Biology, 6*(4), 328.

Canton, I., Cole, D. M., Kemp, E. H., Watson, P. F., Chunthapong, J., Ryan, A. J., et al. (2010). Development of a 3D human in vitro skin co-culture model for detecting irritants in real-time. *Biotechnology and Bioengineering, 106*(5), 794–803.

Cardozo, K. H. M., Guaratini, T., Barros, M. P., Falcão, V. R., Tonon, A. P., Lopes, N. P., Campos, S., Torres, M. A., Souza, A. O., Colepicolo, P., & Pinto, E. (2007). Metabolites from algae with economical impact. *Comparative Biochemistry and Physiology Part C, 146*, 60–78.

Castro, J. P., Jung, T., Grune, T., & Siems, W. (2017). 4–Hydroxynonenal (HNE) modified proteins in metabolic diseases. *Free Radical Biology and Medicine, 111*, 309–315.

Chen, C. L., Liou, S. F., Chen, S. J., & Shih, M. F. (2011). Protective effects of Chlorella-derived peptide on UVB-induced production of MMP-1 and degradation of procollagen genes in human skin fibroblasts. *Regulatory Toxicology and Pharmacology, 60*(1), 112–119.

Cheng, W., Yan-hua, R., Fang-gang, N., & Guo-an, Z. (2011). The content and ratio of type I and III collagen in skin differ with age and injury. *African Journal of Biotechnology, 10*(13), 2524–2529.

Chiang, H.-M., Pan, Y.-Y., Chen, C.-W., & Wen, K.-C. (2011). Fatty acids and their related products modulate melanogenesis. Focus on skin care: Ethnic, whitening & tanning - Supplement to household and personal care today. *Skin Care, 6*(1), 15–19.

Chini Zittelli, G., Lavista, F., Bastianini, A., Rodolfi, L., Vincenzini, M., & Tredici, M. R. (1999). Production of eicosapentaenoic acid by *Nannochloropsis* sp. cultures in outdoor tubular photobioreactors. *Journal of Biotechnology, 70*(1–3), 299–312.

Chiou, W. F., Don, M. J., Liao, J. F., & Wei, B. L. (2011). Psoralidin inhibits LPS-induced iNOS expression via repressing Syk-mediated activation of PI3K-IKK-IκB signaling pathways. *European Journal of Pharmacology, 650*(1), 102–109.

Choi, M. H., Jo, H. G., Kim, M. J., Kang, M. J., & Shin, H. J. (2018). Fruit juice supplementation alters human skin antioxidant levels in vivo: Case study of Korean adults by resonance raman spectroscopy. *Biotechnology and Bioprocess Engineering, 23*(1), 116–121.

Chojnacka, K., & Kim, S. K. (2015). Introduction of marine algae extracts. In S. K. Kim & K. Chojnacka (Eds.), *Marine algae extracts: Processes, products, and applications* (pp. 1–14). New York, NY: John Wiley & Sons.

Christensen, G. J. M., & Brüggemann, H. (2013). Bacterial skin commensals and their role as host guardians. *Beneficial Microbes, 5*(2), 201–215.

Chung, J. H., Seo, J. Y., Choi, H. R., Lee, M. K., Youn, C. S., Rhie, G. E., et al. (2001). Modulation of skin collagen metabolism in aged and photoaged human skin in vivo. *Journal of Investigative Dermatology, 117*(5), 1218–1224.

Curto, E. V., Kwong, C., Hermersdorfer, H., Glatt, H., Santis, C., Virador, V., Hearing, V. J., & Dooley, T. P. (1999). Inhibitors of mammalian melanocytes tyrosinase: In vitro comparisons of alkyl esters of gentisic acid and other putative inhibitors. *Biochemical Pharmacology, 15*, 663–672.

D'costa, A. M., & Denning, M. F. (2005). A caspase-resistant mutant of PKC-δ protects keratinocytes from UV-induced apoptosis. *Cell Death and Differentiation, 12*(3), 224.

Danilenko, D. M., Phillips, G. D. L., & Diaz, D. (2016). In vitro skin models and their predictability in defining normal and disease biology, pharmacology, and toxicity. *Toxicologic Pathology, 44*(4), 555–563.

Darvin, M. E., Sterry, W., Lademann, J., & Vergou, T. (2011a). The role of carotenoids in human skin. *Molecules, 16*(12), 10491–10506.

Darvin, M. E., Fluhr, J. W., Meinke, M. C., Zastrow, L., Sterry, W., & Lademann, J. (2011b). Topical beta-carotene protects against infra-red-light-induced free radicals. *Experimental dermatology, 20*(2), 125–129.

De Pauw, N., Morales, J., & Persoone, G. (1984). Mass culture of microalgae in aquaculture systems: Progress and constraints. *Hydrobiologia, 116*(1), 121–134.

DeAngelis, Y. M., Gemmer, C. M., Kaczvinsky, J. R., Kenneally, D. C., Schwartz, J. R., & Dawson, T. L., Jr. (2005). Three etiologic facets of dandruff and seborrheic dermatitis: *Malassezia* fungi, sebaceous lipids, and individual sensitivity. *Journal of Investigative Dermatology Symposium Proceedings, 10*(3), 295–297.

Del Campo, J. A., García-González, M., & Guerrero, M. G. (2007). Outdoor cultivation of microalgae for carotenoid production: Current state and perspectives. *Applied Microbiology and Biotechnology, 74*(6), 1163–1174.

Dewson, R., & Kluck, M. (2010). Bcl-2 family-regulated apoptosis in health and disease. *Cell Health and Cytoskeleton, 2*, 9–22.

Eckhart, L., Lippens, S., Tschachler, E., & Declercq, W. (2013). Cell death by cornification. *Biochimica et Biophysica Acta (BBA) - Molecular Cell Research, 1833*(12), 3471–3480.

Eisfeld, W., & Mehling, A. (2003). *Use of astaxanthin*. WO0305791 A1.

EU (EC). (2009). EU (EC) Regulation No. 1223/2009 of the European Parliament and of the Council of 30 November 2009 on cosmetic products. *Official Journal of the European Union, L342*, 59.

Fabbrocini, G., & Saint Aroman, M. (2014). Cosmeceuticals based on Rhealba® Oat plantlet extract for the treatment of acne vulgaris. *Journal of the European Academy of Dermatology and Venereology, 28*, 1–6.

Faé Neto, W. A., Borges Mendes, C. R., & Abreu, P. C. (2018). Carotenoid production by the marine microalgae *Nannochloropsis oculata* in different low-cost culture media. *Aquaculture Research, 49*(7), 2527–2535.

Fasshauer, M., & Blüher, M. (2015). Adipokines in health and disease. *Trends in Pharmacological Sciences, 36*(7), 461–470.

Feingold, K. R. (2007). The role of epidermal lipids in cutaneous permeability barrier homeostasis. *Journal of Lipid Research, 48*(12), 2531–2546.

Fernandez, F. G. A., Sevilla, J. M. F., & Grima, E. M. (2017). Microalgae: The basis of mankind sustainability. In B. Llamas (Ed.), *Case study of innovative projects-successful real cases* (pp. 123–140). Rijeka: IntechOpen.

Fields, K., Falla, T. J., Rodan, K., & Bush, L. (2009). Bioactive peptides: Signaling the future. *Journal of Cosmetic Dermatology, 8*(1), 8–13.

Flaim, G., Obertegger, U., Anesi, A., & Guella, G. (2014). Temperature-induced changes in lipid biomarkers and mycosporine-like amino acids in the psychrophilic dinoflagellate. *Freshwater Biology, 59*(5), 985–997.

Forján Lozano, E., Garbayo Nores, I., Casal Bejarano, C., & Vílchez Lobato, C. (2007). Enhancement of carotenoid production in *Nannochloropsis* by phosphate and sulphur limitation. In A. Méndez-Vilas (Ed.), *Communicating current research and educational topics and trends in applied microbiology. Microbiology series No. 2* (Vol. 2, pp. 356–364). Badajoz: FORMATEX.

Fujitani, N., Sakaki, S., Yamaguchi, Y., & Takenaka, H. (2001). Inhibitory effects of microalgae on the activation of hyaluronidase. *Journal of Applied Phycology, 13*(6), 489–492.

Fujitani, N., Hori, M., Takenaka, H., & Yamaguchi, Y. (2002). *Testosterone-5alpha-Reductase inhibitor and hair-growing agent containing the same*. JP20020684943(A).

García, J. L., de Vicente, M., & Galán, B. (2017). Microalgae, old sustainable food and fashion nutraceuticals. *Microbial Biotechnology, 10*(5), 1017–1024.

Gasser, P., Lati, E., Peno-Mazzarino, L., Bouzoud, D., Allegaert, L., & Bernaert, H. (2008). Cocoa polyphenols and their influence on parameters involved in ex vivo skin restructuring. *International Journal of Cosmetic Science, 30*(5), 339–345.

Gersh, S., Mamontov, A., & Weinstein, J. (2002). Sulfation of extracellular polysaccharides of red microalgae: Preparation, characterization and properties. *Journal of Biochemical and Biophysical Methods, 50*, 179–187.

Ghahary, A., & Ghaffari, A. (2007). Role of keratinocyte-fibroblast cross-talk in development of hypertrophic scar. *Wound Repair and Regeneration, 15*, S46–S53.

Gilchrest, B. A. (1983). In vitro assessment of keratinocyte aging. *Journal of Investigative Dermatology, 81*(1), S184–S189.

Gilchrest, B. A. (1996). A review of skin aging and its medical therapy. *The British Journal of Dermatology, 135*, 867–875.

Godin, B., & Touitou, E. (2007). Advanced transdermal skin delivery: Predictions for humans from in vivo, ex vivo and animal models. *Drug Delivery Reviews, 59*, 1152–1161.

Goiris, K., Muylaert, K., Fraeye, I., Foubert, I., De Brabanter, J., & De Cooman, L. (2012). Antioxidant potential of microalgae in relation to their phenolic and carotenoid content. *Journal of Applied Phycology, 24*(6), 1477–1486.

Goiris, K., Muylaert, K., Voorspoels, S., Noten, B., De Paepe, D., Baart, G. J. E., & De Cooman, L. (2014). Detection of flavonoids in microalgae from different evolutionary lineages. *Journal of Phycology, 50*(3), 483–492.

Grewe, M., Vogelsang, K., Ruzicka, T., Stege, H., & Krutmann, J. (2000). Neurotrophin-4 production by human epidermal keratinocytes: Increased expression in atopic dermatitis. *Journal of Investigative Dermatology, 114*(6), 1108–1112.

Griffith, L. G., & Swartz, M. A. (2006). Capturing complex 3D tissue physiology in vitro. *Nature Reviews Molecular Cell Biology, 7*(3), 211.

Guarnieri, M. T., & Pienkos, P. T. (2015). Algal omics: Unlocking bioproduct diversity in algae cell factories. *Photosynthesis Research, 123*(3), 255–263.

Guedes, A. C., Amaro, H. M., & Malcata, F. X. (2011). Microalgae as sources of carotenoids. *Marine Drugs, 9*(4), 625–644.

Guerin, M., Huntley, M. E., & Olaizola, M. (2003). *Haematococcus* astaxanthin: Applications for human health and nutrition. *Trends in Biotechnology, 21*(5), 210–216.

Guillerme, J. B., Couteau, C., & Coiffard, L. (2017). Applications for marine resources in cosmetics. *Cosmetics, 4*(3), 35.

Halliwell, B. (2006). Oxidative stress and neurodegeneration: Where are we now? *Journal of Neurochemistry, 97*(6), 1634–1658.

Hartmann, K. B., Karsten, U., Remias, B., & Ganzera, M. (2015). Analysis of mycosporine-like amino acids in selected algae and cyanobacteria by hydrophilic interaction liquid chromatography and a novel MAA from the red alga *Catenella repens*. *Marine Drugs, 13*, 6291–6305.

Hasegawa, T., Ito, K., Ueno, S., Kumamoto, S., Ando, Y., Yamada, A., et al. (1999). Oral administration of hot water extracts of *Chlorella vulgaris* reduces IgE production against milk casein in mice. *International Journal of Immunopharmacology, 21*(5), 311–323.

Havlickova, B., Bíró, T., Mescalchin, A., Tschirschmann, M., Mollenkopf, H., Bettermann, A., et al. (2009). A human folliculoid microsphere assay for exploring epithelial–mesenchymal interactions in the human hair follicle. *Journal of Investigative Dermatology, 129*(4), 972–983.

Herrmann, M., Zanella, L., Pertile, P., Gaebler, S., Joppe, H., Knupfer, M., Meyer, I., & Vielhaber, G. (2012a). *A novel biological skin tanner from microalgae*. Paper presented at 27th IFSCC Congress, Johannesburg, South Africa, pp. 330–331.

Herrmann, M., Zanella, L., Pertile, P., Gaebler, S., Joppe, H., Vielhaber, G., & Schmaus, G. (2012b). *Microalgae derived extract with promising anti-hair loss potential*. Paper presented at 27th IFSCC Congress, Johannesburg, South Africa, pp. 364–366.

Herrmann, H., Joppe, H., Pertile, P., & Zanella L. (2013). *Extracts of Isochrysis sp*. EP2168570B1.

Hildebrand, M., Manandhar-Shrestha, K., & Abbriano, R. (2017). Effects of chrysolaminarin synthase knockdown in the diatom *Thalassiosira pseudonana*: Implications of reduced carbohydrate storage relative to green algae. *Algal Research, 23*, 66–77.

Hirobe, T. (2014). Keratinocytes regulate the function of melanocytes. *Dermatologica Sinica, 32*(4), 200–204.

Höhn, A., Jung, T., Grimm, S., Catalgol, B., Weber, D., & Grune, T. (2011). Lipofuscin inhibits the proteasome by binding to surface motifs. *Free Radical Biology and Medicine, 50*(5), 585–591.

Horváth, G., Kemény, Á., Barthó, L., Molnár, P., Deli, J., Szente, L., et al. (2015). Effects of some natural carotenoids on TRPA1-and TRPV1-induced neurogenic inflammatory processes in vivo in the mouse skin. *Journal of Molecular Neuroscience, 56*(1), 113–121.

Hugues, N., & Joel, L. (2012). *Utilisation en cosmetique d'une composition a base d'algues unicellulaires*. FR2894473B1.

Imokawa, G. (2019). The xanthophyll carotenoid astaxanthin has distinct biological effects to prevent the photoaging of the skin even by its postirradiation treatment. *Photochemistry and Photobiology, 95*, 490. https://doi.org/10.1111/php.13039.

Imokawa, G., Nakajima, H., & Ishida, K. (2015). Biological mechanisms underlying the ultraviolet radiation-induced formation of skin wrinkling and sagging II: Over-expression of neprilysin plays an essential role. *International Journal of Molecular Sciences, 16*(4), 7776–7795.

Islam, M. N., Alsenani, F., & Schenk, P. M. (2017). Microalgae as a sustainable source of nutraceuticals. In V. K. Gupta, H. Treichel, V. Shapaval, L. A. de Oliveira, & M. G. Tuohy (Eds.), *Microbial functional foods and nutraceuticals* (1st ed., pp. 1–19). Hoboken NJ: Wiley & Sons Ltd.

Jedrzejczak-Silicka, M. (2017). History of cell culture. In S. J. T. Gowder (Ed.), *New insights into cell culture technology* (pp. 1–41). Rijeka: IntechOpen.

Jin, E. S., & Melis, A. (2003). Microalgal biotechnology: Carotenoid production by the green algae *Dunaliella salina*. *Biotechnology and Bioprocess Engineering, 8*(6), 331–337.

Jin, E. S., Polle, J. E., Lee, H. K., Hyun, S. M., & Chang, M. (2003). Xanthophylls in microalgae: From biosynthesis to biotechnological mass production and application. *Journal of Microbiology and Biotechnology, 13*(2), 165–174.

Joshi, S., Kumari, R., & Upasani, V. N. (2018). Applications of algae in cosmetics: An overview. *The International Journal of Innovative Research in Science, Engineering and Technology, 7*, 1269–1278.

Jung, T., Bader, N., & Grune, T. (2007). Lipofuscin: formation, distribution, and metabolic consequences. *Annals of the New York Academy of Sciences, 1119*(1), 97–111.

Juturu, V., Bowman, J. P., & Deshpande, J. (2016). Overall skin tone and skin-lightening-improving effects with oral supplementation of lutein and zeaxanthin isomers: A double-blind, placebo-controlled clinical trial. *Clinical, Cosmetic and Investigational Dermatology, 9*, 325.

Kansanen, E., Kuosmanen, S. M., Leinonen, H., & Levonen, A. L. (2013). The Keap1-Nrf2 pathway: Mechanisms of activation and dysregulation in cancer. *Redox Biology, 1*(1), 45–49.

Katsuyama, M., & Obata, K. (1988). *Hair nourishing cosmetic*. JPS63135315(A).

Kee, H. I., Hill, K., Young, M. E., Soo, K. H., & Koo, L. J. (2018). *Cosmetic composition with microalgae extract for anti-UV and skin-irritation alleviation effect*. KR101856480B1.

Kim, S. K. (2014). Marine cosmeceuticals. *Journal of Cosmetic Dermatology, 13*(1), 56–67.

Kim, E. J., Kim, Y. K., Kim, M. K., Kim, S., Kim, J. Y., Lee, D. H., & Chung, J. H. (2016). UV-induced inhibition of adipokine production in subcutaneous fat aggravates dermal matrix degradation in human skin. *Scientific Reports, 6*, 25616.

Kinkelin, I., Bröcker, E. B., Koltzenburg, M., & Carlton, S. M. (2000). Localization of ionotropic glutamate receptors in peripheral axons of human skin. *Neuroscience Letters, 283*(2), 149–152.
Kligman, A. M. (2005). Cosmeceuticals: A broad-spectrum category between cosmetics and drugs. In P. Elsner & H. I. Maibach (Eds.), *Cosmeceuticals and active cosmetics drug versus cosmetics* (2nd ed., pp. 1–9). Boca Raton, FL: Taylor & Francis.
Kligman, A. M., & Willis, I. (1975). A new formula for depigmenting human skin. *Archives of Dermatology, 111*(1), 40–48.
Klotz, L. O., Sánchez-Ramos, C., Prieto-Arroyo, I., Urbánek, P., Steinbrenner, H., & Monsalve, M. (2015). Redox regulation of FoxO transcription factors. *Redox Biology, 6*, 51–72.
Kobayashi, A., Kang, M.-I., Okawa, H., Ohtsuji, M., Zenke, Y., Chiba, T., Igarashi, K., & Yamamoto, M. (2004). Oxidative stress sensor Keap1 functions as an adaptor for Cul3-based E3 ligase to regulate proteasomal degradation of Nrf2. *Molecular and Cellular Biology, 24*(16), 7130–7139.
Koller, M., Muhr, A., & Braunegg, G. (2014). Microalgae as versatile cellular factories for valued products. *Algal Research, 6*, 52–63.
Kopal, C., Deveci, M., Öztürk, S., & Sengezer, M. (2007). Effects of topical glutathione treatment in rat ischemic wound model. *Annals of Plastic Surgery, 58*(4), 449–455.
Kralovec, J. A., Power, M. R., Liu, F., Maydanski, E., Ewart, H. S., Watson, L. V., et al. (2005). An aqueous *Chlorella* extract inhibits IL-5 production by mast cells in vitro and reduces ovalbumin-induced eosinophil infiltration in the airway in mice in vivo. *International Immunopharmacology, 5*(4), 689–698.
Kruglikov, I. L., & Scherer, P. E. (2016). Dermal adipocytes: From irrelevance to metabolic targets? *Trends in Endocrinology and Metabolism, 27*(1), 1–10.
Kumar, S. (2005). Exploratory analysis of global cosmetic industry: Major players, technology and market trends. *Technovation, 25*(11), 1263–1272.
Kurfurst, R., Nizard, C., Schnebert, S., Perrier, E., Tobin, D., & Singh, S. K. (2010). *Methods useful in studying or modulating skin or hair pigmentation, plant extracts for use in compositions and cosmetic care methods*. WO2010029115A1.
Lee, A., Kim, J. Y., Heo, J., Cho, D.-H., Kim, H.-S., An, I.-S., An, S., & Bae, S. (2018). The inhibition of melanogenesis via the PKA and ERK signaling pathways by *Chlamydomonas reinhardtii* extract in B16F10 melanoma cells and artificial human skin equivalents. *Journal of Microbiology and Biotechnology, 28*(12), 2121–2132.
Letsiou, S., Kalliampakou, K., Gardikis, K., Mantecon, L., Infante, C., Chatzikonstantinou, M., et al. (2017). Skin protective effects of *Nannochloropsis gaditana* extract on H_2O_2-stressed human dermal fibroblasts. *Frontiers in Marine Science, 4*, 221.
Li, M., Rezakhanlou, A. M., Chavez-Munoz, C., Lai, A., & Ghahary, A. (2009). Keratinocyte-releasable factors increased the expression of MMP1 and MMP3 in co-cultured fibroblasts under both 2D and 3D culture conditions. *Molecular and Cellular Biochemistry, 332*(1-2), 1–8.
Lim, J. T. (1999). Treatment of melasma using kojic acid in a gel containing hydroquinone and glycolic acid. *Dermatologic Surgery, 25*, 282–284.
Llewellyn, C. A., & Airs, R. L. (2010). Distribution and abundance of MAAs in 33 species of microalgae across 13 classes. *Marine Drugs, 8*, 1273–1291.
Lubián, L. M., Montero, O., Moreno-Garrido, I., Huertas, I. E., Sobrino, C., González-del Valle, M., & Parés, G. (2000). *Nannochloropsis* (Eustigmatophyceae) as source of commercially valuable pigments. *Journal of Applied Phycology, 12*(3-5), 249–255.
Maas-Szabowski, N., Shimotoyodome, A., & Fusenig, N. E. (1999). Keratinocyte growth regulation in fibroblast cocultures via a double paracrine mechanism. *Journal of Cell Science, 112*(12), 1843–1853.
Mackenna, R. B., Wheatley, V. R., & Wormall, A. (1950). The composition of the surface skin fat ('sebum') from the human forearm. *Journal of Investigative Dermatology, 15*(1), 33–47.
Maiz, D. (2007). The underwater world: A source of inexhaustible inspiration. *Parfums Cosmétiques Actualités, 194*, 136–160.

Marquardt, D., Williams, J. A., Kučerka, N., Atkinson, J., Wassall, S. R., Katsaras, J., & Harroun, T. A. (2013). Tocopherol activity correlates with its location in a membrane: A new perspective on the antioxidant vitamin E. *Journal of the American Chemical Society, 135*(20), 7523–7533.
Martin, G. M., Sprague, C. A., & Epstein, C. J. (1970). Replicative life-span of cultivated human cells. *Laboratory Investigation, 23*(1), 86–92.
Martins, A., Vieira, H., Gaspar, H., & Santos, S. (2014). Marketed marine natural products in the pharmaceutical and cosmeceutical industries: Tips for success. *Marine Drugs, 12*(2), 1066–1101.
Matsui, M. S., Muizzuddin, N., Arad, S., & Marenus, K. (2003). Sulfated polysaccharides from red microalgae have antiinflammatory properties in vitro and in vivo. *Applied Biochemistry and Biotechnology, 104*(1), 13–22.
Mattoli, L., Cangi, F., Maidecchi, A., Ghiara, C., Ragazzi, E., Tubaro, M., et al. (2006). Metabolomic fingerprinting of plant extracts. *Journal of Mass Spectrometry, 41*(12), 1534–1545.
Mattoli, L., Cangi, F., Ghiara, C., Burico, M., Maidecchi, A., Bianchi, E., et al. (2011). A metabolite fingerprinting for the characterization of commercial botanical dietary supplements. *Metabolomics, 7*(3), 437–445.
McGrath, J. A., Eady, R. A. J., & Pope, F. M. (2010). Anatomy and organization of human skin. In *Rook's textbook of dermatology* (Vol. 1, 8th ed., pp. 3–1). Chichester: Wiley.
Megata, H. (2006). *Hair papilla cell growth promoter, vascular endothelial growth factor (VEGF) production promoter and hair-restoring or growing agent*. JP2006282597(A).
Mendes, A., Reis, A., Vasconcelos, R., Guerra, P., & da Silva, T. L. (2009). *Crypthecodinium cohnii* with emphasis on DHA production: A review. *Journal of Applied Phycology, 21*(2), 199–214.
Mewes, K. R., Fuhrmann, G., Heinen, G., Hoffmann-Döhr, S., Reisinger, K., Förster, T., & Petersohn, D. (2017). Reconstructed 3D tissues for efficacy and safety testing of cosmetic ingredients. *IFSCC Magazine, 20*(2), 55–64.
Mishima, Y., Oyama, Y., & Kurimoto, M. (1993). *Skin-whitening agent*. US5,262,153A.
Mishra, N., & Mishra, N. (2018). Exploring the biologically active metabolites of *Isochrysis galbana* in pharmaceutical interest: An overview. *International Journal of Pharmaceutical Sciences and Research, 9*(6), 2162–2174.
Mitsui, T. (1997). *New cosmetic science* (pp. 3–9). Amsterdam: Elsevier Science.
Molina Grima, E., Belarbi, E.-H., Acién Fernández, F. G., Robles Medina, A., & Chisti, Y. (2003). Recovery of microalgal biomass and metabolites: process options and economics. *Biotechnology Advances, 20*(7-8), 491–515.
Moon, S. H., Seo, K. I., Han, W. S., Suh, D. H., Cho, K. H., Kim, J. J., & Eun, H. C. (2001). Pathological findings in cumulative irritation induced by SLS and croton oil in hairless mice. *Contact Dermatitis, 44*(4), 240–245.
Morvan, P., & Vallee, R. (2007). Effects of *Chlorella* extract on skin. *Personal Care, 2007*, 57–64.
Mourelle, M., Gómez, C., & Legido, J. (2017). The potential use of marine microalgae and cyanobacteria in cosmetics and thalassotherapy. *Cosmetics, 4*(4), 46.
Muthuirulappan, S., & Francis, S. P. (2013). Anti-cancer mechanism and possibility of nanosuspension formulation for a marine algae product fucoxanthin. *Asian Pacific Journal of Cancer Prevention, 14*(4), 2213–2216.
Nakamura, M., Haarmann-Stemmann, T., Krutmann, J., & Morita, A. (2018). Alternative test models for skin aging research. *Experimental Dermatology, 27*(5), 495–500.
Nakano, A., Lourenço, C., & Paula, M. (2019). *Uses of extracts of Isochrysis sp*. WO 2019/037843A1.
Nemes, Z., & Steinert, P. M. (1999). Bricks and mortar of the epidermal barrier. *Experimental & Molecular Medicine, 31*(1), 5–19.
Nizard, C., Poggioli, S., Heusèle, C., Bulteau, A. L., Moreau, M., Saunois, A., et al. (2004). Algae extract protection effect on oxidized protein level in human stratum corneum. *Annals of the New York Academy of Sciences, 1019*(1), 219–222.
Nordlund, J. J., Abdel-Malek, Z. A., Boissy, R. E., & Rheins, L. A. (1989). Pigment cell biology: An historical review. *Journal of Investigative Dermatology, 92*(4), S53–S60.

Oh, H. Y., Jin, X., Kim, J. G., Oh, M. J., Pian, X., Kim, J. M., et al. (2007). Characteristics of primary and immortalized fibroblast cells derived from the miniature and domestic pigs. *BMC Cell Biology, 8*(1), 20.

Oyewole, A. O., & Birch-Machin, M. A. (2015). Sebum, inflammasomes and the skin: Current concepts and future perspective. *Experimental Dermatology, 24*(9), 651–654.

Park, S. H., Choi, Y. J., Lee, S. S., & Hong, Y. M. (2014). *Composition for improving condition of hair and preventing hair loss*. KR20140062249A.

Park, S. H., Choi, Y. J., & Lee, S. S. (2015). *Composition comprising Tetraselmis tetrathele extract for improving condition of hair and preventing hair loss*. KR20150100302A.

Park, J. H., Yeo, I. J., Han, J. H., Suh, J. W., Lee, H. P., & Hong, J. T. (2018). Anti-inflammatory effect of astaxanthin in phthalic anhydride-induced atopic dermatitis animal model. *Experimental Dermatology, 27*(4), 378–385.

Paus, R., & Peker, S. (2003). Biology of hair and nails. In J. L. Bolognia, J. L. Jorizzo, & R. P. Rapini (Eds.), *Dermatology* (Vol. 1, pp. 1007–1032). St. Louis, MO: Mosby.

Pavlovic, S., Daniltchenko, M., Tobin, D. J., Hagen, E., Hunt, S. P., Klapp, B. F., et al. (2008). Further exploring the brain–skin connection: Stress worsens dermatitis via substance P-dependent neurogenic inflammation in mice. *Journal of Investigative Dermatology, 128*(2), 434–446.

Paye, M., Barel, A. O., & Maibach, H. I. (2009). Introduction. In A. O. Barel, M. Paye, & H. I. Maibach (Eds.), *Handbook of cosmetic science and technology* (3rd ed., pp. 1–3). New York, NY: Informa Healthcare.

Pertile, P., Zanella, L., Herrmann, M., Joppe, H., & Gaebler, S. (2010). *Extracts of Tetraselmis sp. for cosmetic and therapeutic purposes*. EP2193785A2.

Picardo, M., Ottaviani, M., Camera, E., & Mastrofrancesco, A. (2009). Sebaceous gland lipids. *Dermato-Endocrinology, 1*(2), 68–71.

Pillai, S., Singh, S., & Oresajo, C. (2016). Percutaneous delivery of cosmetic actives to the skin. In Z. D. Draelos (Ed.), *Cosmetic dermatology: Products and procedures* (2nd ed., pp. 65–74). Chichester: Wiley and Sons.

Pillaiyar, T., Manickam, M., & Jung, S.-H. (2017). Recent development of signaling pathways inhibitors of melanogenesis. *Cellular Signalling, 40*, 99–115.

Pisal, D. S., & Lele, S. S. (2005). Carotenoid production from microalga, *Dunaliella salina*. *Indian Journal of Biotechnology, 4*, 476–483.

Pittayapruek, P., Meephansan, J., Prapapan, O., Komine, M., & Ohtsuki, M. (2016). Role of matrix metalloproteinases in photoaging and photocarcinogenesis. *International Journal of Molecular Sciences, 17*(6), 868.

Plaza, M., Herrero, M., Cifuentes, A., & Ibanez, E. (2009). Innovative natural functional ingredients from microalgae. *Journal of Agricultural and Food Chemistry, 57*(16), 7159–7170.

Poeggeler, B., Schulz, C., Pappolla, M. A., Bodó, E., Tiede, S., Lehnert, H., & Paus, R. (2010). Leptin and the skin: A new frontier. *Experimental Dermatology, 19*(1), 12–18.

Poumay, Y., & Coquette, A. (2007). Modelling the human epidermis in vitro: Tools for basic and applied research. *Archives of Dermatological Research, 298*(8), 361–369.

Priyadarshani, I., & Rath, B. (2012). Commercial and industrial applications of micro algae–A review. *Journal of Algal Biomass Utilization, 3*(4), 9–100.

Pulz, O., & Gross, W. (2004). Valuable products from biotechnology of microalgae. *Applied Microbiology and Biotechnology, 65*(6), 635–648.

Queiroz, M. L., da Rocha, M. C., Torello, C. O., de Souza Queiroz, J., Bincoletto, C., Morgano, M. A., et al. (2011). *Chlorella vulgaris* restores bone marrow cellularity and cytokine production in lead-exposed mice. *Food and Chemical Toxicology, 49*(11), 2934–2941.

Raposo, M., de Morais, R., & Bernardo de Morais, A. (2013). Bioactivity and applications of sulphated polysaccharides from marine microalgae. *Marine Drugs, 11*(1), 233–252.

Reis, A., Gouveia, L., Veloso, V., Fernandes, H. L., Empis, J. A., & Novais, J. M. (1996). Eicosapentaenoic acid-rich biomass production by the microalga *Phaeodactylum tricornutum* in a continuous-flow reactor. *Bioresource Technology, 55*, 83–88.

Reus, A. A., Usta, M., & Krul, C. A. (2012). The use of ex vivo human skin tissue for genotoxicity testing. *Toxicology and Applied Pharmacology, 261*(2), 154–163.

Rhyter, J. H., & Goldman, J. C. (1975). Microbes as food in mariculture. *Annual Review of Microbiology, 29*, 429–433.

Rinnerthaler, M., Bischof, J., Streubel, M., Trost, A., & Richter, K. (2015). Oxidative stress in aging human skin. *Biomolecules, 5*(2), 545–589.

Rodrigues, E., Mariutti, L. R., & Mercadante, A. Z. (2012). Scavenging capacity of marine carotenoids against reactive oxygen and nitrogen species in a membrane-mimicking system. *Marine Drugs, 10*(8), 1784–1798.

Rodríguez-Luna, A., Talero, E., Terencio, M., González-Rodríguez, M., Rabasco, A. M., de los Reyes, C., et al. (2017). Topical application of glycolipids from *Isochrysis galbana* prevents epidermal hyperplasia in mice. *Marine Drugs, 16*(1), 2.

Rodríguez-Luna, A., Ávila-Román, J., González-Rodríguez, M. L., Cózar, M. J., Rabasco, A. M., Motilva, V., & Talero, E. (2018). Fucoxanthin-containing cream prevents epidermal hyperplasia and UVB-induced skin erythema in mice. *Marine Drugs, 16*, 378.

Roessler, P. G. (1987). UDPglucose pyrophosphorylase activity in the diatom *Cyclotella cryptica*. Pathway of chrysolaminarin biosynthesis. *Journal of Phycology, 23*(3), 494–498.

Ryu, B., Himaya, S. W. A., & Kim, S.-K. (2015). Applications of microalgae-derived active ingredients as cosmeceuticals. In S.-K. Kim (Ed.), *Handbook of marine microalgae* (pp. 309–316). London: Academic Press.

Safafar, H., Van Wagenen, J., Møller, P., & Jacobsen, C. (2015). Carotenoids, phenolic compounds and tocopherols contribute to the antioxidative properties of some microalgae species grown on industrial wastewater. *Marine Drugs, 13*(12), 7339–7356.

Saladin, K. S. (2007). Part Two, Support and movement — The skin and subcutaneous tissue. In *Human anatomy, International Edition 2007* (pp. 129–150). Singapore: McGraw-Hill Education (Asia).

Sansone, C., Galasso, C., Orefice, I., Nuzzo, G., Luongo, E., Cutignano, A., et al. (2017). The green microalga *Tetraselmis suecica* reduces oxidative stress and induces repairing mechanisms in human cells. *Scientific Reports, 7*, 41215.

Sathasivam, R., & Ki, J.-S. (2018). A review of the biological activities of microalgal carotenoids and their potential use in healthcare and cosmetic industries. *Marine Drugs, 16*(1), 26.

Scandolera, A., Hubert, J., Humeau, A., Lambert, C., De Bizemont, A., Winkel, C., et al. (2018). GABA and GABA-alanine from the red microalgae rhodosorus marinus exhibit a significant neuro-soothing activity through inhibition of neuro-inflammation mediators and positive regulation of TRPV1-related skin sensitization. *Marine Drugs, 16*(3), 96.

Schiff-Deb, C., & Sharma, S. (2015). *Personal care products containing microalgae or extracts thereof*. US20150352034A1.

Schwartz, J. R., DeAngelis, Y. M., & Dawson, T. L., Jr. (2012). Chapter 12. Dandruff and seborrheic dermatitis: A head scratcher. In T. Evans & R. Randall (Eds.), *Practical modern hair science* (pp. 389–414). Darmstadt: Wissenschaftliche.

Shankar, K., Godse, K., Aurangabadkar, S., Lahiri, K., Mysore, V., Ganjoo, A., et al. (2014). Evidence-based treatment for melasma: Expert opinion and a review. *Dermatology and Therapy, 4*(2), 165–186.

Sharma, K., Sharma, D., Sharma, M., Sharma, N., Bidve, P., Prajapati, N., et al. (2018). Astaxanthin ameliorates behavioral and biochemical alterations in in-vitro and in-vivo model of neuropathic pain. *Neuroscience Letters, 674*, 162–170.

Shen, C. T., Chen, P. Y., Wu, J. J., Lee, T. M., Hsu, S. L., Chang, C. M. J., et al. (2011). Purification of algal anti-tyrosinase zeaxanthin from *Nannochloropsis oculata* using supercritical anti-solvent precipitation. *The Journal of Supercritical Fluids, 55*(3), 955–962.

Shih, M. F., & Cherng, J. Y. (2012). Protective effects of *Chlorella*-derived peptide against UVC-induced cytotoxicity through inhibition of caspase-3 activity and reduction of the expression of phosphorylated FADD and cleaved PARP-1 in skin fibroblasts. *Molecules, 17*(8), 9116–9128.

Sidgwick, G. P., McGeorge, D., & Bayat, A. (2016). Functional testing of topical skin formulations using an optimised ex vivo skin organ culture model. *Archives of Dermatological Research, 308*(5), 297–308.

Singh, S. K., Nizard, C., Kurfurst, R., Bonte, F., Schnebert, S., & Tobin, D. J. (2008). The silver locus product (Silv/gp100/Pmel17) as a new tool for the analysis of melanosome transfer in human melanocyte-keratinocyte co-culture. *Experimental Dermatology, 17*(5), 418–426.

Singh, A. K., Ganguly, R., Kumar, S., & Pandey, A. K. (2017). Microalgae: A source of nutraceutical and industrial products. In A. A. Mahdi, M. Abid, M. M. Abid Ali Khan, M. I. Ansari, & R. K. Maheshwari (Eds.), *Molecular biology and pharmacognosy of beneficial plants* (pp. 34–51). Delhi: Lenin Media Private Limited.

Skoczyńska, A., Budzisz, E., Trznadel-Grodzka, E., & Rotsztejn, H. (2017). Melanin and lipofuscin as hallmarks of skin aging. *Advances in Dermatology and Allergology, 34*(2), 97–103.

Solano, F., Briganti, S., Picardo, M., & Ghanem, G. (2003). Hypopigmenting agents: An updated review on biological, chemical and clinical aspects. *Pigment Cell Research, 19*, 550–571.

Spolaore, P., Joannis-Cassan, C., Duran, E., & Isambert, A. (2006). Commercial applications of microalgae. *Journal of Bioscience and Bioengineering, 101*(2), 87–96.

Sri, P., Adimoolam, S., & Mahmud, A. (2013). Percutaneous absorption of triacyglycerols (TAGS), tocols and carotenoids: Comparison studies of crude and refined palm oil. *Malaysian Journal of Pharmaceutical Sciences, 11*(1), 33.

Stark, H. J., Willhauck, M. J., Mirancea, N., Boehnke, K., Nord, I., Breitkreutz, D., et al. (2004). Authentic fibroblast matrix in dermal equivalents normalises epidermal histogenesis and dermo-epidermal junction in organotypic co-culture. *European Journal of Cell Biology, 83*(11-12), 631–645.

Stark, H. J., Boehnke, K., Mirancea, N., Willhauck, M. J., Pavesio, A., Fusenig, N. E., & Boukamp, P. (2006). Epidermal homeostasis in long-term scaffold-enforced skin equivalents. *Journal of Investigative Dermatology Symposium Proceedings, 11*(1), 93–105.

Steinert, P. M., & Marekov, L. N. (1995). The proteins elafin, filaggrin, keratin intermediate filaments, loricrin, and small proline-rich proteins 1 and 2 are isodipeptide cross-linked components of the human epidermal cornified cell envelope. *Journal of Biological Chemistry, 270*(30), 17702–17711.

Stolz, P., & Obermayer, B. (2005). Manufacturing microalgae for skin care. *Cosmetics & Toiletries, 120*(3), 99–106.

Suh, S. S., Hwang, J., Park, M., Seo, H. H., Kim, H. S., Lee, J. H., Moh, S. H., & Lee, T. K. (2014). Anti-inflammation activities of mycosporine-like amino acids (MAAs) in response to UV radiation suggest potential anti-skin aging activity. *Marine Drugs, 12*(10), 5174–5187.

Sun, W., Xing, L., Lin, H., Leng, K., Zhai, Y., & Liu, X. (2016). Assessment and comparison of in vitro immunoregulatory activity of three astaxanthin stereoisomers. *Journal of Ocean University of China, 15*(2), 283–287.

Suzuki, R., Umishio, K., Ifuku, O., Ota, O., Kobayashi, K., & Moro, G. (2002). *Gray hair preventing agent*. JP2002212039A.

Svobodova, A., Walterova, D., & Vostalova, J. (2006). Ultraviolet light induced alteration to the skin. *Biomedical Papers-Palacky University in Olomouc Czech Repub., 150*(1), 25–38.

Terazawa, S., Mori, S., Nakajima, H., Yasuda, M., & Imokawa, G. (2015). The UVB-stimulated expression of transglutaminase 1 is mediated predominantly via the NFκB signaling pathway: New evidence of its significant attenuation through the specific interruption of the p38/MSK1/NFκBp65 ser276 axis. *PLoS One, 10*(8), e0136311.

Terman, A., & Brunk, U. T. (2004). Lipofuscin. *The International Journal of Biochemistry & Cell Biology, 36*(8), 1400–1404.

Tfayli, A., Farhane, Z., Bonnier, F., & Byrne, H. (2014). Comparison of Structure and organization of cutaneous lipids in a reconstructed skin model and human skin: Spectroscopic imaging and chromatographic profiling. *Experimental Dermatology, 23*, 441–443.

Thiele, J. J., Weber, S. U., & Packer, L. (1999). Sebaceous gland secretion is a major physiologic route of vitamin E delivery to skin. *Journal of Investigative Dermatology, 113*(6), 1006–1010.

Tredici, M. R., Rodolfi, L., Biondi, N., Bassi, N., & Sampietro, G. (2016). Techno-economic analysis of microalgal biomass production in a 1-ha Green Wall Panel (GWP®) plant. *Algal Research, 19*, 253–263.
Truzzi, F., Marconi, A., & Pincelli, C. (2011). Neurotrophins in healthy and diseased skin. *Dermato-Endocrinology, 3*(1), 32–36.
Tsai, T. C., & Hantash, B. M. (2008). Cosmeceutical agents: A comprehensive review of the literature. *Clinical Medicine: Dermatology, 2008*, 1–20.
Van Laethem, A., Claerhout, S., Garmyn, M., & Agostinis, P. (2005). The sunburn cell: Regulation of death and survival of the keratinocyte. *The International Journal of Biochemistry & Cell Biology, 37*(8), 1547–1553.
Vert, G., & Chory, J. (2011). Crosstalk in cellular signaling: Background noise or the real thing? *Developmental Cell, 21*(6), 985–991.
Waller, J. M., & Maibach, H. I. (2006). Age and skin structure and function, a quantitative approach (II): Protein, glycosaminoglycan, water, and lipid content and structure. *Skin Research and Technology, 12*, 145–154.
Wang-Michelitsch, J., & Michelitsch, T. M. (2015). Development of age spots as a result of accumulation of aged cells in aged skin. arXiv preprint arXiv:1505.07012.
Watanabe, T., Kitajima, D., & Fujita, S. (1983). Nutritional values of live organisms used in Japan for mass propagation of fish: A review. *Aquaculture, 34*(1-2), 115–143.
Wertz, P. W. (2009). Human synthetic sebum formulation and stability under conditions of use and storage. *International Journal of Cosmetic Science, 31*(1), 21–25.
Weylandt, K. H., Chiu, C. Y., Gomolka, B., Waechter, S. F., & Wiedenmann, B. (2012). Omega-3 fatty acids and their lipid mediators: Towards an understanding of resolvin and protectin formation. *Prostaglandins & Other Lipid Mediators, 97*(3-4), 73–82.
Winget, R. R. (1994). *Anti-inflammatory compositions containing eicosapentaenoic acid bearing monogalactosyldiacylglycerol and methods relating thereto*. WO1994/024984.
Wlaschek, M., Bolsen, K., Herrmann, G., Schwarz, A., Wilmroth, F., Heinrich, P. C., Goerz, G., & Scharffetter-Kochanek, K. (1993). UVA-induced autocrine stimulation of fibroblast-derived-collagenase by IL-6: A possible mechanism in dermal photodamage? *Journal of Investigative Dermatology, 101*(2), 164–168.
Wobbe, L., & Remacle, C. (2015). Improving the sunlight-to-biomass conversion efficiency in microalgal biofactories. *Journal of Biotechnology, 201*, 28–42.
Xia, S., Gao, B., Li, A., Xiong, J., Ao, Z., & Zhang, C. W. (2014). Preliminary characterization, antioxidant properties and production of chrysolaminarin from marine diatom *Odontella aurita*. *Marine Drugs, 12*, 4883–4897.
Xu, W., Hong, S. J., Jia, S., Zhao, Y., Galiano, R. D., & Mustoe, T. A. (2012). Application of a partial-thickness human ex vivo skin culture model in cutaneous wound healing study. *Laboratory Investigation, 92*, 584.
Yano, S., Banno, T., Walsh, R., & Blumenberg, M. (2008). Transcriptional responses of human epidermal keratinocytes to cytokine interleukin-1. *Journal of Cellular Physiology, 214*(1), 1–13.
Yousef, H., & Sharma, S. (2019). *Anatomy, skin (integument), epidermis*. Treasure Island, FL: StatPearls Publishing. Retrieved from https://www.ncbi.nlm.nih.gov/books/NBK470464/.
Zanella, L, & Pertile, P. (2016). *Extracts of Nannochloropsis sp. and their applications*. WO2016/026723A2.
Zanella, L., Pertile, P., Massironi, M., Massironi, M., & Caviola, E. (2012). *Extracts of microalgae and their application*. WO2012/052356A2.
Zanella, L., Pertile, P., & Massironi, M. (2016). *Extracts of microalgae and plants for regulating sebum production*. WO2016/020339A2.
Zhang, J., Xu, X., Rao, N. V., Argyle, B., McCoard, L., Rusho, W. J., et al. (2011). Novel sulfated polysaccharides disrupt cathelicidins, inhibit RAGE and reduce cutaneous inflammation in a mouse model of rosacea. *PLoS One, 6*(2), e16658.
Zhang, J., Sun, Z., Sun, P., Chen, T., & Chen, F. (2014). Microalgal carotenoids: Beneficial effects and potential in human health. *Food & Function, 5*(3), 413–425.

Ziboh, V. A., Miller, C. C., & Cho, Y. (2000). Metabolism of polyunsaturated fatty acids by skin epidermal enzymes: generation of antiinflammatory and antiproliferative metabolites. *The American Journal of Clinical Nutrition, 71*(1), 361s–366s.

Zmijewski, M. A., & Slominski, A. T. (2011). Neuroendocrinology of the skin: An overview and selective analysis. *Dermato-Endocrinology, 3*(1), 3–10.

Part IV
Other High Value Application

Chapter 10
Microalgae as a Vaccine Delivery System to Aquatic Organisms

Aisamuddin Ardi Zainal Abidin, Mohanrajh Suntarajh, and Zetty Norhana Balia Yusof

Abstract Aquaculture is one of the fastest growing food producing sector, as global aquaculture produces about 65 million metric tons of seafood valued at more than US$78 billion annually and supplies 50% of all the fish consumed in the world (Turchini et al., Fish oil replacement and alternative lipid sources in aquaculture feeds, CRC Press, Boca Raton, FL, 2010). On top of that, the aquaculture industry displayed an annual percentage of growth rate (APR) of 9.4% compared to other food producing sectors such as pigs farming (3.1%), poultry (5.1%), beef (1.2%), and mutton and lamb (1.0%). The aquaculture sector, especially fish, contributed up to 17% of animal proteins consumed worldwide and can reach up to 50% in some countries. In 2002, it was reported that the total world aquaculture production was worth 60 billion USD by value. One of the major and primary constraints in the aquaculture system production is disease outbreaks. They could be caused by bacteria, viruses, parasites and fungi. In the catfish industry, a loss of up to 60–80 million USD was caused by pathogenic bacteria *Edwardsiella ictaluri* and *Flavobacterium culumnare*. Apart from that, it was reported that a 50–100 million Euro annual loss in the salmon industry is caused by parasitic lice. Many strategies have been attempted to gauge and control this situation, but there is still an urgent need for better alternatives and also to explore the potential use of genetically modified organisms instead of antibiotics and chemical control. In this chapter, we focus on the potential and application of transgenic microalgae on aquaculture, as it has been dubbed as the organism of the future in terms of its utility, flexibility and, most importantly, sustainability.

A. A. Z. Abidin · M. Suntarajh · Z. N. B. Yusof (✉)
Department of Biochemistry, Faculty of Biotechnology and Biomolecular Science, Universiti Putra Malaysia, Serdang, Selangor, Malaysia

Laboratory of Marine Biotechnology, Institute of Bioscience, Universiti Putra Malaysia, Serdang, Selangor, Malaysia
e-mail: zettynorhana@upm.edu.my

M. A. Alam et al. (eds.), *Microalgae Biotechnology for Food, Health and High Value Products*, https://doi.org/10.1007/978-981-15-0169-2_10

Keywords Microalgae · Vaccine · Expression system · Transformation · Aquaculture

1 Introduction

The aquaculture sector is highly dependent on microalgae as a natural feed for aquatic organisms as well as zooplankton. However, for big hatcheries, mass cultivation of microalgae is a high production constraint and so an alternative feed was developed. One of the highly utilized feeds is the fishmeal feed, as it is cheap and provides sufficient nutrients to the aquatic organisms. The increasing growth of aquaculture has led to the increase in production and price of fishmeal (Tacon and Metian 2009). On top of that, disease prevention measures in fish hatcheries require high cost as well. Due to demand, fish farmers will always prefer cheap and easy measures, without considering the adverse effects they could bring. This chapter aims to explore the potential utilization of transgenic microalgae as feed for the aquaculture sector. With the advancement of microalgae biotechnology, commercial culture of microalgae has been well developed and established. Microalgae species have been widely grown to be utilized in the food health sector (*Spirulina* sp.) and carotenoid production (*Dunaliella salina* and *Haematococcus* pluvialis) (Carvalho et al. 2006; Lee 1997) with the development of large-scale photobioreactors that can be operated under defined, optimal conditions. Apart from that, this chapter discusses the advancement of genetic modification of microalgae for production of natural products or other novel products. These technologies will be further discussed to highlight the significance and potential of transgenic microalgae in the aquaculture sector.

2 Microalgae—A Versatile Microorganism

Microalgae are unicellular or simple multicellular organisms which include prokaryotic organisms such as cyanobacteria (blue green algae) and a broad span of eukaryotic algae. They are a polyphyletic group of autotrophic and heterotrophic microorganisms which can be found in freshwater and marine habitats (Alam and Wang 2019; Gangl et al. 2015). They are a diverse group of organisms with 200,000–800,000 estimated algal species; only approximately 35,000 have been classified and described (Ebenezer et al. 2012). Microalgae are also regarded as a unicellular model for plants (Ball 2005). Compared to plants, the growth and productivity of algae are higher (Lu et al. 2014). Moreover, marine algal species can grow under high saline conditions, which reduces the risk of contamination (Bacellar Mendes and Vermelho 2013). Figure 10.1 shows how versatile microalgae can be.

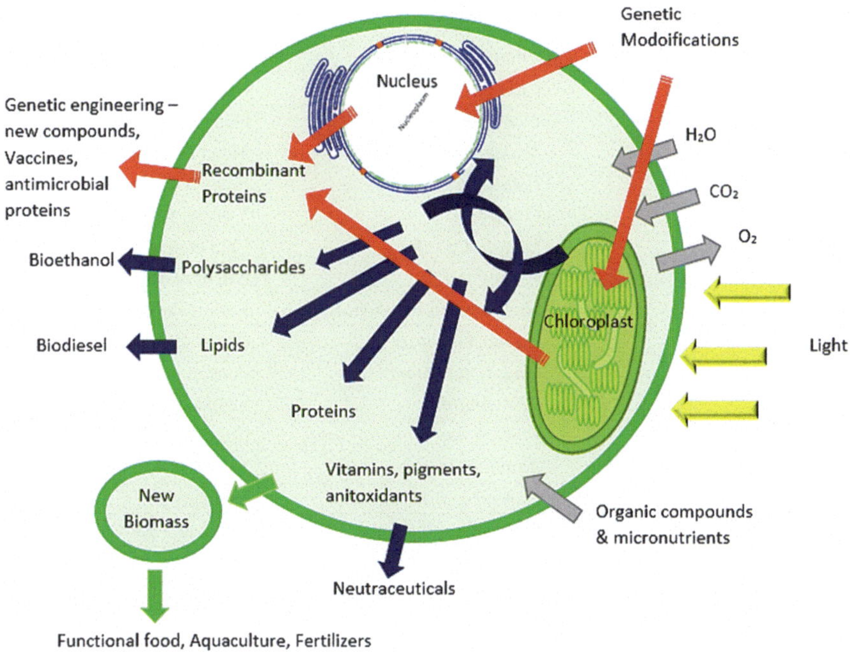

Fig. 10.1 An overview of the natural processes in microalgae with potential for genetic modifications (red arrows)

In the aquaculture industry, microalgae are mainly utilized as live feed for early or larval stage of fish, crustaceans and abalone. On top of that, they are also the main diet for zooplanktons prior to feeding to fish. They are also the key diet for molluscs at all growth stages (Brown 2002). Normally, microalgae are being fed to aquaculture organisms either as monospecies or as mixed species, as their nutritional value can vary significantly between species. The biochemical constituents of microalgae that contribute to their nutritional value are polyunsaturated fatty acids (PUFAs) (such as docosahexaenoic acid (DHA), eicosapentaenoic acid (EPA) and arachidonic acid (AA)), vitamins (such as thiamine, riboflavin, pantothenic acid, pyridoxine, cobalamin and many more), sterols, minerals such as silica (major diet for crustacean) and pigments (Brown et al. 1999; Knauer et al. 1999; Brown 2002).

Increased interest in microalgae studies have led to the identification of numerous new microalgal species, and this highlights their potential as a source of various compounds useful in the feed, food, nutrition, cosmetics and pharmaceutical industries (Gong et al. 2011). Most studies manipulate different parameters (environment) to induce stress in the microalgae interest at a degree it is not fatal to increase the production of the biomolecules in the microalgae (Bjerkeng 2008; Fern et al. 2017; Azim et al. 2018). Most of the recent studies have been focusing on the genetic engineering of microalgae to enhance the productivity of natural compounds through metabolic engineering (Steinbrenner and Sandmann 2006) and expression of useful foreign genes for recombinant protein production (Doron et al. 2016).

Table 10.1 Transformation history of microalgae species

Transformation method	Microalgae species	Reference
Electroporation	*Chlorella* sp.	He et al. (2018)
	C. reinhardtii	Dong et al. (2018)
	N. oculata	Ma et al. (2016)
PEG mediated	*Pleurochrysis carterae*	Endo et al. (2016)
	C. vulgaris	Yang et al. (2015)
	Dunaliella salina	Chai et al. (2013)
Agrobacterium tumefaciens	*C. vulgaris*	Lau et al. (2017)
	Haematococcus pluvialis	Kathiresan et al. (2015)
	Nannochloropsis sp.	Cha et al. (2011)
Glass beads	*Dunaliella salina*	Feng et al. (2014)
	C. reinhardtii	Sizova et al. (2013)
	Platymonas subcordiformis	Cui et al. (2012)
Silicon carbide fibres	*Amphidinium* sp.	Te and Miller (1998)
	C. reinhardtii	Dunahay (1993)
Microprojectile bombardment	*Chlorella zofingiensis*	Liu et al. (2014)
	Chaetoceros sp.	Miyagawa et al. (2001)
	Phaeodactylum tricornutum	Hempel et al. (2011)

3 Transformation of Microalgae

Previous studies have shown that transformation of microalgae is achievable for both chloroplast and nuclear genomes (Coll 2006). There are several methods that have been developed to deliver foreign genes into microalgae species, which are enzyme mediated, electroporation, PEG mediated, *Agrobacterium tumifaciens* mediated, glass beads, silicon carbide whiskers and microprojectile bombardment. The process for transforming microalgae is to trigger temporal permeability of the cell membrane, enabling DNA molecules to enter the cells while preserving viability (León and Fernández 2007). Table 10.1 shows the summary of transformation methods that have been utilized for microalgae transformation.

3.1 Enzyme-Mediated Transformation

The cell wall of microalgae is the ultimate barrier to achieve transformation efficiency. Thus, most transformation protocols of microalgae involve the use of protoplast formation or enzyme-treated cells. The different compositions of microalgae cell wall of different species have made protoplast generation challenging and complex (Popper and Tuohy 2010). Chen et al. (2008) described that a mixture of 4% hemicellulase and 2% driselase shows efficient digestion of cell wall in *N. oculata*. However, an enzyme mixture of 4.0% (w/v) cellulase R-10, 2% (w/v) macerozyme and 0.1% (w/v) pectinase proved effective in digesting *C. vulgaris* cell wall (Yang et al. 2015).

3.2 Electroporation

Electroporation is the method of emitting an electric pulse to introduce DNA into the cells through temporary pores in the plasmalemma. Temperature, osmolarity, electric field and strength, time of discharge, DNA concentration and enzyme treatment should be optimized to obtain high transformation efficiencies (León and Fernández 2007). The advantage of this method is that it is versatile for all type of cells, but the drawbacks are its non-specific transport and the wrong pulses that cause the membrane pulse to become too large and damage the cells (Chow and Tung 1999). The method could also be combined with enzyme-mediated transformation to remove the cell wall prior to electroporation to enable efficient delivery of the genetic material (Yang et al. 2015).

3.3 PEG-Mediated Transformation

Polyethylene glycol is described as an agent for clumping and fusion of protoplasts, which is assumed to facilitate the trapping of DNA into the cells (Fincham 1989). However, the exact role of PEG remains unclear. A study suggested that PEG probably induces the interaction between DNA and the cell surface, and the fusion of protoplasts is not the direct cause of DNA uptake (Kuwano et al. 2008). Apart from that, PEG-mediated transformation involves simple and inexpensive equipment, and yields highly transformed cells (Potrykus 1991). This method also helps to overcome the host range limitations of Agrobacterium-mediated transformation, where PEG-mediated transformation can be readily adapted to a wide range of plant species and tissue sources (Mathur and Koncz 1998).

3.4 Agrobacterium-Mediated Transformation

Agrobacterium-mediated transformation is a unique tool for genetic modification, in which it transfers its plasmid DNA (Ti plasmid) into the nuclear genome of the host cell (usually plant cells). *Agrobacterium tumefaciens*, is often used for the harbouring of plasmids and mode of transfer. It is a technically challenging method due to the large size and low copy number of Ti plasmids, which lead to difficulties in plasmid isolation and manipulation, and the host range is limited in some species (Meyers et al. 2010; Mathur and Koncz 1998). It is highly and mainly used in plant genetic transformation. However, successful transformations have been reported in *Haematococcus pluvialis*, *Chlorella vulgaris* and *Schizochytrium* sp. (Kathiresan et al. 2009; Cheng et al. 2012; San Cha et al. 2012). The advantage of this method is that it does not need a permeabilization step unlike the other transformation methods (Kim et al. 2014).

3.5 Glass Beads

The glass beads method introduces heterologous genes into cells by agitating cells with glass beads, polyethylene glycol (PEG) and foreign DNA. The ruptured cells caused by the glass beads will then allow the exogenous DNA to be forced into the cells with osmotic support (Costanzo and Fox 1988). It is a cost-effective method, utilizes non-specific equipment and results in higher transformation efficiency in *C. reinhardtii* when compared to particle bombardment (Kindle et al. 1989). The disadvantage of this method somehow requires wall-less cells as low transformation efficiency was observed in *Chlamydomonas* sp. cells with complete cell wall (Quesada et al. 1994).

3.6 Silicon Carbide Whiskers

Silicon carbide whiskers (SiC) function as numerous needles that facilitate gene delivery due to the negative charges which results in low affinity between DNA molecules which are negatively charged (Asad and Arshad 2011). It overcomes the cell wall's restriction as compared to glass beads method, as it is also inexpensive. Previous studies have shown that SiC methods were successful in *Amphidinium* sp. and *Chlamydomonas reinhardtii* (Te and Miller 1998; Dunahay 1993) but it requires strict safeguard to avoid the inhalation hazard which is associated with permanent respiratory dysfunction and lung cancer in human (Qin et al. 2012).

3.7 Microparticle Bombardment

Microprojectile bombardment is also referred to as microparticle bombardment, gene gun transformation, or simply biolistics (Gong et al. 2011). It utilizes a gene gun that shoots a small dense particle (usually gold or tungsten) coated with DNA into the host cell (León and Fernández 2007). It has the ability to penetrate intact cell walls without requiring a protoplast regeneration system, enable diversified usage of vector and is comparatively successful compared to other transformation methods in microalgae (Qin et al. 2012). However, it is not widely applicable due to the need of specialized and high-cost equipment (Heiser 1992). Even though particle bombardment was reported as an effective method in algal transformation, it is unfavourable to produce numerous nuclear transformants because the technique is harsh enough to break the stiff silica cell walls that will lead to impairment of microalgae viability (Kindle et al. 1989).

4 Potential and Application of Transgenic Microalgae for Aquaculture

Currently, the main recombinant protein producers are bacteria, which are utilized for industrial enzyme production or other forms of recombinant proteins at a relatively low cost and at a faster pace (Demain and Vaishnav 2009; Corchero et al. 2013). Furthermore, the post-translational process of protein synthesis is very low, making them ideal for industrial recombinant protein production. The importance and significance of bacteria as cell factories for recombinant proteins are well summarized and reviewed by Ferrer-Miralles and Villaverde (2013). In comparison to the well-established cell factory, microalgae also have the advantage as well as potential to be a recombinant protein factory with a greener theme. The characteristics of microalgae such as solar fuelled, rapid growth, economical cultivation and possessing the ability to be genetically manipulated enable them to be applied in the field of biotechnology. Currently, green microalgae have been considered as a high-potential industrial strain for the biomanufacturing of highly valuable molecules and recombinant proteins for various industries such as agriculture, nutraceuticals, bioenergy and even cosmetics (Rasala and Mayfield 2015).

Microalgae have been transformed and have been expressing numerous recombinant proteins for various applications. The most widely studied microalgae for genetic modification is *Chlamydomonas reinhardtii*, which is also known as a model organism for plants. This species has been modified extensively and has showed success in expressing more than 20 different recombinant proteins (Rasala and Mayfield 2015). *C. reinhardtii* has successfully expressed antibodies and immunotoxins (Rasala et al. 2010), gut-active biologics (Manuell et al. 2007), vaccines subunits (Sun et al. 2003; He et al. 2007), industrial enzymes and feed additives (Georgianna et al. 2013; Rasala et al. 2012) and also nutrient supplements such as selenium (Hou et al. 2013). These are significant and well-established evidences that microalgae have the potential similar to bacteria to be a cell factory for recombinant proteins.

Utilization of microalgae in the industry have always been related with high energy and material cost, which are the main constraints (Vo et al. 2018). However, microalgae have been regarded as a cell factory for the production of recombinant protein which are purely environmental friendly as it utilizes CO_2 from the environment and produces O_2 (Fig. 10.1). Compared to bacteria, microalgae produce highly valuable molecules such as proteins, carbohydrates, fatty acids, pigments and other compounds, which can be recovered and separated from the recombinant proteins neglecting the constraints of costs. In terms of aquaculture, these molecules contribute to a well-balanced diet for aquaculture feed. Furthermore, enclosed bioreactors, similar to industrial bacterial recombinant producers, increase the safety in terms of prevention of transgenic microalgae release to the environment. Enclosed bioreactors also increase biomass production with high cost efficiency by recycling media and water (Franconi et al. 2010). Many species of microalgae have also been generally recognized as safe (GRAS) for human and animal consumption making safety not an issue (Rasala and Mayfield 2015). Transgenic microalgae also do not cross-contaminate and can be grown throughout the year with no growing season restrictions (Specht and Mayfield 2014).

4.1 Transgenic Microalgae for Aquaculture via Metabolic Engineering

As previously mentioned, microalgae play a major part in aquaculture. Lipids are a major source of nutrients for growth and development in aquaculture. One of the potentials and applications of microalgae for aquaculture is by the utilization of microlagae with increased lipid content, specifically PUFAs. However, most studies involving the increase of lipid production in microalgae are majorly for biodiesel production (Radakovits et al. 2010). *C. reinhardtii* and *Chlorella pyrenoidosa* have been studied for increased lipid production via eliminating starch metabolism where this aprroach will produce starchless mutant microalgae but with increased lipid composition (Zabawinski et al. 2001; Posewitz et al. 2004). Most studies on increasing lipid metabolism via metabolic engineering were successful in plants such as *Arabidopsis thaliana*, *Nicotiana tabacum*, soybean and rapeseed, where there was a significant increase in oil content as well as desired fatty acids (Nikolau et al. 2003; Stournas et al. 1995; Lardizabal et al. 2006; Dehesh et al. 2001). Apart from that, metabolic engineering of fatty acid biosynthesis in microalgae is still problematic due to many factors including auto-attenuation of exogenous sequences, codon usage bias, GC content and proteasome-mediated degradation (León-Bañares et al. 2004). Currently, most feeds are formulated with fish oil, which increases the cost of the feed and reduces the profit margin of the farmers (Kim et al. 2012). It is also estimated that 90% of the global supply of fish oil is used to produce aquaculture feeds (Turchini et al. 2010). Transgenic microalgae with increased lipid content may play a major role in reducing the cost of the feeds by replacing fish oil.

Carotenoids are pigments that aid in light harvesting, energy transfer during photosynthesis and scavenging of reactive oxygen species (ROS) (Li et al. 2009). There are varieties of carotenoids that come in different colours. Animals and humans do not biosynthesize carotenoids and therefore they must obtain them from diet. Carotenoids are common additives incorporated into animal feeds or even human foods as a natural source of food colouring (Bjerkeng 2008) and are utilized by the skin care industry as UV-absorbents (Sies and Stahl 1997), food dye such as the yellow colour of the margarine, pigments in scales, feathers or skin of birds, fish, amphibians, and reptiles (Blount and McGraw 2008), attracting pollinators or frugivores for plant reproduction (Blount and McGraw 2008), antioxidants and also nutraceuticals applications (Álvarez et al. 2014). In aquaculture, carotenoids are mainly used to make the feed more appealing, as most aquatic organisms are phototactic, and they enhance the organism's appearance, thereby increasing the market value of the products (aquatic organisms such as salmons and lobsters, whose market value relies on how red they are) (Goodwin 1984). Microalgae are natural resources of carotenoids. Two species of microalgae that are the major contributors to the carotenoids in global market are *Dunalella salina* and *Haematococcus pluvialis*, which produce and accumulate high amounts of β-carotene (*D. salina*) and astaxanthin (*H. pluvialis*) (Lorenz and Cysewski 2000). Transgenic studies on microalgae to produce more carotenoids were successful in *C. reinhardtii* and *H. pluvialis*. Deletion of genes involved in the regulation of the carotenoid biosynthesis in *C. reinhardtii* and *H. pluvialis* have shown increase in accumulation of carotenoids (Liu et al. 2013; Steinbrenner and Sandmann 2006). Insertion of exogenous carotenoid biosynthesis gene originated from *H. pluvialis*, namely *BKT* and *PSY*, was also found to increase carotenoids accumulation in *C. reinhardtii* (Couso et al. 2011). However, both these microalgae are not majorly used as aquaculture feed. More studies on producing transgenic lines of microalgae used in aquaculture feed such as *Chaetoceros calcitrans*, *Chlorella vulgaris*, *Nannochloropsis* spp. and many more for increase in carotenoids production should be put forward, as it may result in not only supplying nutritious feed but also increasing the market value of aquaculture products. Furthermore, carotenoids for the global market is currently being produced synthetically (Bjerkeng 2008) and this is causing many concerns regarding Therefore, naturally synthesized carotenoids (especially from microalgae) may penetrate the global carotenoids market which was estimated in 2010 with market value of 1.2 billion US$ with estimated annual growth of 2.3% (Cutzu et al. 2013).

4.2 *Transgenic Microalgae for Aquaculture via Expression of Recombinant Proteins*

A recombinant protein is encoded by a recombinant DNA which is introduced and expressed in a host or a foreign organism. Most recombinant proteins are foreign to the host, and bacteria and yeast have been the key hosts in recombinant proteins production to date. It is important to note the potential of microalgae as a host for expressing recombinant proteins, as the application would be very vast especially in the aquaculture industry. On top of being a host to express recombinant proteins, transgenic microalgae will still have their useful attributes like their biomass, and this feature will increase their functionality and versatility. Studies have shown that microalgae have the capability to be modified extensively for the production of various useful compounds like food additives and oral vaccines for the aquaculture industry.

Disease is a major constraint in aquaculture, and according to Meyer (1991), preventive measures in hatcheries must be taken to prevent disease outbreaks. This can be carried out by (1) preventing potential pathogens from the environment or the farming animals from contaminating, (2) maintaining good water quality, (3) avoiding or reducing environmental stressors such as low dissolved oxygen, temperature control and density control, (4) providing adequate nutrition, (5) isolating cultured animals from feral stocks and (6) immunization. Currently, vaccination is the most efficient method to increase or stimulate the immune system, and the current practical method to vaccinate them is via injection. However, it is labour-intensive and requires the fish to reach a certain size before injection, making vaccination of fry difficult (Sommerset et al. 2005). Apart from that, it is also stressful to the animal, which reduces the efficiency of the vaccines, and it may cause adhesions to animal tissue (Specht and Mayfield 2014). Oral vaccination is another way of vaccinating the aquatic animal, which is done by incorporating vaccines into the feed or into the water of the culture. However, oral vaccines are regarded as inefficient, as they require high amount of antigens if they are being introduced into the water system, and they need a protective layer as they are generally weak and easily degraded. Reports on the inefficiency of oral vaccination were also linked to the destruction of the gut in fish (Nakanishi and Ototake 1997). Hence, the idea of the utilization of transgenic microalgae for this purpose came into the picture, as it may overcome the shortcomings of both injections and oral vaccinations in order to deliver vaccines to aquaculture organisms. They key feature of microalgae as a carrier is that it possesses a protective layer in the form of its cell wall and cell membrane, which could prevent the degradation of the antigens, which in turn will significantly reduce the amount of antigens needed to be introduced into the water system by a few folds. Furthermore, the microalgae itself is the main feed providing nutrients, making it a living nutritious feed combined with immunostimulatory capabilities. Transgenic microalgae producing antigens or vaccines could also be powdered and administered as solid feed, with little or no purification needed (Walker et al. 2005).

As previously mentioned, *C. reinhardtii* is one of the most exploited species of microalgae for recombinant protein expressions. In terms of vaccines expression, studies done were mainly for the production of vaccines against human-related diseases (Rasala and Mayfield 2015). However, vaccines against the white spot disease (WSD), which affects a wide range of crustaceans, were successfully produced from *C. reinhardtii* and *Dunaliella salina* (Surzycki et al. 2009; Feng et al. 2014). Results have shown that crayfish vaccinated with transgenic *D. salina* had an increase in survival rate up to 59% when compared with the unvaccinated crayfish. This finding could be a revolution in aquaculture disease management, as it may prevent losses due to WSD which could go up to more than 35 million US dollars (Yang et al. 1999; Subasinghe et al. 2000).

Apart from immunization, transgenic microalgae could also be utilized in the production of other recombinant proteins for the aquaculture system. A study by Kim et al., in 2002 successfully produced *Chlorella ellipsoidea* that is able to espress fish growth hormone. Transgenic *C. ellipsoidea* producing flounder growth hormone was fed to brine shrimp and rotifer and then the zooplanktons were fed to the flounder fish, as the fish is strictly carnivorous. The study has shown that fish fed with zooplanktons (fed with transgenic *C. ellipsoidea*) showed an increase of 25% in total length and body weight compared with fish fed with zooplanktons (fed with wild-type *C. ellipsoidea*). This study demonstrates that the recombinant protein can be transferred up the food chain with functionality maintained.

Transgenic microalgae have also been studied to produce feed additives. *C. reinhartii* and *Dunaliella tertiolecta* were successfully genetically modified to produce an industrial enzyme called phytase, which degrades phytate, a form of phosphorus-bound protein naturally found in plants (Yoon et al. 2011; Georgianna et al. 2013).

Table 10.2 Major diseases occurring in the aquaculture sector with no vaccines available (Sommerset et al. 2005; Flegel et al. 2008)

Pathogen/Name of disease	Pathogen class	Major aquatic organism infected
Aeromonas salmonicida (Atypical disease)	Bacterial	Various fresh water and sea water species of fish
Flavobacterium branchiophilum (Bacterial gill disease)	Bacterial	Salmonids, carps and various fresh water fish
Flavobacterium psychrophilum (Rainbow trout fry syndrome disease)	Bacterial	Salmonids, fresh water fish
Edwardsiella ictaluri (Enteric septicaemia of catfish disease)	Bacterial	Catfish species
Edwardsiella tarda (Edwardsiella septicaemia disease)	Bacterial	Channel catfish, eel, Japanese flounder
Renibacterium salmoninarum (Bacterial kidney disease)	Bacterial	Salmonids
Photobacterium damsela subspecies *piscicida* (Pasteurellosis disease)	Bacterial	Sea bream/sea bass, amberjack/yellowtail fish
Vibrio spp. (vibriosis disease)	Bacterial	Grouper, various sea water fish, penaeid shrimps
Streptococcus iniae/*Streptococcus phocae* (Streptococciosis)	Bacterial	Tilapia, Asian sea bass, salmonids
Infectious pancreatic necrosis virus (IPNV) (Infectious pancreatic necrosis disease)	Virus	Salmonids. Various seawater species
Infectious hypodermal and hematopoietic necrosis virus (IHHNV) (Hypodermal and hematopoietic necrosis disease)	Virus	Various shrimps
Viral hemorrhagic septicaemia virus (VHSV) (Viral hemorrhagic septicaemia disease)	Virus	Rainbow and brown trout, turbot, Japanese flounder
Yellow head virus (YHV) (Yellow head virus disease)	Virus	Penaeid shrimps
Viral nervous necrosis virus (SJNNV) (Viral nervous necrosis disease)	Virus	Sea bass, groupers, barramundi, halibut
Channel catfish virus (CCV) (Channel catfish virus disease)	Virus	Channel catfish
Taura syndrome virus (TSV) (Red tail disease)	Virus	Penaeid shrimps
Spring viremia of carp virus (SVCV) (Spring viremia of carp disease)	Virus	Carp species
Paramoeba spp. (Amoebic gill disease)	Protists	Salmonids
Cryptobia salmositica (salmonid cryptobiosis disease)	Protists	Salmonids
Ichthyophthirius multifilis, *Cryptocaryon irritans*, *Trichodina* spp. (White spot disease)	Protists	Various fresh and sea water fish
Myxobolus cerebralis (Whirling disease)	Protists	Salmonids
Tetracapsula bryosalmonae (Proliferative kidney disease)	Protists	Salmonids
Lepeophtheirus salmonis, *Caligus* spp.	Protists	Salmonids and various marine fish

As previously mentioned, fishmeal, rich in amino acids and fatty acids, produced by drying and grinding fish, is a major source of feed in the aquaculture industry. However, the cost is high, and hence is less sustainable in the multi-million dollar aquaculture industry. Due to this, studies on alternative protein sources with lower cost like plants emerged in order to substitute fishmeal as the main feed in the aquaculture industry (Gatlin et al. 2007). However, most proteins that originate from plants like phytate are anti-nutritional to animals, as they are indigestible (Kumar et al. 2012). In this case, studies on transgenic microalgae expressing phytase could overcome this constraint, as an earlier study by Jackson et al. (1996), which utilized purified microbial phytase as feed additives in plant-based meal, has shown an increase in bioaccumulation of phosphorus by catfish specifically in bone and the decrease of phosphorus in fish faeces.

5 Future Perspectives and Recommendations

Transgenic microalgae show plenty of potential for numerous applications in the aquaculture sector. Furthermore, microalgae are a major part of aquaculture especially for molluscs feeding, as no substitution is possible yet. Due to these facts, microalgae will be continuously used in aquaculture and genetic modifications can enhance the quality of these microorganisms further. As previously mentioned, disease is a major constraint in the aquaculture industry. More studies should be emphasizing on the utilization of microalgae to improve disease and health management in the aquaculture sector, which will then reduce losses and increase profits for the aquaculture industry. Table 10.2 shows the available commercial vaccines for the major diseases, but more studies should be carried out in developing transgenic microalgae expressing vaccines against those diseases as an alternative preventive measure.

Advancement in biotechnology has created various ways and outcomes in genetic modification, and one of the current technologies in genetic modification is CRISPR/Cas9 system (clustered regularly interspaced short palindromic repeats (CRISPR)/CRISPR associated protein 9 (Cas9) (Doudna and Charpentier 2014). The advantage of this system is that it could modify the genome sequence through deletion or insertion at specific sites in the genome. Another advantage of using this new technology is that it could introduce up to six genetic modifications simultaneously, as it was successfully conducted in yeast (Mans et al. 2018). Furthermore, *C. reinhardtii, Phaeodactylum tricornutum* and *Nannochloropsis* spp. have been successfully transformed with CRISPR/Cas9 system for gene deletion purposes (Shin et al. 2016; Nymark et al. 2016; Wang et al. 2016). Future studies should consider multiple genetic modifications on microalgae, especially the species that are utilized in aquaculture to express more than one recombinant protein or a combination of metabolic engineering for enhancing accumulation of high-value biomolecules with the ability to express recombinant protein for immunization or feed additives. The utilization of transgenic microalgae in other sectors apart from aqua-

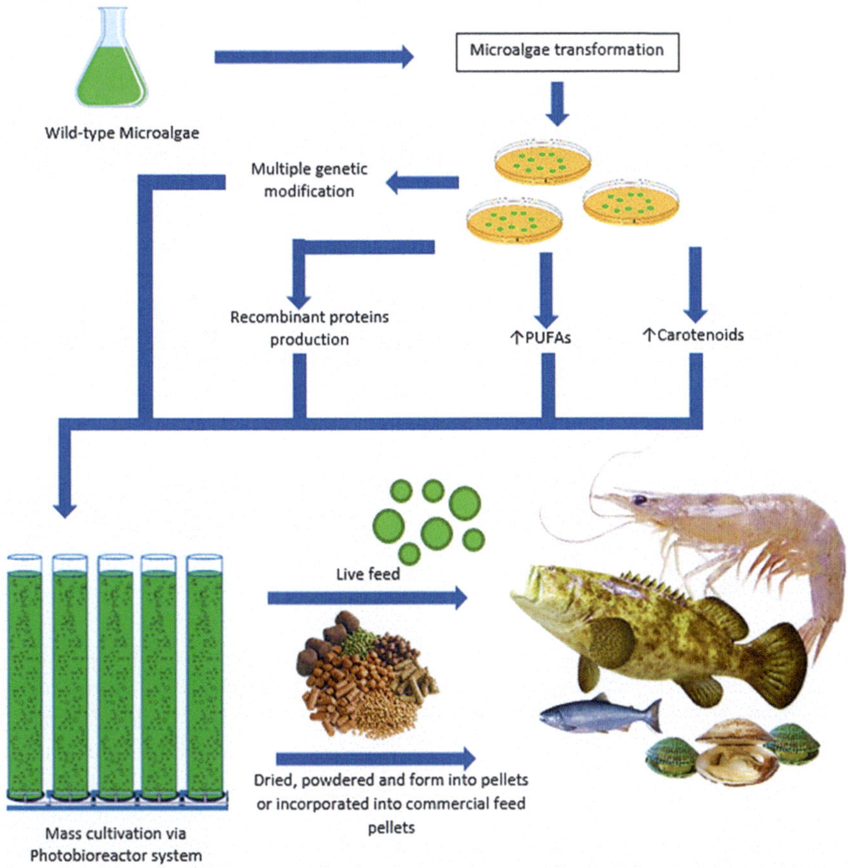

Fig. 10.2 Overall workflow of the transgenic microalgae potential and application in aquaculture system

culture was deeply reviewed by Gong et al. (2011) and Rasala and Mayfield (2015). Figure 10.2 summarizes the overview of the idea of the utilization of transgenic microalgae in the aquaculture industry.

Apart from that, full genome sequencing of microalgae species utilized as feed in the aquaculture industry should be considered, as it is a vital step in the understanding and initiation of successful genetic modifications. Currently, the genome databases are readily available for microalgae such as *Chlamydomonas reinhardtii*, *Nephroselmis olivacea*, *Chaetosphaeridium globosum*, *Chlorella vulgaris*, *Mesostigma viride*, *Guillardia theta*, *Odontella sinensis*, *Cyanophora paradoxal*, *Cyanidium caldarium* and *Euglena gracilis* (Turmel et al. 1999; Turmel et al. 2002; Wakasugi et al. 1997; Lemieux et al. 2000; Douglas et al. 2001; Tada et al. 1999; Chu et al. 2004; Stirewalt et al. 1995; Glöckner et al. 2000; Hallick et al. 1993). More highly utilized species must be well characterized, as it will steer the genetic modifications for various purposes.

6 Conclusion

In summary, what have been accomplished with higher plants and bacteria in terms of over expression of biomolecules and the production of recombinant proteins are applicable to microalgae as well since it has been proven that microalgae are one of the most adaptable and versatile organisms which have really high potential. The utilization of genetic modifications on microalgae should be explored and exploited further for various purposes, with aquaculture being one of the priorities.

Acknowledgements The authors gratefully acknowledge the Higher Institution Centre of Excellence (HICOE) Research Grant (Innovative Vaccines and Therapeutics against Fish Diseases) (Project No. 6369100) and SATREPS (JICA-JST): COSMOS-MOHE G4-B Research Grant (Microalgae for Sustainable Aquaculture Health: Microalgae Vaccine Delivery System) (Project No. 6300866) for the funds to carry out this research.

References

Alam, M. A., & Wang, Z. (Eds.). (2019). *Microalgae biotechnology for development of biofuel and wastewater treatment* (pp. 1–655). Singapore: Springer.

Álvarez, R., Vaz, B., Gronemeyer, H., & de Lera, A. R. (2014). Functions, therapeutic applications, and synthesis of retinoids and carotenoids. *Chemical Reviews, 114*(1), 1–125.

Asad, S., & Arshad, M. (2011). Silicon carbide whisker-mediated plant transformation. In *Properties and applications of silicon carbide* (pp. 345–359). Rijeka: IntechOpen.

Azim, N. H., Subki, A., & Yusof, Z. N. B. (2018). Abiotic stresses induce total phenolic, total flavonoid and antioxidant properties in Malaysian indigenous microalgae and cyanobacterium. *Malaysian Journal of Microbiology, 14*(1), 25–33.

Ball, S. G. (2005). Eukaryotic microalgae genomics. The essence of being a plant. *Plant Physiology, 137*(2), 397–398.

Bjerkeng, B. (2008). Carotenoids in aquaculture: Fish and crustaceans. In *Carotenoids* (pp. 237–254). Basel: Birkhäuser.

Blount, J. D., & McGraw, K. J. (2008). Signal functions of carotenoid colouration. In G. Britton, S. Liaaen-Jensen, & H. Pfander (Eds.), *Carotenoids: Vol. 4: Natural functions* (pp. 213–236). Basel: Birkhäuser.

Brown, M. R. (2002). Nutritional value and use of microalgae in aquaculture. In: *Avances en Nutrición Acuícola VI. Memorias del VI Simposium Internacional de Nutrición Acuícola*, 3 (pp. 281–292).

Brown, M. R., Mular, M., Miller, I., Farmer, C., & Trenerry, C. (1999). The vitamin content of microalgae used in aquaculture. *Journal of Applied Phycology, 11*(3), 247–255.

Carvalho, A. P., Meireles, L. A., & Malcata, F. X. (2006). Microalgal reactors: A review of enclosed system designs and performances. *Biotechnology Progress, 22*(6), 1490–1506.

Cha, T. S., Chen, C. F., Yee, W., Aziz, A., & Loh, S. H. (2011). Cinnamic acid, coumarin and vanillin: Alternative phenolic compounds for efficient Agrobacterium-mediated transformation of the unicellular green alga, *Nannochloropsis* sp. *Journal of Microbiological Methods, 84*(3), 430–434.

Chai, X. J., Chen, H. X., Xu, W. Q., & Xu, Y. W. (2013). Expression of soybean Kunitz trypsin inhibitor gene SKTI in *Dunaliellasalina*. *Journal of Applied Phycology, 25*(1), 139–144.

Chen, H. L., Li, S. S., Huang, R., & Tsai, H. J. (2008). Conditional production of a functional fish growth hormone in the transgenic line of *Nannochloropsis oculata* (Eustigmatophyceae) 1. *Journal of Phycology, 44*(3), 768–776.

Cheng, R., Ma, R., Li, K., Rong, H., Lin, X., Wang, Z., Yang, S., & Ma, Y. (2012). Agrobacterium tumefaciens mediated transformation of marine microalgae *Schizochytrium*. *Microbiological Research, 167*(3), 179–186.

Chow, K. C., & Tung, W. L. (1999). Electrotransformation of *Chlorella vulgaris*. *Plant Cell Reports, 18*(9), 778–780.

Chu, K. H., Qi, J., Yu, Z. G., & Anh, V. O. (2004). Origin and phylogeny of chloroplasts revealed by a simple correlation analysis of complete genomes. *Molecular Biology and Evolution, 21*(1), 200–206.

Coll, J. M. (2006). Methodologies for transferring DNA into eukaryotic microalgae: A review. *Spanish Journal of Agricultural Research, 4*(4), 316–330.

Corchero, J. L., Gasser, B., Resina, D., Smith, W., Parrilli, E., Vázquez, F., Abasolo, I., Giuliani, M., Jäntti, J., Ferrer, P., & Saloheimo, M. (2013). Unconventional microbial systems for the cost-efficient production of high-quality protein therapeutics. *Biotechnology Advances, 31*(2), 140–153.

Costanzo, M. C., & Fox, T. D. (1988). Transformation of yeast by agitation with glass beads. *Genetics, 120*(3), 667–670.

Couso, I., Vila, M., Rodriguez, H., Vargas, M. A., & Leon, R. (2011). Overexpression of an exogenous phytoene synthase gene in the unicellular alga *Chlamydomonas reinhardtii* leads to an increase in the content of carotenoids. *Biotechnology Progress, 27*(1), 54–60.

Cui, Y., Jiang, P., Wang, J., Li, F., Chen, Y., Zheng, G., & Qin, S. (2012). Genetic transformation of Platymonas (*Tetraselmis*) subcordiformis (*Prasinophyceae, Chlorophyta*) using particle bombardment and glass bead agitation. *Chinese Journal of Oceanology and Limnology, 30*(3), 471–475.

Cutzu, R., Coi, A., Rosso, F., Bardi, L., Ciani, M., Budroni, M., Zara, G., Zara, S., & Mannazzu, I. (2013). From crude glycerol to carotenoids by using a *Rhodotorula glutinis* mutant. *World Journal of Microbiology and Biotechnology, 29*(6), 1009–1017.

Dehesh, K., Tai, H., Edwards, P., Byrne, J., & Jaworski, J. G. (2001). Overexpression of 3-ketoacyl-acyl-carrier protein synthase IIIs in plants reduces the rate of lipid synthesis. *Plant Physiology, 125*(2), 1103–1114.

Demain, A. L., & Vaishnav, P. (2009). Production of recombinant proteins by microbes and higher organisms. *Biotechnology Advances, 27*, 297–306.

Dong, B., Cheng, R. Q., Liu, Q. Y., Wang, J., & Fan, Z. C. (2018). Multimer of the antimicrobial peptide Mytichitin-A expressed in *Chlamydomonas reinhardtii* exerts a broader antibacterial spectrum and increased potency. *Journal of Bioscience and Bioengineering, 125*(2), 175–179.

Doron, L., Segal, N. A., & Shapira, M. (2016). Transgene expression in microalgae from tools to applications. *Frontiers in Plant Science, 7*, 505.

Doudna, J. A., & Charpentier, E. (2014). The new frontier of genome engineering with CRISPR-Cas9. *Science, 346*(6213), 1258096.

Douglas, S., Zauner, S., Fraunholz, M., Beaton, M., Penny, S., Deng, L. T., Wu, X., Reith, M., Cavalier-Smith, T., & Maier, U. G. (2001). The highly reduced genome of an enslaved algal nucleus. *Nature, 410*(6832), 1091.

Dunahay, T. G. (1993). Transformation of *Chlamydomonas reinhardtii* with silicon carbide whiskers. *Biotechnology Techniques, 15*(3), 452–460.

Ebenezer, V., Medlin, L. K., & Kei, J. S. (2012). Molecular detection, quantification, and diversity evaluation of microalgae. *Marine Biotechnology, 14*(2), 129–142.

Endo, H., Yoshida, M., Uji, T., Saga, N., Inoue, K., & Nagasawa, H. (2016). Stable nuclear transformation system for the Coccolithophorid alga *Pleurochrysis carterae*. *Scientific Reports, 6*, 22252.

Feng, S., Feng, W., Zhao, L., Gu, H., Li, Q., Shi, K., Guo, S., & Zhang, N. (2014). Preparation of transgenic *Dunaliella salina* for immunization against white spot syndrome virus in crayfish. *Archives of Virology, 159*(3), 519–525.

Fern, L. L., Abidin, A. A. Z., & Yusof, Z. N. B. (2017). Upregulation of thiamine (vitamin B1) biosynthesis gene upon stress application in *Anabaena* sp. and *Nannochloropsis oculata*. *Journal of Plant Biotechnology, 44*(4), 462–471.

Ferrer-Miralles, N., & Villaverde, A. (2013). Bacterial cell factories for recombinant protein production; expanding the catalogue. *Microbial Cell Factories, 12*, 113.

Fincham, J. R. (1989). Transformation in fungi. *Microbiological Reviews, 53*(1), 148–170.

Flegel, T. W., Lightner, D. V., Lo, C. F., & Owens, L. (2008). Shrimp disease control: Past, present and future. In *Diseases in Asian aquaculture VI* (pp. 355–378). Manila, Philippines: Fish Health Section, Asian Fisheries Society.

Franconi, R., Demurtas, O. C., & Massa, S. (2010). Plant-derived vaccines and other therapeutics produced in contained systems. *Expert Review of Vaccines, 9*, 877–892.

Gangl, D., Zedler, J. A. Z., Rajakumar, P. D., Martinez, E. M. R., Riseley, A., & Włodarczyk, A. (2015). Biotechnological exploitation of microalgae. *Journal of Experimental Botany, 66*(22), 6975–6990.

Gatlin, D. M., Barrows, F. T., Brown, P., Dabrowski, K., Gaylord, G. T., Hardy, R. W., Herman, E., Hu, G., Krogdahl, Å., Nelson, R., Overturf, K., Rust, M., Sealey, W., Skonberg, D., Souza, E. J., Stone, D., Wilson, R., & Wurtele, E. (2007). Expanding the utilization of sustainable plant products in aquafeeds: A review. *Aquaculture Research, 38*, 551–579.

Georgianna, D. R., Hannon, M. J., Marcuschi, M., Wu, S., Botsch, K., Lewis, A. J., Hyun, J., Mendez, M., & Mayfield, S. P. (2013). Production of recombinant enzymes in the marine alga *Dunaliellatertiolecta*. *Algal Research, 2*(1), 2–9.

Glöckner, G., Rosenthal, A., & Valentin, K. (2000). The structure and gene repertoire of an ancient red algal plastid genome. *Journal of Molecular Evolution, 51*(4), 382–390.

Gong, Y., Hu, H., Gao, Y., Xu, X., & Gao, H. (2011). Microalgae as platforms for production of recombinant proteins and valuable compounds: Progress and prospects. *Journal of Industrial Microbiology & Biotechnology, 38*(12), 1879–1890.

Goodwin, T. W. (1984). *The biochemistry of the carotenoids. Vol. 2: Animals* (2nd ed.). London: Chapman and Hall.

Hallick, R. B., Hong, L., Drager, R. G., Favreau, M. R., Monfort, A., Orsat, B., Spielmann, A., & Stutz, E. (1993). Complete sequence of *Euglena gracilis* chloroplast DNA. *Nucleic Acids Research, 21*(15), 3537–3544.

He, D. M., Qian, K. X., Shen, G. F., Zhang, Z. F., Yi-Nü, L. I., Su, Z. L., & Shao, H. B. (2007). Recombination and expression of classical swine fever virus (CSFV) structural protein E2 gene in *Chlamydomonas reinhardtii* chroloplasts. *Colloids and Surfaces B: Biointerfaces, 55*(1), 26–30.

He, Y., Peng, H., Liu, J., Chen, F., Zhou, Y., Ma, X., & Wang, K. (2018). *Chlorella* sp. transgenic with Scy-hepc enhancing the survival of *Sparus macrocephalus* and hybrid grouper challenged with *Aeromonas hydrophila*. *Fish & Shellfish Immunology, 73*, 22–29.

Heiser, W. (1992). *Optimization of biolistic transformation using the helium-driven PDS-1000/He system*. BIO-RADUS/EG Bulletin 8–8.

Hempel, F., Lau, J., Klingl, A., & Maier, U. G. (2011). Algae as protein factories: Expression of a human antibody and the respective antigen in the diatom *Phaeodactylum tricornutum*. *PLoS One, 6*(12), e28424.

Hou, Q., Qiu, S., Liu, Q., Tian, J., Hu, Z., & Ni, J. (2013). Selenoprotein-transgenic *Chlamydomonas reinhardtii*. *Nutrients, 5*(3), 624–636.

Jackson, L. S., Li, M. H., & Robinson, E. H. (1996). Use of microbial phytase in channel catfish Ictalurus punctatus diets to improve utilization of phytate phosphorus 1. *Journal of the World Aquaculture Society, 27*(3), 309–313.

Kathiresan, S., Chandrashekar, A., Ravishankar, G. A., & Sarada, R. (2009). Agrobacterium-mediated transformation in the green alga *Haematococcus pluvialis* (Chlorophyceae, Volvocales) 1. *Journal of Phycology, 45*(3), 642–649.

Kathiresan, S., Chandrashekar, A., Ravishankar, G. A., & Sarada, R. (2015). Regulation of astaxanthin and its intermediates through cloning and genetic transformation of β-carotene ketolase in *Haematococcuspluvialis*. *Journal of Biotechnology, 196–197*, 33–41.

Kim, D. K., Kim, K. D., Seo, J. Y., & Lee, S. M. (2012). Effects of dietary lipid source and level on growth performance, blood parameters and flesh quality of sub-adult olive flounder (*Paralichthys olivaceus*). *Asian-Australasian Journal of Animal Sciences, 25*(6), 869.

Kim, S., Lee, Y. C., Cho, D. H., Lee, H. U., Huh, Y. S., Kim, G. J., & Kim, H. S. (2014). A simple and non-invasive method for nuclear transformation of intact-walled *Chlamydomonas reinhardtii*. *PLoS One, 9*(7), e101018.

Kindle, K. L., Schnell, R. A., Fernandez, E., & Lefebvre, P. A. (1989). Stable nuclear transformation of *Chlamydomonas* using the *Chlamydomonas* gene for nitrate reductase. *Journal of Cell Biology, 109*(6), 2589–2601.

Knauer, J., Barrett, S. M., Volkman, J. K., & Southgate, P. C. (1999). Assimilation of dietary phytosterols by Pacific oyster *Crassostrea gigas*spat. *Aquaculture Nutrition, 5*, 257–266.

Kumar, V., Sinha, A. K., Makkar, H. P. S., De Boeck, G., & Becker, K. (2012). Phytate and phytase in fish nutrition. *Journal of Animal Physiology and Animal Nutrition, 96*(3), 335–364.

Kuwano, T., Shirataki, C., & Itoh, Y. (2008). Comparison between polyethylene glycol- and polyethylenimine-mediated transformation of *Aspergillus nidulans*. *Current Genetics, 54*(2), 95–103.

Lardizabal, K. D., Thompson, G. A., & Hawkins, D. (2006). Diacylglycerol acyl transferase proteins. United States. Patent, & Trademark Office. *Official Gazette of the United States Patent and Trademark Office: Patents* (Vol. 1243, No. 3). US Department of Commerce, Patent and Trademark Office.

Lau, C. C., Loh, S. H., Aziz, A., & Cha, T. S. (2017). Effects of disrupted omega-3 desaturase gene construct on fatty acid composition and expression of four fatty acid biosynthetic genes in transgenic *Chlorella vulgaris*. *Algal Research, 26*, 143–152.

Lee, Y. K. (1997). Commercial production of microalgae in the Asia-Pacific rim. *Journal of Applied Phycology, 9*(5), 403–411.

Lemieux, C., Otis, C., & Turmel, M. (2000). Ancestral chloroplast genome in *Mesostigmaviride* reveals an early branch of green plant evolution. *Nature, 403*(6770), 649.

León, R., & Fernández, E. (2007). Nuclear transformation of eukaryotic microalgae: Historical overview, achievements and problems. *Advances in Experimental Medicine and Biology, 616*, 1–11.

León-Bañares, R., González-Ballester, D., Galván, A., & Fernández, E. (2004). Transgenic microalgae as green cell-factories. *Trends in Biotechnology, 22*(1), 45–52.

Li, Z., Wakao, S., Fischer, B. B., & Niyogi, K. K. (2009). Sensing and responding to excess light. *Annual Review of Plant Biology, 60*, 239–260.

Liu, J., Gerken, H., Huang, J., & Chen, F. (2013). Engineering of an endogenous phytoene desaturase gene as a dominant selectable marker for *Chlamydomonas reinhardtii* transformation and enhanced biosynthesis of carotenoids. *Process Biochemistry, 48*(5–6), 788–795.

Liu, J., Sun, Z., Gerken, H., Huang, J., Jiang, Y., & Chen, F. (2014). Genetic engineering of the green alga *Chlorella zofingiensis*: A modified norflurazon-resistant phytoene desaturase gene as a dominant selectable marker. *Applied Microbiology and Biotechnology, 98*(11), 5069–5079.

Lorenz, R. T., & Cysewski, G. R. (2000). Commercial potential for *Haematococcus*microalgae as a natural source of astaxanthin. *Trends in Biotechnology, 18*(4), 160–167.

Lu, Y., Tarkowska, D., Turečková, V., Luo, T., Xin, Y., Li, J., & Xu, J. (2014). Antagonistic roles of abscisic acid and cytokinin during response to nitrogen depletion in oleaginous microalga *Nannochloropsis oceanica* expand the evolutionary breadth of phytohormone function. *Plant Journal, 80*(1), 52–68.

Ma, X., Pan, K., Zhang, L., Zhu, B., Yang, G., & Zhang, X. (2016). Genetic transformation of *Nannochloropsis oculata* with a bacterial phleomycin resistance gene as dominant selective marker. *Journal of Ocean University of China, 15*(2), 351–356.

Mans, R., Wijsman, M., Daran-Lapujade, P., & Daran, J. M. (2018). A protocol for introduction of multiple genetic modifications in *Saccharomyces cerevisiae* using CRISPR/Cas9. *FEMS Yeast Research, 18*(7), foy063.

Manuell, A. L., Beligni, M. V., Elder, J. H., Siefker, D. T., Tran, M., Weber, A., McDonald, T. L., & Mayfield, S. P. (2007). Robust expression of a bioactive mammalian protein in *Chlamydomonas* chloroplast. *Plant Biotechnology Journal, 5*(3), 402–412.

Mathur, J., & Koncz, C. (1998). PEG-mediated protoplast transformation with naked DNA. *Methods in Molecular Biology, 82*, 267–276.

Bacellar Mendes, L. B., & Vermelho, A. B. (2013). Allelopathy as a potential strategy to improve microalgae cultivation. *Biotechnology for Biofuels, 6*(1), 152.

Meyer, F. P. (1991). Aquaculture disease and health management. *Journal of Animal Science, 69*(10), 4201–4208.

Meyers, B., Zaltsman, A., Lacroix, B., Kozlovsky, S. V., & Krichevsky, A. (2010). Nuclear and plastid genetic engineering of plants: Comparison of opportunities and challenges. *Biotechnology Advances, 28*(6), 747–756.

Miyagawa, Y., Tamoi, M., & Shigeoka, S. (2001). Overexpression of a cyanobacterial fructose-1,6-sedoheptulose-1,7-bisphosphatase in tobacco enhances photosynthesis and growth. *Nature Biotechnology, 19*(10), 965–969.

Nakanishi, T., & Ototake, M. (1997). Antigen uptake and immune responses after immersion vaccination. *Developments in Biological Standardization, 90*, 59–68.

Nikolau, B. J., Ohlrogge, J. B., & Wurtele, E. S. (2003). Plant biotin-containing carboxylases. *Archives of Biochemistry and Biophysics, 414*(2), 211–222.

Nymark, M., Sharma, A. K., Sparstad, T., Bones, A. M., & Winge, P. (2016). A CRISPR/Cas9 system adapted for gene editing in marine algae. *Scientific Reports, 6*, 24951.

Popper, Z. A., & Tuohy, M. G. (2010). Beyond the green: Understanding the evolutionary puzzle of plant and algal cell walls. *Plant Physiology, 153*(2), 373–383.

Posewitz, M. C., Smolinski, S. L., Kanakagiri, S., Melis, A., Seibert, M., & Ghirardi, M. L. (2004). Hydrogen photoproduction is attenuated by disruption of an isoamylase gene in Chlamydomonas reinhardtii. *The Plant Cell, 16*(8), 2151–2163.

Potrykus, I. (1991). Gene transfer to plants: Assessment of published approaches and results. *Annual Review of Plant Physiology and Plant Molecular Biology, 42*, 205–225.

Qin, S., Lin, H., & Jiang, P. (2012). Advances in genetic engineering of marine algae. *Biotechnology Advances, 30*(6), 1602–1613.

Quesada, A., Galvan, A., & Fernandez, E. (1994). Identification of nitrate transporter genes in *Chlamydomonas reinhardtii. The Plant Journal, 5*(3), 407–419.

Radakovits, R., Jinkerson, R. E., Darzins, A., & Posewitz, M. C. (2010). Genetic engineering of algae for enhanced biofuel production. *Eukaryotic Cell, 9*(4), 486–501.

Rasala, B. A., Lee, P. A., Shen, Z., Briggs, S. P., Mendez, M., & Mayfield, S. P. (2012). Robust expression and secretion of Xylanase1 in *Chlamydomonas reinhardtii* by fusion to a selection gene and processing with the FMDV 2A peptide. *PLoS One, 7*(8), e43349.

Rasala, B. A., & Mayfield, S. P. (2015). Photosynthetic biomanufacturing in green algae; production of recombinant proteins for industrial, nutritional, and medical uses. *Photosynthesis Research, 123*(3), 227–239.

Rasala, B. A., Muto, M., Lee, P. A., Jager, M., Cardoso, R. M., Behnke, C. A., Kirk, P., Hokanson, C. A., Crea, R., Mendez, M., & Mayfield, S. P. (2010). Production of therapeutic proteins in algae, analysis of expression of seven human proteins in the chloroplast of *Chlamydomonas reinhardtii. Plant Biotechnology Journal, 8*(6), 719–733.

San Cha, T., Yee, W., & Aziz, A. (2012). Assessment of factors affecting Agrobacterium-mediated genetic transformation of the unicellular green alga, *Chlorella vulgaris. World Journal of Microbiology and Biotechnology, 28*(4), 1771–1779.

Shin, S. E., Lim, J. M., Koh, H. G., Kim, E. K., Kang, N. K., Jeon, S., Kwon, S., Shin, W. S., Lee, B., Hwangbo, K., & Kim, J. (2016). CRISPR/Cas9-induced knockout and knock-in mutations in Chlamydomonas reinhardtii. *Scientific Reports, 6*, 27810.

Sies, H., & Stahl, W. (1997). Carotenoids and intercellular communication via gap junctions. *International Journal for Vitamin and Nutrition Research. Internationale Zeitschrift fur Vitamin-und Ernahrungsforschung. Journal international de vitaminologie et de nutrition, 67*(5), 364–367.

Sizova, I., Greiner, A., Awasthi, M., Kateriya, S., & Hegemann, P. (2013). Nuclear gene targeting in *Chlamydomonas* using engineered zinc-finger nucleases. *Plant Journal, 73*(5), 873–882.

Sommerset, I., Krossøy, B., Biering, E., & Frost, P. (2005). Vaccines for fish in aquaculture. *Expert Review of Vaccines, 4*(1), 89–101.

Specht, E. A., & Mayfield, S. P. (2014). Algae-based oral recombinant vaccines. *Frontiers in Microbiology, 5*, 60.

Steinbrenner, J., & Sandmann, G. (2006). Transformation of the green alga *Haematococcuspluvialis* with a phytoene desaturase for accelerated astaxanthin biosynthesis. *Applied and Environmental Microbiology, 72*, 7477–7484.

Stirewalt, V. L., Michalowski, C. B., Löffelhardt, W., Bohnert, H. J., & Bryant, D. A. (1995). Nucleotide sequence of the cyanelle genome from *Cyanophora paradoxa*. *Plant Molecular Biology Reporter, 13*(4), 327–332.

Stournas, S., Lois, E., & Serdari, A. (1995). Effects of fatty acid derivatives on the ignition quality and cold flow of diesel fuel. *Journal of the American Oil Chemists' Society, 72*(4), 433–437.

Subasinghe, R. P., Arthur, J. R., Phillips, M. J., & Reantoso, M. B. (2000). *Thematic review on management strategies for major diseases in shrimp aquaculture*. Cebu, Philippines: FAO, UN.

Sun, M., Qian, K., Su, N., Chang, H., Liu, J., & Shen, G. (2003). Foot-and-mouth disease virus VP1 protein fused with cholera toxin B subunit expressed in *Chlamydomonas reinhardtii* chloroplast. *Biotechnology Letters, 25*(13), 1087–1092.

Surzycki, R., Greenham, K., Kitayama, K., Dibal, F., Wagner, R., Rochaix, J. D., Ajam, T., & Surzycki, S. (2009). Factors effecting expression of vaccines in microalgae. *Biologicals, 37*(3), 133–138.

Tacon, A. G., & Metian, M. (2009). Fishing for feed or fishing for food: Increasing global competition for small pelagic forage fish. *Ambio, 38*, 294–302.

Tada, N., Shibata, S., Otsuka, S., Namba, K., & Oyaizu, H. (1999). Comparison of gene arrangements of chloroplasts between two centric diatoms, *Skeletonema costatum* and *Odontella sinensis*. *DNA Sequence, 10*(4–5), 343–347.

Te, M. R., & Miller, D. J. (1998). Genetic transformation of dinoflagellates (*Amphidinium* and *Symbiodinium*): Expression of GUS in microalgae using heterologous promoter constructs. *The Plant Journal, 13*(3), 427–435.

Turchini, G. M., Ng, W. K., & Tocher, D. R. (2010). *Fish oil replacement and alternative lipid sources in aquaculture feeds*. Boca Raton, FL: CRC Press.

Turmel, M., Otis, C., & Lemieux, C. (1999). The complete chloroplast DNA sequence of the green alga *Nephroselmisolivacea*: Insights into the architecture of ancestral chloroplast genomes. *Proceedings of the National Academy of Sciences of the United States of America, 96*(18), 10248–10253.

Turmel, M., Otis, C., & Lemieux, C. (2002). The chloroplast and mitochondrial genome sequences of the charophyte Chaetosphaeridium globosum: insights into the timing of the events that restructured organelle DNAs within the green algal lineage that led to land plants. *Proceedings of the National Academy of Sciences of the United States of America, 99*(17), 11275–11280.

Vo, H. N. P., Ngo, H. H., Guo, W., Nguyen, T. M. H., Liu, Y., Liu, Y., Nguyen, D. D., & Chang, S. W. (2018). A critical review on designs and applications of microalgae-based photobioreactors for pollutants treatment. *Science of the Total Environment, 651*(1), 1549–1568.

Wakasugi, T., Nagai, T., Kapoor, M., Sugita, M., Ito, M., Ito, S., Tsudzuki, J., Nakashima, K., Tsudzuki, T., Suzuki, Y., & Hamada, A. (1997). Complete nucleotide sequence of the chloroplast genome from the green alga *Chlorella vulgaris*: The existence of genes possibly involved in chloroplast division. *Proceedings of the National Academy of Sciences of the United States of America, 94*(11), 5967–5972.

Walker, T. L., Purton, S., Becker, D. K., & Collet, C. (2005). Microalgae as bioreactors. *Plant Cell Reports, 24*(11), 629–641.

Wang, Q., Lu, Y., Xin, Y., Wei, L., Huang, S., & Xu, J. (2016). Genome editing of model oleaginous microalgae Nannochloropsis spp. by CRISPR/Cas9. *The Plant Journal, 88*(6), 1071–1081.

Yang, B., Liu, J., Liu, B., Sun, P., Ma, X., Jiang, Y., Wei, D., & Chen, F. (2015). Development of a stable genetic system for *Chlorella vul*garis—A promising green alga for CO_2biomitigation. *Algal Research, 12*, 134–141.

Yang, Y. G., Shariff, M., Lee, L. K., & Hassan, M. D. (1999, November). Malaysia: national review on management strategies for major diseases in shrimp aquaculture. In *WB/NACA/WWF/FAO. Thematic review on management strategies for major diseases in shrimp aquaculture. Proceedings of a workshop held in Cebu, Philippines on* (pp. 28–30).

Yoon, S. M., Kim, S. Y., Li, K. F., Yoon, B. H., Choe, S., & Kuo, M. M. C. (2011). Transgenic microalgae expressing *Escherichia coli* AppA phytase as feed additive to reduce phytate excretion in the manure of young broiler chicks. *Applied Microbiology and Biotechnology, 91*(3), 553–563.

Zabawinski, C., Van Den Koornhuyse, N., D'Hulst, C., Schlichting, R., Giersch, C., Delrue, B., Lacroix, J. M., Preiss, J., & Ball, S. (2001). Starchless mutants of *Chlamydomonas reinhardtii* lack the small subunit of a heterotetrameric ADP-glucose pyrophosphorylase. *Journal of Bacteriology, 183*(3), 1069–1077.

Chapter 11
Microalgae as Sustainable Producers of Bioplastic

D. Tharani and Muthusamy Ananthasubramanian

Abstract Extensive exploitation of nonrenewable resources for various products has showed its undesirable consequence on the environment following its limited availability. Plastics have impacted the environment in a negative way polluting terrestrial and marine life to a great extent. Biodegradable products from renewable resources are the possible alternative solution. Bioplastics are one such promising replacements with characteristics similar to fossil fuel-derived polymers, with increased biodegradability and blending property. Bacterial and algal systems accumulate polyhydroxyalkanoates (PHA) as part of their metabolic processes depending on the availability of carbon source. PHA from bacterial systems has proved to be efficient in accumulation, but their commercialization is challenging due to the high cost involved. Bacterial systems demand critical process parameters which makes scaling-up expensive. The shortcomings of bacterial PHA commercialization can be overcome by employing microalgal biomass that accumulates PHA. The versatility in carbon source utilization enables cultivation on different resources. This eliminates the dependence on single substrate, thus aiding in mixed growth population. Current studies indicate an accumulation of 27% PHA by *Chlorella pyrenoidosa*. Taking blending properties into consideration, the whole biomass of *Chlorella* to glycerol (4:1) showed improved plasticity with polyolefins. Interestingly, direct incorporation of 50% *Spirulina platensis* biomass into polyolefin shows properties comparable to petroleum-derived plastics. Further developments in the strain could be achieved through metabolic engineering by directing flux toward PHA accumulation. But the maintenance of genetically modified microalgae for continuous production is critical, taking the doubling time of the organism into account. In order to have a sustainable bioplastic recovery from microalgal biomass, an integrated biorefinery approach should be adopted.

Keywords *Chlorella* · *Spirulina* · Bioplastic · Microalgae · Polyhydroxyalkanoates · Sustainability · Wastewater

D. Tharani · M. Ananthasubramanian (✉)
Department of Biotechnology, PSG College of Technology, Coimbatore, India
e-mail: mas.bio@psgtech.ac.in

M. A. Alam et al. (eds.), *Microalgae Biotechnology for Food, Health and High Value Products*, https://doi.org/10.1007/978-981-15-0169-2_11

1 Introduction

Plastics derived from fossil fuels are versatile polymers used in day-to-day life of humans. The physical and mechanical properties of plastics along with their resistance to hydrolysis puts them into flexible, fine tailored polymer material. Today, a majority of plastic products are disposable. End-users view them as economical, user-friendly, and tactile to be used in electronic devices, household products, and even in medical appliances. Being a highly advantageous material, petroleum-based plastics account for only 4% fossil fuel derivatives (British Plastics Federation 2019). The drastic increase in plastic production from 0.5 million tons to 260 million tons in six decades shows its diversity in application (Hopewell et al. 2009). If plastic production continues at the current rate, it is estimated to use up 20% of fossil fuel in 2050 (Plastic Pollution Coalition 2017). Used plastic materials find their way to either landfilling sites or incinerators (Hopewell et al. 2009). In addition to plastic disposal, the additives used in plastic materials pose a major threat to the environment. Bisphenol-A (BPA) and phthalates are commonly used in polycarbonate plastics, polyvinyl compounds, and in epoxy resins. These compounds can leach out in landfilling sites becoming a threat through biomagnification. Inside the human body, BPA mimics estrogen, while phthalates act similar to anti-androgenic compounds. Identified as endocrine disruptors, they have a direct link in altering the homeostasis. These additives play a role ranging from obesity, reproductive problems, to cancer (Rawsthorne et al. 2011; Lind et al. 2011; Manikkam et al. 2013).

Aquatic life is also at risk due to plastic-toxicity and additive leachates. The lesser toxic Bisphenol-S (BPS) has shown detrimental effect on the biotransformation pathways in the marine organism *Chironomusriparius* (Dong et al. 2018; Herrero et al. 2018). Both physical entrapment and ingestion affect the marine organism, negatively impacting the eco-system. The persistence of plastics in aquatic system leads to fragmentation with time, forming microplastics. Microplastics are 100 nm–5 mm particles or fibers of synthetic and semi-synthetic origin, which are easily taken up by marine organism during predation. The incidence of ingestion rate varies with location and natural availability. Studies on biomagnification of organic pollutants from microplastics (Steer et al. 2017; Bessa et al. 2018) emphasize the need for alternatives to conventional petroleum-derived nonbiodegradable plastics.

To eliminate the dependency on fossil fuels, many biomass-derived nonbiodegradable plastics are being commercialized. One breakthrough in biomass-derived bioplastics is polythioesters (Bögershausen et al. 2002). Though formed through bacterial fermentation, they are nonbiodegradable. The starting materials for polyethylene terephthalate (PET) are obtained from renewable sources followed by chemical conversion (Wang and Tong 2016). With added materialistic properties, PET, polystyrene (PS), and polypropylene (PP) have broader market range. Differing only in the raw material source, these nonbiodegradable bioplastics occupy nearly 80% of the total bioplastic market. Packaging industry is the major trendsetter in driving the commercialization of bioplastic (European Bioplastics, 15th April, 2019). The main advantage of nonbiodegradable bioplastic is its production from renewable resources with lesser carbon footprint when compared to petroleum-derived plastics. (Fig. 11.1) Although resourceful, their usage adds to the negative impact on the environment due to their persistence to biodegradation. To overcome this drawback, new chemical modifications to the polymers, enzyme-based degradation, thermal treatment, and photosensitive degradation behaviors of PS, PP, and PET are studied. Implementing these strategies for individual polymers in day-to-day life is complicated. Segregation of wastes, especially plastic, at the source of generation is still a challenging task in highly populated cities. As a replacement, biodegradable biopolymers are gaining significant insights recently. PLAs, PHAs, and a variety of blends have shown promising durability as successful biopolymers with biodegradability.

Polyhydroxyalkanoates are biological polyesters stored as energy reserve during microbial metabolism. They are produced by a wide variety of microbes under nutrient limiting conditions (Anderson and Dawes 1990; Müller and Seebach 1993). They are categorized as biodegradable bioplastics, requiring limited amendments in properties before commercial implementation. Yet as blends, they prove to be competitive to conventional oil-based plastics in their application. Based on the number of carbon, they are classified as either short chain length (scl—up to 5 carbon monomers) or medium chain length (mcl—6 to 14 monomers) polymers. The polymer properties are enhanced through block and copolymerization, blending with compatible polymers of methacrylate, polyethylene glycol (PEG), and other raw sources like cellulose (Li et al. 2016; Chen et al. 2016). Although flexible in its modification, marketing polyhydroxyalkanoates is still a risk factor due to its high production cost.

Microbial synthesis of plastics requires good-quality carbon substrates, controlled environment, and high energy input for polymer extraction. Bacteria such as

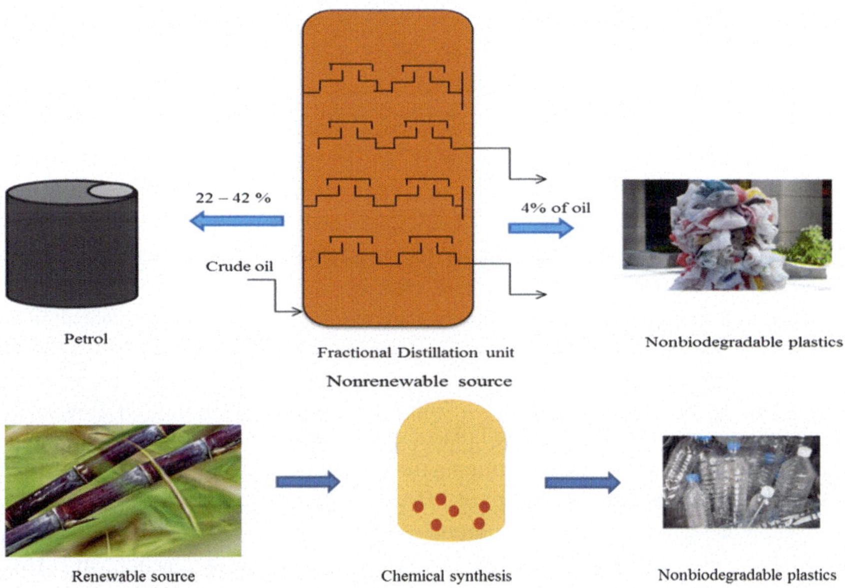

Fig. 11.1 Nonbiodegradable plastic production from nonrenewable and renewable sources

Ralstonia eutropha, *Pseudomonas*, *E. coli*, *Bacillus*, and *Halomonas* show higher accumulation of PHA granules. Usually carbon:nitrogen (C/N) ratios, type of carbon source—monomers or polymers, solid retention time (SRT), and hydraulic retention time (HRT) of the culture conditions have a great impact on efficient biopolymer accumulation (Johnson et al. 2010). With the advancements in biotechnology, yield and productivity of PHA have seen an increase. Modulating the carbon flux toward storage and insertion of *Pha* genes are the basic strategies employed by synthetic biologists (David et al. 2014; Levin et al. 2018; Beckers et al. 2016). Recently, the dependence on single neat carbon source for PHA production is replaced by a mixture of sources such as the hydrolysates and wastewaters. This reduces the cost burden where carbon source alone accounts for the major part of the total capital (Choi and Lee 2000). Besides bacteria, members of algae are also being studied for bioplastic accumulation. *Synechocystis* (Khetkorn et al. 2016; Hellingwerf et al. 2017) and *Chlorella* (Cassuriaga et al. 2018; Druzian et al. 2018) both act as model organisms in tuning the metabolic pathways.

Concurrent research adapting algae as axenic or mixed cultures shows promising results pertaining to bioplastic accumulation. As mixed cultures, algal consortium is capable of utilizing diverse carbon substrates. This type of consortium could be enriched and taken up for wastewater treatment. Nitrogen present in the wastewater provides the necessary limiting conditions for product accumulation (Uggetti et al. 2018). This way, the dual problem of wastewater treatment with nutrient recovery and bioplastic accumulation can be resolved simultaneously. The major bottleneck concerning microalgal PHA production is its recovery. Separation of algal cells, dewatering, and solvent extraction are the overall downstream processes involved (Fasaei et al. 2018). Instead of using the extractable fraction, the whole cells of microalgae can be substituted to form plastic blends with different biodegradable materials (Yan et al. 2016). Working with wastewater reduces the costs on carbon source. Sufficient biomass productivity can be achieved with industrial effluents rather than utilizing the fresh water reserve. The positive aspects of microalgal productivity employing wastewater as a sole source have gained new interests in obtaining value-added products.

2 Structure and Properties of Bioplastics

2.1 PHA-Homopolymer

Polyhydroxyalkanoates are polyesters with elastomeric or thermoplastic properties comparable to present-day oil-based plastics. PHA is versatile with homo-, co-, and heteropolymer combination, determining the material characteristics of the polymer (Fig. 11.2). One fascinating feature of PHA is its variety in polymer composition, which can be altered based on the carbon substrate provided to the microorganism. Considering scl polymer, PHB is the simplest with uniform stereochemistry. This enables the polymer to be crystallized in an orthorhombic pattern. However, the

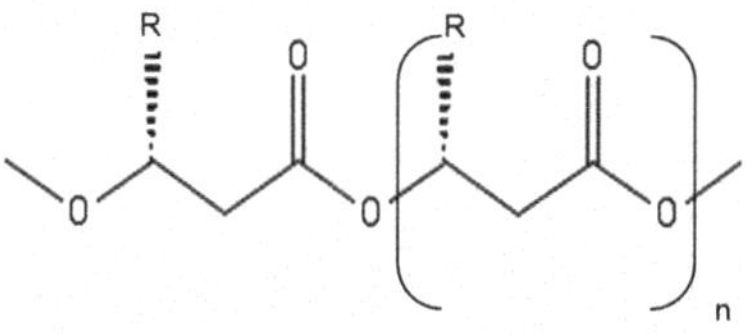

Fig. 11.2 Structural backbone of Polyhydroxyalkanoates. *n* repeating unit of polyme, *R* alkyl group

crystallization rate of PHB is relatively low compared to that of synthetic polymers, which can be overcome by the addition of nucleating agents. This crystallization tendency limits the application of scls. To extend the usage of PHA, mcls are employed, wherein the elastomeric properties are enhanced. The two major drawbacks in using mcl-PHAs are their melting temperature ranging from 40 to 60 °C making it vulnerable even at 40 °C and lower crystallization time (Koning 1995). Different copolymers with hydroxyvalerate prove promising in real-world application. In order to make use of their properties, many blends of synthetic and natural polymers are investigated for increasing the rigidity and elasticity.

2.2 *PHA or Microalgal Whole Cell—Blends and Composites*

Blending of polymers gives an added advantage to the synergistic material property of individual polymers employed. Both synthetic and biodegradable blends are available. mcl-PHA with natural and synthetic rubber showed increased thermal stability (Bhatt et al. 2008). PHA/PLA blend from injection molding has higher miscibility and crystallinity index with increasing amount of PHA (Loureiro et al. 2015).The crystallinity of the PHA blend is reduced with ethyl cellulose without compromising the material property (Chan et al. 2011). Similarly, poly(3-hydroxybutyrate) [P(3HB)] polymers also suffer from higher crystallinity rendering them unable to be used in many tissue engineering applications. As a means of overcoming the drawback, an oligomeric plasticizer, which is also an mcl-PHA, is used. Since both are from microbial sources and biocompatible, their use in soft tissue engineering will aid in decreasing the toxicity caused by external plasticizers in vitro (Lukasiewicz et al. 2018).

Zeller et al. (2013) utilized the whole cell biomass of *Spirulina* and *Chlorella* as alternatives for traditional plastics. Blends of algal biomass with glycerol and polyethylene were studied for their mechanical strength and thermal degradation. The strength of the blends obtained is determined by the pressure, temperature, and protein content of the cells along with their stability. It is necessary that the actual material durability is obtained from the cells rather than from the plasticizing agents (Savenkova et al. 2000). Torres et al. (2015) studied the biocomposites of residual microalgal biomass from biodiesel production with poly(butylene adipate-co-terephthalate). The biocomposites underwent mechanical studies including tensile strength, flexural modulus, elongation at break, and thermal stability. The green composite with 20% residual biomass, 30% glycerol, and 7.5 phr (parts per hundred rubber) urea showed excellent properties on extrusion molding.

On the contrary, Bulota and Budtova (2015) proposed that increasing the algal concentration in composites decreases the mechanical properties; and the proper optimization of composites with respect to end usage is necessary when algae is used as a filler. Mixing the residual *Nanochloropsis* (80%) with cellulose (20%) using ionic liquids indicates higher hydrolytic degradation of the film with an increase in biomass. But the tensile strength and elongation at break with 20% biomass and 80% cellulose are better compared to the other composites. Thus for any composite preparation with biomass, the nature of the cells, kind of bonding it forms with the blend, temperature, composition of biomass, processing time, and molding type play a major role. Tran et al. (2016) used surface-modified microalgal ash to be incorporated as a filler in poly (vinyl alcohol) (PVA) film. A solution casting of the above has uniform tensile strength over an ash content of 15–20%. Zhao et al. (2017) provided another example in improving the composites of PHA with 2 wt% of ribbon-like hexagonal boron nitride. Additives as nanofillers improved the mechanical and thermal properties.

Lipid-extracted microalgal biomass of *Nannochloropsis salina* finds its usage as a composite material with PVA. As demonstrated by others, higher biomass concentration decreased mechanical properties of the composites while increasing its thermal properties. In order to overcome the reduced mechanical properties, another polymeric material, poly(diallyldimethylammonium chloride) (PD), was introduced. The resulting composite with a composition of PVA 68%, Biomass 20%, and PD 12% had a melting point of 216.5 ± 1.9. These types of composites with higher thermal stability could be used for 3D-printing (Tran et al. 2018). Utilizing *Spirulina* as filler, PVA films were casted with glycerol as a co-filler. In obtaining the film, the composite with *Spirulina*–PVA–Glycerol (SPG) 10–7–3 showed higher tensile strength and elastic modulus. SPG 10–9–3 showed greater elongation at break. Glycerol being a plasticizer, aids in molecular interaction with *Spirulina* proteins contributing to the elasticity of the film produced. With increased water resistance, the produced film is proposed for packaging applications (Shi et al. 2017).

Moghaddas Kia et al. (2018) produced an edible antioxidant-enriched film using *Spirulina*, sodium caseinate, and Zedo gum. The water solubility of the film was tested, as it corresponds to the slow release of antioxidants. It is found that on increasing the concentration of microalgae, the water solubility also increases. The presence of microalgal cells in the bioplastic affects the properties to a greater extent. The type of cell used, with or without cell wall, affects the crystallinity of the produced film. A comparative study with intact cell and cell-disrupted microalgal strains as composites of corn starch showed greater difference in water vapor and oxygen permeability properties. Although intact cells reduce the elasticity of the biocomposites, elongation at break could be maintained. This is because the base polymeric crystallinity of starch backbone is reduced when incorporating with *Nanochloropsis* (Fabra et al. 2018). Another study on the usage of intact or disrupted cells of *Nannochloropsis* in developing a hybrid film with starch shows that ultrasound treatment has a significant influence on the tensile strength and elongation at break. Prolonging the treatment has a negative impact on tensile strength, but it raises the elongation at break. Furthermore, the presence of surfactant employed in disrupting the cell wall had a negative impact in obtaining a continuous film (Fabra et al. 2017). Composite nanofibers are produced from *Scenedesmus almeriensis* and polyethylene. When scaled up, the nanofibers showed bead formation (Sankaranarayanan et al. 2018).

In response to stress (centrifugation), biomass accumulates triacylglycerol (TAG). The TAG-accumulated whole cells can be used for molding bioplastics. *Chlamydomonas* being highly exploited for various industrial production processes is also used for making bioplastic beads. This again becomes an alternative to the petroleum-derived plastic beads. Although the tensile strength is only 14% of that of fossil fuel-derived plastic beads, the property could be enhanced by using different plasticizers. Similarly, carotenoid-extracted biomass of *Chlamydomonas* is also tested for the strength of plastic beads. The presence of different cross-linking agents inside the cell enhances the relative strength of the biomass-derived beads (Kato 2019) (Table 11.1).

3 PHA from Microalgae

3.1 Bioaccumulation of PHA in Microalgae

Microalgal cultures can be adapted to accumulate value-added molecules into their cells as part of their regular metabolism. One such kind of storage molecule is the PHB, which is a carbon reserve aiding the organism in nutrient-deprived conditions. Microalgal and cyanobacterial species of *Chlorella*, *Botyrococcus*, *Spirulina*, *Synechocystis*, and *Synechococcus* have shown good amount of PHB storage along with higher lipid accumulation. Since PHB is stored during nutrient limiting conditions, there is always a trade-off between biomass yield and PHB productivity. Lower the cellular growth higher the PHB accumulation. *Chlorella fusca* LEB 111 cultures when grown on pentoses with nitrogen-deficient condition and shorter light period (6 h) showed 17.4% w/w of polymer. The same strain gave 10.7% w/w of PHB when exposed to a photoperiod of 12 h. With xylose as the carbon source and light cycles of 18 h, 16.2% w/w of polymer accumulated. Factors like the light intensity, carbon source, and nitrogen content all determine the PHB accumulation (Cassuriaga et al. 2018). *Botyrococcus braunii* also showed efficient PHB storage capability. But the yield of PHB is very low, 0.382 mg/g, which was observed only on the 25th day with a purity of 16.4% (Kavitha et al. 2016).

3.1.1 Metabolic Engineering of Microalgae for Bioplastic Production

With a view of improvising the yield and productivity, many modifications at the genetic level controlling the genes and directing the carbon flux toward product formation are adopted. Many new tools like the clustered regularly interspaced short palindromic repeats (CRISPR-cas9), transcription activator-like effector nucleases (TALEN), and zinc finger nucleases are employed to bring out the necessary changes

Table 11.1 Properties of various blends, composites of PHA, and microalgae as biodegradable bioplastics

Composites or blends	Type of molding/ polymerization	Melting temperature (°C)	Glass transition temperature (°C)	Young's modulus (MPa)	Elongation at break (%)	Tensile strength (MPa)	Reference
Neat PHB	–	175	5	2950	2	40	Koning (1995)
Cellulose/microalgae (with 19.14% microalgae)	Composite film	nd	nd	3300 ± 500	3.7 ± 0.8	117 ± 12	Yan et al. (2016)
Cellulose/microalgae (with 78.83% microalgae)	Composite film	nd	nd	1500 ± 100	1.0 ± 0.1	15 ± 22	
PVA/PASH	Solution casting	226.7–229.8	nd	940 ± 1100	180 ± 6.4	4.6 ± 2.8	Tran et al. (2016)
PVA/GASH	Solution casting	226.9–229.0	nd	980 ± 140	240 ± 28	40.7 ± 1.7	
Flax/PHB	Compression molding	nd	nd	40,000	1.5 ± 0.5	40	Barkoula et al. (2010)
Flax/PHB/12%HV	Injection molding	nd	nd	6000	2.5	40	
PLLA-PHB-PLA copolymers (36L-28BBL-36L)	Triblockcopolymers	166.7	38.7	338 ± 2	21 ± 1.1	20 ± 1.7	Aluthge et al. (2013)
Blend PHB/PHHx (PHHx42mol%)	Polymers from recombinant strains of *P. putida* KTOYO6ΔC (phaPCJAc) *P. putida*KTQQ20 [Diblock copolymer]	173.8	−8.2, −27.3	80.21 ± 5.23	10.14 ± 1.12	4.32 ± 0.45	Chen et al. (2013)
Blend PHB-b-PHHx (PHHx42mol%)		172.1	2.7, −16.14	7.58 ± 2.70	207.31 ± 15.38	1.42 ± 0.24	
Random P(3HB-co-3HHx) (HHx 21mol%)		55.4	−18.1	23.58 ± 4.10	75.29 ± 9.25	1.84 ± 0.36	

PLA-15PHA	Injection molding—flexible film	148.9 ± 0.7	54.7 ± 0.2	1220 ± 140	100 ± 40	31 ± 5	Armentano et al. (2015)
PLA-15PHB-10Carv		149.5 ± 0.1 168.2 ± 0.2	54.4 ± 0.6	1130 ± 160	105 ± 26	24.3 ± 1.7	
PLA-15PHB-10Carv-15OLA		137.1 ± 1.4 145.7 ± 0.4 153.8 ± 0.6 163.1 ± 1.3	36.7 ± 1.8	330 ± 60	150 ± 30	14.8 ± 1.7	
PHB/starch (70/30)	Films (solvent casting)	167	7.3	949	9.4	19.23	Godbole et al. (2003)
PHB/PIP-g-PVAc (80/20)	Copolymer blend films	175	6	711	13	14.3	Yoon et al. (1999)
PHB/PLC (77/23)	Film (solvent casting)	60; 168	−60; 4	730	9	21	Kumagai and Doi (1992)
PHB/PHO (75/25)	Film (solvent casting)	172	−35	370	30	6.2	Dufresne and Vincendon (2000)
PHB/PHBV (25/75)	Electrospun fiber mats	152;163	nd	150	7	2	Sombatmankhong et al. (2006)
PHBV/a-PHB (50/50)	Film casting	133	2	240	33	7	Scandola et al. (1997)
SPG-10-7-3	Film casting	nd	nd	25	12	15	Shi et al. (2017)
SPG-10-9-3		nd	nd	880	45	12.5	

nd not determined

directed toward product yield. Both nuclear and chloroplast transformation is possible with advanced techniques. The possibilities of genetic transformation in *Chlamydomonas reinhardtii* was shown for the first time by Rochaix and van dillewijn (1982) with the plasmid pYe*arg*4. Since then, various studies are being carried out in enhancing the strain with added advantages. Besides the positive side, one major issue with the microalgal transgenic system is its genetic instability after modification. Stable growth in few strains of *Chlamydomonas*, *Chlorella* (Chow and Tung 1999), *Porphyridium* (Shapira et al. 2002), and *Euglena* could be observed with DNA transformation. Nevertheless, few strains show relatively stable nature making them amenable for modification and concurrent application.

PHB accumulation in *Chlamydomonas reinhadtii*cc-849 (cell wall deficient) transgenic cells was observed in the cytoplasm using acetyl co-A as the substrate. This transformation required two genes, *pha*B and *pha*C, for successful PHB accumulation as confirmed by transmission electron microscopy (TEM) (Chaogang et al. 2010). However, these cotransformants suffered low growth rate, which might be attributed to the presence of foreign gene. The bacterial enzymes from *R.eutropha* H16 necessary for PHB synthesis (PhaA, PhaB, PhaC) were successfully expressed into the diatom *Phaeodactylum tricornutum* by Hempel et al. (2010). The sequences were put under the control of a nitrate-inducible promoter that showed larger accumulating granules in the cytosol when grown in nitrate-containing medium. PHB of about 10.6% of cell dry weight was obtained. The authors also claim a 100-fold increase in PHB levels as compared to the plant-based systems (Fig. 11.3).

The PHB production in cyanobacteria is well established through genetic engineering approach. But the necessity of genome sequence, ease of genetic manipulation, stability of transgenic strains, and their maintenance are still a challenge. The need for sophisticated conditions and proper maintenance will definitely add to the cost of the product. And in those strains wherein the techniques of synthetic biology worked, further improvement could be directed toward achieving higher CO_2 conversion, efficient light utilization, effortless harvesting, and smooth extraction of products. Additionally, the cost incurred with the supply of high-purity sugars in growing engineered algal cells becomes a hurdle for its commercialization. In order to have a successful market value, the marginal cost should be reduced, which in turn relies on the processing parameters.

3.2 *Extraction of Polyhydroxyalkanoates*

PHAs, being intracellular polymeric reserves, are to be extracted using specific solvents. The techniques involved in the extraction process affect the polymer composition, their properties, and final purity. In general, the extraction involves biomass harvesting, cell disruption, polymer recovery, and its purification (da Silva et al. 2018). Extraction of PHAs is done by mechanical, chemical, or biological means of cell disruption followed by the use of organic solvents or supercritical fluids for the recovery of PHA (Fig. 11.4). Majority of PHA processing involves the usage of fos-

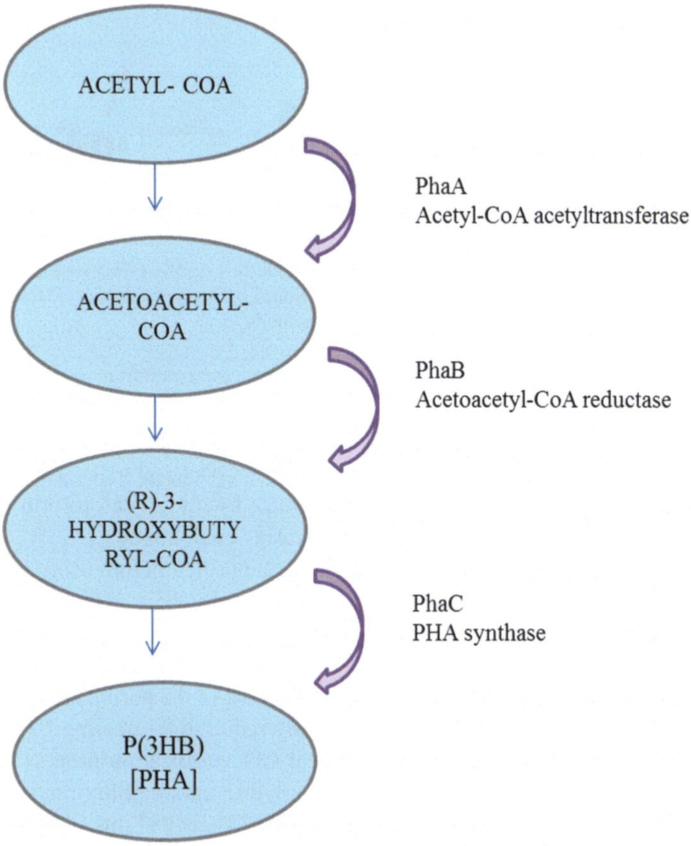

Fig. 11.3 Biopolymer accumulation pathway in microbial cell

sil fuel-derived organic solvents. In the chemical method for cell disruption, varying concentrations of sodium hypochlorite are employed. Treatment with sodium hypochlorite allows the cell content to be let out while retaining the sticky biomass with polymer. Repeated washing with water or methanol and treating the residue with solvents like acetone facilitate PHA recovery. Martins et al. (2014) followed the above protocol, obtaining polymer after drying in oven at 32 °C for 2 h. Yield is given as gram polymer obtained per gram of biomass. Purity is checked with FTIR analysis of both biomass and polymer content. FTIR analysis of biomass after extraction is done to verify the extraction efficiency of the solvent in getting the entire polymer from biomass.

Alternatively, Samrot et al. (2011) employed freeze-dried biomass for processing. The freeze-dried biomass was treated with 15 ml chloroform and 30% sodium hypochlorite. With continuous agitation for 1 h at 37 °C followed by centrifugation, the polymer was extracted using chloroform:ethanol at 1:9 ratio following evaporation. Using only methanol and chloroform, PHA was extracted as a translucent film.

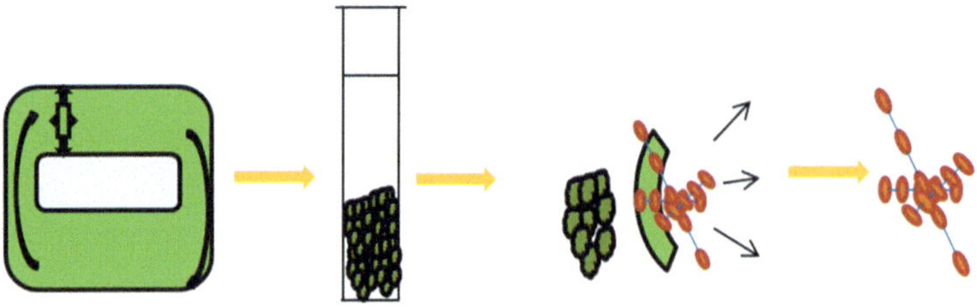

Fig. 11.4 General process involved in Polyhydroxyalkanoate extraction from microalgal cells

The harvested biomass was treated overnight with methanol. This treatment helped in removing the pigment molecules in case of cyanobacterial cells. The methanol was then evaporated at 60 °C and PHA is extracted with hot chloroform immediately precipitating into chilled methanol (Shrivastav et al. 2010).

PHA obtained through sequential steps is to be purified before using it as a raw material for thermoplastic production. Higher grade PHA without impurities increases the market value of the polymer. Applications in medical field require polymer of standard grade and high quality. Traditionally, organic solvents are used for purification purposes. This in turn adds to the cost of the purification process. It is estimated that around 50–165 L of solvents are required for treating 1 kg of polymer. Alternative methods involve supercritical CO_2 purification process, wherein pressurized CO_2 is allowed to react with the polymer until equilibrium is reached. After which the treated polymer is recovered. The efficiency of this system could be increased by the addition of ethanol along with CO_2. These methods are helpful in eliminating oils from PHBs (Daly et al. 2018).

3.3 Challenges in PHA Production and Recovery from Microalgae

The disadvantages in downstream processing of microalgal strains toward PHB production have paved the way for using whole cell biomass in bioplastics. Rather than extracting PHB for bioplastics, algal biomass as blends and composites has proven to be commercially competitive. Producing large amount of biomass could be made sustainable by employing wastewater as a source of nutrients. Microalgal biomass growing on wastewater can be harvested, followed by PHA extraction and the resulting residual biomass can be used as composite for bioplastics.

In the process of PHA recovery and purification from algal biomass, great deal of organic solvents are utilised. These solvents will not find use through recycling or reuse due to the impurities present. With PHA recovery and purity, none of the methods has proven tracks of 100% recovery or purity. New techniques involving fewer amount of solvent with larger partition coefficient between solvents should be developed. Facilities like supercritical fluid extraction should be experimented for large-scale production systems. By overcoming these shortcomings, PHA production can be seen as cost-effective and environment friendly.

4 Production Technologies in Bioplastic from Microalgae and PHA

Polyhydroxyalkanoates are short-chain polymers that require other polymeric materials or fillers to enhance the mechanical and physical properties of the resulting product (Li et al. 2016). Before blending two components as copolymers, their compatibility should be checked. Polarity of the polymers, stability of the components (thermal and chemical), the solvent required during molding, and ease of flow are a few properties to be determined prior to selecting a production methodology. For PHA, a vast variety of techniques from simple film casting to extrusion molding are being tested.

4.1 Film Casting

Film casting is the basic production process that comes handy for laboratory-scale experiments. Basically the polymer of study is pretreated, dissolved in a suitable solvent system, and poured onto die or molds. Necessary curing time is provided for stable crystallization after which the polymeric casts are taken out. Film casting, as simple as they seem, requires careful study on the rheological properties of each components. The particle size, molecular weight, and uniform distribution in a solution are few properties relating to polymer. Polarity, optimum volatility, lesser phase separation, and resistance to vapor absorption from atmosphere are a few properties to be noted in choosing a suitable solvent system. For film casting, usually high molecular weight polymers with moderate volatile solvents are preferred (Siemann 2005). Currently, the technology to even recover the used solvent after casting has been developed. The common solvent system used for dissolving PHA is chloroform (Godbole et al. 2003). Solubility of chloroform in PHA is well established. Usually the material formed through film casting exhibits isotropic nature. Pretreatment of the polymer to remove residual moisture is necessary before its dissolution with solvent. Thermal treatment is adopted to have a homogenously mixed solution. Following the casting, a controlled environment with constant humidity will aid in the development of a more uniform, flexible, and lesser crystallized film.

Solvent casting is the most common method to study the properties of polymer, its characterization, and for determining the mechanism behind drug release.

Akhtar et al. (1991) showed the relation between copolymer composition, its crystallinity, and types of matrix formed during the processing on drug release pattern. Many different p(HB-HV) copolymers with varying HV composition were casted. Interestingly, copolymers with high hydroxyvalerate showed increased drug release when compared to neat PHB films. The processing temperature involved in solvent casting and the crystallization rate of the copolymers play a major role in trapping the drug.

4.2 *Compression Molding*

Compression molding is used with thermosets and thermoplastics. In compression molding, the material is casted into the desired shape at a specific temperature, usually 190 °F, and at a constant pressure. During the process, the material is either preheated to its melting temperature or poured into heated molds. Either way, the polymer is made to fill the molds following compression provided by constant pressure for a period of time. After treatment, the material is cured and stripped off from the molds. This type of molding helps in attaining homogenous blend as the polymers are preheated. It also provides enough tensile strength to the products. One major drawback in extending this technology for bioplastics is the pretreatment process involved. Involvement of high temperature denatures or degrades many heat-labile bioplastic polymers. This also restricts the usage of fillers that are unstable in higher temperature. When mass production is targeted, compression molding is the economical production technology. Influence of plasticizers in compression molding of PHA is shown by Requena et al. (2016).

Microalgal whole cells are also tested for compression molding. Zeller et al. (2013) made thermomechanical molding of *Spirulina* and *Chlorella* with 57% and 58% of protein content, respectively. The molding process includes treatment of samples under 150 °C for 20 min followed by cooling for 10 min. The relation between different plasticizers and microalgal cells is analyzed by determining their mechanical properties.

4.3 *Injection Molding*

Injection molding involves complete polymer melting before casting into shapes. This process involves the solidification of injected molten material inside the required molds. During molding process, the molten material is to be introduced into the molds. For this, the molds are made with runners and gates. During the curing process, the material present in the runners and gates also solidifies resulting in wastage. Thermoplastics can be reused through another cycle of melting and can be incorporated into next the molding cycle. The design complexity of the product and product quantity influence the economic value. To test the possibility of carbon dioxide fixation inside a polymer, *Spirulina* and *Nannochloropsis* were injection molded into other thermoplastics. The mechanical properties of the blends varied based on the composition but proved competitive for packaging application (Shi et al. 2017).

4.4 Electro Spinning

Electrospinning refers to the production of fibers with nano- or microscale diameter. The polymeric solution dissolved in a suitable volatile solvent is drawn as fibers by applying voltage between the tip of the container and the collector plate. As the polymeric solution hits the collector plate, the solvent evaporates, solidifying the polymer as it deposits. This results in a nonwoven mat with controlled porosity and rigidity (Doshi and Reneker 1995). Electrospinning finds application in rigid packing, tissue engineering for scaffolds, drug delivery system, and wound healing patches. Electrospinning may be of either solvent spinning or melt spinning type. Electrospun fibers have a high surface-to-volume ratio. The ideality of any polymer to be adapted in electrospinning depends on various parameters like the solvent volatility, concentration of polymers, surface tension, surface charge density, and viscosity.

The innate mechanical property of electrospun polymeric fiber is also influenced by the collection method used. With neat poly[(*R*)-3-hydroxybutyrate-*co*-(*R*)-3-hydroxyvalerate], electrospun fibers from counter electrode collector and rotating disk collector showed variations in their tensile properties. This is due to the molecular alignment of polymers with fiber axis aiding in increasing the tensile strength and tensile modulus (Chan et al. 2009). With the increasing usage of electrospinning and polyhydroxyalkanoates, its application toward tissue engineering has increased. The possibility of PHA as biocompatible implants with other monomer compositions and its effect in tissue compatibility are shown by Ying et al. (2008). The biosorption of the subcutaneous scaffold increases in vivo with their mechanical property mimicking the skin. The diameter of the electrospun fibers greatly affects the biosorption capability that depends on the molecular weight and the monomer composition. Researchers found that P(3HB-co-97mol%-4HB) contributes to minimal fibrous encapsulation as the implant gets degraded with time (12 weeks) in vivo.

The percentage of polymer in the starting solution plays a major role in determining the thickness, porosity, diameter, and uniformity of fibers. As the polymer composition increases, thickening of fibers occurs due to slower evaporation of solvent. Increasing the diameter increases the elasticity of the implants but reduces their tensile property. The applied voltage also has an effect on the mechanical properties and the morphology. When lowering the voltage, beaded fibers are formed; hence selection of optimum voltage becomes necessary (Volova et al. 2014).

PHA as a potential packaging material is well known. Improvement of packaging material containing proteins like gluten is evaluated by Fabra et al. (2015). A three-layered film is developed to check for barrier properties. Incorporation of PHB and poly hydroxybutyrate-co-valerate copolymer (PHB3V—3% valerate content) on two sides of gluten is evaluated for oxygen and water barrier properties.

Water vapor barrier properties for PHB3V depended on the processing temperature used for three-layer assembly rather than the deposition time for fibers. The water vapor permeability and oxygen permeability values for PHB3V obtained are 3.14 ± $0.2e^{-11}$ kg m Pa^{-1} s^{-1} m^{-2} and 4.36 ± $0.05e^{-1}5$ m^3 m m^{-2} s^{-1} Pa^{-1}, respectively. However, water barrier properties for PHB are directly dependent on the deposition time and on the layer thickness. The oxygen permeability of the material solely depends on the polymer type used.

The application of PHA films as anti-bacterial surface is studied by Kehail and Brigham (2018). They produced Electrospun poly(3-hydroxybutyrate-*co*-3-hydroxyhexanoate).

[P(HB-*co*—30 mol%HHx)] fibers; the functionalization of the surface is done through sodium chloride pretreatment. This functionalization could produce negatively charged carboxylic acid, sufficient to have bonding with lysozyme. The fibers are then loaded with lysozyme by immersing the sheets in lysozyme overnight. This resulted in a maximum of 5.1 ± 0.8 μg of enzyme embedded onto Electrospun sheets. They exhibited about 42% inhibition on biofilm formation of *Rhodococcus opacus* PD630.

5 Patents and Intellectual Property Rights from Microalgal Bioplastic

Microalgal bioplastic is still under research, and very few companies have initiated to commercialize the product. Taking the economic burden of commercialization into consideration, both researchers as well as industrial research and development groups try out various strategies in making the process less currency-intensive. As a result, starting from feedstock development to product composition, every step could be patented or turned into a trade secret. With microalgal bioplastic, metabolically engineered strain could be patented while the process of obtaining biopolymer from cells could be an IPR right to a particular company. After obtaining the biopolymer, the composition involved in thermoplastic manufacturing, its fine-tuning, and its production process could also be patented (Fig. 11.5).

Unique patent (Shi and James 2013—US 8,524,811 B2) on manufacturing of thermoplastic blends from algae deals with types of microalgae, maximum allowable algal content, and powdered microalgal particle size along with the mechanical property of the obtained polymer. The claim states that the plasticized algal polymer could be made from either a single type of *Nannochloropsis*, *Spirulina* and *Chlorella* or as compositions. And the melting temperature can range from 60 to 190 °C. In this patent, one specific aspect is that the blends from microalgal powder could have a particle size of only 115μ.

A patent by Lavoisier et al. (2017)—WO 2017/046356 Al, discussed the process development for incorporating micro and microalga powder with reduced protein content for bioplastic. This patent includes the claims from culturing the algae, its harvesting, and the unique process in reducing the protein content by treatment with enzymes or citric acid or other chemicals. It also covers the method for getting the powdered form of algae and its subsequent incorporation as packaging films).

6 Commercialized Microalgal Bioplastic

Global bioplastic producers from algal source are limited. Need for more sustained process development, efficient cost-cutting technology, and reducing the cost involved with energy input may be stated as few reasons. Kimberly-Clark Corporation and Algix® are major players in microalgal bioplastic manufacturing. They use algae from natural habitat to be molded into products, 3D printed, or made into sheets. Many commercial materials are marketed under the brand name Solaplast (*Solaplast Material data centre*, 27th April, 2019). Algix® uses algae as flexible foam after proper preprocessing. This foam finds application from mats, backpacks to footwear and toys. Algix® along with 3D fuel has extended its market in producing more sustainable 3D filaments from algae. Solaplast (40% algae) has four different materials; three of which (Solaplast 2112, 1222, 1312) are made of food-grade algae and one (Solaplast 2020) with industrial algae. Of the four materials, Solaplast 1222 and Solaplast 1312 are durable resins, while Solaplast 2112 and Solaplast 2020 are compostable resins (Table 11.2).

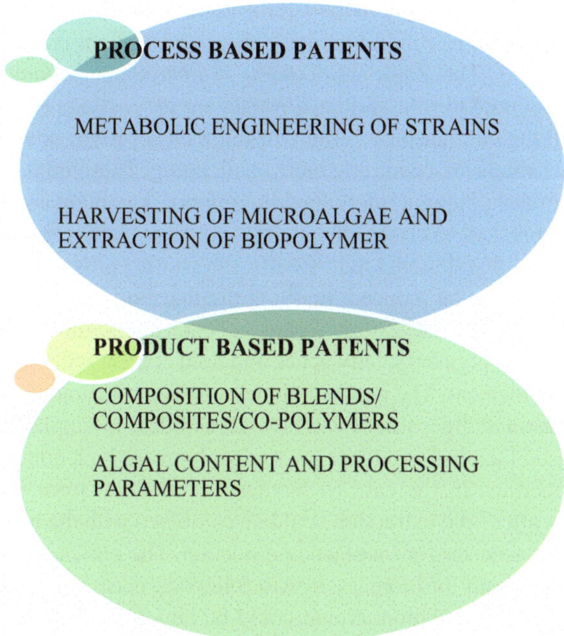

Fig. 11.5 Patenting options employing microalgal cells, bioplastic, and PHA

Table 11.2 Material properties of Solaplast from Algix®

Solaplast material	Composition	Nature of the polymer	Tensile strength (MPa)	Tensile modulus (MPa)	Elongation (%)	Flexural strength (MPa)
2020	Industrial algae (40%) Hyperbranched PLA (60%)	Compostable resin	28.7	1765	3.3	8280
2112	Food-grade algae (40%) + polybutylene adipate terephthalate (60%)	Compostable resin	5.9	145	24	1450
1222	Food-grade algae (40%) + copolymer PP (60%)	Durable resin	16.4	876	10.4	5390
1312	Food-grade algae (40%) + ethylene vinyl acetate (60%)	Durable resin	3.57	46.9	180	549

7 Circular Bioeconomy for Sustainable Microalgal Bioplastic Production

Microalgae are proving their effectiveness to be used as alternatives for the nonrenewable resource-derived products (Alam and Wang 2019). From fuels to polymers, feedstocks to medicinal applications, the multitude of microalgal biorefinery is constantly expanding. The flexibility in their cultivation, utilization of waste as a source, and natural environmental conditions make them as excellent competitors to their bacterial counterparts. The major shortcoming of microalgal technology transfer is the capital cost endured during various downstream processing. This could also be reduced by adapting circular bioeconomy of microalgae. Right now, treatment and disposal of wastewaters are becoming more challenging. They could be used as low-cost substrates in developing a microalgal biorefinery. The biomass obtained could be considered equivalent to crude oil subjected to fractionation, gaining new products at each stage of distillation. On this basis, the cost in cultivating microalgae for different purpose would be reduced, marking the feasibility of biorefinery. Energy for biomass growth and harvesting would become a one-time investment. The extraction of different pigments, fatty acids, and lipid content will make the basis for further product refinement. The residual biomass from extraction could be used as a substrate for anaerobic digestion or can be used as fillers in bioplastic manufacturing. Thermoplastic blends from direct microalgal biomass look competitive in their mechanical properties. In the case of biodiesel production from microalgae, the residual biomass after lipid extraction could be combined with the by-product, glycerol, acting as a plasticizer for bioplastic production. The remaining glycerol could be adopted as a medium for biomass growth following necessary purification. This way each microalgal production stream could be closed, where the output of one process becomes the starting material for the other.

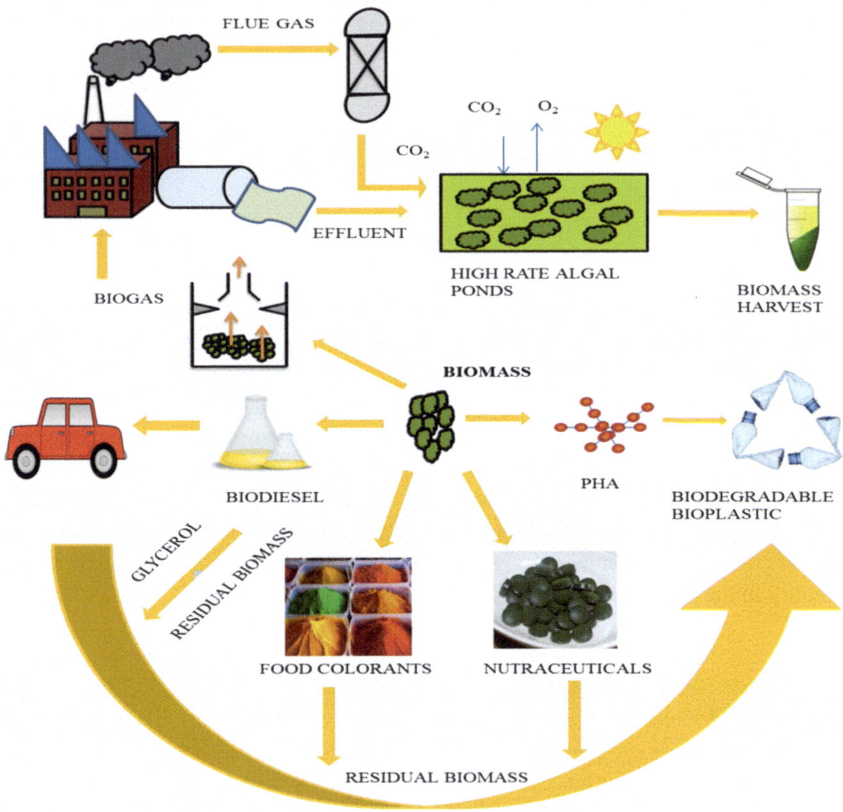

Fig. 11.6 A possible sustainable bioeconomic process in bioplastic production from microalgae

Implementing a process with circular economy could be achieved easily where the side streams and applications of each compound is identified with its recycle ratio. The main challenge vests in converting a traditional processing step into a more sustainable one. In such a process, the raw materials are from renewable biological sources leading to sequential extraction of useful products along with reuse, thereby closing the loop. In these steps it is not only the process parameters or the product that is concentrated but the minimization of waste stream exiting the system is also checked. Maximum waste valorization with microalgal technology and minimal waste production in the process line marks the true circular bioeconomy (Fig. 11.6).

Microalgae, like other plants, can be cultivated on nonarable lands. In many countries with abundant natural solar energy, arable land—a direct requirement for feedstock development—is limited or underutilized. The Indian subcontinent being the second largest democracy has approximately 40% of nonarable land (The WorldBank Data, 15th April, 2019). In such a case, encouraging microalgal cultivation aids in improving the lifestyle and economy of people living in those countries. From getting high-value products to the bulk chemicals, algal biomass could also be used to address the reduced energy input slowly getting us at energy-neutral solutions. Through streamlining the carbon flux toward product formation, yield and productivity could be increased. Optimised downstream processing with compound separation at each stage would play a critical role in establishing a strong sustainable bioeconomy from microalgae in bioplastic production.

8 Conclusion and Future Perspective

Microalgae are promising feedstock as a renewable source. Simple cultivation with minimal nutrient requirements makes them an interesting raw material for fractionating valuable products. High-value to low-value products could be obtained through customized downstream processing. Many a replacements for fossil fuel-derived produce are tried, but the major bottleneck lies in improving the biomass productivity and economic sustainability. The downstream processing established now at industrial scale could be made more sustainable by incorporating a circular economy with microalgal cultivation and processing. A thorough utilization of algal biomass through sequential valorization alone can make the microalgal technology a reality in the near future.

References

Akhtar, S., et al. (1991). The influence of crystalline morphology and copolymer composition on drug release from solution cast and melt-processed P(HB-HV) copolymer matrices. *Journal of Controlled Release, 17*, 225–234.

Alam, M. A., & Wang, Z. (Eds.). (2019). *Microalgae biotechnology for development of biofuel and wastewater treatment* (pp. 1–655). Singapore: Springer.

Aluthge, D. C., et al. (2013). PLA-PHB-PLA triblock copolymers: Synthesis by sequential addition and investigation of mechanical and rheological properties. *Macromolecules, 46*(10), 3965–3974. https://doi.org/10.1021/ma400522n.

Anderson, A., & Dawes, E. (1990). Occurrence, metabolism, metabolic role, and industrial uses of bacterial polyhydroxyalkanoates. *Microbiological Reviews American Society for Microbiology, 54*(4), 450–472.

Armentano, I., et al. (2015). Bio-based PLA_PHB plasticized blend films: Processing and structural characterization. *LWT – Food Science and Technology, 64*(2), 980–988. https://doi.org/10.1016/j.lwt.2015.06.032.

Barkoula, N. M., Garkhail, S. K., & Peijs, T. (2010). Biodegradable composites based on flax/polyhydroxybutyrate and its copolymer with hydroxyvalerate. *Industrial Crops and Products, 31*(1), 34–42. https://doi.org/10.1016/j.indcrop.2009.08.005.

Beckers, V., et al. (2016). Integrated analysis of gene expression and metabolic fluxes in PHA-producing *Pseudomonas putida* grown on glycerol. *Microbial Cell Factories, 15*(1), 1–18. https://doi.org/10.1186/s12934-016-0470-2.

Bessa, F., et al. (2018). Occurrence of microplastics in commercial fish from a natural estuarine environment. *Marine Pollution Bulletin, 128*, 575–584. https://doi.org/10.1016/j.marpolbul.2018.01.044.

Bhatt, R., et al. (2008). PHA-rubber blends: Synthesis, characterization and biodegradation. *Bioresource Technology, 99*(11), 4615–4620. https://doi.org/10.1016/j.biortech.2007.06.054.

British Plastics Federation (2019). Retrieved November 15, 2019 from https://www.bpf.co.uk/press/oil_consumption.aspx.

Bögershausen, A., et al. (2002). Biosynthesis of novel thermoplastic polythioesters by engineered *Escherichia coli*. *Nature Materials, 1*(4), 236–240. https://doi.org/10.1038/nmat773.

Bulota, M., & Budtova, T. (2015). PLA/algae composites: Morphology and mechanical properties. *Composites Part A: Applied Science and Manufacturing, 73*, 109–115. https://doi.org/10.1016/j.compositesa.2015.03.001.

Cassuriaga, A. P. A., et al. (2018). Innovative polyhydroxybutyrate production by *Chlorella fusca* grown with pentoses. *Bioresource Technology, 265*(October), 456–463. https://doi.org/10.1016/j.biortech.2018.06.026.

Chan, K. H. K., et al. (2009). Effect of molecular orientation on mechanical property of single electrospun fiber of poly[(R)-3-hydroxybutyrate-co-(R)-3-hydroxyvalerate]. *Journal of Physical Chemistry B, 113*(40), 13179–13185. https://doi.org/10.1021/jp905820h.

Chan, R. T. H., et al. (2011). Manipulation of polyhydroxybutyrate properties through blending with ethyl-cellulose for a composite biomaterial. *International Journal of Polymer Science, 2011*, 1–8. https://doi.org/10.1155/2011/651549.

Chaogang, W., et al. (2010). Biosynthesis of poly-3-hydroxybutyrate (PHB) in the transgenic green alga *Chlamydomonas reinhardtii*. *Journal of Phycology, 46*(2), 396–402. https://doi.org/10.1111/j.1529-8817.2009.00789.x.

Chen, G. Q., et al. (2013). *Psedomonas putida* KT2442 as a platform for the biosynthesis of polyhydroxyalkanoate with adjustable monomer contents and compositions. *Bioresource Technology, 142*, 225–231. https://doi.org/10.1016/j.biortech.2013.05.027.

Chen, W., et al. (2016). Modification and potential application of short-chain-length polyhydroxyalkanoate (SCL-PHA). *Polymers, 8*(8), 273. https://doi.org/10.3390/polym8080273.

Choi, J., & Lee, S. Y. (2000). Economic considerations in the production of poly(3-hydroxybutyrate-co-3-hydroxyvalerate) by bacterial fermentation. *Applied Microbiology and Biotechnology, 53*(6), 646–649. https://doi.org/10.1007/s002530000326.

Chow, K. C., & Tung, W. L. (1999). Electrotransformation of Chlorella vulgaris. *Plant Cell Reports, 18*(9), 778–780. https://doi.org/10.1007/s002990050660.

da Silva, C. K., Costa, J. A. V., & de Morais, M. G. (2018). Polyhydroxybutyrate (PHB) synthesis by *Spirulina* sp. LEB 18 using biopolymer extraction waste. *Applied Biochemistry and Biotechnology, 185*(3), 822–833. https://doi.org/10.1007/s12010-017-2687-x.

Daly, S. R., et al. (2018). A green process for the purification of biodegradable poly(β-hydroxybutyrate). *The Journal of Supercritical Fluids, 135*, 84–90. https://doi.org/10.1016/j.supflu.2018.01.007.

David, Y., et al. (2014). Metabolic engineering of *Ralstonia eutropha* for the production of polyhydroxyalkanoates from sucrose. *Biotechnology and Bioengineering, 112*(3), 638–643. https://doi.org/10.1002/bit.25469.

Dong, X., et al. (2018). Proteomic profile and toxicity pathway analysis in zebrafish embryos exposed to bisphenol A and di-n-butyl phthalate at environmentally relevant levels. *Chemosphere, 193*, 313–320. https://doi.org/10.1016/j.chemosphere.2017.11.042.

Doshi, J., & Reneker, D. H. (1995). Electrospinning process and applications of electrospun fibers. *Journal of Electrostatics, 35*, 151–160.

Druzian, J. I., et al. (2018). Influence of nitrogen on growth, biomass composition, production, and properties of polyhydroxyalkanoates (PHAs) by microalgae. *International Journal of Biological Macromolecules, 116*, 552–562. https://doi.org/10.1016/j.ijbiomac.2018.05.064.

Dufresne, A., & Vincendon, M. (2000). Poly(3-hydroxybutyrate) and poly(3-hydroxyoctanoate) blends: Morphology and mechanical behavior. *Macromolecules, 33*(8), 2998–3008. https://doi.org/10.1021/ma991854a.

European Bioplastics. (2019). Retrieved April 15, 2019, from https://www.european-bioplastics.org/market/.

Fabra, M. J., López-Rubio, A., & Lagaron, J. M. (2015). Three-layer films based on wheat gluten and electrospun PHA. *Food and Bioprocess Technology, 8*(11), 2330–2340. https://doi.org/10.1007/s11947-015-1590-0.

Fabra, M. J., et al. (2017). Development and characterization of hybrid corn starch-microalgae films: Effect of ultrasound pre-treatment on structural, barrier and mechanical performance. *Algal Research, 28*, 80–87. https://doi.org/10.1016/j.algal.2017.10.010.

Fabra, M. J., et al. (2018). Structural and physicochemical characterization of thermoplastic corn starch films containing microalgae. *Carbohydrate Polymers, 186*, 184–191. https://doi.org/10.1016/j.carbpol.2018.01.039.

Fasaei, F., et al. (2018). Techno-economic evaluation of microalgae harvesting and dewatering systems. *Algal Research, 31*, 347–362. https://doi.org/10.1016/j.algal.2017.11.038.

Godbole, S., et al. (2003). Preparation and characterization of biodegradable poly-3-hydroxybutyrate-starch blend films. *Bioresource Technology, 86*(1), 33–37. https://doi.org/10.1016/S0960-8524(02)00110-4.

Hellingwerf, K. J., et al. (2017). Genetic engineering of *Synechocystis* sp. PCC6803 for poly-β-hydroxybutyrate overproduction. *Algal Research, 25*, 117–127. https://doi.org/10.1016/j.algal.2017.05.013.

Hempel, F., et al. (2010). Microalgae as bioreactors for bioplastic production. *Microbial Cell Factories, 10*(81), 2–7.

Herrero, Ó., et al. (2018). The BPA-substitute bisphenol S alters the transcription of genes related to endocrine, stress response and biotransformation pathways in the aquatic midge *Chironomus riparius* (Diptera, Chironomidae). *PLoS One, 13*(2), 1–17. https://doi.org/10.1371/journal.pone.0193387.

Hopewell, J., Dvorak, R., & Kosior, E. (2009). Plastics recycling: Challenges and opportunities. *Philosophical Transactions of the Royal Society B: Biological Sciences, 364*(1526), 2115–2126. https://doi.org/10.1098/rstb.2008.0311.

Johnson, K., Kleerebezem, R., & van Loosdrecht, M. C. M. (2010). Influence of the C/N ratio on the performance of polyhydroxybutyrate (PHB) producing sequencing batch reactors at short SRTs. *Water Research, 44*(7), 2141–2152. https://doi.org/10.1016/j.watres.2009.12.031.

Kato, N. (2019). Production of crude bioplastic-beads with microalgae: Proof-of-concept. *Bioresource Technology Reports, 6*, 81–84. https://doi.org/10.1016/j.biteb.2019.01.022.

Kavitha, G., et al. (2016). Biosynthesis, purification and characterization of polyhydroxybutyrate from *Botryococcus braunii* kütz. *International Journal of Biological Macromolecules, 89*, 700–706. https://doi.org/10.1016/j.ijbiomac.2016.04.086.

Kehail, A. A., & Brigham, C. J. (2018). Anti-biofilm activity of solvent-cast and electrospun polyhydroxyalkanoate membranes treated with lysozyme. *Journal of Polymers and the Environment, 26*(1), 66–72. https://doi.org/10.1007/s10924-016-0921-1.

Khetkorn, W., et al. (2016). Enhancement of poly-3-hydroxybutyrate production in *Synechocystis* sp. PCC 6803 by overexpression of its native biosynthetic genes. *Bioresource Technology, 214*, 761–768. https://doi.org/10.1016/j.biortech.2016.05.014.

Koning, G. (1995). Physical properties of bacterial poly(R-3-hydroxyalkanoate). *Canadian Journal of Microbiology, 41*, 303–309.

Kumagai, Y., & Doi, Y. (1992). Enzymatic degradation and morphologies of binary blends of microbial poly(3-hydroxy butyrate) with poly(ε-caprolactone), poly(1,4-butylene adipate) and poly(vinyl acetate). *Polymer Degradation and Stability, 36*(3), 241–248. https://doi.org/10.1016/0141-3910(92)90062-A.

Lavoisier, P., Pierre, R., & Benoit, M. (2017). Method for preparing an algae powder with reduced protein content and bioplastic composition formulated from such a powder. WO 2017/046356 Al.

Levin, D., et al. (2018). Polyhydroxyalkanoate (PHA) polymer accumulation and *pha* gene expression in Phenazine (phz$^-$) and Pyrrolnitrin (prn$^-$) defective mutants of *Pseudomonas chlororaphis* PA23. *Polymers, 10*(11), 1203. https://doi.org/10.3390/polym10111203.

Li, Z., Yang, J., & Loh, X. J. (2016). Polyhydroxyalkanoates: Opening doors for a sustainable future. *NPG Asia Materials, 8*(4), e265–e220. https://doi.org/10.1038/am.2016.48.

Lind, P. M., et al. (2011). Circulating levels of bisphenol A (BPA) and phthalates in an elderly population in Sweden, based on the prospective investigation of the vasculature in Uppsala seniors (PIVUS). *Ecotoxicology and Environmental Safety, 75*, 242–248. https://doi.org/10.1016/j.ecoenv.2011.09.004.

Loureiro, N. C., et al. (2015). Thermal characterization of polyhydroxyalkanoates and poly(lactic acid) blends obtained by injection molding. *Polymer – Plastics Technology and Engineering, 54*(4), 350–356. https://doi.org/10.1080/03602559.2014.935422.

Lukasiewicz, B., et al. (2018). Binary polyhydroxyalkanoate systems for soft tissue engineering. *Acta Biomaterialia, 71*, 225–234. https://doi.org/10.1016/j.actbio.2018.02.027.

Manikkam, M., et al. (2013). Plastics derived endocrine disruptors (BPA, DEHP and DBP) induce epigenetic transgenerational inheritance of obesity, reproductive disease and sperm epimutations. *PLoS One, 8*(1), e55387. https://doi.org/10.1371/journal.pone.0055387.

Martins, R. G., et al. (2014). Bioprocess engineering aspects of biopolymer production by the cyanobacterium *Spirulina* strain LEB 18. *International Journal of Polymer Science, 2014*, 1–7. https://doi.org/10.1155/2014/895237.

Moghaddas Kia, E., Ghasempour, Z., & Alizadeh, M. (2018). Fabrication of an eco-friendly antioxidant biocomposite: Zedo gum/sodium caseinate film by incorporating microalgae (*Spirulina platensis*). *Journal of Applied Polymer Science, 135*(13). https://doi.org/10.1002/app.46024.

Müller, H.-M., & Seebach, D. (1993). Poly(hydroxyalkanoates): A fifth class of physiologically important organic biopolymers? *Angewandte Chemie International Edition in English, 32*(4), 477–502. https://doi.org/10.1002/anie.199304771.

Plastic pollution Coalition. (2017). Retrieved April 15, 2019, from https://www.plasticpollutioncoalition.org/pft/2017/9/29/fueling-plastics-new-research-details-fossil-fuel-role-in-plastics-proliferation.

Rawsthorne, T. W., et al. (2011). Food packaging and bisphenol A and bis(2-ethyhexyl) phthalate exposure: Findings from a dietary intervention. *Environmental Health Perspectives, 119*(7), 914–920. https://doi.org/10.1289/ehp.1003170.

Requena, R., et al. (2016). Effect of plasticizers on thermal and physical properties of compression-moulded poly[(3-hydroxybutyrate)-co-(3-hydroxyvalerate)] films. *Polymer Testing, 56*, 45–53. https://doi.org/10.1016/j.polymertesting.2016.09.022.

Rochaix, J. D., & van Dillewijn, J. (1982). Transformation of the green alga *Chlamydomonas reinhardii* with yeast DNA. *Nature, 296*, 70–72. https://doi.org/10.1038/296070a0.

Samrot, A. V., et al. (2011). Accumulation of poly[(R)-3-hydroxyalkanoates] in *Eenterobacter cloacae* SU-1 during growth with two different carbon sources in batch culture. *Applied Biochemistry and Biotechnology, 163*(1), 195–203. https://doi.org/10.1007/s12010-010-9028-7.

Sankaranarayanan, S., et al. (2018). The influences of solvents on the electrospun of whole *Scenedesmus almeriensis* and poly(ethylene oxide) for the preparation of composite nanofibers. *Composites Communications, 10*, 18–24. https://doi.org/10.1016/j.coco.2018.05.003.

Savenkova, L., et al. (2000). Mechanical properties and biodegradation characteristics of PHB-based films. *Process Biochemistry, 35*(6), 573–579. https://doi.org/10.1016/S0032-9592(99)00107-7.

Scandola, M., et al. (1997). Polymer blends of natural poly(3-hydroxybutyrate-co-3-hydroxyvalerate) and a synthetic atactic poly(3-hydroxybutyrate). Characterization and biodegradation studies. *Macromolecules, 30*(96), 2568–2574.

Shapira, M., et al. (2002). Stable chloroplast transformation of the unicellular red alga *Porphyridium* species. *Plant Physiology, 129*(1), 7–12.

Shi, B., & James, H. (2013). Algae blended compositions for thermoplastic compositions. US 8,524,811 B2.

Shi, B., et al. (2017). Glycerol-plasticized *spirulina*–poly(vinyl alcohol) films with improved mechanical performance. *Journal of Applied Polymer Science, 134*(20), 1–10. https://doi.org/10.1002/app.44842.

Shrivastav, A., Mishra, S. K., & Mishra, S. (2010). Polyhydroxyalkanoate (PHA) synthesis by *Spirulina subsalsa* from Gujarat coast of India. *International Journal of Biological Macromolecules, 46*(2), 255–260. https://doi.org/10.1016/j.ijbiomac.2010.01.001.

Siemann, U. (2005). Solvent cast technology – A versatile tool for thin film production. *Progress in Colloid and Polymer Science, 130*, 1–14. https://doi.org/10.1007/b107336.

Solaplast. (2019). Retrieved April 27, 2019, from https://www.materialdatacenter.com/ms/en/Solaplast/ALGIX%2C+LLC/7025.

Sombatmankhong, K., et al. (2006). Electrospun fibre Mtas of poly(3-hydroxybutyrate), poly(3-hydroxybutyrate-co-3-hydroxyvalerate) and their blends. *Journal of Polymer Science, Part B: Polymer Physics, 44*, 2923–2933. https://doi.org/10.1002/polb.20915.

Steer, M., et al. (2017). Microplastic ingestion in fish larvae in the western English Channel. *Environmental Pollution, 226*, 250–259. https://doi.org/10.1016/j.envpol.2017.03.062.

The WorldBank Data. (2019). Retrieved April 15, 2019, from https://data.worldbank.org/indicator/AG.LND.AGRI.ZS?locations=IN.

Torres, S., et al. (2015). Green composites from residual microalgae biomass and poly(butylene-adipate-co-terephthalate): Processing and plasticization. *ACS, Sustainable Chemistry and Engineering, 3*, 614–624. https://doi.org/10.1021/sc500753h.

Tran, D. T., et al. (2016). Preparation and characterization of poly(vinyl alcohol) biocomposites with microalgae ash. *Journal of Applied Polymer Science, 133*(26), 1–12. https://doi.org/10.1002/app.43599.

Tran, D. T., et al. (2018). Lipid-extracted algal biomass based biocomposites fabrication with poly(vinyl alcohol). *Algal Research, 31*, 525–533. https://doi.org/10.1016/j.algal.2016.08.016.

Uggetti, E., et al. (2018). Production of polyhydroxybutyrate and crabohydrates in amixed cyanobacterial culture: Effect of nutrients limitation and photoperiods. *New Biotechnology, 42*, 1–11. https://doi.org/10.1016/j.nbt.2018.01.001.

Volova, T., et al. (2014). Electrospinning of polyhydroxyalkanoate fibrous scaffolds: Effects on electrospinning parameters on structure and properties. *Journal of Biomaterials Science, Polymer Edition, 25*(4), 370–393. https://doi.org/10.1080/09205063.2013.862400.

Wang, F., & Tong, Z. (2016). Solely biomass-derived polyethylene terephthalate (PET): Conversion of bio-based isoprene and acrolein to p-xylene and terephthalic acid. *ChemistrySelect, 1*(17), 5538–5541. https://doi.org/10.1002/slct.201600055.

Yan, C., et al. (2016). Cellulose/microalgae composite films prepared in ionic liquids. *Algal Research, 20*, 135–141. https://doi.org/10.1016/j.algal.2016.09.024.

Ying, T. H., et al. (2008). Scaffolds from electrospun polyhydroxyalkanoate copolymers: Fabrication, characterization, bioabsorption and tissue response. *Biomaterials, 29*(10), 1307–1317. https://doi.org/10.1016/j.biomaterials.2007.11.031.

Yoon, J. S., et al. (1999). Toughening of poly(3-hydroxybutyrate) with poly(*cis*-1,4-isoprene). *European Polymer Journal, 35*(5), 781–788. https://doi.org/10.1016/S0014-3057(98)00068-8.

Zeller, M. A., et al. (2013). Bioplastics and their thermoplastic blends from *Spirulina* and *Chlorella* microalgae. *Journal of Applied Polymer Science, 130*(5), 3263–3275. https://doi.org/10.1002/app.39559.

Zhao, Y., et al. (2017). Self-sacrificed template synthesis of ribbon-like hexagonal boron nitride nano-architectures and their improvement on mechanical and thermal properties of PHA polymer. *Scientific Reports, 7*(1), 1–6. https://doi.org/10.1038/s41598-017-08524-7.

Chapter 12
Microalgae as Biofertilizer in Modern Agriculture

Suolian Guo, Ping Wang, Xinlei Wang, Meng Zou, Chunxue Liu, and Jihong Hao

Abstract Microalgae are a kind of widespread photosynthetic organisms including eukaryotic green algae and prokaryotic blue algae. They have great potential to be used as biological resources in the fields of medicine, health products, feed, fuel and so on. These fascinating organisms can also be used in modern agriculture for their ability to enrich soil nutrients and enhance utilization of macro and micronutrients. In addition to improving soil fertility and quality, microalgae can also produce plant growth hormones, polysaccharides, antimicrobial compounds and other metabolites to promote plant growth. This section focuses on the effects of cyanobacteria and green algae as biofertilizers on improving fertility and quality of soil and promoting plant growth. Recent research developments and future prospects of their application in modern agriculture are also discussed.

Keywords Microalgae · Soil fertility · Plant growth · Modern agriculture

1 Introduction

The application of chemical fertilizer plays an important role in ensuring high yield of food crops, and it also brings serious environmental and land pollution problems.

The excessive application of chemical fertilizer has resulted in a series of problems, such as serious imbalance of N, P and K ratio, soil hardening, salinization, nutrient reduction, groundwater pollution, which hinder the sustainable development of modern agriculture. Therefore, how to reduce the amount of chemical

S. Guo (✉) · M. Zou · C. Liu · J. Hao
College of Pharmaceutical and Biological Engineering, Shenyang University of Chemical Technology, Shenyang, Liaoning, China

P. Wang
Marine Science and Technology, Zhejiang Ocean University, Zhoushan, Zhejiang, China

X. Wang
Ports Administration of Jiaxing Municipality, Jiaxing, Zhejiang, China

M. A. Alam et al. (eds.), *Microalgae Biotechnology for Food, Health and High Value Products*, https://doi.org/10.1007/978-981-15-0169-2_12

fertilizer and find environmentally friendly alternatives is an important subject of current scientific research.

Biofertilizers are green, healthy and pollution-free, which are considered as the best substitute for chemical fertilizers. Many studies have shown that biofertilizers can effectively improve the utilization of nitrogen, phosphorus, potassium and other elements in soil, enhance the stress resistance of crops, improve the quality and yield of agricultural products, and significantly inhibit the toxicity of plant pathogens in soil.

Among the various kinds of biofertilizers, formulations based on photosynthetic organisms, including prokaryotic cyanobacteria and eukaryotic microalgae, have been paid much attention, owing to their excellent capacity of increasing soil fertility and crop yields. Heterocystous cyanobacteria contribute towards nutrient enrichment by nitrogen fixation and mineralization, while non-heterocystous cyanobacteria and green algae mainly contribute towards nutrient fertilization through the release of insoluble or immobile nutrients in soil. Photosynthetic capacity of microalgae and their application in soil leads to carbon enrichment, which improves soil organic matter, promotes other mineralization processes and enhances the utilization of macro and micronutrients in soil and rhizosphere. In addition, microalgae can produce metabolites such as phytohormones, polysaccharides and so on, which can bring additional benefits to agricultural production. This fascinating group of organisms can be applied in modern agriculture because it helps to improve nutrient availability, maintain soil organic carbon and fertility, and enhance plant growth and crop yield by stimulating soil microbial activity.

Therefore, in agronomic practice, the use of algae biofertilizer can completely or partially replace chemical fertilizer, reduce nutrient loss, protect agricultural soil fertility and meet the needs of sustainable development.

2 Microalgae as Biofertilizers Improving Soil Fertility and Quality

Chemical fertilizers contain a large number of ineffective components, which accumulate in the soil after long-term use and form water-insoluble substances with metal ions in the soil, reducing soil fertility. Long-term use of chemical fertilizers results in unbalanced proportion of nitrogen, phosphorus and potassium in soil, decrease of nutrient composition, salinization, hardening, etc. Besides that, overtillage and frequent use of heavy machinery change the structure of the soil, resulting in difficulties in water maintenance and nutrient mobilization.

Microalgae can not only fix CO_2 and N_2 by assimilation to increase soil fertility but also secrete EPS and form symbiotic system to improve soil structure. Different traits pertinent to cyanobacteria and green microalgae that make them promising biofertilizing options in modern agriculture are discussed in the following sections.

2.1 *Increase in Soil Fertility*

2.1.1 Carbon Sequestration

Carbon dioxide (CO_2) is the major contributor (68%) in greenhouse gases (GHGs) emission. The use of chemical fertilizers has increased greenhouse gas emissions in many ways, such as extraction of raw materials, production process, transportation and mechanized fertilization, all of which have increased carbon dioxide emissions. GHGs are also emitted from the soil in the form of CO_2 and NO_2 when chemical fertilizers are applied.

Cyanobacteria and green microalgae are important sources of organic matter in the agro-ecosystem, as they are directly involved in the assimilation of atmospheric carbon dioxide into the organic algal biomass through photosynthesis. As the primary producers, algae are responsible for 50% of the total photosynthesis on the earth. The algal assimilation of carbon dioxide can significantly raise the soil organic carbon content. Besides that, algae can increase the organic carbon pool in soil through the algal excretion of carbon (exopolysaccharides), and promote the growth of other microflora and fauna. Their remains can be degraded by algae to further fortify the soil organic content. Reports on the inoculation of green algae and cyanobacteria in various crops showed enhancement in the microbial activity, microbial biomass and overall organic carbon of soil. The pot experiment conducted by Yilmaz and Sönmez was to evaluate the potential of different biofertilizers on soil organic carbon. Results showed that compared to the control treatments, the algal biofertilizer soil amendments significantly increased soil organic carbon.

2.1.2 Nitrogen Fixation

Besides carbon, nitrogen is the most important nutrient in biomass production. Nitrogen content of biomass ranged from 1% to 10%, according to the population. The typical reaction of nitrogen deficiency is cell discoloration (chlorophyll reduction, humus increase) and accumulation of organic carbons such as polysaccharides and some oils (PUFA). Biological nitrogen fixation refers to the process in which nitrogen in the atmosphere can be converted into ammonia by nitrogen-fixing microorganisms. In relation to little soluble inorganic ammonium salt in the earth's crust, biological nitrogen fixation plays an extremely important role in maintaining the nitrogen cycle in nature. Almost all living beings rely on the organic nitrogen fixed by organisms. However, biological nitrogen fixation occurs only through a few special microorganisms such as bacteria and algae, which have the particular nitrogenase system in their bodies. The research and utilization of nitrogen-fixing organisms can open up a source of fertilizer for agriculture and is of great significance in maintaining and improving soil fertility.

Cyanobacteria have specialized cells called heterocysts, which are capable of fixing atmospheric N and thereby catering to the demands of soil micro- and macrofauna and flora and plants. Various studies have shown that the soil nitrogen content of agricultural crops inoculated with cyanobacteria or cyanobacteria consortium increased significantly. Cyanobacteria inoculation can save 25–40% of chemical nitrogen fertilizer. Perera proved that inoculating filamentous nitrogen-fixing cyanobacteria to rice crops could reduce the amount of fertilizer by 50%, and the grain yield and quality were not affected. The application of nitrogen fixation by Isosporic cyanobacteria in paddy fields is not new, but recent reports have broadened the scope of application in a variety of vegetables, cotton and food crops.

Swarnalakshmi et al. (2013) showed that the inoculation of Anabaena sp. biofilm in wheat crop increases soil N content to 57.42% and 40%, compared to 50% and 100% doses of chemical fertilizer control group, respectively. Similarly, Osman et al. (2010) evaluated the potential of two cyanobacteria, *Nostoc entophytum* and *Oscillatoria augustissima*, as biofertilizers for pea plant. Their research showed that the cyanobacterial inoculation reduces 50% use of chemical fertilizer and improves the nutritional value of pea seeds. The use of cyanobacterial biofertilizer not only reduces the use of chemical fertilizers but also increased the crop yields of rice and various other crops. Jha and Prasad (2006) showed that the inoculation of N-fixing cyanobacterial strains in rice fields increased the straw and grain yield to a significant extent, and saved 25% N of chemical fertilizer. In another study, Singh and Datta (2007) reported that inoculation of N-fixing *Anabaena variabilis* enhanced the plant growth and yield of rice compared to the field application of chemical fertilizers. There could be environmental concern on leaching of excessive N that is biologically fixed in the soil; however, the extent could be very low compared to that caused by the use of chemical fertilizers. Most of the fixed nitrogen is in complex chemical forms. Leaching may occur due to erosion of soil caused by natural calamities like heavy rainfall. However, exopolysaccharide producing cyanobacteria form biological soil crusts and prevent the leaching of soil nitrogen. Overall, cyanobacterial inoculation can provide 25–40 kg N/ha and save the cost of fertilizers for the farming community significantly, and also reduce the enviromental pollution by preventing the leakage of biofertilizer nutrient from farmland.

2.1.3 Mineralization and Solubilization of Nutrients

In addition to the primary contribution of algae to soil organic matter, algae also play a role in the mineralization and solubilization of macro- and micronutrients in soil, which are important for plant growth. Algae, mainly cyanobacteria, are also reported to mineralize compounds by the production of organic acids or siderophore.

Organic acids including humic acid play an important role in mineral weathering. Cyanobacteria can secrete humic-like substances, which are of great importance to agriculture. It is reported that the exopolysaccharides secreted by the cyanobacterium *Microcystis aeruginosa* act as biological pumps and facilitate bio-absorption of phenanthrene. Yandigeri found that the cyanobacteria *Westiellopsis prolifica* and *Anabaena variabilis* could solubilize the insoluble tricalcium phosphate and Mussorie rock phosphate efficiently.

Siderophores are organic compounds produced by microbes, which are helpful to chelate ferric iron under iron deficiency, and make them available to microorganisms and plants. Cyanobacteria such as *Anabaena flos-aquae*, *Anabaena cylindrica* and *Anabaena* spp., have the ability to synthesize siderophores to chelate Fe, Cu and other trace elements. Several reports showed that the combined action of cyanobacteria, green algae and bacteria could increase the content of trace elements such as Fe, Mn, Cu and Zn in plants, which is of great significance to the growth of crops such as grain. However, the transport mechanisms from soil to rhizosphere and to aboveground parts of plants remain to be further studied.

2.2 *Promotion of Soil Quality*

2.2.1 Improvement of Soil Structure

Soil is eroded by physical forces such as water, wind and agricultural activities, which affect the fertility and productivity of agricultural soil. It is reported that many green algae and cyanobacteria can secrete EPS to the surrounding environment. EPS can not only increase soil organic carbon but also prevent soil erosion and improve soil structure. EPS produced by green algae and cyanobacteria is so adhesive, that it can help to aggregate soil particles providing the minimum pore size required for crop growth and root penetration.

Another aspect of maintaining soil structure is that the stable soil aggregates formed can resist rainfall and water flow to a certain extent. Malam Issa et al. (2007) showed that after 6 weeks of inoculation, the organic mineral soil aggregates consisting of filaments and EPS were formed, and the stability of the aggregates was enhanced compared with the control. Recent studies have shown that green algae can also improve soil structure and aggregate stability. Yilmaz and Sönmez (2017) studied the effects of different biofertilizers on the stability of soil aggregates. The results showed that the stability of soil microaggregates (0.25–0.050 mm) was improved by inoculating *Chlorella* alone or in combination with vermiculite, compared with the control treatments with and without chemical fertilizer application.

Cyanobacteria facilitate an increase in the start time of soil runoff and reduce soil erosion. The effect of cyanobacteria such as *Nostoc*, *Oscillatoria* and *Lyngbya* on run off prevalence in degraded soil under laboratory conditions was studied by Sadeghi et al. (2017). The results showed that the beginning time of soil runoff increased by 38.2–168.8% when cyanobacteria were inoculated into degraded soil. In another study, Kheirfam et al. (2017). showed that soil loss decreased by 99% after 60 days inoculated with cyanobacteria.

2.2.2 Reclamation of Wasteland

Green microalgae and cyanobacteria are ubiquitous organisms that can tolerate extreme environmental conditions. Their ability to survive in the drought, salt affected areas and oil- and metal-contaminated areas makes it possible to reclaim these wastelands. Trejo (2012) studied the potential of immobilized beads of *Chlorella aeruginosa* and *Spirillum Azotobacter* from three-stage sewage treatment for desert eroded soil reclamation. The results showed that three consecutive applications of microalgae–bacterium consortium could significantly increase soil organic matter, organic carbon and microbial biomass carbon. According to Acea (2003), the application of *Oscillators, Candida* and *Filaria* contributes to the recovery of heat-damaged soil. It has been proved that the inoculation of cyanobacteria is helpful to the recovery of microbial crust in heated soil (350 °C), due to the increase in number of nitrate and nitrite producers, with the increase in starch and cellulose mineralized microorganisms. Abed (2010) reported that cyanobacteria and bacteria have a unique relationship in the repairing of oil-contaminated areas, and they support their mutual growth. A study by Chaillan et al. (2006) also showed that cyanobacteria may not directly involve in the degradation of petroleum products, and their interaction with other related microorganisms can improve their activity of remediation and restore fertility in oil and oil-contaminated areas. Subhashini and Kaushik (1981) studied the possibility of using cyanobacteria to improve saline soil. It has been reported that cyanobacteria can trap excessive sodium ions into EPS matrix and limit the absorption of sodium ions by plants. However, with the degradation of cyanobacteria biofilm, sodium ions are released back into the environment. This cultivation method will not affect the effective use of nutrients by crops in saline environment.

Cyanobacteria and green microalgae have good removal ability of heavy metals and have been applied in metal-contaminated sites. The potential of nitrogen-fixing cyanobacteria in improving field application of fly ash was studied. After inoculating cyanobacteria with fly ash, the content of nitrogen and phosphorus in fly ash increased and the content of heavy metals decreased. Tripathi's research (Tripathi et al. 2008) shows that using cyanobacteria as biofertilizer in mixed fly ash soil can improve the stress ability of rice crops to fly ash. They reported that cyanobacteria inoculated with fly ash (under high metal pollution) prevented plants from accumulating heavy metals and promoted plant growth. Therefore, cyanobacteria can be used as effective inoculants and biofertilizers in metal-contaminated sites.

3 Microalgae as Biofertilizers Promoting Plant Growth

Microalgae also have important applications in promoting plant growth. It can directly secrete plant-derived hormones such as growth hormone and cytokinin to stimulate crop growth. Microalgae can also induce the immune system of plants by producing antibiotics and improve their disease resistance. In addition, microalgae can improve the root microbial system of crops and promote crop growth together with other microorganisms.

3.1 Production of Plant Growth Hormones

Phytohormones play an important role in plant growth and development. External supplementation of phytohormones (synthetic or natural) in agriculture has become a method to increase crop productivity and yields and to control weed. However, the potential risk of their leakage to adjacent areas and water bodies is of great environmental concern. Many strains of green microalgae and cyanobacteria can produce intracellular hormones, and some strains can even secrete hormones to growth media and surrounding environment. Growth hormones such as auxin, cytokinin and jasmonic acid produced by algae can be used as agricultural biological stimulants. At present, there are reports on the use of cyanobacterial hormones for in vitro regeneration of valuable plants and as plant growth promoters for useful crops. The androgen response of anther culture and maize regeneration was enhanced by the inoculation of green algae and cyanobacteria. The growth hormone extracted from cyanobacteria could promote the germination and growth of rice seedlings. Hussein and Hassner studied the effects of hormones secreted by cyanobacteria on plant growth under sterile and field conditions. The results showed that cyanobacterial hormone levels (cytokinin and auxin) were positively correlated with plant growth parameters (such as stem length, root length, spike length and seed weight). The increase of hormone levels in plants is due to the interaction between cyanobacteria and plant roots. Mazhar et al. (2013) found that the levels of endogenous and exogenous auxins increased in the cyanobacteria–wheat symbiotic system, indicating the signal transduction between plants and cyanobacteria. Therefore, the use of potential phytohormone-excreting cyanobacterial strains as biofertilizers in agricultural practice can be an environmentally friendly way to promote plant growth. However, there are few studies on field scale evaluation of algae hormone application, which need further exploration.

3.2 Elicitation of Plant Defense Mechanisms

It is reported that cyanobateria can regulate plant defense mechanisms by activing of β-1,3 endoglucanase. chitinase, catalase, peroxidase, polyphenol oxidase, phenylalanine ammonia lyase, etc., to active the antioxidant and pathogenic mechanisms. Priya et al. (2015) showed that inoculation with cyanobacteria significantly increased the activities of immune-related enzymes such as peroxidase, polyphenol oxidase and phenylalanine ammonia-lyase in rice roots and stems. Kumar et al. (2013) pointed out that the inoculation of cyanobacteria significantly enhanced the activity of β-1,3-endoglucanase in root and shoot of seed spice crops. Babu et al. (2015) studied the effect of inoculating different cyanobacteria (*Anabaena laxa*

RPAN8, *Calothrix* sp. and *Anabaena* sp. CW2) on the activity of defense enzymes in wheat plant. Different cyanobacteria help to improve plant immunity. The activities of peroxidase, polyphenol oxidase and phenylalanine ammonia lyase were significantly increased with high sugar (mannose) treatment.

Results have shown that plant microalgae/cyanobacteria interactions directly and indirectly contribute to the improvement of plant immunity and the robustness to abiotic and biological stresses. However, the practical mechanism of algae–plant interaction in improving defense enzyme activity and immunity still needs to be studied comprehensively, by exploring the various beneficial aspects of different green algae and cyanobacteria for their use as biofertilizer. Then, microalgae species with multiple agricultural values can be used as bio-options for sustainable agriculture.

3.3 Colonization of Plant Tissues

Cyanobacteria inoculation regulates rhizosphere microbial community, resulting in changes in microbial community structure and quantity. It has been reported that cyanobacteria can form symbiotic relationships with algae, fungi, gymnosperms, ferns and vascular plants. Krings et al. (2009) studies have shown that cyanobacteria can enter through stomata and colonize in stomatal cavity, intercellular space, parenchyma cells and arbuscular mycorrhizal zone, forming loops or intracellular coils. Karthikeyan (2009) observed the short cyanobacterial filaments in root hairs and cortical areas in wheat seed imbibition test. Through scanning electron microscopy and DNA fingerprint analysis, it was found that cyanobacteria colonized different parts of wheat. Cyanobacteria inoculation promoted growth and enhanced nitrogen fixation ability, and enhanced plant defense ability by inducing IAA production, with improving plant growth and nutritional status. Priya et al. (2015) reported that *Elenkinii* was able to colonize rice roots and stems and stimulate microbial populations of nitrogen and phosphorus-solubilizing microorganisms in the colonized sites. Cyanobacteria agent improved the symbiosis of chickpea, made beneficial changes in rhizosphere soil and rhizobium microbial community profile, and then improved crop growth and soil fertility. A thorough study of the diversity of cyanobacteria and the interaction between partners in symbiotic combinations, especially with important agricultural plants, will be conducive to the sustainable development of agriculture.

4 Microalgae as Biofertilizers Controlling Disease and Pest

In agricultural practice, the use of chemical fungicides to control pests and diseases is harmful to the sustainability of agricultural ecosystems. It is an urgent need to explore sustainable alternatives for pathogens control. Utilization of biological options represents an effective environmental strategy to control soil-borne pathogens. The most common microorganisms for biological control are bacteria and fungi. In the last decades, microalgae, mainly cyanobacteria, have also been recognized as the potential biocontrol for pathogens and pests. In addition to reducing the use of chemicals harmful to the environment, cyanobacteria also provide additional benefits of increased nutrition, thereby improving plant disease resistance and enhancing crop yields.

4.1 Disease Control

It is reported that cyanobacteria could produce a large number of antimicrobial compounds, including hydrolase, ambigol A benzoic acid, carbamidocyclophane A and majusculonic acid. These antimicrobial compounds inhibit or kill pathogenic bacteria, fungi or nematodes or other microbial groups/fauna by modifying and destroying the structure and function of cytoplasmic membranes, inactivating enzymes and inhibiting protein synthesis in targeted organisms. Algae extracts contain bioactive substances including polyphenols, tocopherols, as well as antimicrobial pigments, which are helpful for the prevention and control of soil-borne diseases. Some compounds in cyanobacteria are found to exhibit biological control/insecticidal/insecticidal properties, such as benzoic acid from *Calothrix* sp., chlorine-containing antibiotics from *Scytonema* and majusculamide-like compound from *Anabaena laxa*, by inducing the activity of plant defense enzymes. Extracts from cyanobacteria *Anabaena* spp. and *Oscillatoria* spp. presented antibiotic activity against *Bacillus subtilis*, *Micrococcus flavus* and *Staphylococcus aureus*.

Cyanobacteria can directly or indirectly inhibit cotton root rot by promoting beneficial changes in rhizosphere microbial communities. The antifungal activities of *Oscillator*, *Anabaena*, *Nostoc*, *Nodularia* and *Calotrix* against *Alternaria alternate*, *Botrytis cinerea*, *Rhizopus stolonifera* varied with temperature, and reached the highest at 35 °C. It is showed that extracts from cyanobacteria *Nodularia harveyana* and *Nostoc insulare* anti-cyanobacterial, have antibacterial (*Staphylococcus aureus*, *Bacillus cereus*) and antifungal (*Candida albicans*) activities. Some cyanobacterial strains such as *Anabaena spp.* have a dual role in bio-control of plant pathogens and bio- stimulation of plant. Chaudhary et al. (2012) compared the effects of formulations based on *Anabaena variabilis* and *A. oscillarioides* on disease resistance of tomato seedlings and their biological fertility. The disease severity of tomato seedlings was reduced by 10–15% and plant growth was significantly increased by using the variant of *Anabaena preparation*. Singh and Datta (2007) found that in addition to increasing crop yields and soil fertility in paddy fields, the variant *Anabaena* had herbicide resistance.

Cyanobacteria are also involved in the production of cyanotoxins that can have direct negative effect on humans and animals. There have been reports about the ability of plants, including food crops, to accumulate cyanobacterial toxins. Therefore, strict biological control of cyanobacterial toxins and their effects on different model animals is needed. Therefore, critical verification of cyanobacterial toxins and their effect on different model animal systems is essential. An Issa (1999) study reported that the metabolites produced by cyanobacteria *Oscillatoria angustissima* and *Calothrix parietina* had antimicrobial activity against some cyanobacterial and bacterial strains, but did not affect model mice. However, more studies are needed to study the functions of cyanobacterial antibiotics and their indirect effects on micro- and macroflora and fauna of aquatic and terrestrial environments.

4.2 Pest Control

It is reported that cyanobacteria-dominated microalgae can also reduce the number of plant pests such as nematodes by producing peptide toxins and nematocides. Khan et al. (2005) found that inoculating cyanobacterium *Microcystis cyanobacteria* in soil could not only increase tomato yield but also reduce the number of nematodes and the formation of eggs in tomato crops. Khan showed that tomato seedlings soaked in *M. vaginalis* culture filtrate reduced the population of rhizobium and nematode by 65.9% and 97.5%, respectively. The cyanobacterium *Oscillatoria chlorina* is also found to have nematicidal activity against *Meloidogyne incognita*. Compared with untreated soil, the formula of 1% *Oscillatoria chlorina* powder in soil reduced the number of eggs and nematodes by 68.9% and 97.6%, respectively. The effect of inoculating cyanobacteria before plant seedling is better in nematocidal activity and promoting plant growth. Youssef and Ali compared the nematicidal activity of *Anabaena oryzae*, *Nostoc calcicola* and *Spirulina* sp. against *Meloidogyne incognita* infecting cowpea plants. Among these microalgae *N. calcicola* is the most effective, reducing the formation of eggs and nodules and increasing plant productivity. Chandel's study showed that the culture filtrate of *Aulosira fertilissima* could inhibit the hatching of root-knot nematode *Meloidogyne triticoryzae*. Peptide toxins extracted from cyanobacterium MKU 106 showed insecticidal activity against *Helicoverpa armigera* at a concentration of 0.01%. Cyanotoxin can be used as an anti-feed compound for style larvae when applied to cotton leaves at a concentration of 0.1%. Studies have shown that compared to organic fertilizer treatment, the number of mosquito larvae in the field is higher when chemical fertilizer (urea) is applied. However, inoculating cyanobacteria as biofertilizer increased grain yield without increasing mosquito population. Biondi et al. 2004 reported that the methanolic extracts of the cyanobacterium *Nostoc strain* ATCC 53789 have nematicidal activity (against *Caenorhabditis elegans*), antifungal activity (against *Armillaria* sp., *Penicillium expansum*, *Phytophthora cambivora*, *Rhizoctonia solani*, *Rosellinia* sp., *Sclerotinia sclerotiorum* and *Verticillium albo-atrum*), insecticidal activity (against *Helicoverpa armigera*), herbicidal property (against some grasses) and cytotoxicity (against *Artemia salina*). Therefore, it is necessary to determine and formulate suitable strategies for microalgae/cyanobacteria fertilizers implementation in order to prevent toxicity to non-target organisms and successfully apply them into conventional agronomic practices.

5 Economics of Microalgae Fertilizer

The environmental and economic feasibility of algae fertilizer is questioned because it requires a large amount of water and nutrients for its growth. Therefore, the combination of microalgae biofertilizer production with non-drinking water sources, low-cost nutrient sources or value chains and by-products of the production technologies seems to be a feasible trend in the future.

5.1 Integration of Microalgal Cultivation and Nutrient Recycling in Wastewaters

Due to the large requirements of chemical nutrients for microalgal growth, it is not economically feasible for low-value products such as biofuels and biofertilizers to cultivate microalgae on an industrial scale. This challenge can be met by using inexpensive sources of nutrients. Using different types of wastewater as nutrient source to cultivate microalgae can not only treat wastewater but also produce microalgae biomass at the same time, which has broad application prospects. Generally speaking, wastewater contains abundant organic and inorganic nutrients such as C, N and P, which can be effectively utilized by microalgae. Phosphorus is a non-renewable resource, which is rich in wastewater. It is very important to recycle this nutrient. Microalgae can be used to accumulate phosphorus from wastewater and then be used as biofertilizer. Microalgae were successfully cultivated from pig farm wastewater, aquaculture wastewater, soybean processing wastewater, potato processing wastewater, carpet factory wastewater and domestic wastewater.

Solid waste can also be used as a nutrient source for microalgae culture. Agricultural activities and other agricultural industries produce a large amount of solid waste. In recent years, it has been found that various solid wastes, such as food wastes, poultry manure and dairy wastes, have also been studied for the cultivation of microalgae. Microalgae culture using nutrients from anaerobic-digested agricultural wastes (such as agricultural wastes, cow dung) is also an emergency choice. Microalgae cultured from these wastes can be used as biofertilizer. Cabanelas reported that *Chlorella* cultured in different types of wastewater could be used as biofertilizer. Renuka et al. (2016) found that the formulations of microalgae cultured in wastewater as biofertilizer could improve yield and nutritional characteristics of wheat and grain. Similarly, Wuang et al. (2016) cultivated *Spirulina filamentosa* using aquaculture wastewater, subsequently the biomass was used as biofertilizer for leaf vegetables *Arugula* (*Erucasativa*), *Bayam Red* (*Ameranthusgangeticus*) and *Pak Choy* (*Brassica rapa* ssp. *chinensis*). Compared to chemical fertilizer (Triple-Pro 15-15-15), the iron (Fe), magnesium (Mg), calcium (Ca) and zinc (Zn) content of algae biomass was higher. These trace elements (iron, magnesium, calcium, zinc) play an important physiological role in plant growth. Therefore, these wastewaters can be used as a nutrient source of microalgae biomass. Agricultural wastewater may contain pesticides, pharmaceutical compounds and other substances that may affect the growth of algae. Therefore, as the growth matrix of microalgae, different wastewaters need to be thoroughly evaluated to meetthe quality requirements of microalgae as different products such as food, feed, biological fertilizer and so on.

5.2 *Microalgae Residue of Value-Added Products as Biofertilizer*

Cyanobacteria and microalgae are widely used as important sources of commercialized biomolecules (lipids, carbohydrates and proteins) and bioactive compounds. The remaining algae biomass after extraction of interesting compounds is usually rich in essential macro-and microelements, which can be used as feedstocks for other applications such as protein source and biofertilizer. Different studies have shown that lipid biomass extracted from algae is rich in nutrients and can be used as a nutrient source in agriculture. In Zhu's research (2013), it was found that the algae biomass extracted from oil contained abundant C (49.0%), H (6.96%), N (5.76%) and O (26.3%) with energy value of 22.9 MJ kg^{-1}. Similarly, Ehimen et al. (2011) showed that lipid-extracted biomass of *Chlorella* sp. contained 44.7–47.4% of C, 7.05–7.46% of H, 9.39–10.13% of N and 34.57–36.22% of O. Maurya et al. studied the effect of lipid-extracted microalgae biomass on corn biofertilizer. Their research shows that the use of algae biomass extracted from oil as fertilizer in corn crops not only reduces the use of chemical fertilizers but also improves the productivity of crops. Another sustainable approach is to use anaerobic digestion of algae wastes to generate biomethane and biohydrogen for additional benefits. Recently, Solé-Bundó et al. (2017) found that digestate from the untreated microalgae and anaerobically co-digested microalgae biomass in primary sludge was rich in organic matter and nutrients, which can be used for agricultural soil improvement. The application of co-digestive algae biomass studies on *Lepidium sativum* in vivo showed that plant growth stimulation and minimum phytotoxicity. The contents of *E. coli* and heavy metal were lower than European legislative standards. In agricultural practice, using algae waste as biological fertilizer is an economical and effective method to utilize algae biomass.

6 Prospect of Microalgae as Fertilizer in Agriculture

As a fertilizer, it needs a lot of nutrients to meet the needs of commercial agriculture. To provide these nutrients in the form of microalgae biofertilizer a large amount of microalgae biomass is required. The nitrogen in anhydrous ammonia of chemical fertilizer reaches up to 82%, while nitrogen (N) in the microalgae biomass is about 1–10%. Therefore, it can be assumed that microalgae need about 15 times as much material to achieve similar fertilization levels. According to Plastina (2017), if 186 lbs. ac^{-1} or 208 kg ha^{-1} of nitrogen is recommended for corn production, 1.4 tons ac^{-1} or 3.1 tons ha^{-1} of microalgae biomass is needed to achieve a similar level of benefit. In this regard, using in vivo culture as inoculant is a possible solution to this problem. This requires only small amounts of cultured algae as inoculant, and the growth of microalgae and cyanobacteria provides nutrients and valuable compounds for crops in a continuous manner.

The use of living microalgae and cyanobacterial-based inoculants is beneficial, because they provide the benefits of continuous nutrient sequestration throughout the phases of plant growth, as well as preventing soil erosion, nutrient leaching and maintaining soil structure and fertility. It is well known that the use of cyanobacteria as growth cells can save 25–75% of nitrogen fertilizer in addition to providing other essential elements (macronutrients and micronutrients) and metabolites useful for plant growth. Moreover, microalgae biomass also contains 40–60% carbon, 1–4% phosphorus and other essential elements. The most important beneficial attribute of microalgae biofertilizer is the improvement of soil organic carbon, which cannot be achieved by chemical fertilizer.

In recent years, great progress has been made in the commercial utilization of algae as biofertilizer. Algae biofertilizer products are in-market, ensuring their effectiveness in improving plant productivity and soil fertility. Delmont Fresh Agricultural Products Company conducted field trails on algae biofertilizer in Arizona's pristine desert. The algae fertilization was helpful for the reclamation of abandoned wasteland, reduced fossil inputs and increased crop yield.

There is a growing awareness of the use of algae biofertilizers and their benefits. The market potential of algae biofertilizers is enormous, but the challenges associated with commercialization need to be addressed through extensive field surveys and cost-effectiveness development for fertilizer production technology.

7 Conclusions

Recent advances have shown that the utilization of consortia/biofilm of green algae and cyanobacteria with different beneficial microorganisms as biofertilizer in agriculture is promising. In addition to being used as a nutritional supplement, algal biofertilizers also bring other benefits such as biocontrolling plant pathogens, reducing the use of chemicals and slumping greenhouse gas emissions. However, the success of algae fertilizer depends largely on the economic feasibility of biomass production. Utilization of waste substrates for the cultivation of algal fertilizers is an economical and feasible strategy with other environmental benefits. Whereas, practical waste-related issues still need in-depth study and field assessment.

In order to further understand algae biofertilizer and promote its commercialization process, we should focus on the mechanism of algae fertilizer, the accumulation of algae bio-active molecules and their effects on plant growth, biological system and soil structure. The use of modern molecular tools, such as genomics and proteomics, provides a new technical means to elucidate the mechanism of bio-alliance and plant synergy, which could be helpful to further verify the efficacy of algae biofertilizer and its commercial application.

References

Abed, R. M. M. (2010). Interaction between cyanobacteria and aerobic heterotrophic bacteria in the degradation of hydrocarbons. *International Biodeterioration and Biodegradation, 64*(1), 58–64.

Acea, M. (2003). Cyanobacterial inoculation of heated soils: Effect on microorganisms of C and N cycles and on chemical composition in soil surface. *Soil Biology and Biochemistry, 35*(4), 513–524.

Babu, S., Bidyarani, N., Chopra, P., et al. (2015). Evaluating microbe-plant interactions and varietal differences for enhancing biocontrol efficacy in root rot challenged cotton crop. *European Journal of Plant Pathology, 1*(42), 345–362.

Biondi, N., Piccardi, R., Margheri, M. C., Rodolfi, L., Smith, G. D., & Tredici, M. R. (2004). Evaluation of Nostoc strain ATCC 53789 as a potential source of natural pesticides. *Applied and Environmental Microbiology, 70*(6), 3313–3320.

Chaillan, F., Gugger, M., Saliot, A., Coute, A., & Oudot, J. (2006). Role of cyanobacteria in the biodegradation of crude oil by a tropical cyanobacterial mat. *Chemosphere, 62*(10), 1574–1582.

Chaudhary, V., Prasanna, R., Nain, L., et al. (2012). Bioefficacy of novel cyanobacteria-amended formulations in suppressing damping off disease in tomato seedlings. *World Journal of Microbiology and Biotechnology, 28*(12), 3301–3310.

Ehimen, E. A., Sun, Z. F., Carrington, C. G., Birch, E. J., & Eaton-Rye, J. J. (2011). Anaerobic digestion of microalgae residues resulting from the biodiesel production process. *Applied Energy, 88*(10), 3454–3463.

Issa, A. A. (1999). Antibiotic production by the cyanobacteria *Oscillatoria angustissima* and *Calothrix parietina*. *Environmental Toxicology and Pharmacology, 8*(1), 33–37.

Jha, M. N., & Prasad, A. N. (2006). Efficacy of new inexpensive cyanobacterial biofertilizer including its shelf-life. *World Journal of Microbiology and Biotechnology, 22*(1), 73–79.

Karthikeyan, N., Prasanna, R., Sood, A., Jaiswal, P., Nayak, S., & Kaushik, B. D. (2009). Physiological characterization and electron microscopic investigations of cyanobacteria associated with wheat rhizosphere. *Folia Microbiologica, 54*, 43–51.

Khan, Z., Park, S. D., Shin, S. Y., Bae, S. G., Yeon, I. K., & Seo, Y. J. (2005). Management of *Meloidogyne incognita* on tomato by root-dip treatment in culture filtrate of the bluegreen alga, *Microcoleus vaginatus*. *Bioresource Technology, 96*(12), 1338–1341.

Kheirfam, H., Sadeghi, S. H., Zarei Darki, B., & Homaee, M. (2017). Controlling rainfall-induced soil loss from small experimental plots through inoculation of bacteria and cyanobacteria. *Catena, 152*, 40–46.

Koutra, E., Grammatikopoulos, G., & Kornaros, M. (2017). Microalgal post-treatment of anaerobically digested agro-industrial wastes for nutrient removal and lipids production. *Bioresource Technology, 224*, 473–480.

Krings, M., Hass, H., Kerp, H., Taylor, T. N., Agerer, R., & Dotzler, N. (2009). Endophytic cyanobacteria in a 400-million-yr-old land plant: a scenario for the origin of a symbiosis? *Review of Palaeobotany and Palynology, 153*(1-2), 62–69.

Kumar, M., Prasanna, R., Bidyarani, N., Babu, S., Mishra, B. K., Kumar, A., Adak, A., Jauhari, S., Yadav, K., Singh, R., & Saxena, A. K. (2013). Evaluating the plant growth promoting ability of thermotolerant bacteria and cyanobacteria and their interactions with seed spice crops. *Scientia Horticulturae, 164*, 94–101.

Malam Issa, O., Défarge, C., Le Bissonnais, Y., Marin, B., Duval, O., Bruand, A., D'Acqui, L. P., Nordenberg, S., & Annerman, M. (2007). Effects of the inoculation of cyanobacteria on the microstructure and the structural stability of a tropical soil. *Plant and Soil, 290*(1), 209–219.

Mazhar, S., Cohen, J. D., & Hasnain, S. (2013). Auxin producing non-heterocystous cyanobacteria and their impact on the growth and endogenous auxin homeostasis of wheat. *Journal of Basic Microbiology, 53*, 996–1003.

Osman, M. E. H., El-Sheekh, M. M., El-Naggar, A. H., & Gheda, S. F. (2010). Effect of two species of cyanobacteria as biofertilizers on some metabolic activities, growth, and yield of pea plant. *Biology and Fertility of Soils, 46*(8), 861–875.

Plastina, A. (2017). Estimated Costs of Crop Production in Iowa. Ag Decision Maker File A1-20.

Priya, H., Prasanna, R., Ramakrishnan, B., et al. (2015). Influence of cyanobacterial inoculation on the culturable microbiome and growth of rice. *Microbiological Research, 171*, 78–89.

Renuka, N., Prasanna, R., Sood, A., et al. (2016). Exploring the efficacy of wastewater-grown microalgal biomass as a biofertilizer for wheat. *Environmental Science and Pollution Research, 23*(7), 6608–6620.

Renuka, N., Guldhe, A., Prasanna, R., Singh, P., & Bux, F. (2018). Microalgae as multi-functional options in modern agriculture: Current trends, prospects and challenges. *Biotechnology Advances, 36*(4), 1255–1273.

Sadeghi, S. H., Kheirfam, H., Homaee, M., Darki, B. Z., & Vafakhah, M. (2017). Improving runoff behavior resulting from direct inoculation of soil micro-organisms. *Soil and Tillage Research, 171*, 35–41.

Singh, S., & Datta, P. (2007). Outdoor evaluation of herbicide resistant strains of Anabaena variabilis as biofertilizer for rice plants. *Plant and Soil, 296*(1-2), 95–102.

Subhashini, D., & Kaushik, B. (1981). Amelioration of sodic soils with blue-green algae. *Soil Research, 19*(3), 361–366.

Swarnalakshmi, K., Prasanna, R., Kumar, A., et al. (2013). Evaluating the influence of novel cyanobacterial biofilmed biofertilizers on soil fertility and plant nutrition in wheat. *European Journal of Soil Biology, 55*, 107–116.

Trejo, A., De-Bashan, L. E., Hartmann, A., et al. (2012). Recycling waste debris of immobilized microalgae and plant growth-promoting bacteria from wastewater treatment as a resource to improve fertility of eroded desert soil. *Environmental and Experimental Botany, 75*, 65–73.

Tripathi, R. D., Dwivedi, S., Shukla, M. K., Mishra, S., Srivastava, S., Singh, R., Rai, U. N., & Gupta, D. K. (2008). Role of blue green algae biofertilizer in ameliorating the nitrogen demand and fly-ash stress to the growth and yield of rice (Oryza sativa L.) plants. *Chemosphere, 70*(10), 1919–1929.

Wuang, S. C., Khin, M. C., Chua, P. Q. D., et al. (2016). Use of *Spirulina* biomass produced from treatment of aquaculture wastewater as agricultural fertilizers. *Algal Research, 15*, 59–64.

Yilmaz, E., & Sönmez, M. (2017). The role of organic/bio-fertilizer amendment on aggregate stability and organic carbon content in different aggregate scales. *Soil and Tillage Research, 168*, 118–124.

Part V
Biomass Production

Chapter 13
Microalgae Biomass Production: An Overview of Dynamic Operational Methods

Gul Muhammad, Md. Asraful Alam, Wenlong Xiong, Yongkun Lv, and Jing-Liang Xu

Abstract Microalgae biomass is a promising and sustainable feedstock with wide range of applications for biofuel, cosmetics, pharmaceutical, functional foods, aquaculture, and nutraceutical. The production scheme for microalgae products comprises several stages of processing. It starts with microalgal strain development and cultivation, followed by microalgal biomass harvesting or separation from the culture media and consequent thickening, dewatering, drying, and target product extraction. Efficient and cost-effective harvesting and drying approaches severely affect the overall energy consumption and production cost of microalgal-based products in a large scale. There are plenty of reviews and research articles available on various cultures and harvesting technologies of microalgae biomass generation. However very few articles are available on the drying and storage methods of biomass. Thus, in this chapter we provide a short overview of strain selection, cultivation, and harvesting, with additional importance in drying and storage advancement along with the advantages and disadvantages of various methods documented till now.

Keywords Microalgae · Biomass · High-moisture · Drying · Storage condition · Stabilization

1 Introduction

Microalgae are microscopic photosynthetic organisms that synthesize different organic substances like lipids, carbohydrates, protein, and vitamins through carbon fixation. Due to the wide range of applications such as biofuel, cosmetics, pharmaceutical, functional foods, aquaculture, and nutraceutical, microalgae gained

G. Muhammad · M. A. Alam · W. Xiong · Y. Lv · J.-L. Xu (✉)
School of Chemical Engineering, Zhengzhou University, Zhengzhou, China
e-mail: xujl@zzu.edu.cn

M. A. Alam et al. (eds.), *Microalgae Biotechnology for Food, Health and High Value Products*, https://doi.org/10.1007/978-981-15-0169-2_13

more attention at industrial and academic scale (Choi et al. 2019). One of the major challenges to the development of economical microalgal industry is the variability in microalgal biomass productivity year-round due to seasonal variations in temperature and insolation. The cultivation can be performed in closed photobioreactors or in open raceway ponds (Yew et al. 2019). Culture conditions like temperature, nutrients, light intensity, and pH need to optimized to stimulate the accumulation of valuable compounds in the cells. Microalgae biomass harvesting is a costly process as microalgae cells are tiny in size and contain much water in the culture system. However numerous research works have been carried out in the last few years on lab scale those need to optimize in pilot scale (Alam et al. 2017). Harvesting methods for microalgae include sedimentation, flotation, filtration, flocculation, and centrifugation, which are applied in either a single or a combined process. However, there is no all-purpose harvesting technique that could treat all types of microalgae suspensions, considering both energy consumption and cost. Harvested biomass contains high moisture and it leads to the damage of the quality of biomass slurry in room temperature within a few hours. Thus upon harvesting, it is necessary to dry the biomass immediately to make it stable and storable for further use. Different methods are applied like solar, oven, microwave, spray, freeze, cross flow air-drying, and incinerator. In this chapter we provide a short overview of microalgae, from strain collection to biomass generation and subsequently drying and storage.

2 Strain Selection

Microalgae are diversified high photosynthetic organisms mostly found in fresh or marine water. They have gained more attention since last decade due to the high content of valuable molecules such as 30–70% lipid, 40–65% carbohydrates, and 20–40% proteins (Chew et al. 2017). Lipids can be utilized as potential feedstock for biodiesel, whereas microalgal carbohydrates can be used as an alternative carbon source for conventional carbohydrates (treated lignocellulosic biomass or sugar) in fermentation industries. In addition, long-chain fatty acids present in microalgae possess major functions like food health supplements. Apart from carbohydrates and proteins, microalgae also contain valuable pigments and other macromolecules that can be used in pharmaceutical industries to develop various health products (Yen et al. 2013). Various reports indicate that there are about 150,000 microalgal species present in the world according to algaebase.org (Yew et al. 2019). But only a few species have been investigated in terms of valuable application. Most of the algal culture collection centres perform isolation and identification for their own use. However, many researchers are also focusing on the selection and breeding of microalgae strains such as *Chlorella*, *Dunaliella*, *Spirulina*, and *Haematococcus* for commercial interests (Xin et al. 2009). Strains such as *Dunaliella salina* with rich β-carotene early 1960s were used as nutritional sources. *Spirulina* has been reported for medical applications (i.e. viral infections, cancer, and cardiovascular diseases) because of their immunogenic characteristics, *Chlorella*, a freshwater microalga, is also used as a sugar and blood cholesterol reducer.

Currently, many research groups are trying to improve the yield of the required molecules (such as lipid) and valuable compounds (such as carotenoids, lutein) in microalgae under stressful conditions. Most of the studies only focus on the single strategy at lab scale. Therefore, it is necessary to focus on to multiple stresses for robost microalgal strain development to use wastewater treatment plus biomass production at industrial scale for biofuel and high value-added products production.

Omics approaches are widely adopted to enhance the lipid biosynthesis of various species under different stresses (i.e. nitrogen limitation), whereas in this technology analysis of bioactive compounds (carotenoids, lutein etc.) are limited as compared to the lipid. Genetic engineering technology is used to modified the metabolic pathways for improving the yield of targeted molecules. Adaptive laboratory evolution (ALE) has been extensively applied for promoting innovative biological and phenotypic functions in robust strains construction. Nutrient and environmental stress are very popular in ALE. For example, *Chlorella* sp. AE10 was examined with 10% CO_2, 3.7 g L^{-1} biomass concentration obtained after 31 cycles, which was 2.9 times higher than control (Li et al. 2015). ALE is attempted to generate sufficient nutrients in a robust *C. reinhardtii*, with double increment in lipid content resulting in a slight growth found under nitrogen starvation stress (Yu et al. 2013). The above-mentioned approaches such as omics technology, genetic engineering, and ALE for development of robust microalgae strains are discussed in detail in our book (*Microalgae Biotechnology for Development of Biofuel and Wastewater Treatment*).

The cost of microalgae biofuel is higher when compared to the traditional fuel, which requires an integration of lipid production and high value-added products to overcome the cost. Similarly, robust microalgae species with sufficient lipid and value-added products can also compress the expense during industrialization.

3 Microalgae Culturing Techniques

The major requirements for microalgae biomass are light (naturally or artificially), nutrients (carbon, nitrogen, and phosphorus), and preservation of adequate culture (temperature and pH) (Menegazzo and Fonseca 2019). The procedures to cultivate microalgae are autotrophic, heterotrophic, or mixotrophic conditions (Menegazzo and Fonseca 2019). The most common procedure for microalgae cultivation is autotrophic cultivation, where light is needed and carbon dioxide is used as energy and carbon source. Autotrophic cultivation returns slow growth and lowers biomass productivity due to the daily and seasonal variations in the form of light. Moreover, biomass produced during the day time may be lost due to the respiration at night. This can be mitigated by mixotrophic cultivation, where combination of organic and inorganic are used and cells consume CO_2 during photosynthesis and organic carbon during respiration. Therefore, the cells are engaged in the photosynthesis–respiration cycle to produce new cells and biomass (Rashid et al. 2019). Some species are capable to grow in dark conditions, and in this type of cultivation organic carbon sources such as glucose, acetate, and wastewater are used as substrates to reproduce microalgae (Tan et al. 2018). This type of cultivation is termed as heterotrophic cultivation. This procedure for microalgae cultivation is independent of light or solar energy. Each one of them has advantages and disadvantages.

3.1 Raceway Pond

Generally there are two types of cultivation systems to grow microalgae; one is the open raceway pond and the other is the closed bioreactor (Tan et al. 2018). The depth of raceway pond is usually about 0.3–0.5 m to ensure high growth rate. A paddle wheel is installed for mixing in the raceway pond and the motion is monitored at the flow channel as shown in Fig. 13.1 (Chisti 2007). These types of ponds are made up of concrete and sometimes lined with white plastic. During the day, the cultivation medium is fed permanently to the raceway pond before circulation starts. The paddle wheel is operated continuously to avoid settling of microalgae, and the broth is harvested on the completion of the circular loop (Brennan and Owende 2010; Chisti 2007; Pierre et al. 2011). Open raceway pond systems are practiced in China, USA, and Israel, which produce a microalgae yield of 0.5 g L^{-1} (Richmond 1990).

The main challenges for open raceway ponds are evaporation of water, less efficient utilization of CO_2 due to the water loss, temperature fluctuation, and seasonal variation. Additionally, other organisms which may contaminate or consume microalgae have a great effect on the microalgae biomass yield.

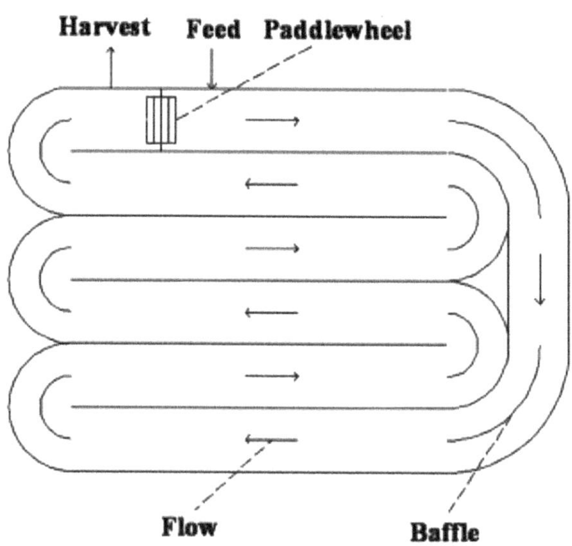

Fig. 13.1 Raceway pond (Chisti 2007)

3.2 Closed Photobioreactor

Photobioreactor is a closed type reactor that uses light as an energy source for photobiological reaction. This type of reactor produces more biomass compared to raceway ponds. Contamination can be avoided under specific conditions (Tan et al. 2018). For commercial scale, photobioreactors are not a good option for photoremediation of wastewater, as a huge volume of wastewater is generated daily (Rawat et al. 2011). From an economic point of view, the closed photobioreactor is costly compared to the open raceway pond. On the other hand, less land is required for closed photobioreactors to cultivate microalgae. For instance, it was estimated that about 19,000–57,000 lipid content per acre per year can be obtained from microalgae at optimum conditions on using closed photobioreactors. This yield of microalgae oil is 200 times more compared to oil-bearing crops (Tan et al. 2018).

A variety of closed reactors such as flat plate reactor, tubular airlift and bubble column bioreactor, helical tubular bioreactor, and alfa-shaped bioreactor have been developed according to reactor geometry. Table 13.1 gives an idea about the aforementioned bioreactors. The pros and cons of open pond and closed bioreactors, advances in technologies, automation, and manufacturer information are discussed in detail in our previous book (*Microalgae Biotechnology for Development of Biofuel and Wastewater Treatment*). Especially, closed photobioreactors are suggested for high-value long-chain fatty acids or proteins, whereas open pond is suitable only for biofuel production (Zeng et al. 2011).

4 Harvesting of Biomass

4.1 Centrifugation

Centrifugation is considered as the fastest and widely used harvesting method. Particle size and density are the major factors in centrifugation (Bux 2013). This method works on the basis of density difference which makes it fast and stable. Centrifugation offers several advantages compared to other commonly used processes, for instance, chemical (i.e. flocculants)-free biomass and 100% recovery efficiency. The quality of the biomass remains same because of the short duration. The high quality and chemical-free biomass are basic concerns for food products, cosmetic (Levine and Fleurence 2018), and pharmaceutical industries (Alam and Wang 2019). Centrifugation was reported to be very effective but energy-intensive and expensive to be applied on commercial scale. Undoubtably this is the most effective, efficient, and reliable technology, but one should keep in mind its high operational cost and energy consumption (Bux 2013).

Table 13.1 Details for different photobioreactors

Bioreactor type	Detail
Flat plate bioreactor 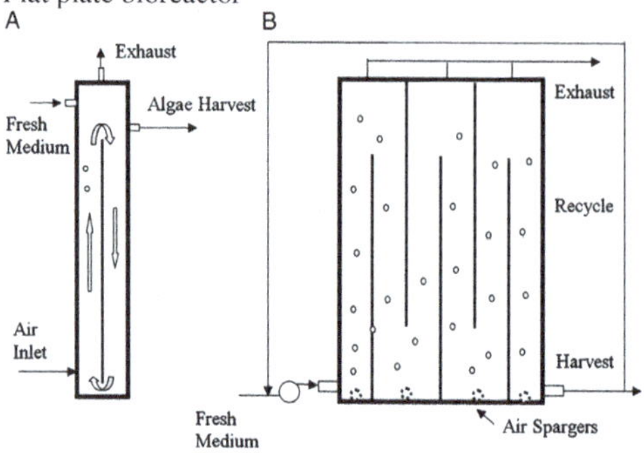 	Working: In this type of reactor light path is characterized by large illuminated solid/volume ratio. Pump-driven reactor works based on the liquid flow that is created by the pump and the turbulence causes the mixing. On the other hand, in airlift reactor compressed air is used for mixing
Bubble column and tubular airlift bioreactor  	Working: Vertical transparent tubes (glass or polyethylene) are used to utilize the maximum available sunlight, and CO_2 is supplied via bubbling. Their fabrication is not expensive Limitations: Due to the small area of the tubes gas does not transfer efficiently and photosynthesis does not take place well
Helical tubular bioreactor 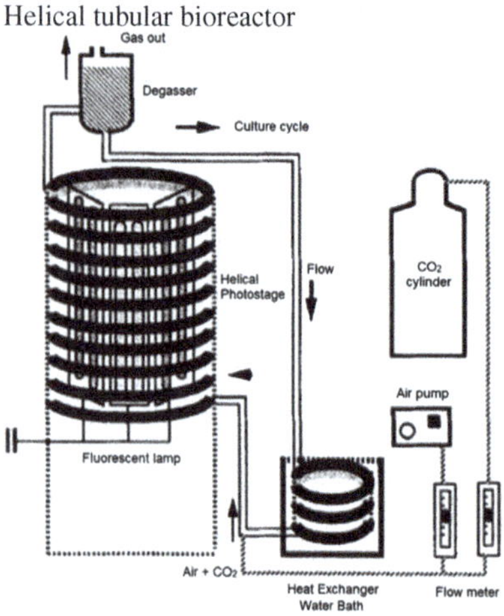 	Working: Flexible tubular pipes with coiled shape, heat exchanger, and gas exchanger tower. Due to the coil-shaped structure high algal biomass is produced Limitations: Remove the deposited algal biomass inside the tube

(continued)

Table 13.1 (continued)

Bioreactor type	Detail
Alfa-shaped bioreactor 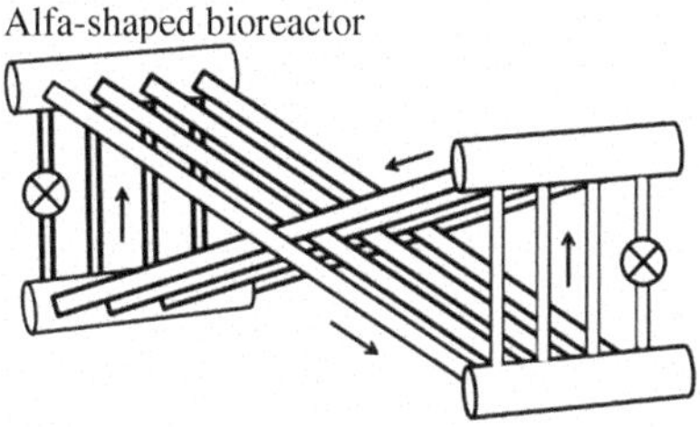	Working: Two airlift pumps are used to aerate and to mix the culture. High liquid flow and mass transfer can be maintained with low air supply. Surface-to-volume ratio is high Limitation: Foam is formed due to the high cell density

4.2 Sedimentation

Sedimentation is considered as an economical and most preferable microalgae harvesting method applied in water and wastewater treatment (Chun-Yen et al. 2011). During the sedimentation process the particles in a suspension settle down due to gravity and form concentrated slurry with clear liquid above (Mathimani and Mallick 2018; Pragya et al. 2013). Sedimentation depends upon the size of the cell or density. Larger the microalgae cell size, faster the sedimentation. This strategy to harvest microalgae is an attractive and less-energy-demanding strategy, and it requires very low cost for infrastructure (Roselet et al. 2019). However, floating-type microalgae cannot be harvested in this way, and it takes long time to settle down the algae cells.

4.3 Filtration

This method is known as a solid–liquid separation process where the algal culture runs through the filters, algal culture sticks with the membrane, and water passes through it (Menegazzo and Fonseca 2019). So it is the inverse of sedimentation, and separation occurs due to the difference in solids and membrane pores. To keep the flow fluent, pressure is required across the membrane like conventional filtration. Different types of filtrations are used for microalgae harvesting, namely microfiltration, vacuum filtration, tangential flow filtration, and dead-end filtration. However, filtration is not economical due to the high cost of filters and filters can be blocked quickly with cells' adhesion and frequent washing is needed during harvesting, which is not feasible for large production.

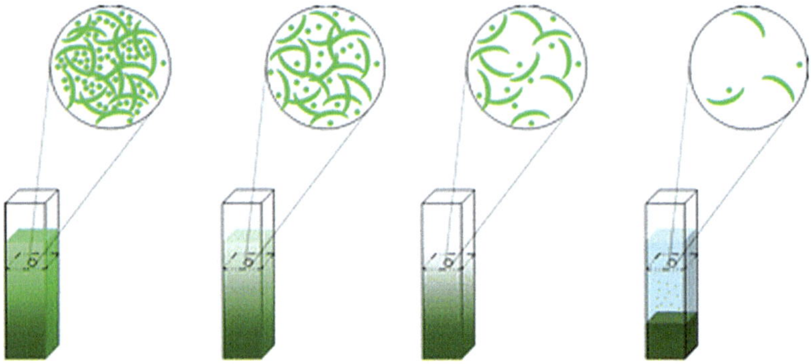

Fig. 13.2 Flocculation technique to harvest microalgae (Salim et al. 2011)

4.4 Flocculation

4.4.1 Chemical Flocculation

Flocculation is a microalgae harvesting method which is used to make aggregates called as algal flocs (Chun-Yen et al. 2011). This method is applied as a pre-concentrated to destabilize algal cells from water to increase the cell density. Microalgae harvesting via flocculation is often performed using chemical, physical, and bio-based flocculants. When flocculation takes place automatically with the increases of pH to a certain level in the solution, and is known as auto flocculation (Wan et al. 2015). In recent years cell flocculation for microalgae harvesting has also been of great interest.

Chemical flocculants are applied to a wide range of microalgae species (Wan et al. 2015). These methods works on the basis of coagulation, which is followed by settling at the bottom of the cultivating apparatus while the density of the cell is increased as shown in Fig. 13.2 (Salim et al. 2011). Mostly the chemical flocculants are classified as organic and inorganic flocculants (ammonia, metal salts, etc.). Successful application of chemical flocculation to various microalgae species (*Scenedesmus* sp., *Chlorella* sp., *N. salina*, *Neochloris* sp.) have been reported in (Wan et al. 2015; Alam et al. 2017).

Four common mechanisms are used in chemical flocculation: (a) charge neutralization, (b) electrostatic patch, (c) bridging, and (d) sweeping. In charge neutralization and electrostatic patch the positively charged ions adsorb on the negative surface of the microalgae cell. As a result, particles come together because of the van der walls attraction as shown in (Fig. 13.3a). While in electrostatic surfaces, charges are reversed with the particles, and patches are formed with each other (Fig. 13.3b). Whereas in bridging, microalgae cell attached to the segment of polymer to form bridges that causes aggregation and flocculation takes place (Fig. 13.3c). In sweeping process, cells or particles are entrapped by a massive aggregation, which results in flocculation (Fig. 13.3d).

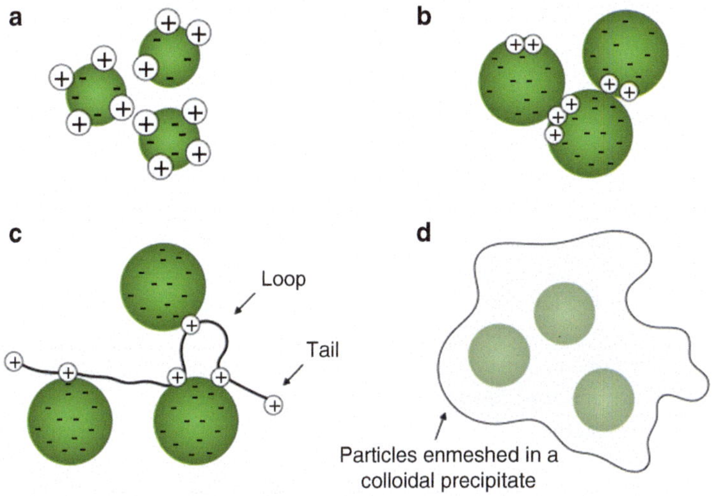

Fig. 13.3 Flocculation mechanism charge neutralization (**a**), electrostatic patch mechanism (**b**) bridging mechanism (**c**), and sweeping flocculation (**d**) (Roselet et al. 2019)

4.4.2 Bio-flocculation

The flocculation process is induced by microorganisms, extracellular polysaccharides (EPS), or other substances (proteins) from flocculating organisms. Currently it is considered as an innovative and economical approach to harvest the microalgae (Alam et al. 2017). When compared to other approaches bio-flocculation is an environment-friendly, cheap, and sustainable approach to harvest microalgae in a bulk quantity (Ummalyma et al. 2017). Bio-flocculation can be categorized into three types (1) microbial flocculation, (2) microorganism-associated flocculation, (3) microalgal self-flocculation.

Most of the recent studies confirmed the advantages of microbial bio-flocculation in wastewater treatment. For example 90% bio-flocculation efficiency was achieved of *Nannochloropsis* oceanic microalgae using the bacterial strain *Solibacillus silvestris* without an effect on growth (Wan et al. 2013). In another study the feasibility of microbial flocculant poly-c-glumatic acid (c-PGA) to harvest fresh water *Chlorella protothecoides* and marine *Chlorella vulgaris* were optimized via response surface methodology (RSM) (Zheng et al. 2012). The underlying mechanism is poorly understood and needs further research due to its chemical-free approach (Ummalyma et al. 2017).

An innovative type of flocculation is microalgal-self flocculation freely flocculated cells flocculate the other microalgal cells with them. This type of flocculation is effective, chemical-free, and can be used in a wide range of applications from high value-added products to low value-added products (Alam et al. 2017). Only a few examples of the self-flocculating microalgae have been reported like

Scenedesmus obliquus, *C. vulgaris* JSC-7, and *Ankistrodesmus falcatus*. Most of the studies are performed on a lab scale. For example, when CNW11 is co-cultured with JSC-7, the harvesting efficiency increased three times (25.6–68.3%) (Alam et al. 2014). This group also investigated and found that the polysaccharide molecules present in the cell wall of *C. vulgaris* JSC-7 are mainly responsible for the patching and bridging with other non-flocculating cells and thereby enhancing the flocculation rate (Alam et al. 2014).

4.4.3 Electrolytic Flocculation

The electrolytic flocculation is free from flocculants. The main principle behind electrolytic flocculation is the movement of negatively charged algal cells towards anode where they lose charge and form aggregates (Mubarak et al. 2019) as shown in Fig. 13.4. Poelman et al. (1997) reported the effectiveness of electrolytic flocculation with 80–95% harvesting efficiency in 35 min with different species involved: green algae (*Staurastrum* sp., *Closterium* sp., *Cryptomonas sp.*, *Pediastrum sp.*, etc.), diatoms (*Melosira* sp., *Asterionella* sp., *Cyclotella sp.*, etc.), and blue green algae (*Asphanizomenon* sp., *Coelosphaerium* sp., etc.). This technique consumes very little amount of energy, 0.3 kWh m^{-3}. It was also noted that when the voltage decreased, the removal efficiency was also decreased, and less energy was consumed when the distance between the electrodes was reduced (Poelman et al. 1997). In another study 93.6% recovery efficiency was achieved in 30 min using electrolytic flocculation with 6 W power supply (Xu et al. 2010). Moreover, this type of flocculation is free

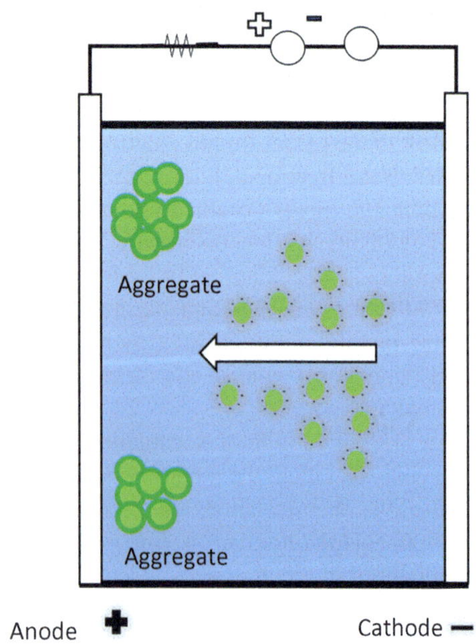

Fig. 13.4 Electrolytic flocculation

of risk from contamination and less expensive but has some problems such as reactor design, periodic replacement of electrodes, and initial investment cost.

4.5 *Floatation*

At laboratory scale, floatation is applicable for harvesting algae (small unicellular). Floatation is a separation process in which solid particles (microalgae) are carried to the liquid surface with the help of gas bubbles. This is achieved by the adhesion of gas bubbles to the cells by increasing buoyancy and decreasing their density. Compared to sedimentation, floatation is considered to be more effective in microalgae harvesting (Menegazzo and Fonseca 2019; Singh et al. 2011). Many factors can affect the adhesion of gas bubbles to microalgae, such as the size of the microalgae, collision and adhesion between microalgae, and gas bubbles. It is easier to bring small-sized microalgae particles to the surface compared to larger. But due to decrease in size also there is a less chance of collision and adhesion with bubbles. Furthermore, both (bubbles and microalgae) are negatively charged surfaces so it results in electrostatic repulsion. Therefore, it is required to add additives (i.e. surfactants) (Garg et al. 2012; Granados et al. 2012; Zhou et al. 2017). In addition, floatation can be categorized with regard to bubble size technology, such as dissolved air floatation, dispersed floatation, or electrolytic floatation (Laamanen et al. 2016; Menegazzo and Fonseca 2019).

5 Drying Process of Microalgae Biomass

After harvesting microalgae through separation from the media, the biomass is often moist and contains high amount of water which need to dry before further process depends upon the type of the product which is will be extracted from the next step, i.e., microalgal lipid, carbohydrates, pigments, or others. Various studies reported the drying methods (Chatsungnoen and Chisti 2016), Such as solar, oven, microwave, spray, freeze, cross flow air-drying, and incinerator. Common methods are discussed below.

5.1 *Solar Drying*

Solar drying is the sustainable source for drying algae, but it requires a long time and more surface area (Brennan and Owende 2010). Furthermore, it is difficult to control the quality of the desired product with conventional open drying solar method; because of the slow drying rate, low temperature can degrade the biomass quality and causes bacterial contamination. A closed solar drying device was developed for drying the microalgae biomass, which carries the temperature from

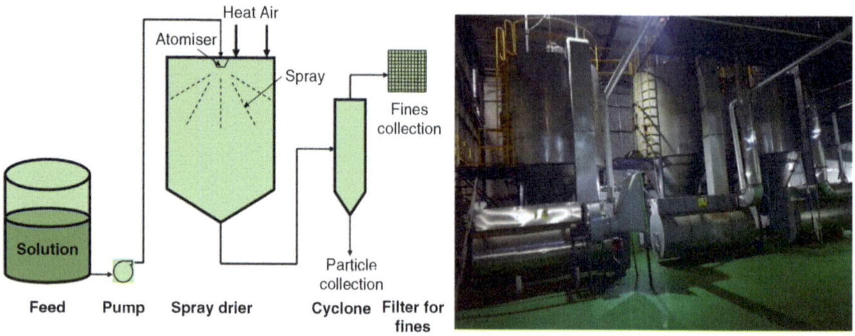

Fig. 13.5 Spray dryer; schematic of spray drying process (left), atomize spray dryer in Fuqing King Dnarmsa Spirulina Co., Ltd, China (right). The microalgae mud can be atomized through an atomizer on the top of the tower and be dried by hot air

35 to 60 °C; 90% of the moisture content was removed within 3–5 h. In this experiment, two microalgae species namely *Spirulina* and *Scenedesmus* were tested and found that the dried biomass contained only 10% moisture (Prakash et al. 1997). It was claimed that sun drying is an acceptable solution when biomass will be used as animal feed.

5.2 Spray Drying

Spray drying system (Fig. 13.5) includes liquid atomization, mixing of gas or droplets, and drying from liquid droplets. In a vertical tower, atomized droplets are sprayed in which hot gases pass in downward direction. Drying is achieved within few seconds, product is removed from the bottom of tower, and a cyclonic dust separator is used to remove exhausted.

Spray drying is supposed to be an appropriate method for high value-added products and can produce green dark or green microalga powder. The colour depends upon the spray drying and temperature (Chen et al. 2015). Furthermore, spray dried biomass can also be used for human consumption. In an evaluation of two different methods, drum drying and spray drying, the first one was recommended due to its advantages like better digestibility, less energy, and lower investment.

5.3 Rotary Drying

In rotary drying method a sloped rotating cylinder known as rotary dryer is used. The material (algae) is dried from one end to other by using gravity. In the past, different types of dryers were developed for drying, such as direct heating (drying material is in contact with the hot gases) and indirect heating type (hot gases are separated from the material using steel shell). Rotary kiln dryers and drum drying are widely used for drying wastewater sludge. Drum drying is the most common method used to dry microalgae (Mohn 1978). Thin-layered drum dryer used to dry *Scenedesmus* algae showed excellent yield product. Drying microalgae via drum drier has dual advantages; the sterilizing of the sample and the breaking of the cell wall.

5.4 *Freeze Drying*

Freeze drying is a common method used in the food industry and also in research, because the cell constituents are protected without cell wall disruption (Chen et al. 2015). Compared to spray drying, freeze drying retains more amount of protein in dry microalgae, with less than 10% protein loss (Desmorieux and Hernandez 2004). During the freeze drying process samples freeze slowly and larger intracellular crystals are formed.

Drying methods are selected on the basis of scale and operation speed. Merits and demerits of some drying methods are summarized in Table 13.2.

Among the aforementioned methods, spray drying is considered as a promising method to extract high value-added products (high protein content). However, spray drying is expensive and can destroy certain pigments (Tan et al. 2018). So in future, attention should be given to improve drying device and need to give important to develop device and method which can be used for all types of product derived from microalgae biomass.

6 Advancement in Storage Process of Microalgae Biomass

Algal biomass is promising and attractive, and is increasing as a feedstock for high-value-added products and biofuels. But due to seasonal variations, it is hard to produce enough biomass year-round for delivering a constant supply of feedstock to the conversion facility. Generally, herbaceous biomass is harvested annually and stock for the use of whole year for biofuel production or agriculture use.

Table 13.2 Summary of various drying methods

Method	Advantages	Disadvantages
Solar	Sustainable and no energy consumption	Dependence on the weather
Spray	Fast and economical method, suitable for algae production for human consumption	Degradation in the quality, operational cost
Freeze	High energy intensive	Applicable for small scale
Oven drying	Less energy intensive	Suitable for small scale
Cross flow air-drying	Economical and fast drying	Energy cost
Incinerator	Algal biomass burning can be avoided	High cost and complicated

Eventhough, microalgal biomass production occurs many times throughout the year in many areas, but biomass yield and growth rate varies because of the seasonal variations like temperature and solar irradiations thus need to store for continue supply to the conversion facility (Coleman et al. 2014; Moody et al. 2014; Wigmosta et al. 2011). There are two types of microalgae storage; one is called dry route and other is the wet route. The first one is expensive, as the natural gas-fed dryers cost about 150$ per ton biomass. So the second one (wet anaerobic storage) is an alternative to the first one, which is used for a long time to store herbaceous biomass for livestock use (Wendt et al. 2017). For example, in the past ensiling was used to store the different kinds of herbaceous biomass (wheat straw, corn stover, switch grass, and other grasses) to produce bioenergy (Linden et al. 2010; Oleskowicz-Popiel et al. 2011; Shinners et al. 2007, 2010, 2011).

Ensiling was successfully applied to preserve the microalgae biomass as well. It has been reported that if wet *Scenedesmus obliquus* biomass is stored with 80% water content in acidified anaerobic conditions for one month, about 6–14% dry matter is lost (Wendt et al. 2017). In another study, the effect of different parameters like drying, storage, and air exposure on the stability of astaxanthin was found. *Haematococcus pluvialis* was dried by freeze or spray drying under vacuum air −20 °C to 37 °C for 5 months. Freeze drying showed 41% higher astaxanthin recovery than spray drying and at −20 °C and 4 °C showed maximum astaxanthin stability with a little 12.3 ± 3.1% degradation during 5 months. Regarding the economic point of view almost AUD$600 higher profit can be obtained compared to 100 kg of *H. pluvialis*. So, freeze drying is recommended as a cost-effective method for longer storage (Ahmed et al. 2015). The effect of spray drying and storage on astaxanthin content on *H. pluvialis* at different conditions, inlet temperature (220 and 180 °C) and outlet temperature (80, 110, 120 °C) for 9 weeks period was evaluated. A reasonable preservation of astaxanthin obtained (180 °C/80 °C, −21 °C nitrogen, 180 °C/110 °C, 21 °C nitrogen and 220/80 °C, 21 °C). Preventing the degradation of astaxanthin content in *H. pluvialis* biomass, spray dried carotenized cells (180/110 °C) under nitrogen and −21 °C was recommended for storage by Raposo et al. (2012).

From the above studies it is clear that wet anaerobic microalgae powder can be stored from 1 month to 4 months. An alternative drying method like co-storage of microalgae with corn stover would also be a valid method for stability during seasonal fluctuations. So anaerobic wet storage can be applied on industrial scale for a short-term stabilization of microalgae industry. Advanced research focusing on the long-term storage and mechanism behind abolish biomass quality is needed to understand the better approach to preserve algae biomass. Personal communication with the expert from Fuqing King Dnarmsa Spirulina Co., Ltd, China suggested that microalgal pigments, such as algal blue protein, carotene, chlorophyll, etc., can easily decompose under high temperature, light, or oxygen during the storage period. So dry microalgae powder needs to be preserved in a temperature- and humidity-controlled environment to comply with the requirements of standard regulations and customer's needs. Generally, the control temperature of the microalgae powder warehouse needs to be set below 20 °C and the humidity should be below 75%. Microalgae dry powder needs packing in double-PE plastic bag and needs to be air-free/vacuum before the bag is filled with biomass. In addition, the microalgae powder bags should be placed on the tray 20 cm above the ground and 20 cm distance from the side wall and roof, and an adequate control device needs to be installed to protect the biomass from insects, flies, mosquitoes, etc.

7 Conclusions

Microalgae is a promising eco-friendly feedstock for food, feed, chemicals and biofuel. A robust strain with high yield of target product is beneficial for the industry. It is necessary to optimize culture conditions to get sufficient yield and avoid contamination of biomass. A combination of flocculation and centrifugation is proven better than flotation and filtration for harvesting and dewatering microalgae. Spray drying is the best option for industry, and even solar drying is cheap considering the energy consumption. To understand the drying and storage approaches, more fundamental study is needed to fulfil the present knowledge gap.

References

Ahmed, F., Yan, L., Fanning, K., Netzel, M., & Schenk, P. M. (2015). Effect of drying, storage temperature and air exposure on astaxanthin stability from *Haematococcus pluvialis*. *Food Research International, 74*, 231–236.

Alam, M. A., Wan, C., Guo, S.-L., Zhao, X.-Q., Huang, Z.-Y., Yang, Y.-L., Chang, J.-S., & Bai, F.-W. (2014). Characterization of the flocculating agent from the spontaneously flocculating microalga *Chlorella vulgaris* JSC-7. *Journal of Bioscience and Bioengineering, 118*(1), 29–33.

Alam, M. A., & Wang, Z. (Eds.). (2019). *Microalgae biotechnology for development of biofuel and wastewater treatment*. Singapore: Springer.

Alam, M. A., Wang, Z., & Yuan, Z. (2017). *Generation and harvesting of microalgae biomass for biofuel production. Prospects and challenges in algal biotechnology* (pp. 89–111). Singapore: Springer.

Brennan, L., & Owende, P. (2010). Biofuels from microalgae—A review of technologies for production, processing, and extractions of biofuels and co-products. *Renewable & Sustainable Energy Reviews, 14*(2), 557–577.

Bux, F. (2013). Biotechnological applications of microalgae. *International e-Journal of Science, Medicine & Education, 6*(Suppl 1), S24–S37.

Chatsungnoen, T., & Chisti, Y. (2016). Oil production by six microalgae: Impact of flocculants and drying on oil recovery from the biomass. *Journal of Applied Phycology, 28*(5), 2697–2705.

Chen, C.-L., Chang, J.-S., & Lee, D.-J. (2015). Dewatering and drying methods for microalgae. *Drying Technology, 33*, 443.

Chew, K. W., Jing, Y. Y., Show, P. L., Suan, N. H., Juan, J. C., Ling, T. C., Lee, D. J., & Chang, J. S. (2017). Microalgae biorefinery: High value products perspectives. *Bioresource Technology, 229*, 53–62.

Chisti, Y. (2007). Biodiesel from microalgae. *Biotechnology Advances, 25*(3), 294–306.

Choi, S.-A., Oh, Y.-K., Lee, J., Sim, S. J., Hong, M. E., Park, J.-Y., Kim, M.-S., Kim, S. W., & Lee, J.-S. (2019). High-efficiency cell disruption and astaxanthin recovery from *Haematococcus pluvialis* cyst cells using room-temperature imidazolium-based ionic liquid/water mixtures. *Bioresource Technology, 274*, 120–126.

Chun-Yen, C., Kuei-Ling, Y., Rifka, A., Duu-Jong, L., & Jo-Shu, C. (2011). Cultivation, photobioreactor design and harvesting of microalgae for biodiesel production: A critical review. *Bioresource Technology, 102*(1), 71–81.

Coleman, A. M., Abodeely, J. M., Skaggs, R. L., Moeglein, W. A., Newby, D. T., Venteris, E. R., & Wigmosta, M. S. (2014). An integrated assessment of location-dependent scaling for microalgae biofuel production facilities. *Algal Research, 5*(1), 79–94.

Desmorieux, H., & Hernandez, F. (2004). Biochemical and physical criteria of *Spirulina* after different drying processes. In *Proceedings of the 14th international drying symposium (IDS 2004)*, Sào Paulo, 22–25 August 2004 (pp. 900–907).

Garg, S., Li, Y., Wang, L., & Schenk, P. M. (2012). Flotation of marine microalgae: Effect of algal hydrophobicity. *Bioresource Technology, 121*(10), 471–474.

Granados, M. R., Acién, F. G., Gómez, C., Fernández-Sevilla, J. M., & Molina Grima, E. (2012). Evaluation of flocculants for the recovery of freshwater microalgae. *Bioresource Technology, 118*, 102–110.

Laamanen, C. A., Ross, G. M., & Scott, J. A. (2016). Flotation harvesting of microalgae. *Renewable and Sustainable Energy Reviews, 58*, 75–86.

Levine, I., & Fleurence, J. (2018). *Microalgae in health and disease prevention*. London: Academic Press.

Li, D., Wang, L., Zhao, Q., Wei, W., & Sun, Y. (2015). Improving high carbon dioxide tolerance and carbon dioxide fixation capability of *Chlorella* sp. by adaptive laboratory evolution. *Bioresource Technology, 185*, 269–275.

Linden, J. C., Henk, L. L., Murphy, V. G., Smith, D. H., Gabrielsen, B. C., Tengerdy, R. P., & Czako, L. (2010). Preservation of potential fermentables in sweet sorghum by ensiling. *Biotechnology and Bioengineering, 30*(7), 860–867.

Mathimani, T., & Mallick, N. (2018). A comprehensive review on harvesting of microalgae for biodiesel – Key challenges and future directions. *Renewable and Sustainable Energy Reviews, 91*, 1103–1120.

Menegazzo, M. L., & Fonseca, G. G. (2019). Biomass recovery and lipid extraction processes for microalgae biofuels production: A review. *Renewable and Sustainable Energy Reviews, 107*, 87–107.

Mohn, F. H. (1978). Improved technologies for the harvesting and processing of microalgae and their impact on production costs. *Archiv fur Hydrobiologie, Beihefte Ergebnisse der Limnologie, 1*, 228–253.

Moody, J. W., Mcginty, C. M., & Quinn, J. C. (2014). Global evaluation of biofuel potential from microalgae. *Proceedings of the National Academy of Sciences of the United States of America, 111*(23), 8691–8696.

Mubarak, M., Shaija, A., & Suchithra, T. V. (2019). Flocculation: An effective way to harvest microalgae for biodiesel production. *Journal of Environmental Chemical Engineering, 7*(4), 103221.

Oleskowicz-Popiel, P., Thomsen, A. B., & Schmidt, J. E. (2011). Ensiling – Wet-storage method for lignocellulosic biomass for bioethanol production. *Biomass & Bioenergy, 35*(5), 2087–2092.

Pierre, C., Arnaud, H., Laurent, L., Monique, R., Romy-Alice, G., & Jean-Philippe, S. (2011). Lifecycle assessment of microalgae culture coupled to biogas production. *Bioresource Technology, 102*(1), 207–214.

Poelman, E., De Pauw, N., & Jeurissen, B. (1997). Potential of electrolytic flocculation for recovery of micro-algae. *Resources, Conservation and Recycling, 19*(1), 1–10.

Pragya, N., Pandey, K. K., & Sahoo, P. K. (2013). A review on harvesting, oil extraction and biofuels production technologies from microalgae. *Renewable & Sustainable Energy Reviews, 24*(10), 159–171.

Prakash, J., Pushparaj, B., Carlozzi, P., Torzillo, G., Montaini, E., & Materassi, R. (1997). Microalgal biomass drying by a simple solar device. *International Journal of Solar Energy, 18*, 303.

Raposo, M. F. J., Morais, A. M. M. B., & Rui, M. S. C. M. (2012). Effects of spray-drying and storage on astaxanthin content of *Haematococcus pluvialis* biomass. *World Journal of Microbiology & Biotechnology, 28*(3), 1253–1257.

Rashid, N., Lee, B., & Chang, Y.-K. (2019). Recent trends in microalgae research for sustainable energy production and biorefinery applications. In *Microalgae biotechnology for development of biofuel and wastewater treatment* (pp. 3–20).

Rawat, I., Kumar, R. R., Mutanda, T., & Bux, F. (2011). Dual role of microalgae: Phycoremediation of domestic wastewater and biomass production for sustainable biofuels production. *Applied Energy, 88*(10), 3411–3424.

Richmond, A. (1990). Quantitative assessment of the major limitations on productivity of *Spirulina platensis* in open raceways. *Journal of Applied Phycology, 2*(3), 195–206.

Roselet, F., Vandamme, D., Muylaert, K., & Abreu, P. (2019). Harvesting of microalgae for biomass production. In *Microalgae biotechnology for development of biofuel and wastewater treatment, Alam M. A. and Wang Z. (eds.)* (pp. 211–244). Singapore: Springer.

Salim, S., Bosma, R., Vermuë, M. H., & Wijffels, R. H. (2011). Harvesting of microalgae by bioflocculation. *Journal of Applied Phycology, 23*(5), 849–855.

Shinners, K., Wepner, A. D., Muck, R. E., & Weimer, P. J. (2011). Aerobic and anaerobic storage of single-pass, chopped corn stover. *BioEnergy Research, 4*(1), 61–75.

Shinners, K. J., Binversie, B. N., Muck, R. E., & Weimer, P. J. (2007). Comparison of wet and dry corn stover harvest and storage. *Biomass & Bioenergy, 31*(4), 211–221.

Shinners, K. J., Boettcher, G. C., Muck, R. E., Weimer, P. J., & Casler, M. D. (2010). Harvest and storage of two perennial grasses as biomass feedstocks. *Transactions of the ASABE, 53*(2), 359–370.

Singh, A., Nigam, P. S., & Murphy, J. D. (2011). Mechanism and challenges in commercialisation of algal biofuels. *Bioresource Technology, 102*(1), 26–34.

Tan, X., Man, K. L., Uemura, Y., Lim, J. W., Wong, C. Y., & Lee, K. T. (2018). Cultivation of microalgae for biodiesel production: A review on upstream and downstream processing. *Chinese Journal of Chemical Engineering, 26*(1), 17–30. S1004954117304305.

Ummalyma, S. B., Gnansounou, E., Sukumaran, R. K., Sindhu, R., Pandey, A., & Sahoo, D. (2017). Bioflocculation: An alternative strategy for harvesting of microalgae – An overview. *Bioresource Technology, 242*, 227.

Wan, C., Alam, M. A., Zhao, X. Q., Zhang, X. Y., Guo, S. L., Ho, S. H., Chang, J. S., & Bai, F. W. (2015). Current progress and future prospect of microalgal biomass harvest using various flocculation technologies: Biomass, bioenergy, biowastes, conversion technologies, biotransformations, production technologies. *Bioresource Technology, 184*, 251–257.

Wan, C., Zhao, X. Q., Guo, S.-L., Alam, M. A., & Bai, F. W. (2013). Bioflocculant production from *Solibacillus silvestris* W01 and its application in cost-effective harvest of marine microalga Nannochloropsis oceanica by flocculation. *Bioresource Technology, 135*, 207–212.

Wendt, L. M., Wahlen, B. D., Li, C., Kachurin, G., Ogden, K. L., & Murphy, J. A. (2017). Evaluation of a high-moisture stabilization strategy for harvested microalgae blended with herbaceous biomass: Part I—Storage performance. *Algal Research, 25*, 567–575, S2211926416307664.

Wigmosta, M. S., Coleman, A. M., Skaggs, R. J., Huesemann, M. H., & Lane, L. J. (2011). National microalgae biofuel production potential and resource demand. *Water Resources Research, 47*(3), 289–306.

Xin, M., Yang, J. M., Xin, X., Lei, Z., Nie, Q. J., & Mo, X. (2009). Biodiesel production from oleaginous microorganisms. *Renewable Energy, 34*(1), 1–5.

Xu, L., Wang, F., Li, H.-Z., Hu, Z.-M., Guo, C., & Liu, C.-Z. (2010). Development of an efficient electroflocculation technology integrated with dispersed-air flotation for harvesting microalgae. *Journal of Chemical Technology & Biotechnology, 85*(11), 1504–1507.

Yen, H. W., Hu, I. C., Chen, C. Y., Ho, S. H., Lee, D. J., & Chang, J. S. (2013). Microalgae-based biorefinery – From biofuels to natural products. *Bioresource Technology, 135*(2), 166–174.

Yew, G. Y., Lee, S. Y., Show, P. L., Tao, Y., Law, C. L., Nguyen, T. T. C., & Chang, J.-S. (2019). Recent advances in algae biodiesel production: From upstream cultivation to downstream processing. *Bioresource Technology Reports, 7*, 100227.

Yu, S., Zhao, Q., Miao, X., & Shi, J. (2013). Enhancement of lipid production in low-starch mutants *Chlamydomonas reinhardtii* by adaptive laboratory evolution. *Bioresource Technology, 147*(7), 499–507.

Zeng, X., Danquah, M. K., Chen, X. D., & Lu, Y. (2011). Microalgae bioengineering: From CO_2 fixation to biofuel production. *Renewable & Sustainable Energy Reviews, 15*(6), 3252–3260.

Zheng, H., Gao, Z., Yin, J., Tang, X., Ji, X., & Huang, H. (2012). Harvesting of microalgae by flocculation with poly (γ-glutamic acid). *Bioresource Technology, 112*, 212–220.

Zhou, Y., Lai, Y. J. S., Eustance, E., Straka, L., & Rittmann, B. E. (2017). How myristyltrimethylammonium bromide enhances biomass harvesting and pigments extraction from *Synechocystis* sp. PCC 6803. *Water Research, 126*, 189–196.

Chapter 14
Microalgal Carbohydrates and Proteins: Synthesis, Extraction, Applications, and Challenges

Ayesha Shahid, Fahad Khan, Niaz Ahmad, Muhammad Farooq, and Muhammad Aamer Mehmood

Abstract Microalgae are a promising feedstock for renewable energy, nutraceuticals, pharmaceuticals, and other high-value industrial products. The major components of algal biomass are carbohydrates, lipids, and proteins, whose concentration depends upon cultivation conditions, composition of growth media, light intensity/duration, and CO_2 supplies. Microalgae can also be exploited as an alternative "protein crop" based on amino-acid composition, protein quality, and digestibility. Algal carbohydrates are mainly in the form of starch and cellulose, which can be used to produce bioethanol and degradable bioplastics. Although use of algal biomass for various products looks attractive, yet its commercial demonstration is hindered by the slow growth, low product yields, unavailability of high-throughput extraction procedures, and the product-refining processes. This book chapter comprehends the cultivation conditions to enhance the algal protein and carbohydrate content along with extraction techniques, and associated challenges for the recovery, separation, and characterization of these metabolites. Likewise, the potential applications of the microalgae-based carbohydrates and proteins in energy, food, pharmaceutical, and cosmetic industries along with future opportunities are also discussed to devise a roadmap for designing robust algal biorefineries.

A. Shahid · F. Khan · M. Farooq
Department of Bioinformatics and Biotechnology, Bioenergy Research Center, Government College University Faisalabad, Faisalabad, Pakistan

N. Ahmad
Agricultural Biotechnology Division, National Institute for Biotechnology & Genetic Engineering, Faisalabad, Pakistan

M. A. Mehmood (✉)
Department of Bioinformatics and Biotechnology, Bioenergy Research Center, Government College University Faisalabad, Faisalabad, Pakistan

School of Bioengineering, Sichuan University of Science & Engineering, Zigong, China
e-mail: draamer@gcuf.edu.pk

M. A. Alam et al. (eds.), *Microalgae Biotechnology for Food, Health and High Value Products*, https://doi.org/10.1007/978-981-15-0169-2_14

Keywords Algal carbohydrates · Algal proteins · Bioactive compounds · Product enrichment · Extraction technologies

1 Introduction

Microalgae have gained immense attention as an alternative feedstock to obtain bioactive metabolites with a special focus on "minimum waste generation" (El-Dalatony et al. 2019). It has potential to address the persisting challenges of global warming, alternative renewable energy resources, food security, and human health (Salla et al. 2016). In addition, microalgae have also the inherent ability of treating wastewater and accumulation of carbon-based compounds in the forms of carbohydrates, proteins, and lipids while harnessing the solar energy into biomass (Afzal et al. 2017; Shahid et al. 2017). According to an estimate, the share of microalgae-based products in the global market will reach to ~1143 million $ by 2024 (Mehta et al. 2018).

In microalgae, 15–60% of biomass normally accounts for carbohydrate content—a direct product of CO_2 fixation through the Calvin cycle during photosynthesis (Fig. 14.1). They are mainly present in the form of starch (present in plastids), cellulose, glycogen, polysaccharides (present in the cell-wall), agar, etc. (Khan et al. 2018b). Cyanobacteria normally accumulate carbohydrates as glycogen, while microalgae do as starch (amylopectin-like polysaccharide). The carbohydrates present in cell walls provide structural support, while the intracellular carbohydrates act as storage molecules, which are used as energy sources to drive the metabolic processes or act as protectants for the survival under stress (Markou et al. 2012). Although the microalgal carbohydrates have low energy content (15.7 kJ g^{-1}), they are preferred choice for the production of biohydrogen, bioethanol, and biobutanol (Markou et al. 2012) due to the availability of high levels of fermentable sugars, low-hemicellulose, and zero-lignin content. However, based on the microalgal species, carbohydrate content and composition may vary (Chen et al. 2013)

Green microalgae are rich source of proteins, which accounts for 60–70% of the algal-biomass composition with the energy content of 16.7 kJ g^{-1} (Markou et al. 2012), having productional and nutritional benefits over traditional protein sources such as legumes, egg, milk, meat, and soybean. Depending on habitat, season, and algal species, the protein composition may also vary (Fig. 14.1). They normally contain high amounts of essential amino acids like glutamic acid and aspartic acids. As compared to terrestrial plants, higher amounts of methionine, lysine, histidine, tryptophan, threonine, and cysteine have been reported in algal proteins (Guedes et al. 2019). Microalgae-based proteins are gaining interest as an alternative protein source due to their balanced amino-acid proportions. Microalgae are of commercial importance due to their structural, functional, and nutritional value and higher protein yields (4–15 metric

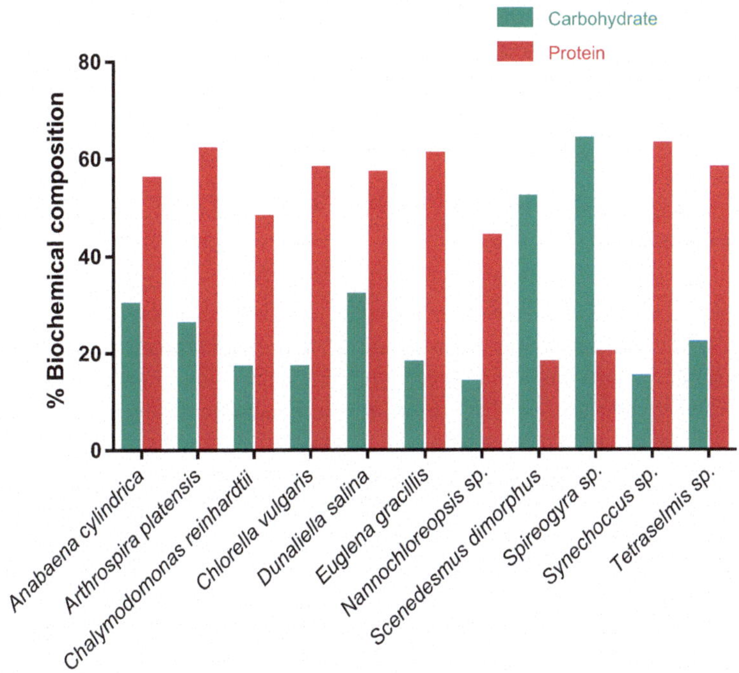

Fig. 14.1 Microalgae-based carbohydrate and protein content based on algal species

tons ha^{-1} $year^{-1}$) when compared to terrestrial plants (0.6–2 metric tons ha^{-1} $year^{-1}$) (Bleakley and Hayes 2017). They are mainly used as animal feed, and as food and/or food supplement for human consumption (Guedes et al. 2019). Algal proteins also have pharmaceutical importance as they are known to boost immune system, digestion, and energy levels, reduce fatigue, improve kidney, liver and cardiovascular functions, and detoxify the antinutrients in the body (Mehta et al. 2018).

However, wild-type strains normally contain low levels of carbohydrates or the biomass production is compromised under optimized conditions. In order to obtain the high amounts of targeted product, the first and important step is the selection of suitable microalgal strain, which must be rich in the target product under optimized conditions. Next step is to optimize culturing conditions (abiotic factors) to enhance the biomass productivity and accumulation of the product of interest (Yen et al. 2013).

This chapter discusses the importance of microalgae-based carbohydrates and proteins. While, accumulation of large amounts of metabolites especially carbohydrate and proteins are challenging. Because, either the wild-type strains are not capable to accumulate these metabolites or the hyperaccumulation of one of the metabolites may compromise the growth productivity. This chapter briefly discusses stress manipulation as a suitable strategy to enhance growth and production of desired products. A variety of pre-treatment and extraction methods are available to convert microalgal metabolites in useful industrial products. An overview of the most suitable strategies along with their possible merits and demerits is also provided to help the reader choose the most appropriate method according to the need. Moreover, possible applications of microalgae-based carbohydrates and proteins at commercial scales are also highlighted.

2 Manipulation of the Culturing Parameters for Product Enrichment

Large-scale cultivation decisively contributes to a sustainable industry for the economically viable production of biomass and other high-value products. Identification of robust microalgal strains and process engineering are some considerations which must be taken into account for the improved metabolite production (Chia et al. 2018). Microalgae tend to accumulate large amounts of proteins during exponential phase (de Carvalho et al. 2019), so, to attain the biofuel potential of microalgae the need is to enhance the carbohydrate content of microalgae. Carbohydrate content can be improved by decreasing starch degradation and by increasing glucan storage. Manipulation of the environmental factors including pH, light, nutrients, temperature, carbon source, etc. is the most common and affordable approach that directly influences the biochemical composition. However, in most cases, the enhancement in metabolite content occurs at the expense of biomass production (Chen et al. 2017). It is therefore crucial to understand and manipulate the factors to enhance the carbohydrate content without compromising the biomass production (Chen et al. 2013). Following section will elaborate the impact of various physicochemical factors to improve the targeted biomass and product production.

2.1 Impact of Nutrients on Carbohydrate Accumulation

Macronutrients like nitrogen, phosphorus, sulfur, potassium, etc., are essential for the optimum growth of microalgae (Table 14.1). However, nutrient-limited or starved conditions are feasible and commonly employed for altering metabolite composition (Markou et al. 2012). Limiting nitrogen and phosphorus can shift the carbon fixed during the Calvin cycle toward the production of non-nitrogenous compounds like polysaccharides (mainly starch and glycogen) and lipids (de Farias Silva et al. 2019a).

Increase in carbohydrate content from 12% to 54% was reported for *Chlorella* as a result of protein reduction by 60–20% when cultivation conditions were shifted from nitrogen-sufficient to nitrogen-starved conditions (de Farias Silva et al. 2019b). Similar results were observed in *Chlorella* cultivated under average nitrogen and limited phosphorus conditions, where an increase in carbohydrate content from 10% to 60% with a reduction in protein content from 57% to 7% (Samiee-Zafarghandi et al. 2018). Interestingly, contrary to a popular belief, when *Scenedesmus* sp. was

Table 14.1 Impact of nutrients on the carbohydrate content of microalgae

Nutrient limitation/ starvation	Microalgae	Carbohydrate production (mg L^{-1})	Metabolite improvement (%)	Biomass production (mg L^{-1})	References
Nitrogen	*Arthrospira platensis*	4.3	9.36	192	Lai et al. (2019)
Phosphorus	*Arthrospira platensis*	6.31	59.7	195	Lai et al. (2019)
N, P, vitamins, and metal	*Tetraselmis*	420 mg g^{-1}	130 (1.3-fold)	5720	Dammak et al. (2017)
Calcium and magnesium	*Chlorella sorokiniana*	450	50	–	Hanifzadeh et al. (2018)
Nitrogen	*Microcystic aeruginosa*	–	20	2.25×10^7 cells/mL	Hanifzadeh et al. (2018)
Sulfur	*Chlamydomonas reinhardtii*	5070	~51	–	Mathiot et al. (2019)
Multiple nutrients	*Desmodesmus* sp.	400 mg g^{-1}	64	1950	Hernández-García et al. (2019)

cultivated in high nitrate concentration, an increase in carbohydrate from 80 to 160 mg g^{-1} and protein from 208 to 524 mg g^{-1} was observed (Ranadheer et al. 2019).

Though nutrient starvation seems a feasible option especially to enhance the carbohydrates, the process viability is compromised due to low biomass yield. In batch cultivation, the extreme nitrogen starvation may negatively influence the carbohydrate content. Therefore, nutrient limitation is considered a more attractive strategy as compared to the starvation approach, favoring the carbohydrate accumulation; however, it takes place at the cost of protein reduction. Two-stage cultivation or continuous culturing may also be suitable in this regard.

2.2 Impact of Irradiance and Temperature on Carbohydrate Accumulation

Temperature greatly influences the sucrose formation; normally high temperature improves the carbohydrate production which may be due to the involvement of thermophilic enzyme system in sucrose production (Barati et al. 2019). Irradiance is another important factor, which is a major energy provider for the photosynthesis and its quantity and quality directly influences the growth rate, cell composition, and CO_2 fixation rate. Although the impact of light intensity is species-dependent, a linear correlation with the increase in irradiance intensity has been reported with the carbohydrate content and biomass productivity. Further increase in irradiance (more than the optimum) may damage the photosystem or biomass saturation may reduce the light penetration due to self-shading (Ho et al. 2014; Markou et al. 2012). Similarly, low light intensities (below 275 μmol m^{-2} s^{-1}) may gradually decline starch synthesis. Varying irradiance can regulate key enzyme (phosphoglucomutase) for starch synthesis (Ho et al. 2014).

Scenedesmus and *Desmodesmus* sp. showed cold-stress tolerant ability when cultivated in Nordic conditions. Natural strains are capable to tolerate the stress by producing high biomass (>1000 mg L^{-1}) and accumulating carbohydrate and lipids as major storage molecules under low temperature (5 °C) and continuous light. Under these conditions, *Scenedesmus* produced ~1400 mg L^{-1} biomass with carbohydrate to amide ratio of 1.5, whereas *Desmodesmus* produced ~700 mg L^{-1} of biomass with carbohydrate to amide ratio of 1.6 (Ferro et al. 2018). Similarly, *Parachlorella kessleri* exhibited excellent biomass and starch productivities of 1040 and 220 mg L^{-1} day^{-1}, respectively, under a continuous high-light intensity of 600 μmol m^{-2} s^{-1} (Takeshita et al. 2014). Study of simultaneous effect of nutrient, irradiance, and temperature on *Pseudoneochloris marina* suggested an increase in biomass (260 mg L^{-1} day^{-1}) under high nutrient (74.1 mg L^{-1}), high-light intensity (252–364 μmol m^{-2} s^{-1}), and moderate temperature (28 °C). However, these conditions were not shown to be favorable for the protein, lipid, and pigment productivities. While, the maximum protein content of 236 mg g^{-1} was observed at low light (140 μmol m^{-2} s^{-1}) and low temperature (20 °C) (Gonçalves et al. 2019).

The impact of high irradiance is studied widely to promote the biomass production and carbohydrate content of microalgae (Table 14.2). In response to high-light intensities, microalgae tend to accumulate high-energy compounds like carbohydrate and lipids to protect the cells against stress conditions. On the other hand, limited information is available on the effects of temperature on the carbohydrate accumulation because most of the studies focused on lipid accumulation under high temperature. Another reason maybe the availability of few thermotolerant microalgal strains. However, there is a need to study the impact of temperature to enhance the carbohydrate and protein contents of microalgae.

2.3 *Impact of Organic Carbon Source on Carbohydrate Accumulation*

Carbon is essential for regular processing and to maintain the cultures of microalgae. Sucrose (as carbon source) is crucial for growth and development, cell signaling, energy storage, and stress assimilation. Inorganic carbon in the form of HCO_3^- and CO_2 is important for photosynthesis, CO_2 sequestration, and accumulation of desired products. Though, atmospheric CO_2 is easily available to microalgae but can be leaked easily, the more stable form of inorganic carbon is provided by the carbon concentration mechanism (CCM) in the form of HCO_3^- ions, which are provided by the conversion of atmospheric CO_2 through the carbonic anhydrase (CA) activity which takes place in protein shell of carboxysome that acts as unidi-

Table 14.2 Impact of irradiance on the biomass and carbohydrate production in microalgae

Irradiance stress (μmol m^{-2} s^{-1})	Microalgae	Carbohydrate productivity (mg L^{-1} day^{-1})	Metabolite improvement (%)	Biomass production (mg L^{-1})	References
896	*Scenedesmus obtusiusculus*	280	31	978	Toledo-Cervantes et al. (2018)
300	*Chlorella sorokiniana*	170	–	2800	Gifuni et al. (2018)
650	*Tetradesmus obliquus*	800	~30	1700	de Farias Silva et al. (2018)
310	*Desmodesmus* sp. P5	–	13.4	2380	Coronel et al. (2019)
2000	*Botryococcus braunii*	900	–	1300 mg L^{-1} day^{-1}	García-Cubero et al. (2018)

rectional membrane to enhance carboxylation efficiency by accumulating CO_2 around *Rubisco* (major enzyme involves in CO_2 fixation) (Tourang et al. 2019). Another way to enhance dissolved inorganic carbon is the transport of HCO_3^- ions by membrane transporters or ATP activity from extracellular fluid to inside the cell. The addition of sodium bicarbonate as HCO_3^- ions source is the most suitable approach, increasing the availability of dissolved inorganic carbon increases media pH and promotes cell growth as well as the production of energy-rich molecules (Choi et al. 2019; Pancha et al. 2015). However, some algae are sensitive to high pH changes, resulting in a negative influence on biomass productivity and metabolite content.

In a study, which was conducted to evaluate the impact of carbon source and nutrient concentrations on the biomass and carbohydrate productivities of *A. platensis*, it was observed that media supplementation with up to 16 g L^{-1} of $NaHCO_3$ enhanced the biomass productivity. A further increase in $NaHCO_3$ concentration did not increase biomass rather resulted in its depletion. Since higher bicarbonate concentrations had a negative impact on carbohydrate accumulation, therefore, a concentration of 9.8 g L^{-1} was found optimum to promote the biomass and carbohydrate production in *A. platensis* (Tourang et al. 2019). Addition of bicarbonates in the media affects the growth and metabolite content by directly influencing its pH. The two cyanobacterial strains *Lyngbya limnetica* and *Oscillatoria obscura* when cultivated at pH 9.0 produced maximum biomass of 1196 and 1226 mg L^{-1} with a carbohydrate content of 219 and 192 mg g^{-1}, respectively (Kushwaha et al. 2018). In addition to bicarbonates, the direct addition of CO_2 and the use of other carbon sources like pentose, sucrose have proven to be beneficial for the improvement of carbohydrates accumulation. However, there are also cases where addition of the bicarbonates improved the production of both the proteins and carbohydrates. The impact of various carbon sources on microalgal production of bioactive compounds is summarized in Table 14.3.

3 Genetic Modification of Microalgae for Product Enrichment

Industrial scale-up of microalgae-based products can also be achieved by improving the algal metabolism. In recent years, metabolic/genetic engineering in combination with synthetic biology came forth as promising strategies to develop the so-called "super strains" by genome editing to manipulate multiple attributes of interest. These characteristics could be: (1) improvement in biomass/metabolite production yield, (2) resistance to toxins or abiotic factors, (3) enhanced photosynthetic efficiency, (4) ability to consume various carbon sources, and (5) reduction in the formation of unfavorable by-products (Naghshbandi et al. 2019). The OMICS approaches including genomics, transcriptomics, proteomics, metabolomics, and glycomics are now routinely being used to identify the genetic targets, their regulatory elements, and to study the impact of external factors on the gene involved in

Table 14.3 Impact of carbon source on the growth and metabolite content of microalgae

Microalgae	Culture media	Stress conditions	Biomass productivity (mg L^{-1} day^{-1})	Carbohydrate content (%)		Protein content (%)		References
				Before	After	Before	After	
Scenedesmus	BG11	0.9 g L^{-1} $NaHCO_3$	28.32	18.5	30.9	47	49.5	Pancha et al. (2015)
Chlorella minutissima	BBM	5% pentose	60	32.5	58.6	15.4	14.1	de Freitas et al. (2019)
Coelastrum sp.	Dairy waste-water	12% CO_2	267	–	58.45	–	–	Mousavi et al. (2018)
Asterarcys quadricellulare	BBM	5% CO_2	900 mg L^{-1}	31.2	71.4	20	14	Varshney et al. (2018)
Chlorella sorokiniana	BBM	5% CO_2	960 mg L^{-1}	30.2	~53	24	10	Varshney et al. (2018)

enzyme synthesis or metabolites production (Khan et al. 2018a). Regulating the gene expression through random or targeted mutagenesis and genetic disruption, nuclear/chloroplast-based transformation, heterologous-protein expression, RNAi silencing, and CRISPR-based insertion/deletion of selected genes have been implied in recent years to achieved robustness (Ghag et al. 2019). However, these studies are still at infancy as compared to other well-established expression systems like bacteria, yeast, or plants. Most of these studies are limited to the model algal organisms such as *Chlorella, Chlamydomonas,* and *Synechococcus*, should be expanded to other valuable algal species. A possible reason for the limited work done in this area is the complex metabolic pathways, diverse nature, and feedback inhibition which renders the reconstruction of these pathways in microalgae extremely difficult. Particularly, the improvement in carbohydrate or protein content of microalgae by genetic engineering is lacking thus far and more studies are required for this purpose.

3.1 Improvement in Production and Quality of Carbohydrates

As mentioned in the introduction, microalgae store carbohydrates mainly as starch. The production of carbohydrates through genetic engineering can be enhanced by: (1) overexpressing the enzymes involved in starch biosynthesis, (2) blocking the starch degradation or by-product formation, and (3) altering the secretion of soluble sugars. The ADP-glucose pyrophosphorylase is a rate-limiting enzyme in starch anabolism pathway; modifications in allosteric and catalytic properties of this enzyme can enhance starch synthesis (Ho et al. 2014). Genetic modification of *Synechococcus* could enhance the sucrose production by many-folds even higher when compared to sugarcane. Overexpression of sucrose-phosphate synthase (SPS) in *Anabaena* enhanced the intracellular sucrose level by 10% (based on dry-weight) under NaCl stress induction (Smachetti et al. 2019). Similarly, co-overexpression of aldolase with alcohol dehydrogenase and pyruvate decarboxylase in *Synechocystis* under *PnrsB* promoter enhanced the ethanol production by 69% with 10% increase in biomass production (Liang et al. 2018).

Additionally, carbohydrate accumulation can also be enhanced by the overexpression of glucosyltransferase (the enzyme responsible for the conversion of sugar forms) to prevent feedback-inhibition or by inactivating starch-catabolic enzymes (Ghag et al. 2019). The overexpression of phycobilisome response regulator (RpaB) in *Synechococcus*, resulting in >twofold enhanced productivity of sucrose secretion by inducing cell-growth arrest, improved the bioproduct formation and photosynthetic efficiencies (Abramson et al. 2018). In order to improve the bioethanol conversion efficiency, manipulation of polysaccharidal sugars has been explored. Though these studies are limited to plants where cellulases from prokaryotic/eukaryotic origin have been expressed (Bhalla et al. 2013), it could also be applied to microalgae to improve the carbohydrate quality and bioethanol production.

3.2 Improvement in Recombinant Protein Production

Microalgae are potential sources for recombinant protein production. The algal strains can be engineered for enhanced protein synthesis via nuclear, mitochondrial, and chloroplast transformations. The eukaryotic nature of microalgae enables them to produce glycosylated proteins due to the presence of post-translational pathways in them. Though microalgae-based recombinant proteins are yet to be commercialized, it offers the cost-effective production of industrial or therapeutically important recombinant proteins. *C. reinhardtii* has been extensively studied for therapeutic protein production and >100 such proteins have been expressed successfully in this system (Gong et al. 2011).

Microalgae-based biomanufacturing is preferred due to effectiveness in terms of cost and energy, fewer contamination chances, and simplified downstream processing. Recently, human lactoferrin (hLF) has been produced in *C. reinhardtii* by transforming an optimized version of the lactoferrin gene. The transgenic strain produced 0.73 μg of lactoferrin/40 μg of soluble protein. The extract from transgenic alga showed antibacterial activity against *Klebsiella variicola* and *Escherichia coli* (Pang et al. 2019). Similarly, transgenic *C. pyrenoidosa* produced ~53 mg L^{-1} of hLF under optimized conditions which were fourfold higher when compared to the control (Wang et al. 2019). In another study, malaria detection protein (used in ELISA testing) and cell-traversal protein (antigen) from sporozoites and ookinetes of mosquitoes have been successfully expressed and produced in *C. reinhardtii* chloroplast (Shamriz and Ofoghi 2019).

C. *pyrenoidosa* (K05) was modified through irradiation-mediated mutagenesis, which resulted in a 31.8% improvement in protein content and 11.6% improvement

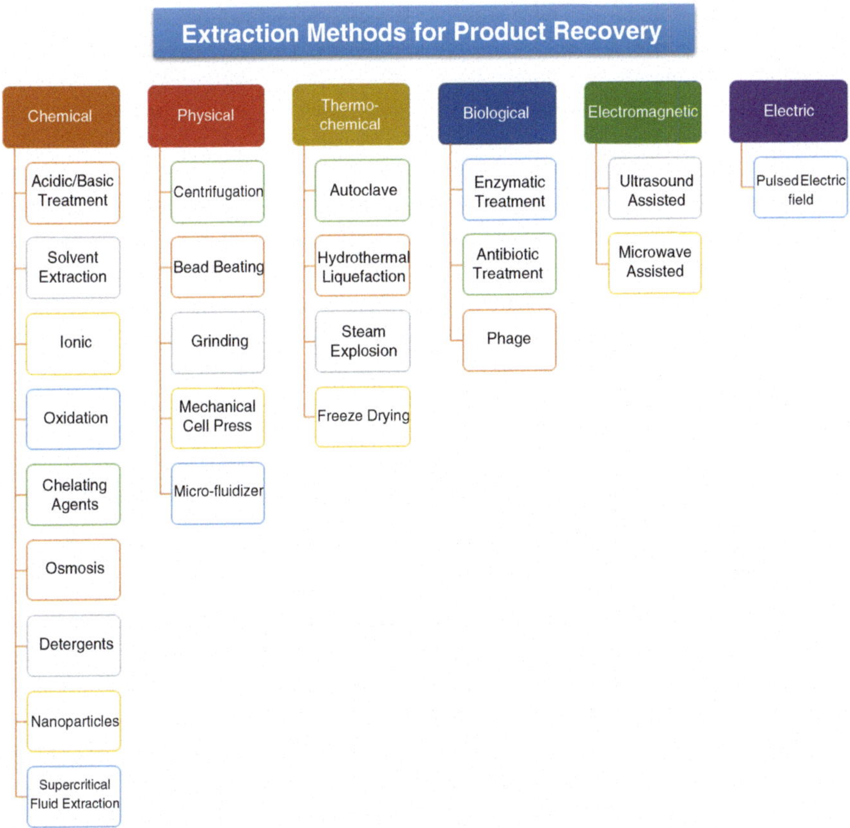

Fig. 14.2 Various extraction methods for product recovery from microalgae

in biomass production (Song et al. 2018). Additionally, efforts were made to improve the secretion and purification of recombinant proteins by using C-terminal protein-fusion with a two-phase alternative aqueous technique (based on amphiphilic-hydrophobic protein tag) respectively (Baier et al. 2018). In general, genetic engineering and synthetic biology tools seem an attractive option to reconstruct the metabolic pathways for improved protein and carbohydrate content and quality. Further studies are required to enhance the protein content, quality, secretion, and purification by using transgenic microalgae.

4 Extraction Methods for Product Recovery

The efficient extraction of microalgal biomass is a necessary prerequisite to produce high-value commodities. An ideal strategy of extraction should have the ability to selectively extract the desired products while minimizing the production of byproducts and toxic compounds. There are several methods that can be employed for the cell wall disruption to isolate the desired component. However, its suitability for a particular strain mainly depends on the rigidity of the cell wall and the component under consideration (Kapoore 2014). Generally, these extraction methods can be divided into chemical, physical, thermal/thermochemical, biological, electromagnetic, and electric as shown in Fig. 14.2. Each method has its pros and cons which are summarized in Table 14.4 and only selected advanced methods are discussed here.

The conventional extraction methods like solvent extraction are quite a labor-intensive, involve the use of highly toxic organic solvents, can lead to changes in the stereochemistry and decomposition of important bioactive components via excessive heat, light, and oxygen. The use of toxic solvents like chloroform and methanol on an industrial scale can pose serious environmental hazards and human health issues (Denery et al. 2004). On the other hand, traditional approaches like grinding and cryogenic grinding (involving the use of liquid nitrogen) are found to be efficient at lab scale but they are expensive and impractical to be used at a larger scale.

The bead milling (bead size of 0.5 mm) has shown to be an effective approach for microalgal cell disruption when compared to nine other mechanical methods employed for some microalgae but it was found to be ineffective for some species like *Chlorophytes* and *Chlorella* sp. that made its consistency objectionable (Lee et al. 2010). It is a commonly employed technique at lab scale and its modified form namely "agitated bead beating" is usually used at an industrial scale. However, in techniques such as bead beating and homogenizers, excessive heat production is a major barrier in their application at a large scale. Moreover, an additional step of bead separation is involved. Mechanical cell press has shown to be effective for plants but ineffective for microalgae due to smaller cell size. Autoclaving and homogenization are also commonly used at lab-scale techniques but their use does not seem feasible at an industrial scale in both the technical and economic terms (Gong and Bassi 2016; Lee et al. 2010). Soxhlet extraction is generally considered as the most suitable method for the extraction of important biochemical components from plant sources but it is time-consuming (up to 15 h) and involves the use of toxic solvents. Moreover, much lower yield is reported from Soxhlet extraction in comparison to other conventional approaches (Cravotto et al. 2008). To overcome the toxicity and heat production, some other chemicals and biological methods are devised but the contamination is another major concern that makes the downstream processing difficult (Jaki et al. 2006). Biological methods which involve the use of enzymes such as cellulases, xylanases, and pectinases are environment-friendly and mild in nature yet expensive and have limited efficiency. Similarly, the recovery of enzymes is also an important concern (Gong and Bassi 2016).

Physical techniques like ultrasound-assisted extraction (Cravotto et al. 2008), microwave-assisted extraction (Biller et al. 2013; Cravotto et al. 2008), high-pressure homogenization (Gilbert-López et al. 2017) (microfluidizer), supercritical fluid extraction (Gong and Bassi 2016), and pulse electric field lysis (Goettel et al. 2013) (PEF) are developed to enhance the efficiency of extraction process. In ultrasound-mediated extraction, ultrasound waves of 20–100 MHz are utilized to create localized cavitation bubbles that lead to cell wall disruption, when expanded. Among the

Table 14.4 Comparison of various extraction methods for microalgal product recovery

Category	Method	Strains used	Advantages	Drawbacks	References
Chemical	Acidic/basic treatment	*Chlorococcum* sp., *Scenedesmus* sp.	Energy-efficient	Loss of carotenoids need for an acid/base to be disposed of after the process	Günerken et al. (2015)
	Solvent extraction	*Crypthecodinium cohnii*	Cost-effective, can be used at a larger scale	Delayed process, often uses large amounts of solvents, sometimes toxic	Cravotto et al. (2008)
	Ionic method	*Chlorococcum* sp., *Chlorella*	Cost-effective	Toxicity issues, not a fully developed technique	Shankar et al. (2017)
	Osmotic shock	*Thraustochytrid strains*	Energy-efficient, can be used at a large scale	Time-taking, wastewater is produced, cell disruption is not much efficient	Byreddy et al. (2015)
	Detergents				
	Nanoparticles	*Chlorella* sp.	Environment-friendly	Under the development phase, costly, difficult removal of nanoparticles after the process	Seo et al. (2016)
	Supercritical fluid extraction	*Haematococcus pluvialis*, *Nannochloropsis*, *Scenedesmus*, *Haematococcus*, *Chlorella*	Suitable for carotenoid extraction, rapid, non-toxic	Costly and cannot be employed at an industrial scale	Nobre et al. (2006) and Sikkema et al. (2010)

(continued)

Table 14.4 (continued)

Category	Method	Strains used	Advantages	Drawbacks	References
Physical	*Centrifugation*				
	Horn sonication	*Crypthecodinium cohnii*, *Nannochloropsis gaditana*	Non-toxic, efficient cell disruption, rapid and low maintenance cost	Localized cavitation, high energy requirements, expensive	Al hattab and Ghaly (2015)
	Bath sonication	*Nannochloropsis oculata*, *Chlorella* sp.	Non-toxic, efficient cell disruption, rapid and low maintenance cost	High-energy requirements and operational cost	Al hattab and Ghaly (2015)
	Bead beating	*Chlorella* sp., *Nostoc* sp., *Tolypothrix* sp., *Nannochloropsis* sp., *Scenedusmus* sp., *Chlorococcum* sp., *Botryococcus* sp.	Efficient cell disruption, fast extraction	Efficiency depends upon cell nature, removal of beads, energy input, and maintenance cost	Al hattab and Ghaly (2015)
	Grinding	*Arthrospira platensis*, *Chlorella vulgaris*, *Haematococcus pluvialis*, *Porphyridium cruentum*, *Nannochloropsis oculata*	Rapid and easy lab-scale technique	Degrades some important cell components, time taking	Safi et al. (2014)
	Mechanical cell press	*Scenedesmus dimorphus*, *Chlorella* sp., *Nannochloropsis*, *Botryococcus* sp. *MCC31*, *Ankistrodesmus falcatus*	Industrial-scale technique for oil recovery	Energy inefficient, cell disruption is not good	Aarthy et al. (2018)
	Micro-fluidizer	*Nannochloropsis* sp., *Chlorella* sp., *Tetraselmis suecica*	Good for lipid extraction, cell disruption at room temperature	Not good for protein extraction, high-energy requirements	Spiden et al. (2013)

(continued)

Table 14.4 (continued)

Category	Method	Strains used	Advantages	Drawbacks	References
Thermo-chemical	Autoclave	*Haematococcus pluvialis*	Easy to use, low maintenance cost	Not suitable for pigments, energy requirements are high	Mendes-Pinto et al. (2001)
	Hydrothermal liquefaction	*Monoraphidium* sp., *Stigeoclonium* sp., *Nannochloropsis occulata*, *Chlorogloeopsis fritschii*, *Pseudochoricystis eliipsoidea*	Does not require drying of biomass	Energy requirements are high, heat production, variable recovery rate, expensive catalysts are used	Biller et al. (2013) and Passos et al. (2015)
	Steam explosion	*Nannochloropsis gaditana*	Relatively easy maintenance and lower energy requirements, efficient cell disruption	Variable results among different species	Al hattab and Ghaly (2015)
	Freeze drying	*Dunaliella tertiolecta*, *Scenedesmus* sp., *P. tricornutum*, *Nannochloropsis* sp.	Single-step drying and extraction procedure, does not disturb the bioactivity of cell components	Inefficient cell disruption, high-energy requirements and difficult to maintain, time taking	Al hattab and Ghaly (2015)
	Hydrodynamic cavitation	*Nannochloropsis oculata*	Low energy requirements	Limited area of cavitation	Ali and Watson (2015)
Biological	Enzymatic treatment	*Chlorella vulgaris*, *Scenedesmus dimorphus*, *Nannochloropsis* sp.	Selective, delicate and efficient cell wall hydrolysis without affecting carotenoid bioactivity	Enzymes are expensive, difficult to recover after each reaction, delayed process	Al hattab and Ghaly (2015)

(continued)

Table 14.4 (continued)

Category	Method	Strains used	Advantages	Drawbacks	References
Electromagnetic	Ultrasound-assisted	*Scenedesmus dimorphus*, *Nannochloropsis oculata*	Relatively energy-efficient	Variable results depending on species	Wang et al. (2014a)
	Microwave-assisted	*S. obliquus*, *Botryococcus* sp., *Chlorella vulgaris*, *Scenedesmus* sp.	Relatively energy efficient with good product recovery ratio, efficient cell disruption, reaction time is fast	Maintenance cost is higher, heat generation	Al hattab and Ghaly (2015)
Electric	Pulsed electric field	*Auxenochlorella protothecoides*	Selective, delicate and energy-efficient, does not affect carotenoids	Under the development phase	Goettel et al. (2013)

sonication techniques, bath sonication is much suitable which can be used at a large scale when compared with horn sonication, due to a single operational unit and lesser energy input. Some recent methods such as supercritical fluid extraction (SFE) and pressurized liquid extraction (PLE) are also suggested as green technology. SFE makes the use of non-toxic CO_2 in supercritical conditions for the extraction of lipids from cells. Whereas PLE is generally used for the extraction of polar and non-polar components from corn and oats by using different solvents. Moreover, this technique has been used to extract bioactive compounds like antioxidants from two microalgae species *Arthrospira platensis* and *Dunaliella salina*. However, the constant need for maintenance of high temperature and pressure conditions, and high-energy inputs make both the techniques impractical for industrial applications. Moreover, the use of high temperature compromises the yield of temperature-sensitive components such as carotenoids. Pulse electric field lysis (PEF) is also employed to carry out the extraction of desired products from microalgae. In this technique, the cells are subjected to pulses of a strong electric field that leads to the formation of pores in the cell wall as happens in the process of electroporation (DNA transfer process). The formation of temporary pores in the cell wall causes the biochemical components to leach out of the cell. This method is not applicable to the extraction of carotenoids due to the requirements of some essential organic solvents.

Current extraction methods are constrained by time consumption, use of large amounts of toxic solvents, high-energy inputs, and expensive operational procedures leading to difficulty in scaling up the process (Balasubramanian et al. 2011). The use of microwave-assisted extraction (MAE) for microalgae has been primarily used not only for lipid extraction but also for proteins, carbohydrates, and pigments. Microwaves have an advantage of quick heating, unidirectional flow of heat and mass, selective energy dissipation, faster, increase the purity and yield of the desired product. The microwave-assisted extraction resulted in up to 94.92% cell disruption in *Nannochloropsis oculate* (Ali and Watson 2015). Conclusively, MAE is an environment-friendly approach with minimum use of toxic solvents and diverse applicability to the extraction of various bioactive metabolites but still, the problem of heat generation and maintenance cost are some of the issues that need to be addressed.

5 Product Purification

As discussed earlier, microalgae have recently been employed as an alternative production platform for recombinant protein expression. *C. reinhardtii* genome is well studied and thus serves as a suitable model and expression system for the expression of recombinant proteins. Moreover, it is easy and cost-effective to culture it on a large scale (Shamriz and Ofoghi 2016). The optimization of a sustainable and efficient process of protein isolation for industrial applications from the broth of microalgae is a big challenge. Therapeutic recombinant proteins and bioactive molecules derived from *C. reinhardtii* require purification up to 99% (Rasala et al. 2010). The diatom *Phaeodactylum tricornutum* was genetically engineered to express a recombinant human antibody (Hempel et al. 2011). A-Sepharose beads column was used to purify the expressed recombinant protein and the purity was confirmed by performing gel electrophoresis. Various chromatography techniques such as affinity, size exclusion, hydrophobic interaction, and ion-exchange chromatography are generally used for lab-scale purification of the desired protein from microalgae (Verdel et al. 2000; Wang et al. 2014b). An enzyme of *Betaphycus gelatinus* D-galactose-6-sulfurylase, a monomeric protein (65 kDa) was purified using hydrophobic interaction and the exchange chromatography with an increased purification (4.9-fold) and 3.7% yield of cell lysate (Wang et al. 2014b). A monomeric haloperoxidases enzyme (43 kDa) was isolated using gel filtration and ion exchange chromatography from a freshwater microalgae *Cladophora glomerate* (Verdel et al. 2000). Similarly, a protein named *N*-acetyl-D-galactosamine-specific protein was isolated by affinity chromatography from red alga *Aglaothamnion oosumiense* (Han et al. 2012).

In addition to chromatography-based purification, membrane filtration strategies such as cross-flow filtration and ultrafiltration are also used to purify microalgal proteins (Henderson et al. 2010; Her et al. 2004; Zhang et al. 2010). High product yield, greater volumetric throughput, fewer requirements, and cost-effectiveness are some of the advantages of using filtration methods over chromatography. The major concern in using membrane filtration is the availability of required size membrane which is crucial for isolation of a specific protein (Buyel et al. 2015).

The cost of microalgal metabolite production depends on multiple factors including the cost involved in biomass production, the yield of desired metabolites, and the cost of purification. Moreover, it has been found that the contribution of the recovery process is 60% greater than the cost of biomass, i.e., 40% in the production of valuable metabolite like eicosapentaenoic acid (EPA) (Molina Grima et al. 2003).

6 Challenges with Microalgal-Bioactive Compounds

Although the use of microalgal biomass to produce bioactive compounds appears promising, their economic feasibility for the large-scale biomass production is a major problem in the commercialization of microalgae-based commodities. Few algal species including *Chlorella*, *Spirulina*, *Botryococcus*, *Haematococcus*, *Nannochloropsis*, *Isochrysis*, *Porphyridium*, *Dunaliella*, and *Tetraselmis* have shown commercial potential. Unfortunately, large-scale production of these strains is still at infancy (Alam et al. 2019) mainly due to their slow growth rate which makes them susceptible to contamination chances. Algal biomass market is around 20,000 metric tons of microalgal biomass/year with the price based on the consumption source ranging from 0.4 € kg^{-1} to up to 100 € kg^{-1} for biofuel production and human consumption, respectively. Protein-rich algal biomass could be sold for 0.75 € kg^{-1} (Sathasivam et al. 2017). However, the market share of algal carbohydrate especially for bioethanol production is quite promising and has to be increased so that it can replace the other conventional sources (Khan et al. 2018b).

There is a need to identify the robust potential microalgae species having the ability to cultivate at large scales to combat the problems related to contamination, reduced growth, and reduced metabolite production. Another attractive option could be the process optimization, to enhance the biomass and metabolite production in response to various biotic and abiotic factors (Shahid et al. 2019) or the culturing conditions. Usually, algal strains can adapt themselves to various environmental conditions and start producing bioproduct based on their habitat. Moreover, robust algal strains are usually extremophiles and can grow exceptionally at a diverse range of factors like carbon source, temperature, light, media composition, nutrient, and salinity levels and modify their composition to adapt those conditions and to reduce contamination chances (Rashid et al. 2019). This feature of microalgae could be exploited to raise the bioprocessing economics.

Harvesting and drying of algal biomass are other challenging processes. Harvesting alone accounts for 20–30% of the production cost. Microbe-induced flocculation or self-flocculating algal species like *C. vulgaris* JSC7 could be an interesting option (Alam et al. 2014). However, the selection of harvesting method may influence the quality of the desired product. Flocculation followed by sedimentation is suggested as an economical method but is most suitable to obtain low-grade products. High-grade products require large amounts of biomass which can only be processed using continuous operational centrifugation. In order to decrease production cost and to increase shelf-life, it is mandatory to obtain the biomass with high solid content. Sun drying could be an excellent choice due to its high efficiency, and low-maintenance and operation cost especially in the countries with abundant sunlight (Rizwan et al. 2018).

Integrated extraction of the desired product (fuel and food) in a cost-effective and efficient way is a major constraint in the commercialization of algal commodities. Development of a generalized extraction method is tricky as its selection depends on the desired product and algal species under consideration. Moreover, the selected method should not damage other products of interest (Rizwan et al. 2018; Rösch et al. 2019). The use of microwave, ultrasounds, and selective solvents is an attractive option. In protein extraction, polysaccharide content in the cell wall of algae hinders the process and has a determinantal effect on the purity and yield of the product. Moreover, the cell wall acts as a protective covering and reduces the protein digestibility. Various methods like air drying, sun drying, fermentation, etc. are reported to increase protein functionality and digestibility. Similar is the case with algal peptides, which remain inactive until treated with either exogenous enzymes or subjected to enzymatic or protease hydrolysis during fermentation. Mostly algal proteins and peptides are bitter in taste but are excellent fortifying agents especially in extreme conditions of high temperature and pH (Sahni et al. 2019). In order to reduce the protein impartment, it is suggested to extract protein content along with its polysaccharide constituent followed by fatty-acid content. The obtained polysaccharides then can be fermented to bioethanol and biobutanol (Rösch et al. 2019).

A plausible solution to reduce the production in the open-pond system is the use of wastewater as nutrient media. A case study suggested the price reduction from 4.5 to 3.6 € kg^{-1} for raceway ponds and from 2.3 to 1.4 € kg^{-1} for thin-layer cascade (TLC). A further reduction in production cost (2.1 and 0.6 € kg^{-1} for raceway and TLC) is possible if the wastewater treatment cost is considered as an input for biomass production (Fernández et al. 2019). Process optimization in combination with integrated biorefinery is the best option to obtain a variety of bioproducts (food and fuel) in a cost-effective, efficient, and environment-friendly manner.

7 Applications of Microalgal-Bioactive Compounds

Microalgae are an untapped source of valuable biochemicals which can be exploited to produce energy, edibles, and other health products by employing efficient and cost-effective approaches (Fig. 14.3). Microalgae are also capable of producing many valuable bioactive compounds that are yet to be explored. Apart from bioethanol production, the carbohydrate and protein component of microalgae can be used on a commercial scale for various applications such as human health and nutrition, cosmetics, pharmaceuticals, feed for animal and aquatic life, and as biofertilizers (Table 14.5).

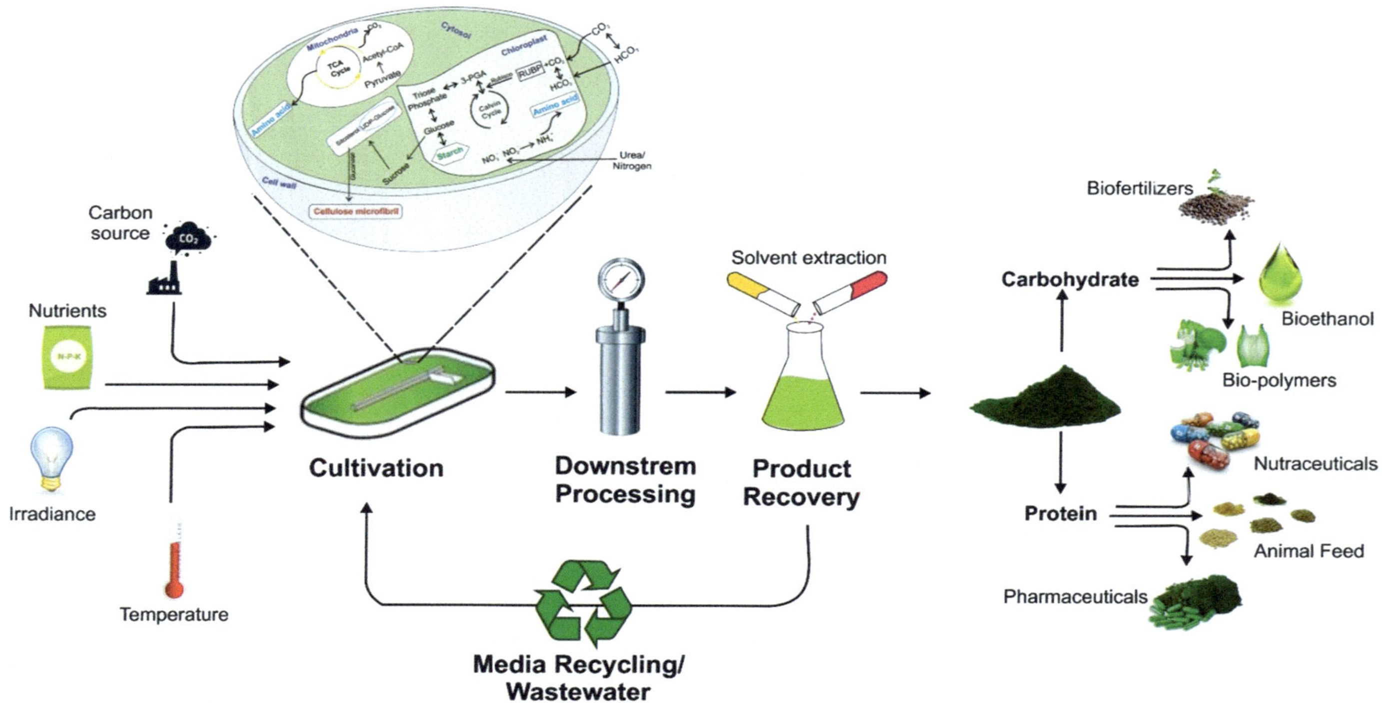

Fig. 14.3 Schematic diagram for the cost-effective and efficient production of microalgal carbohydrate and protein-based products

Table 14.5 Protein- and carbohydrate-based industrial products of microalgae

Type	Product	Species	Utility	References
Protein	Protein powder/ tablets	*Chlorella* sp. *Athrospira* sp.	• As health food products • As an animal and poultry feedstock	Bleakley and Hayes (2017)
	Phycobiliproteins	*Spirulina platensis* *Porphyridium* sp.	• Used extensively in food production • Used in pharmacology and medicine industry • In the cosmetics industry	Manirafasha et al. (2016)
	Peptides	*Chlorella pyrenoidosa* *Cyanobacteria*	• In the pharmaceutical industry as anti-cancer, antioxidant, and anti-inflammatory • In the production of functional foods for cardiovascular health	Ejike et al. (2017) and Kim and Kang (2011)
	Mycosporin-like amino acids	*Cyanobacteria* sp., *Dinophyta*, etc.	• Used as sunscreens	Llewellyn and Airs (2010)
Carbohydrate	Polysaccharides	*Porphyridium* sp., *Rhodella* sp., various *Cyanobacteria*	• As a solidifying agent in the preparation of culture media for the growth of microorganisms • As a thickening agent in place of gelatin, pectin, and starch • Cosmaceuticals	Buono et al. (2014) and Venugopal (2016)
	Sulfated polysaccharides	*C. stiamatophora*, *C. pyrenoidosa*, *Phaeodactylum tricornutum*, *Porphyridium* sp.	• Anti-inflammatory, antioxidant, immunomodulatory	Buono et al. (2014)

7.1 Bioethanol

Microalgae are renowned as third-generation feedstock to produce biodiesel. However, some microalgae species are a rich source of carbohydrates in the form of cellulose and starch thus contributing in the production of bioethanol. Bioethanol production from such carbohydrate-rich microalgal species is advantageous in comparison to plants due to higher growth rate and CO_2 fixation ability of microalgae. Moreover, microalgal carbohydrate content is mainly composed of cellulose and starch which makes it easier to convert into monosaccharides in comparison to lignocellulosic biomass. *Chlorella, Scenedesmus, Dunaliella, Chlamydomonas,* and

Tetraselmis have found to be rich in carbohydrates (above 40% of the dry weight). Various species of genus *Chlorella* possess high carbohydrate content, for example, *C. vulgaris* has 37–55% of carbohydrates (dry weight) (Hossain et al. 2019).

Prior to the fermentation of these components by microorganisms, they need to be hydrolyzed into simple fermentable sugars. Chemical (acid and alkali) and enzymatic hydrolysis are the two commonly used procedures for this purpose. Acid hydrolysis is comparatively cheaper and faster but also results in the decomposition of important components and produce toxic compounds that usually interfere with the fermentation. Enzymatic hydrolysis is a costly and slow process, yet it is a mild and environment-friendly procedure and can yield higher glucose without producing inhibitory byproducts. Enzymatic hydrolysis often requires costly pretreatment procedures in order to enhance the hydrolysis efficiency. The conversion of hydrolyzed microalgal biomass is carried out by two major ways, either by separated hydrolysis and fermentation (SHF) or simultaneous saccharification and fermentation (SSF) (Khan et al. 2018b).

7.2 Human Health and Nutritional Products

In the past few years, especially after the identification of probiotics, the aquatic organisms particularly microalgae have been intensively investigated for their health benefits. The cardiovascular diseases can be treated by sterols which are produced by microalgae. The *Spirulina* sp. has been reported to produce clionasterol, that is helpful in preventing cardiovascular diseases (Munir et al. 2019). Likewise, many antioxidant compounds (e.g., mycosporines, dimethyl-sulfoniopropionate, β-carotene, astaxanthin, and some other carotenoids) are also obtained from microalgae. The oxidative stress damage can be compensated with these antioxidants. The *Spirulina maxima* and *Spirulina platensis* strains have been frequently used as a source of human food. The *Spirulina* sp. has been grown at commercial level as a food source because it can improve the immune system. It positively affects lactic acid bacteria of the gastrointestinal tract that improves body hormones. It is also used in the treatment of other diseases like cancer, diabetes, arthritis, cardiovascular diseases, and anemia (Mani et al. 2007).

The *Chlorella* sp. is also reported to be used as a food source. It is rich in different vitamins, proteins (51–58% of dry weight), and carotenoids (Richmond 2004). It is also involved in the production of "*Chlorella* Growth Factor" that enhances lactic acid bacteria's growth in the body. This microalga sp. also has β-glucan that performs its functions as immune stimulator such as, it reduces blood lipids and free radicals (Iwamoto 2003). It has many health benefits like it decreases cholesterol levels, reduces blood sugar, and increases hemoglobin levels in the blood. During ethionine intoxication and hunger, it acts as a hepatoprotective agent, increases the growth of the intestinal bacteria, and protects from kidney failure. It can stimulate the immune system to hinder the growth of *Listeria monocytogenes* and *Candida albicans*. In mammals, extracts obtained from *Chlorella* are used for increasing the production of cytokines and splenocytes and for activating some other immune responses. The Japanese have been utilizing *C. ellipsoidea* for making various food products, e.g., cookies, powdered noodles, bread rolls, green tea, soy sauce, soups, and ice cream (Barrow and Shahidi 2008).

The biomass of microalgae is composed of enzymes, fiber contents, carbohydrates, and proteins. It can also produce different vitamins (e.g., A, B1, B2, B6, and C) and could be a source of various minerals like calcium, iodine, potassium, magnesium, niacin, and iron. Because all essential nutrients are present in microalgae, so it is used as a primary food source in different Asian countries particularly Japan, China, and Korea. Due to their high nutritional value, it is also used in other continents. Only a limited number of microalgae species can be used as human dietary supplements because of the market demand and strict food safety regulations. Microalgae are marketed in the form of liquids, tablets, and capsules for the human's usage and are also added in beverages, gums, pastas, candy bars, and snacks. They are utilized as natural food colorants and in nutritional supplements. These are some important nutritional and medicinal benefits of microalgae that make them beneficial for humans (Koyande et al. 2019).

7.3 Cosmetics

Chlorella sp. and *Arthrospira* sp. have some important components that are used in skincare products. Some international companies like Daniel Jouvance and LVMH have their own microalgae production units. The extracts of microalgae are usually used in skin and face care, sun protection, and haircare products. The *Dunaliella salina, Alaria esculenta, Spirulina platensis, Mastocarpus stellatus, Chlorella vulgaris, Ascophyllumnodosum, Nannochloropsis oculata,* and *Chondrus crispus* are some of the microalgal species that are usually being utilized in cosmetics. The microalgal extracts are commonly used in cosmetics because of some important properties such as high level of antioxidants, water binding capacity, and slimy texture. The *Arthrospira* sp. is used in the production of a protein-rich extract that has an anti-aging agent. The extracts obtained from the *C. vulgaris* are involved in tissues regeneration and reduction of wrinkles. The extract of *Nannochloropsis oculata* has skin tightening properties and the extract of *D. salina* can enhance skin energy metabolism and significantly increase cell growth (Ariede et al. 2017; Couteau and Coiffard 2018). So, microalgae have many imperative properties to be used in cosmetics production.

7.4 Pharmaceuticals

The bioactive molecules are naturally obtained from microalgae by a biological process that cannot be synthesized easily by using chemical methods. The antibiotics of chemically diverse types like tannins, polysaccharides, fatty acids, bromophenol, terpenoids, and alcohols are produced by microalgae. Some neurotoxic and hepatotoxic compounds are also produced by several microalgae. In the pharmaceutical industry, these compounds have some potential applications (Patel et al. 2019). The blue-green algae, *Prymnesiumparvum* sp., and *Ochromonas* sp. can produce toxins that can be utilized in the pharmaceutical industry. *Spirulina*, *Scenedesmus*, and *Chlorella* species are used as human supplements. The extracts of microalgae improve skin, fertility, immune responses, and control weight. However, high concentrations of these extracts can be harmful particularly when cyanobacteria are utilized (Katircioglu et al. 2012).

7.5 Aquaculture

Aquaculture animals obtain nutrients from the food chain. The primary producers in food chain are microalgae. The survival and growth of adults as well as their larvae depend upon the nutrients produced from microalgae. Microalgae including *Skeletonema*, *Pavlova*, *Nannochloropsis*, *Chlorella*, *Isochrysis*, *Phaeodactylum*, *Tertraselmis*, *Chaetoceros*, *Thalassiosira*, and *Tetraselmis* play a vital role in external appearance and physiological growth of aquatic animals (Apandi et al. 2019). Both for the marine and fresh aquatic animals, microalgae are used as feed additives and food source until now. The cladocerans, rotifers, and shrimp have been cultured by different types of zooplankton that are further used in finfish farming and crustacean. *Haematococcus pluvialis*, *Dunaliella salina*, and *Spirulina* sp. are utilized for culturing ornamental fish, salmonid fish, and prawns (Shah et al. 2018). The usage of microalgae has been encouraged as a food source. However, risk of toxicological contamination and high manufacturing costs are some limitations for using microalgae as food product.

7.6 Animal Feed (Pets and Farming)

Microalgae have been reported as a feed supplement in several studies. A wide range of animals can use *Arthrospira* sp. as their feed, for example, cats, dogs, cows, aquarium fish, ornamental birds, breeding bulls, and horses. It puts a positive effect on the physiology of animals. It can be used in the feed of poultry as a source of protein (5–10%) (Richmond 2004). But it also has hazardous effects on poultry if applied for a longer time at high concentrations. It effects usually color of broiler shanks, skin, and egg yolk. Almost 30% of microalgal production is utilized for preparing animal feed and 50% of *Arthrospira* sp. is produced for making feed supplements worldwide (Molino et al. 2018).

7.7 *Biofertilizers*

In the absence of air, the microalgal biomass can be converted to charcoal, syngas, and bio-oil via pyrolysis at a higher temperature (350–700 °C) (Yang et al. 2019). The biochar produced by this mechanism can be utilized as a source of sequestration of carbon and as a biofertilizer. The emission of CO_2 can be reduced by 84% by applying biochar for carbon sequestration (Chen et al. 2019). In the field of agriculture, microalgae can be used as a source of biofertilizer and as soil conditioners (Ronga et al. 2019). The blue-green algae can reduce the nitrogen fertilizer usage, improve physicochemical properties of soil, and enhance the yield of biomass. It can also improve electrical conductivity, pH, residual carbon and nitrogen of soil. Moreover, grain quality is improved regarding protein content. In low-land and up-land conditions, microalgae of following genera *Anabaena*, *Tolypothrix*, *Aulosira*, and *Nostoc* can fix atmospheric nitrogen to improve the growth of paddy crops (Esteves-Ferreira et al. 2018). The microalgae production is usually carried out by four methods that are (1) field method, (2) nursery cum, (3) tank method, and (4) pit method. The latter two are more beneficial for individual farming while former two are better for bulk production of microalgae. Additional income can be generated by this technology by selling algal biofertilizer.

8 Commercialization of Microalgal Products

Various patents have been filed in recent years targeting the efficient extraction of pigments, proteins, carbohydrates, and other valuable products from the cell lysate (Table 14.6). For instance, a patent was granted to Sepal Technologies Ltd., for developing a novel method of harvesting microalgal cells from water without rupturing (Borodyanski and Konstantinov 2003). This method involves successive steps of flocculation, flotation followed by dehydration to harvest microalgal biomass in concentrated form without cell rupturing. Another patent was granted to Green Extraction Technologies for developing an instrument for fractionation of microalgal biomass. Katz and colleagues hold another patent for the novel strategy of flocculation-deflocculation used for harvesting the microalgal biomass produced in various forms of waters including freshwater, saltwater, brackish water, and treated wastewater (Katz et al. 2013). Another method of isolating microalgae using externally applied magnetic field has been established using a composite of paramagnetic nanoparticles (Tohver et al. 2001).

A multi-step protein extraction method for isolating highly purified protein (less than 5 kDa) from *Chlorella* species was patented involving washing, thermal permeabilization (at 50–150 °C), removal of permeabilized biomass by combined pro-

Table 14.6 Recent patents in industrial process optimization and utilization of microalgal biomass

Process	Patent date	Assignee	Patent number	Target strains	References
Microalgae separation method without cell rupturing	25-02-2003	Sepal Technologies Ltd.	US-09748249/ US-6524486 B2	Generally applicable	Borodyanski and Konstantinov (2003)
Microalgae fractionation	14-04-2011	Valicor Inc.	US-20110086386A1	–	Czartoski et al. (2016)
Flocculation-deflocculation method for improved harvesting	25-04-13	The University of Texas System (Board of Regents)	WO 2013059754A1/ US-20130102055	*Chlorella*, *Nannochloropsis*, *Spirulina*, *Dunaliella*, *Oscillatoria*, *Scenedesmus*, *Amphora*, *Phormidium*, *Ochromonas*, *Selenastrum*	Katz et al. (2013)
Microalgae fractionation for proteins and lipids	13-01-2013	Old Dominion University Research Foundation	WO2013086302A1	*Scenedesmus* sp., *Chlamydomonas* sp., *Spirulina* sp., *Chlorella* sp., *Porphyridium* sp., *Euglena* sp.	Kumar and Hatcher (2013)
Enzymatic digestion for production of lipid, protein, and carbohydrate from microalgae	05-03-15	The University of Toledo	WO2015031762A1	*Schizochitrium*, *chlorella* sp., *Limacium*	Vadlamani et al. (2016)
Method for extraction of sugars and lipids	28-05-15	Eni S.P.A.	WO2015075630A1	–	Massetti et al. (2016)
Albumin protein extraction using selective heating from freshwater microalgae	05-11-13	Heliae Dev LLC	US-8574587B2	–	Aniket (2013)

(continued)

Table 14.6 (continued)

Process	Patent date	Assignee	Patent number	Target strains	References
Globulin protein extraction using selective heating from freshwater microalgae	03-06-14	Heliae Dev LLC	US-8741629B2	*Nannochloropsis* sp.	Aniket (2014)
Microalgae harvesting process	04-01-2015	The Colorado School of Mines	US-9464268/ US-20150152376A1/ US-20110201076	Generally applicable	Tohver et al. (2001)
Extraction of soluble proteins from microalgae	01-06-17	Roquette Freres	US-20170152294	*Chlorella* sp.	Patinier (2017)
Production of protein-rich edible product from microalgae	02-02-2017	Synthetic Genomics Inc.	WO2017019125A1	–	Rutt et al. (2017)
Isolation of protein from solid biomass of microalgae, macroalgae, vegetables, and their combinations	27-03-14	Wageningen Universiteit	WO2014046543A1	–	Zhang et al. (2014)

cess of filtration, flocculation, multi-step centrifugation followed by ultrafiltration using specific sized membrane with a cutoff value less than 5 kDa (Patinier 2017). The isolation process of globulin protein from freshwater microalgae *Nannocchloropsis* sp. was invented by the Heliae Development LLC. This method involves mixing of freshwater algae with saltwater and heated at a temperature below the boiling point of extract results in a liquid fraction with an enriched content of globulin proteins. The separation of globulin-enriched liquid fraction was separated from biomass fraction by a series of steps involving treatment with various solvents like ethanol, methanol, acetone, isopropanol, acetonitrile, and ethyl acetate (Aniket 2013). A patent was filed in 2013, regarding protein fractionation along with lipids from microalgae. This study used a bioreactor having a temperature of 200–350 °C which led to the separation of cell lysate into a solid and liquid phase which resulted in the extraction of 60% of the total protein (Aniket 2014).

An efficient method of sugar extraction from microalgal biomass was patented by a Colombian company (Ecopetrol S. A.) in 2012. This study proposed the use of sulfuric acid treatment at high temperature (between 100–200 °C) and pressure

(101.14–303.42 kPa) that led to the disruption of cell wall of microalgal cells of *C. vulgaris*, *S. bassilensis* and *Chlamydomonas* sp., and thus facilitated the process of carbohydrate extraction (Improved method for obtaining fermentable sugars based on microalgae and macroalgae 2011). Another process of carbohydrate and lipid extraction from microalgae was patented in 2014. This method utilized the extracted carbohydrates for the purpose of fermentation to produce alcohols like ethanol and butanol (Massetti et al. 2014). Moreover, a strategy of flocculating the microalgae using an anionic flocculant (polyacrylamides, polycarboxylates, polyacrylates, and polymethacrylates) and an elevated pH (equal to or higher than 10) was used. The carbohydrate and lipid content were then extracted by solvent extraction and acid hydrolysis of the concentrated semi-solid biomass. A patent, suggesting the use of enzymatic hydrolysis (using fungal acid protease) of *Chlorella sp., Schizochitrium limacium* was allotted to the University of Toledo for improved extraction of carbohydrate, lipid, and protein content (Choi et al. 2010).

9 Conclusion and Prospects

Microalgal biorefineries have immense potential in terms of product diversity and sustainability and could become multi-billion dollars industry in near future. Continuous technological developments are being made to increase the efficiency of microalgal biorefineries but still a multi-product commercial biorefinery for large-scale production of value-added products is not feasible. Current studies show that a multi-product commercial biorefinery is too costly to operate. The standard downstream processing of the bulk commodities accounts for 30% to the total cost whereas in current biorefineries it is up to 50–60%. So, in order to reduce the overall expense, simplified downstream processing is required with a reduced number of multi-operational units. The complexity of the design is much crucial mainly in the cell disruption and extraction part of the biorefinery section. Certainly, we need to develop more efficient procedures of cell disruption and extraction along with the overall optimization of the process. We need a better understanding of cell wall structure and composition of each microalga to employ the best suitable technique for its treatment. The microalgae-based products require extraction of the metabolic content for which various cell wall disruption techniques have been developed. However, the microwave-assisted extraction (MAE) and pulsed electric field method (PEF) have shown to be the most promising approaches. In order to enable the scaling up of any technique, further optimization of the operating costs and energy consumption are needed to reach the ultimate goals of high yielding quality products and easy recovery. Development of new and efficient techniques is a prerequisite to the advancement of the overall process of microalgal cell disruption and enhanced recovery of bioactive compounds. Conclusively, along with the overall cost reduction of the process, mildness is another important criterion to be considered while designing a microalgal biorefinery approach. Maximum utilization of all components of the microalgal biomass in the biorefinery will lead to the development of large-scale production of valuable commodities from the current small-scale procedures in the future.

References

Aarthy, A., Kumari, S., Turkar, P., & Subramanian, S. (2018). An insight on algal cell disruption for biodiesel production. *Asian Journal of Pharmaceutical and Clinical Research, 11*(2), 21–21.

Abramson, B. W., Lensmire, J., Lin, Y.-T., Jennings, E., & Ducat, D. C. (2018). Redirecting carbon to bioproduction via a growth arrest switch in a sucrose-secreting cyanobacterium. *Algal Research, 33*, 248–255.

Afzal, I., Shahid, A., Ibrahim, M., Liu, T., Nawaz, M., & Mehmood, M. A. (2017). Microalgae: A promising feedstock for energy and high-value products. In *Algae based polymers, blends, and composites* (p. 55). San Diego: Elsevier.

Al hattab, M., & Ghaly, A. (2015). Microalgae oil extraction pre-treatment methods: Critical review and comparative analysis. *Journal of Fundamentals of Renewable Energy and Applications, 5*(4), 172.

Alam, M. A., Muhammad, G., Rehman, A., Russel, M., Shah, M., & Wang, Z. (2019). Standard techniques and methods for isolating, selecting and monitoring the growth of microalgal strain. In M. A. Alam & Z. Wang (Eds.), *Microalgae biotechnology for development of biofuels and wastewater treatment* (pp. 75–94). Singapore: Springer.

Alam, M. A., Wan, C., Guo, S.-L., Zhao, X.-Q., Huang, Z.-Y., Yang, Y.-L., Chang, J.-S., & Bai, F.-W. (2014). Characterization of the flocculating agent from the spontaneously flocculating microalga *Chlorella vulgaris* JSC-7. *Journal of Bioscience and Bioengineering, 118*(1), 29–33.

Ali, M., & Watson, I. A. (2015). Microwave treatment of wet algal paste for enhanced solvent extraction of lipids for biodiesel production. *Renewable Energy, 76*, 470–477.

Aniket, K. (2013). Selective heated extraction of albumin proteins from intact freshwater algal cells. Google Patents.

Aniket, K. (2014). Selective heated extraction of globulin proteins from intact freshwater algal cells. Google Patents.

Apandi, N. M., Mohamed, R. M. S. R., Al-Gheethi, A., & Kassim, A. H. M. (2019). Microalgal biomass production through phycoremediation of fresh market wastewater and potential applications as aquaculture feeds. *Environmental Science and Pollution Research, 26*(4), 3226–3242.

Ariede, M. B., Candido, T. M., Jacome, A. L. M., Velasco, M. V. R., de Carvalho, J. C. M., & Baby, A. R. (2017). Cosmetic attributes of algae—A review. *Algal Research, 25*, 483–487.

Baier, T., Kros, D., Feiner, R. C., Lauersen, K. J., Müller, K. M., & Kruse, O. (2018). Engineered fusion proteins for efficient protein secretion and purification of a human growth factor from the green microalga *Chlamydomonas reinhardtii*. *ACS Synthetic Biology, 7*(11), 2547–2557.

Balasubramanian, S., Allen, J. D., Kanitkar, A., & Boldor, D. (2011). Oil extraction from *Scenedesmus obliquus* using a continuous microwave system – Design, optimization, and quality characterization. *Bioresource Technology, 102*(3), 3396–3403.

Barati, B., Gan, S.-Y., Lim, P.-E., Beardall, J., & Phang, S.-M. (2019). Green algal molecular responses to temperature stress. *Acta Physiologiae Plantarum, 41*(2), 26.

Barrow, C. J., & Shahidi, F. (2008). *Marine nutraceuticals and functional foods*. Boca Raton, FL: CRC Press.

Bhalla, A., Bansal, N., Kumar, S., Bischoff, K. M., & Sani, R. K. (2013). Improved lignocellulose conversion to biofuels with thermophilic bacteria and thermostable enzymes. *Bioresource Technology, 128*, 751–759.

Biller, P., Friedman, C., & Ross, A. B. (2013). Hydrothermal microwave processing of microalgae as a pre-treatment and extraction technique for bio-fuels and bio-products. *Bioresource Technology, 136*, 188–195.

Bleakley, S., & Hayes, M. (2017). Algal proteins: Extraction, application, and challenges concerning production. *Food, 6*(5), 33.

Borodyanski, G., & Konstantinov, I. (2003). Microalgae separator apparatus and method. Google Patents.

Buono, S., Langellotti, A. L., Martello, A., Rinna, F., & Fogliano, V. (2014). Functional ingredients from microalgae. *Food & Function, 5*, 1669–1685.

Buyel, J. F., Twyman, R. M., & Fischer, R. (2015). Extraction and downstream processing of plant-derived recombinant proteins. *Biotechnology Advances, 33*, 902–913.

Byreddy, A. R., Gupta, A., Barrow, C. J., & Puri, M. (2015). Comparison of cell disruption methods for improving lipid extraction from thraustochytrid strains. *Marine Drugs, 13*(8), 5111–5127.

Chen, B., Wan, C., Mehmood, M. A., Chang, J.-S., Bai, F., & Zhao, X. (2017). Manipulating environmental stresses and stress tolerance of microalgae for enhanced production of lipids and value-added products–A review. *Bioresource Technology, 244*(Pt 2), 1198–1206.

Chen, C.-Y., Zhao, X.-Q., Yen, H.-W., Ho, S.-H., Cheng, C.-L., Lee, D.-J., Bai, F.-W., & Chang, J.-S. (2013). Microalgae-based carbohydrates for biofuel production. *Biochemical Engineering Journal, 78*, 1–10.

Chen, W., Meng, J., Han, X., Lan, Y., & Zhang, W. (2019). Past, present, and future of biochar. *Biochar, 1*(1), 75–87.

Chia, S. R., Chew, K. W., Show, P. L., Yap, Y. J., Ong, H. C., Ling, T. C., & Chang, J. S. (2018). Analysis of economic and environmental aspects of microalgae biorefinery for biofuels production: A review. *Biotechnology Journal, 13*(6), e1700618.

Choi, S. P., Nguyen, M. T., & Sim, S. J. (2010). Enzymatic pretreatment of *Chlamydomonas reinhardtii* biomass for ethanol production. *Bioresource Technology, 101*, 5330–5336.

Choi, Y. Y., Hong, M.-E., Chang, W. S., & Sim, S. J. (2019). Autotrophic biodiesel production from the thermotolerant microalga *Chlorella sorokiniana* by enhancing the carbon availability with temperature adjustment. *Biotechnology and Bioprocess Engineering, 24*(1), 223–231.

Coronel, C. D., Do Nascimento, M., & Curatti, L. (2019). Effect of matching microalgal strains origin and regional weather condition on biomass productivity in environmental photobioreactors. *Bioresource Technology Reports, 5*, 104–112.

Couteau, C., & Coiffard, L. (2018). Microalgal application in cosmetics. In *Microalgae in health and disease prevention* (pp. 317–323). London: Elsevier.

Cravotto, G., Boffa, L., Mantegna, S., Perego, P., Avogadro, M., & Cintas, P. (2008). Improved extraction of vegetable oils under high-intensity ultrasound and/or microwaves. *Ultrasonics Sonochemistry, 15*(5), 898–902.

Czartoski, T. J., Perkins, R., Villanueva, J. L., & Richards, G. (2016). Algae biomass fractionation. Google Patents.

Dammak, M., Hadrich, B., Miladi, R., Barkallah, M., Hentati, F., Hachicha, R., Laroche, C., Michaud, P., Fendri, I., & Abdelkafi, S. (2017). Effects of nutritional conditions on growth and biochemical composition of *Tetraselmis* sp. *Lipids in Health and Disease, 16*(1), 41.

de Carvalho, J. C., Sydney, E. B., Tessari, L. F. A., & Soccol, C. R. (2019). Culture media for mass production of microalgae. In *Biofuels from algae* (pp. 33–50). London: Elsevier.

de Farias Silva, C. E., Barbera, E., & Bertucco, A. (2019a). Biorefinery as a promising approach to promote ethanol industry from microalgae and cyanobacteria. In *Bioethanol production from food crops* (pp. 343–359). London: Elsevier.

de Farias Silva, C. E., Sforza, E., & Bertucco, A. (2018). Stability of carbohydrate production in continuous microalgal cultivation under nitrogen limitation: Effect of irradiation regime and intensity on *Tetradesmus obliquus*. *Journal of Applied Phycology, 30*(1), 261–270.

de Farias Silva, C. E., Sforza, E., & Bertucco, A. (2019b). Enhancing carbohydrate productivity in photosynthetic microorganism production: A comparison between cyanobacteria and microalgae and the effect of cultivation systems. In *Advances in feedstock conversion technologies for alternative fuels and bioproducts* (pp. 37–67). London: Elsevier.

de Freitas, B. C. B., Brächer, E. H., de Morais, E. G., Atala, D. I. P., de Morais, M. G., & Costa, J. A. V. (2019). Cultivation of different microalgae with pentose as carbon source and the effects on the carbohydrate content. *Environmental Technology, 40*(8), 1062–1070.

Denery, J. R., Dragull, K., Tang, C. S., & Li, Q. X. (2004). Pressurized fluid extraction of carotenoids from *Haematococcus pluvialis* and *Dunaliella salina* and kavalactones from *Piper methysticum*. *Analytica Chimica Acta, 501*(2), 175–181.

Ejike, C. E. C. C., Collins, S. A., Balasuriya, N., Swanson, A. K., Mason, B., & Udenigwe, C. C. (2017). Prospects of microalgae proteins in producing peptide-based functional foods for promoting cardiovascular health. *Trends in Food Science & Technology, 59*, 30–36.

El-Dalatony, M. M., Salama, E.-S., Kurade, M. B., Kim, K.-Y., Govindwar, S. P., Kim, J. R., Kwon, E. E., Min, B., Jang, M., & Oh, S.-E. (2019). Whole conversion of microalgal biomass into biofuels through successive high-throughput fermentation. *Chemical Engineering Journal, 360*, 797–805.

Esteves-Ferreira, A. A., Inaba, M., Fort, A., Araújo, W. L., & Sulpice, R. (2018). Nitrogen metabolism in cyanobacteria: Metabolic and molecular control, growth consequences and biotechnological applications. *Critical Reviews in Microbiology, 44*(5), 541–560.

Fernández, F. A., Sevilla, J. M. F., & Grima, E. M. (2019). Costs analysis of microalgae production. In *Biofuels from algae* (pp. 551–566). London: Elsevier.

Ferro, L., Gorzsás, A., Gentili, F. G., & Funk, C. (2018). Subarctic microalgal strains treat wastewater and produce biomass at low temperature and short photoperiod. *Algal Research, 35*, 160–167.

García-Cubero, R., Cabanelas, I. T. D., Sijtsma, L., Kleinegris, D. M., & Barbosa, M. J. (2018). Production of exopolysaccharide by *Botryococcus braunii* CCALA 778 under laboratory simulated Mediterranean climate conditions. *Algal Research, 29*, 330–336.

Ghag, S. B., Vavilala, S. L., & D'Souza, J. S. (2019). Metabolic engineering and genetic manipulation of novel biomass species for biofuel production. In *Advanced bioprocessing for alternative fuels, biobased chemicals, and bioproducts* (pp. 13–34). London: Elsevier.

Gifuni, I., Olivieri, G., Pollio, A., & Marzocchella, A. (2018). Identification of an industrial microalgal strain for starch production in biorefinery context: The effect of nitrogen and carbon concentration on starch accumulation. *New Biotechnology, 41*, 46–54.

Gilbert-López, B., Barranco, A., Herrero, M., Cifuentes, A., & Ibáñez, E. (2017). Development of new green processes for the recovery of bioactives from *Phaeodactylum tricornutum. Food Research International, 99*(Pt 3), 1056–1065.

Goettel, M., Eing, C., Gusbeth, C., Straessner, R., & Frey, W. (2013). Pulsed electric field assisted extraction of intracellular valuables from microalgae. *Algal Research, 2*(4), 401–408.

Gonçalves, C. F., Menegol, T., & Rech, R. (2019). Biochemical composition of green microalgae *Pseudoneochloris marina* grown under different temperature and light conditions. *Biocatalysis and Agricultural Biotechnology, 18*, 101032.

Gong, M., & Bassi, A. (2016). Carotenoids from microalgae: A review of recent developments. *Biotechnology Advances, 34*(8), 1396–1412.

Gong, Y., Hu, H., Gao, Y., Xu, X., & Gao, H. (2011). Microalgae as platforms for production of recombinant proteins and valuable compounds: Progress and prospects. *Journal of Industrial Microbiology & Biotechnology, 38*(12), 1879–1890.

Guedes, A. C., Amaro, H. M., Sousa-Pinto, I., & Malcata, F. X. (2019). Algal spent biomass—A pool of applications. In *Biofuels from algae* (pp. 397–433). London: Elsevier.

Günerken, E., D'Hondt, E., Eppink, M. H. M., Garcia-Gonzalez, L., Elst, K., & Wijffels, R. H. (2015). Cell disruption for microalgae biorefineries. *Biotechnology Advances, 33*(2), 243–260.

Han, J. W., Klochkova, T. A., Shim, J. B., Yoon, K., & Kim, G. H. (2012). Isolation and characterization of a sex-specific lectin in a marine red alga, *Aglaothamnion oosumiense* Itono. *Applied and Environmental Microbiology, 78*, 7283–7289.

Hanifzadeh, M., Garcia, E. C., & Viamajala, S. (2018). Production of lipid and carbohydrate from microalgae without compromising biomass productivities: Role of Ca and Mg. *Renewable Energy, 127*, 989–997.

Hempel, F., Lau, J., Klingl, A., & Maier, U. G. (2011). Algae as protein factories: Expression of a human antibody and the respective antigen in the diatom *Phaeodactylum tricornutum. PLoS One, 6*, e28424.

Henderson, R. K., Parsons, S. A., & Jefferson, B. (2010). The impact of differing cell and algogenic organic matter (AOM) characteristics on the coagulation and flotation of algae. *Water Research, 44*, 3617–3624.

Her, N., Amy, G., Park, H.-R., & Song, M. (2004). Characterizing algogenic organic matter (AOM) and evaluating associated NF membrane fouling. *Water Research, 38*, 1427–1438.

Hernández-García, A., Velásquez-Orta, S. B., Novelo, E., Yáñez-Noguez, I., Monje-Ramírez, I., & Ledesma, M. T. O. (2019). Wastewater-leachate treatment by microalgae: Biomass, carbohydrate and lipid production. *Ecotoxicology and Environmental Safety, 174*, 435–444.

Ho, S.-H., Ye, X., Hasunuma, T., Chang, J.-S., & Kondo, A. (2014). Perspectives on engineering strategies for improving biofuel production from microalgae—A critical review. *Biotechnology Advances, 32*(8), 1448–1459.

Hossain, N., Mahlia, T. M. I., Zaini, J., & Saidur, R. (2019). Techno-economics and sensitivity analysis of microalgae as commercial feedstock for bioethanol production. *Environmental Progress & Sustainable Energy*. https://doi.org/10.1002/ep.13157.

Iwamoto, H. (2003). Industrial production of microalgal cell-mass and secondary products – Major industrial species. In *Chlorella* (pp. 253–263). Oxford, UK: Blackwell Publishing Ltd.

Jaki, B. U., Franzblau, S. G., Cho, S. H., & Pauli, G. F. (2006). Development of an extraction method for mycobacterial metabolome analysis. *Journal of Pharmaceutical and Biomedical Analysis, 41*(1), 196–200.

Kapoore, R. V. (2014). Mass spectrometry based hyphenated techniques for microalgal and mammalian metabolomics. http://etheses.whiterose.ac.uk/8234/.

Katircioglu, H., Beyatli, Y., Aslim, B., Yüksekdag, Z., & Atici, T. (2012). Screening for antimicrobial agent production of some microalgae in freshwater. *The Internet Journal of Microbiology, 2*(2), 1–5.

Katz, L. E., Kinney, K. A., Choi, J., & Chen, E. (2013). Continuous flocculation deflocculation process for efficient harvesting of microalgae from aqueous solutions. Google Patents.

Khan, A. Z., Shahid, A., Cheng, H., Mahboob, S., Al-Ghanim, K. A., Bilal, M., Liang, F., & Nawaz, M. Z. (2018a). Omics technologies for microalgae-based fuels and chemicals: Challenges and opportunities. *Protein and Peptide Letters, 25*(2), 99–107.

Khan, M. I., Shin, J. H., & Kim, J. D. (2018b). The promising future of microalgae: Current status, challenges, and optimization of a sustainable and renewable industry for biofuels, feed, and other products. *Microbial Cell Factories, 17*(1), 36.

Kim, S.-K., & Kang, K.-H. (2011). Medicinal effects of peptides from marine microalgae. *Advances in Food and Nutrition Research, 64*, 313–323.

Koyande, A. K., Chew, K. W., Rambabu, K., Tao, Y., Chu, D.-T., & Show, P.-L. (2019). Microalgae: A potential alternative to health supplementation for humans. *Food Science and Human Wellness, 8*, 16–24.

Kumar, S., & Hatcher, P. G. (2013). Fractionation of proteins and lipids from microalgae. Google Patents.

Kushwaha, D., Upadhyay, S., & Mishra, P. K. (2018). Growth of cyanobacteria: Optimization for increased carbohydrate content. *Applied Biochemistry and Biotechnology, 184*(4), 1247–1262.

Lai, Y. H., Puspanadan, S., & Lee, C. K. (2019). Nutritional optimization of *Arthrospira platensis* for starch and total carbohydrates production. *Biotechnology Progress, 35*(2), e2798.

Lee, J.-Y., Yoo, C., Jun, S.-Y., Ahn, C.-Y., & Oh, H.-M. (2010). Comparison of several methods for effective lipid extraction from microalgae. *Bioresource Technology, 101*(1), S75–S77.

Liang, F., Englund, E., Lindberg, P., & Lindblad, P. (2018). Engineered cyanobacteria with enhanced growth show increased ethanol production and higher biofuel to biomass ratio. *Metabolic Engineering, 46*, 51–59.

Llewellyn, C. A., & Airs, R. L. (2010). Distribution and abundance of MAAs in 33 species of microalgae across 13 classes. *Marine Drugs, 8*, 1273–1291.

Mani, U. V., Iyer, U. M., Dhruv, S. A., Mani, I. U., & Sharma, K. S. (2007). Therapeutic utility of *Spirulina*. In *Spirulina in human nutrition and health* (pp. 85–114). London: Taylor & Francis Group.

Manirafasha, E., Ndikubwimana, T., Zeng, X., Lu, Y., & Jing, K. (2016). Phycobiliprotein: Potential microalgae derived pharmaceutical and biological reagent. *Biochemical Engineering Journal, 109*, 282–296.

Markou, G., Angelidaki, I., & Georgakakis, D. (2012). Microalgal carbohydrates: An overview of the factors influencing carbohydrates production, and of main bioconversion technologies for production of biofuels. *Applied Microbiology and Biotechnology, 96*(3), 631–645.

Massetti, F., Capuano, F., Medici, R., & Miglio, R. (2014). Process for the extraction of lipids and sugars from algal biomass. Google Patents.

Massetti, F., Capuano, F., Medici, R., & Miglio, R. (2016). Process for the extraction of lipids and sugars from algal biomass. Google Patents.

Mathiot, C., Ponge, P., Gallard, B., Sassi, J.-F., Delrue, F., & Le Moigne, N. (2019). Microalgae starch-based bioplastics: Screening of ten strains and plasticization of unfractionated microalgae by extrusion. *Carbohydrate Polymers, 208*, 142–151.

Mehta, P., Singh, D., Saxena, R., Rani, R., Gupta, R. P., Puri, S. K., & Mathur, A. S. (2018). High-value coproducts from algae—An innovational way to deal with advance algal industry. In *Waste to wealth* (pp. 343–363). Singapore: Springer.

Mendes-Pinto, M. M., Raposo, M. F. J., Bowen, J., Young, A. J., & Morais, R. (2001). Evaluation of different cell disruption process on encysted cells of *Haematococcus pluvialis. Journal of Applied Phycology, 13*, 19–24.

Molina Grima, E., Belarbi, E.-H., Acién Fernández, F. G., Robles Medina, A., & Chisti, Y. (2003). Recovery of microalgal biomass and metabolites: Process options and economics. *Biotechnology Advances, 20*, 491–515.

Molino, A., Iovine, A., Casella, P., Mehariya, S., Chianese, S., Cerbone, A., Rimauro, J., & Musmarra, D. (2018). Microalgae characterization for consolidated and new application in human food, animal feed and nutraceuticals. *International Journal of Environmental Research and Public Health, 15*(11), 2436.

Mousavi, S., Najafpour, G. D., & Mohammadi, M. (2018). CO_2 bio-fixation and biofuel production in an airlift photobioreactor by an isolated strain of microalgae *Coelastrum* sp. SM under high CO_2 concentrations. *Environmental Science and Pollution Research, 25*(30), 30139–30150.

Munir, M., Qureshi, R., Bibi, M., & Khan, A. M. (2019). Pharmaceutical aptitude of Cladophora: A comprehensive review. *Algal Research, 39*, 101476.

Naghshbandi, M. P., Tabatabaei, M., Aghbashlo, M., Aftab, M. N., & Iqbal, I. (2019). Metabolic engineering of microalgae for biofuel production. https://doi.org/10.1007/7651_2018_205.

Nobre, B., Marcelo, F., Passos, R., Beirão, L., Palavra, A., Gouveia, L., & Mendes, R. (2006). Supercritical carbon dioxide extraction of astaxanthin and other carotenoids from the microalga *Haematococcus pluvialis. European Food Research and Technology, 223*(6), 787–790.

Pancha, I., Chokshi, K., Ghosh, T., Paliwal, C., Maurya, R., & Mishra, S. (2015). Bicarbonate supplementation enhanced biofuel production potential as well as nutritional stress mitigation in the microalgae *Scenedesmus* sp. CCNM 1077. *Bioresource Technology, 193*, 315–323.

Pang, X., Tong, Y., Xue, W., Y-f, Y., Chen, X., Liu, J., & Chen, D. (2019). Expression and characterization of recombinant human lactoferrin in edible alga *Chlamydomonas reinhardtii. Bioscience, Biotechnology, and Biochemistry, 83*(5), 851–859.

Passos, F., Carretero, J., & Ferrer, I. (2015). Comparing pretreatment methods for improving microalgae anaerobic digestion: Thermal, hydrothermal, microwave and ultrasound. *Chemical Engineering Journal, 279*, 667–672.

Patel, A., Matsakas, L., Rova, U., & Christakopoulos, P. (2019). A perspective on biotechnological applications of thermophilic microalgae and cyanobacteria. *Bioresource Technology, 278*, 424–434.

Patinier, S. (2017). Method for extracting soluble proteins from microalgal biomass. Google Patents.

Ranadheer, P., Kona, R., Sreeharsha, R. V., & Mohan, S. V. (2019). Non-lethal nitrate supplementation enhances photosystem II efficiency in mixotrophic microalgae towards the synthesis of proteins and lipids. *Bioresource Technology, 283*, 373–377.

Rasala, B. A., Muto, M., Lee, P. A., Jager, M., Cardoso, R. M. F., Behnke, C. A., Kirk, P., Hokanson, C. A., Crea, R., Mendez, M., & Mayfield, S. P. (2010). Production of therapeutic proteins in

algae, analysis of expression of seven human proteins in the chloroplast of *Chlamydomonas reinhardtii*. *Plant Biotechnology Journal, 8*, 719–733.
Rashid, N., Lee, B., & Chang, Y.-K. (2019). Recent trends in microalgae research for sustainable energy production and biorefinery applications. In M. A. Alam & Z. Wang (Eds.), *Microalgae biotechnology for development of biofuels and wastewater treatment* (pp. 3–20). Singapore: Springer.
Richmond, A. (2004). *Handbook of microalgal culture: Biotechnology and Applied Phycology*. London: John Wiley & Sons, Inc.. https://doi.org/10.1002/9780470995280.
Rizwan, M., Mujtaba, G., Memon, S. A., Lee, K., & Rashid, N. (2018). Exploring the potential of microalgae for new biotechnology applications and beyond: A review. *Renewable and Sustainable Energy Reviews, 92*, 394–404.
Ronga, D., Biazzi, E., Parati, K., Carminati, D., Carminati, E., & Tava, A. (2019). Microalgal biostimulants and biofertilisers in crop productions. *Agronomy, 9*(4), 192.
Rösch, C., Roßmann, M., & Weickert, S. (2019). Microalgae for integrated food and fuel production. *GCB Bioenergy, 11*(1), 326–334.
Rutt, G. C., Flatt, J. H., Domaille, P., & Toledo, G. V. (2017). Protein rich food ingredient from biomass and methods of preparation. Google Patents.
Safi, C., Ursu, A. V., Laroche, C., Zebib, B., Merah, O., Pontalier, P. Y., & Vaca-Garcia, C. (2014). Aqueous extraction of proteins from microalgae: Effect of different cell disruption methods. *Algal Research, 3*(1), 61–65.
Sahni, P., Aggarwal, P., Sharma, S., & Singh, B. (2019). Nuances of microalgal technology in food and nutraceuticals: A review. *Nutrition & Food Science*. https://doi.org/10.1108/NFS-01-2019-0008.
Salla, A. C. V., Margarites, A. C., Seibel, F. I., Holz, L. C., Brião, V. B., Bertolin, T. E., Colla, L. M., & Costa, J. A. V. (2016). Increase in the carbohydrate content of the microalgae *Spirulina* in culture by nutrient starvation and the addition of residues of whey protein concentrate. *Bioresource Technology, 209*, 133–141.
Samiee-Zafarghandi, R., Karimi-Sabet, J., Abdoli, M. A., & Karbassi, A. (2018). Increasing microalgal carbohydrate content for hydrothermal gasification purposes. *Renewable Energy, 116*, 710–719.
Sathasivam, R., Radhakrishnan, R., Hashem, A., & Abd_Allah, E. F. (2017). Microalgae metabolites: A rich source for food and medicine. *Saudi Journal of Biological Sciences, 26*(4), 709–722.
Seo, J. Y., Praveenkumar, R., Kim, B., Seo, J. C., Park, J. Y., Na, J. G., Jeon, S. G., Park, S. B., Lee, K., & Oh, Y. K. (2016). Downstream integration of microalgae harvesting and cell disruption by means of cationic surfactant-decorated Fe_3O_4 nanoparticles. *Green Chemistry, 18*(14), 3981–3989.
Shah, M. R., Lutzu, G. A., Alam, A., Sarker, P., Chowdhury, M. K., Parsaeimehr, A., Liang, Y., & Daroch, M. (2018). Microalgae in aquafeeds for a sustainable aquaculture industry. *Journal of Applied Phycology, 30*(1), 197–213.
Shahid, A., Khan, A. Z., Liu, T., Malik, S., Afzal, I., & Mehmood, M. A. (2017). Production and processing of algal biomass. In *Algae based polymers, blends, and composites* (pp. 273–299). San Diego: Elsevier.
Shahid, A., Malik, S., Alam, M. A., Nahid, N., & Mehmood, M. A. (2019). The culture technology for freshwater and marine microalgae. In M. A. Alam & Z. Wang (Eds.), *Microalgae biotechnology for development of biofuels and wastewater treatment* (pp. 21–44). Singapore: Springer.
Shamriz, S., & Ofoghi, H. (2016). Outlook in the application of *Chlamydomonas reinhardtii* chloroplast as a platform for recombinant protein production. *Biotechnology and Genetic Engineering Reviews, 32*, 92–106.
Shamriz, S., & Ofoghi, H. (2019). Expression of recombinant PfCelTOS antigen in the chloroplast of *Chlamydomonas reinhardtii* and its potential use in detection of malaria. *Molecular Biotechnology, 61*(2), 102–110.

Shankar, M., Chhotaray, P. K., Agrawal, A., Gardas, R. L., Tamilarasan, K., & Rajesh, M. (2017). Protic ionic liquid-assisted cell disruption and lipid extraction from fresh water *Chlorella* and *Chlorococcum* microalgae. *Algal Research, 25*, 228–236.

Sikkema, R., Junginger, H. M., Pichler, W., Hayes, S., & Faaij, A. P. C. (2010). The international logistics of wood pellets for heating and power production in Europe. *Biofuels, Bioproducts and Biorefining, 4*, 132–153.

Smachetti, M. E. S., Cenci, M. P., Salerno, G. L., & Curatti, L. (2019). Ethanol and protein production from minimally processed biomass of a genetically-modified cyanobacterium overaccumulating sucrose. *Bioresource Technology Reports, 5*, 230–237.

Song, X., Wang, J., Wang, Y., Feng, Y., Cui, Q., & Lu, Y. (2018). Artificial creation of *Chlorella pyrenoidosa* mutants for economic sustainable food production. *Bioresource Technology, 268*, 340–345.

Spiden, E. M., Yap, B. H. J., Hill, D. R. A., Kentish, S. E., Scales, P. J., & Martin, G. J. O. (2013). Quantitative evaluation of the ease of rupture of industrially promising microalgae by high pressure homogenization. *Bioresource Technology, 140*, 165–171.

Takeshita, T., Ota, S., Yamazaki, T., Hirata, A., Zachleder, V., & Kawano, S. (2014). Starch and lipid accumulation in eight strains of six *Chlorella* species under comparatively high light intensity and aeration culture conditions. *Bioresource Technology, 158*, 127–134.

Tohver, V., Smay, J. E., Braem, A., Braun, P. V., & Lewis, J. A. (2001). Nanoparticle halos: A new colloid stabilization mechanism. *Proceedings of the National Academy of Sciences, 98*, 8950–8954.

Toledo-Cervantes, A., Sol6rzano, G. G., Campos, J. E., Martínez-García, M., & Morales, M. (2018). Characterization of *Scenedesmus obtusiusculus* AT-UAM for high-energy molecules accumulation: Deeper insight into biotechnological potential of strains of the same species. *Biotechnology Reports, 17*, 16–23.

Tourang, M., Baghdadi, M., Torang, A., & Sarkhosh, S. (2019). Optimization of carbohydrate productivity of *Spirulina* microalgae as a potential feedstock for bioethanol production. *International journal of Environmental Science and Technology, 16*(3), 1303–1318.

Vadlamani, A. K., Relue, P., Viamajala, S., Shao, H., & Varanasi, S. (2016). Enzymatic digestion of microalgal biomass for lipid, sugar, and protein recovery. Google Patents.

Varshney, P., Beardall, J., Bhattacharya, S., & Wangikar, P. P. (2018). Isolation and biochemical characterisation of two thermophilic green algal species-*Asterarcys quadricellulare* and *Chlorella sorokiniana*, which are tolerant to high levels of carbon dioxide and nitric oxide. *Algal Research, 30*, 28–37.

Venugopal, V. (2016). *Marine polysaccharides*. Boca Raton, FL: CRC Press. https://doi.org/10.1201/b10516.

Verdel, E., Kline, P., Wani, S., & Woods, A. (2000). Purification and partial characterization of haloperoxidase from fresh water algae *Cladophora glomerata*. *Comparative Biochemistry and Physiology, Part B: Biochemistry & Molecular Biology, 125*, 179–187.

Wang, M., Yuan, W., Jiang, X., Jing, Y., & Wang, Z. (2014a). Disruption of microalgal cells using high-frequency focused ultrasound. *Bioresource Technology, 153*, 315–321.

Wang, W., Wu, X., Tan, J., Zhu, L., Mou, Y., Zhang, D., & Gao, J. (2019). Using response surface methodology optimize culture conditions for human lactoferrin production in desert *Chlorella*. *Protein Expression and Purification, 155*, 130–135.

Wang, X., Jin, M., Balan, V., Jones, A. D., Li, X., Li, B.-Z., Dale, B. E., & Yuan, Y.-J. (2014b). Comparative metabolic profiling revealed limitations in xylose-fermenting yeast during co-fermentation of glucose and xylose in the presence of inhibitors. *Biotechnology and Bioengineering, 111*(1), 152–164.

Yang, C., Li, R., Zhang, B., Qiu, Q., Wang, B., Yang, H., Ding, Y., & Wang, C. (2019). Pyrolysis of microalgae: A critical review. *Fuel Processing Technology, 186*, 53–72.

Yen, H.-W., Hu, I.-C., Chen, C.-Y., Ho, S.-H., Lee, D.-J., & Chang, J.-S. (2013). Microalgae-based biorefinery—From biofuels to natural products. *Bioresource Technology, 135*, 166–174.

Zhang, C., Sanders, J., & Bruins, M. (2014). A process for isolating proteins from solid protein-containing biomass selected from vegetable biomass, algae, seaweed and combinations thereof. Google Patents.

Zhang, X., Hu, Q., Sommerfeld, M., Puruhito, E., & Chen, Y. (2010). Harvesting algal biomass for biofuels using ultrafiltration membranes. *Bioresource Technology, 101*, 5297–5304.

Chapter 15
Pretreatment and Lipid Extraction from Wet Microalgae: Challenges, Potential, and Application for Industrial-Scale Application

Md Shamim Howlader and William Todd French

Abstract The production of oil from microalgae has tremendous potential for reducing environmental problems generated using conventional fossil fuels. The present barrier for industrial-scale lipid production from algal biomass for biofuel application comes from the high extraction cost which is usually performed after drying the biomass. The lipid extraction cost can be significantly reduced if the extraction is performed directly on wet biomass. The lipid recovery from the wet biomass at the present state is very low to be competitive at large-scale application. Due to the high moisture content, a pretreatment of wet biomass is needed prior to the lipid extraction to increase the overall oil recovery. There are different pretreatments (e.g., high-pressure homogenization, ultrasound sonication, microwave irradiation, etc.) that can be used to disrupt the robust cell wall of microalgae prior to the oil extraction. Sometimes, both the pretreatment and lipid extraction can be performed using the same apparatus to reduce the overall production cost. The process economy and the cost of lipid extraction of different pretreatment methods need to be assessed carefully before considering its commercial-scale application.

Keywords Biofuel · Microalgae · Lipid extraction · Pretreatment

1 Introduction

We are in a state where we are rapidly running out of time to combat some of the basic survival problems currently we are facing to save our planet from certain obvious threats. Global warming is one of the most debated issues of the twenty-first

M. S. Howlader (✉) · W. T. French
Dave C. Swalm School of Chemical Engineering, Mississippi State University, Mississippi State, MS, USA
e-mail: msh324@msstate.edu; french@che.msstate.edu

M. A. Alam et al. (eds.), *Microalgae Biotechnology for Food, Health and High Value Products*, https://doi.org/10.1007/978-981-15-0169-2_15

century which many people are viewing in different angles; some believe this is happening and causing a serious threat to our existence and some believe this as non-existent. In reality, we are already having drastic changes in our loving planet due to the increasing world surface temperature. The use of fossil fuel is one of the main sources of global warming which generally occurs by the emission of greenhouse gases originated from fossil fuel and other highly industrial sources (Ramakrishnan 2015). To reduce the global warming, we need to shift our focus on producing fuels from greener sources which not only be environmentally friendly but also need to be cost-effective to make it feasible for large-scale application. Generally, biodiesel is a greener fuel which emits less greenhouse gas and renewable produced from different vegetable oil sources such as soybean and rapeseed oil. Biodiesel from these sources is not a long-term solution because they are used as food as well as they are season-dependent for their growth and production (Mazanov et al. 2016; Park et al. 2008; Patel et al. 2017).

Different types of microorganisms are the potential sources to produce biodiesel at large-scale application without much complexity. Generally, the oleaginous microbes are termed as the suitable candidates for the biofuel production because these microbes can produce more than 20% oil (up to 80%) on dry cell weight (DCW) basis (Shields-Menard et al. 2018). Among oleaginous microbes, yeasts are suitable for large-scale biodiesel production because their growth is faster, they can take a wide range of substrates, and their productivity is higher compared to other microbes (Patel et al. 2017; Alfenore and Molina-Jouve 2016; Adrio 2017). Though different bacteria and fungi are used for many applications, they can also be considered for biodiesel production (Mukhopadhyay 2015; Portillo et al. 2018). Among different microbes, algae are most widely studied microbes because of their versatilities such as they have thousands of species to explore, do not compete with food unlike vegetable oil, ability to fix atmospheric CO_2, can be cultivated anywhere in the world, and have high biomass and lipid content when the strain is correctly developed (Singh and Gu 2010; Griffiths and Harrison 2009; Alam et al. 2019; Wan et al. 2015; Lu et al. 2019). Generally for lipid extraction from algae, the process starts with the cultivation of algae either in an open pond with less control or in a photo-bioreactor with more controlled environment abbreviated as PBR. The second step of the algal biofuel process is the harvesting of algae followed by the extraction of lipid from cell biomass using a suitable solvent.

For traditional lipid extraction from algae, the algal biomass is dried after harvesting and the solvent extraction is performed on dried biomass. The drying of wet algae is very energy-intensive that makes the process uneconomical for commercialization of the algae-based biofuels (Sathish and Sims 2012). Lipid extraction from wet algal biomass can eliminate this shortcoming by bypassing the drying of the biomass. Though wet extraction process can reduce the energy requirement, the problem still remains as the lipid recovery is usually very low from the wet biomass. A pretreatment on wet biomass prior to the solvent extraction can improve the lipid recovery which can be conducted in many different ways such as physical, chemical, enzymatic, etc. There have been many publications on algal biomass pretreatment on improving the lipid recovery from wet biomass (Yap et al. 2015; Wang et al. 2015; Dong et al. 2016; Howlader et al. 2018a; Martinez-Guerra et al. 2018). In addition to lipid, the algal biomass can also be treated for protein, carbohydrate, and other bioproduct recovery. In this chapter, we discuss different means of pretreatment techniques applied to treat the biomass for improving the lipid recovery. Additionally, we also briefly discuss findings on protein recovery after algal biomass pretreatment using different cell disruption methods. Here, we have reported the data obtained by researchers mostly in the last 5 years but not beyond 10 years for presenting the most updated findings.

2 Lipid Extraction from Dried Biomass

Traditionally, the lipid is extracted from the algal biomass or other microbes after drying wet cells using different means. For laboratory-scale lipid extraction, freeze-drying is the widely used method of biomass drying where the drying is conducted by applying vacuum at very low temperature. For large-scale lipid extraction and biodiesel production, the algal biomass is dried using drum dryer or spray dryer prior to the lipid extraction. Though the lipid extraction from dried biomass is preferred because of higher lipid recovery, the overall lipid extraction cost increases due to the higher energy requirement of drying making the process unsuitable in the large-scale application. Hence, lipid extraction from wet algal cell biomass is preferred.

3 Lipid Extraction from Wet Route

The lipid extraction from wet biomass is a preferred means for improving the lipid and other bioproducts recovery from microalgae for developing cost-effective bioprocesses. The high moisture content in the wet algae prevents the solvent to recover lipid which can be overcome by pretreating the wet algae. There are different ways to conduct the pretreatment on wet microalgae such as high-pressure homogenization, microwave irradiation, ultrasound sonication, high-pressure gas application, steam explosion, etc. The cell disruption or pretreatment methods applied into microalgae are discussed below.

3.1 High-Pressure Homogenization

High-pressure homogenization (HPH) is the most widely used technique for microbial pretreatment to extract the intracellular compounds for different applications. Unlike many other methods discussed earlier, high-pressure homogenization has been used at industrial scale in dairy processing application for protein extraction and purification (Cano-Ruiz and Richter 1997). Generally for HPH in algae processing, a very high pressure (50–1500 bar) with single or multiple passes is applied to the wet algal cell biomass with a defined biomass concentration to release the intracellular lipid or protein by breaking the thick cell wall (Samarasinghe et al. 2012; Yap et al. 2014; Günerken et al. 2015). The product is then extracted from the homogenized sample using multiple methods such as membrane filtration (mechanical method) and solvent extraction (chemical method). There have been numerous reports on improving lipid recovery after treating with HPH at different pressurized conditions. For example, Yao et al. (2018) found that 57.4% lipid can be recovered using the hexane extraction by HPH treatment which is considerably higher compared to the Bligh & Dyer method where 44% lipid was recovered. Another report by Cho et al. (2012) showed that lipid recovery can be improved from 19.8% to 24.9% when the cell wall was ruptured using HPH compared to untreated algal cell biomass when the lipid extraction time for HPH-treated sample and -untreated samples were 30 min and 5 h, respectively. Since the main drawback for lipid extraction and subsequent biodiesel production is the cost of extraction, the energy requirement for HPH treatment is significantly lower compared to other methods if the starting solid content in microalgae is higher. Having said that, HPH is not suitable for algal biomass with lower cell density because the processing cost of microalgae having the lower cell concentration is very high making the process unsuitable for large-scale application. The energy requirement for HPH treatment can be obtained using the following equation:

$$E\left(\mathrm{MJ}/\mathrm{kg}\right) = PN / C_{\mathrm{m}}\rho \tag{1}$$

where E is the total energy to treat per kg of dried biomass, P is the applied pressure in MPa, C_{m} is the solid percentage (%) of the treated algal cell, t is the time (h), and N is the number of passes used during the treatment. Though having higher solid content in the treated cell is advantageous from the energy consumption point of view, there are several drawbacks for cells with higher density such as high viscosity resulting in additional unit operations to pump the liquid to the HPH facility, emulsion formation is another concern that needs to be considered, and it was reported that lipid recovery using hexane extraction was decreased when HPH was used as the treatment on cell with solid content increase from 20% to 25% (Yap et al. 2015; Dong et al. 2016).

3.2 *Microwave-Assisted Pretreatment*

Microwave irradiation is another promising pretreatment method of algal microbes for improving the lipid recovery from both dry and wet biomass. Microwave irradiation creates an electromagnetic radiation with a frequency of 0.3–300 GHz where the algal cell wall is broken due to the non-contact heat sources, and the intracellular lipid or other metabolites are accessible to extract using solvent(s) (Drira et al.

2016). Microwave irradiation is one of the most popular pretreatment methods applied in laboratory scale for algal lipid extraction. One of the many advantages MW pose is that this method is suitable for both dry and wet biomass. Other notable advantages include a non-contact heat source, the process needs fewer equipment compared to other methods, and the faster energy transfer (Drira et al. 2016). There have been reports on microwave irradiation to improve the lipid recovery from algal biomass. For example, de Moura et al. compared the lipid extraction of MW-treated algal biomass with ultrasound sonication treatment and reported 20% (w/w) increase in lipid recovery from MW compared to the US. They also found that the fatty acid profile of the extracted lipid was not affected by the MW pretreatment (de Moura et al. 2018). Heo et al. (2017) reported a lipid recovery of 83% (w/w) from MW-treated algal cells whereas they found only 51% (w/w) lipid from the untreated cells. Balasubramanian et al. (2011) compared the lipid extraction after treating the algal biomass with MW and water bath control and found 76–77% lipid can be recovered from MW-treated biomass compared to 43–46% (w/w) lipid recovery using water bath control. From the above discussion, it is clear that MW irradiation is applicable to recover lipid from the algal cell with higher success rate compared to traditional methods, but the process has to be cost-effective from the energy consumption perspective for industrial-scale application. The total energy requirement for treating a defined algal cell as well as the energy available from the obtained biodiesel is required for the cost estimation. There have been reports on energy requirement of microwave irradiation method and it can be concluded that MW cannot be used for large-scale processing of microalgae due to its high energy requirement (Ali and Watson 2015; Lee et al. 2012). For example, Lee et al. (2012) performed a careful study on energy requirement for MW treatment for different algae and found that 420 MJ was needed for treating per kg of dried biomass, which is very high making the process unsuitable for commercial-scale application.

3.3 *Ultrasound Sonication*

Ultrasound sonication is a widely used pretreatment technique in algae processing for biodiesel application. The process of ultrasonication and its working principles are discussed elsewhere (Howlader et al. 2018a). We will focus on studying the effect of different parameters, i.e., power, treatment time and the intensity of the sonicator as well as the solid concentration in the algal cell biomass on lipid recovery. Generally, the lipid recovery is expected to be increased with increasing different treatment parameters, but the quality of the final product can be degraded due to longer treatment time, and hence, the production cost also increases. As a result, the treatment parameters need to be chosen wisely considering these for process optimization. There have been many recent publications on increasing lipid recovery by treating the algal biomass using ultrasound sonication. For example, Garoma and Janda reported an increase of 26.4% lipid recovery from *Chlorella vulgaris* after treating with ultrasound sonication compared to the untreated cells. All the treatments were conducted having the biomass concentration of 15% (w/w) because they investigated the effect of biomass concentration on lipid extraction and found that even if the biomass was increased beyond 15% the lipid extraction yield did not improve for a defined solvent system (Garoma and Janda 2016). Adam et al. (2012) found that the percentage of dry matter of biomass is the most significant factor for lipid extraction using ultrasound sonication where they extracted lipid by studying the effect of power, extraction time, and percentage dry matter. In a recent study by Ellison et al., they studied the effect of ultrasonication power and treatment time on lipid recovery using both Bligh and Dyer, and hexane extraction from a mixed algae sample. They found the optimum ultrasonication conditions to be 750 W and 30 min for increasing the lipid recovery from 8.3% to 16.9% when lipid was extracted using the Bligh and Dyer method (Ellison et al. 2019). The ultrasonication treatment can also be applied at the logarithmic growth phase of algae to improve the biomass and lipid content when used at low concentration. For example, Ren et al. (2019) used an ultrasonication power of 20 W and a frequency of 20 kHz to improve the growth and lipid accumulation where it was found that biomass and lipid content reached an optimum value of 2.78 g/L and 890 mg/L compared to control where the biomass and lipid content was 2.00 g/L and 550 mg/L, respectively. The higher lipid and biomass content resulted from greater substrate consumption due to the ultrasonication treatment. This remarkable study paved a new door for biofuels application and if this method can be scaled up to industrial-scale the cost of biodiesel production would be considerably lower making the process feasible.

3.4 *Steam Explosion*

The steam explosion can be applied on wet algal biomass to release the intracellular lipid with or without adding acid at various concentrations. The steam explosion generally utilizes a temperature of 120–240 °C with pressure from 1.03 to 3.45 MPa for a short exposure time (approximate 5 min) (Cheng et al. 2015; Lorente et al. 2015, 2017). The system remains in the pressurized state for the assigned time and then suddenly depressurized to the ambient condition. There have been reports in improving the lipid as well as other metabolites recovery from various microbes using the steam explosion technique. For example, Lorente et al. (2017) found an improvement in lipid recovery from 2.1% (w/w) to a staggering 17.6% (w/w) when treated *Nannochloropsis gaditana* using the steam explosion at 150 °C supplemented with 5% sulfuric acid compared to untreated sample using hexane as the solvent. The same authors also reported an increase in protein content from 1.4% (w/w) to 9.1% (w/w) for *N. gaditana* when treated with the steam explosion compared to untreated sample (Lorente et al. 2018). It is interesting to note that the lipid and protein recovery did not improve for other microbes which signify that steam explosion is dependent on cell wall of different microbes. In another article, Lorente et al. found that addition of acid at different concentration increased the lipid recovery during steam explosion. For example, the lipid recovery using hexane extraction method increased from 2.5% (w/w) to 10.0% (w/w) when the acid concentration was increased from 0% to 10.0% as an additive to steam explosion. Lorente et al. (2015) further compared the effect of lipid recovery using steam explosion with other pretreatment methods and found a significant increase in lipid recovery for steam explosion compared to ultrasound sonication and microwave treatment. From the initial study by several researchers, the steam explosion is found to be a potential pretreatment method which can be applied on microalgae in industrial-scale application. Having many advantages of steam explosion for biofuel application, this method is suitable for treating the microalgae with the addition of acid. The lipid recovery is very low without any acid addition which will pose a barrier for its large-scale application since acid recovery at large scale is very challenging as of cost of recovery and environmental consideration. Unlike other traditional cell disruption techniques such as HPH and microwave irradiation, the steam explosion has not explored thoroughly to understand the process in a more detailed manner for biofuel and protein extraction application which is needed for scaling up this method in commercial scale.

3.5 Pressurized Gas Treatment

Pressurized gases can also be used to treat the wet microalgae to improve the lipid recovery as well as other bioproducts. For pressurized gas treatment, the microbial cell suspension with a defined biomass concentration is pressurized using the treated gas at the desired pressure, temperature, and agitation for a certain exposure time. The system is then suddenly depressurized, and the cell is lysed due to the expansion of the gas during depressurization stage (Howlader et al. 2017a, 2019). The mechanism of cell disruption using pressurized gas is an interesting phenomenon where the solubility of the treated gas in the microbial cell suspension which contains more than 99% (w/w) water plays an important role for efficient cell disruption. Though the exact mechanism of cell disruption using pressurized gas is not available to date, there are several possible explanations discussed in the literature (Garcia-Gonzalez et al. 2007). For example, if the treated gas is reactive to the water present in the cell suspension, the overall pH of the suspension is reduced due to the formation of carbonic acid when carbon dioxide is used as the treated gas. The normal cellular activity of the cell decreases due to the reduction in pH because cells are optimally active for a certain pH. The second explanation is when the pressurized CO_2 is used to treat the cell, some of the unreacted gas pass through the cell membrane where it reacts with metabolites (e.g., Ca^{2+} and Mg^{2+}) present in the cell and precipitate as $CaCO_3$ and $MgCO_3$. Due to the loss of certain important metabolites, cells also lose their normal activity. The third explanation is when the pressure is suddenly reduced at the end of the treatment; the cells are mechanically broken like a popped balloon due to the sudden expansion of the treated gas. These factors are combinedly responsible for cell disruption when a pressurized gas is used for cell disruption. There have been different gases explored for microbial cell disruption, and CO_2 was found to be the most suitable one due to its superior advantages such as reactivity with water present in the cell suspension and precipitation of metabolites which does not occur for other gases (e.g., N_2, O_2, Ar, etc.). Also, CO_2 is inflammable, cheap, safer to use, and non-toxic. Furthermore, as oleaginous microbes contain more than 20% (w/w) lipid on dry cell weight basis, the process would be advantageous if the treated gas is moderate to highly soluble in lipid-rich cell for appropriate solubilization. It has been reported in the literature that the solubility of CO_2 in both lipid (triglyceride and tributyrin) and lipid-rich cell is higher compared to other gases (Howlader et al. 2017a, b, 2018b). Traditionally, pressurized CO_2 has been used extensively in food application for the inactivation of unwanted microbes or pathogens present in the food (Xu et al. 2010). Although heat treatment was also considered for similar purpose, pressurized CO_2 was found to be superior because the quality of food degrades for heat treatment due to the high treatment temperature. Since pressurized CO_2 has been successfully implemented in food industry, this method can also be applied for treating wet microalgae for improving the lipid recovery for biodiesel application. For example, Howlader et al. (2017a) reported an increase of 40% lipid recovery from CO_2-treated *Rhodotorula glutinis* cell when the treatment pressure was 4000 kPa with the treatment time of 5 h. One of the main advantages of this method is the energy requirement for treating per kg of dried biomass which was found to be comparable with other traditional methods (Howlader et al. 2018a). Being at very early stage of the research to explore its full potential, pressurized CO_2 needs further study to consider this as an alternative method for microbial cell disruption at an industrial scale.

3.6 Ionic Liquid-Assisted Method

Ionic liquids (IL) are liquids at room temperature which are gaining popularity for solid-liquid separation in recent years due to their tunable properties such as the viscosity, density, polarity, and hydrophobicity. These properties can be changed by changing their structure. Ionic liquids can be simultaneously used for both algal biomass pretreatment and lipid extraction which have several advantages compared to conventional methods such as non-flammable, low vapor pressure, high thermal and chemical stability, etc. (Desai 2016). Kim et al. (2012) compared the lipid extraction from *Chlorella vulgaris* using the mixture of ionic liquid [Bmim] [CF_3SO_3] and methanol with conventional Bligh and Dyer method and found a lipid recovery of 19% and 11%, respectively. Orr et al. (2016) studied a wide range of ionic liquids for treating the algal biomass to extract the lipid using hexane extraction where they investigated the mass ratio of algae to ionic liquids, incubation time, water content, and co-solvents. They indicated that ionic liquids can be used to disrupt the algal cell wall to improve the lipid recovery having high moisture content in the cell. Additionally, lower energy is needed for treating the algal cell biomass using different ionic liquids making the process an alternative approach for biofuel application. To et al. (2018) also studied the effect of different ionic liquids on lipid extraction and found [Che][ARG] as the most effective ionic liquid for lipid extraction from *C. vulgaris* and *Spirulina platensis* where [Emim][OAc] was found to be the least efficient. The structural differences between different ILs are responsible for varied lipid recovery even if the treatment conditions are the same using the same microbes. Another study was performed by Choi et al. on lipid extraction using the mixture of IL and molten salts where they found the following results: single molten salts ($FeCl_3$•$6H_2O$) extracted 113 mg/g, single IL [Emim][OAc] extracted 218.7 mg/g, and the mixture of molten salts ($FeCL_3$•$6H_2O$) and [Emin][OAc] extracted 227.6 mg/g. The results indicated an increase in lipid recovery when molten salts are mixed with ILs (Choi et al. 2014). Though ionic liquids are considered to be one of the suitable candidates in algal biomass treatment for different application, this method mostly confined in laboratory scale to date and further studies are needed for the feasibility of large-scale application.

3.7 Other Pretreatment Methods

Cell disruption using chlorine (Cl_2) can be a potential source for algal biomass pretreatment for improving the lipid recovery due to the lower energy requirement of the chlorine-treated biomass compared to other methods. It was found by Garoma and Yazdi that the total specific energy requirement for the chlorine-treated algal biomass was 3.73 MJ/kg having the biomass concentration of 0.2 g/L which is significantly lower compared to all of the conventional methods. Though Cl_2 treatment has the superior advantage from the energy requirement consideration, the lipid recovery was lower using Cl_2 treatment compared to even the untreated biomass due to the reactivity of chlorine with residual biomass (Garoma and Yazdi 2019). The chlorine method would be feasible in lipid industry if the reactivity can be diminished and further research is needed to make the process come into reality. Dilute acid treatment is another technique which showed promise for algal biomass pretreatment because up to 79% (w/w) lipid was recovered by treating the wet biomass using this method as reported by Sathish and Sims (2012). Though dilute acid pretreatment has the advantage of requiring lower energy for treating the wet microbes, this method cannot be used in large-scale application due to the environmental concerns the dilute acid poses. For example, if the dilute acid treatment is used for commercial application, tons of acid will be needed to process the biomass which would pose the environmental problems when it needs the disposal. Surfactants can also be used to treat the algal biomass prior to the lipid extraction to improve the lipid recovery. It has been shown that surfactants are suitable for increasing the lipid recovery after treating the wet biomass (Lai et al. 2016). Generally for surfactant-assisted pretreatment, concentration of surfactants, treatment time, and pH are varied to find the extraction efficiency. Though the surfactant-assisted method is suitable for increasing the lipid content, this method would be unsuitable for large-scale application due to the high cost of surfactant. Enzymatic pretreatment can also be applied on microalgae for improving the lipid recovery. When using enzymatic pretreatment, a defined amount of enzyme is dissolved and incubated for a defined time. The parameters generally controlled during enzymatic pretreatments are concentration of enzyme, incubation time, pH of the reaction, and incubation temperature. The enzyme can be used either alone or as a mixture of different enzymes.

4 Protein Recovery Using Cell Disruption

Different cell disruption method can be used to recover protein and other byproducts which have been under study by numerous researchers. It has been reported that the high-pressure homogenization (HPH) can be used to improve the protein release after disintegrating the cell membrane at the desired treatment condition. For example, Safi et al. (2017a) found 95% cell disintegration with more than 50% protein recovery with lower energy requirement compared to some other cell disruption techniques such as bead beating, pulse electric field (PEF), and enzymatic treatment. In another study performed by Safi et al. (2017b), they found that HPH released more protein (49%) during the pretreatment compared to enzymatic treatment (35%), but the overall protein yield after the membrane filtration was higher in case of enzymatic disruption. Sometimes, additional treatment is needed to release the protein content from algal biomass. For example, it was found that ultrasonication alone is not effective for improving the protein release from algal thick cell wall as reported by Phong et al. They found the protein recovery increased when the ultrasonication was combined with alkali pretreatment. They found that the protein recovery of *C. vulgaris* increased from 20% to 25% when cell was treated using the combination of sonication with alkali treatment compared to sonication alone (Phong et al. 2018). Lee et al. investigated the protein recovery from microalgae *C. vulgaris* by disrupting the microalgal cell using ultrasonication where they found ultrasonication method was superior in terms of protein recovery when compared with freeze-thawing and non-ionic detergent triton X-100. The optimum ultrasonic power and time were 400 W and 30 min having the biomass of 6.0 g/L where the protein recovery was 25.3% dry weight (Lee et al. 2017). Safi et al. performed another study on protein recovery from five microalgae named as *H. pluvialis*, *N. oculate*, *C. vulgaris*, *A. platensis*, and *P. cruentum* using five different cell disruption techniques which were control, manual grinding, ultrasonication, chemical method (added 2 N NaOH to maintain a pH of 12 to perform protein extraction), and HPH. They found a remarkable protein yield using high-pressure homogenization (HPH) with the reported 90% protein yield from *P. cruentum*. The other methods recovered lower protein as the protein yield was 24.8%, 49.5%, 67.0%, and 73.5% from control, manual grinding, ultrasonication, and chemical method, respectively. It is interesting to note that the protein yield using chemical method was significantly higher compared to other methods except for HPH and the processing cost of chemical method is lower compared to ultrasonication and HPH, which proves that chemical method using NaOH has a potential to be used for large-scale application for protein recovery from algal biomass (Safi et al. 2014).

Lupatini et al. (2017) also reported similar results for protein recovery using ultrasonication cell disruption on *S. platensis*. They reported that up to 76% protein can be recovered using the ultrasonication for 33–40 min treatment.

5 Future Directions

Microalgal cell disruption using different techniques is currently under study for different applications and there has been a decent success in regard to lipid recovery from both wet and dried biomass. Single disruption method sometimes cannot recover the product at the desired productivity, and combination of different techniques can be applied to eliminate that hurdle. For example, the protein recovery was improved when ultrasonication was combined with alkali treatment compared to the ultrasonication alone (Phong et al. 2018). Martinez-Guerra et al. (2018) produced fatty acid methyl ester (FAME) using the combination of microwave irradiation and ultrasound sonication. Liang et al. (2012) found that up to 49% lipid can be recovered when enzymatic-assisted extraction is combined with ultrasonication. Among all methods studied, the high-pressure homogenization (HPH) was found to be the most suitable method for industrial-scale cell disruption to date and other methods showed promise in small-scale or laboratory-scale application, but yet to be feasible in large-scale process. Although there have been several disadvantages of HPH treatment such as solid content in the algal biomass needs to be higher for reducing the energy content, having high viscosity in the sample requires additional unit operations, and emulsion formation during the process hinders the solvent extraction, the algal biofuels can be made feasible utilizing appropriate study of this method because HPH can completely rupture the tough cell wall of the microalgae as well as improve the recovery of metabolites. Other methods are still in the early stage of the development and need further research to consider the feasibility of algae processing for recovering different bioproducts.

6 Conclusion

Different cell disruption methods have been discussed along with the lipid recovery as well as advantages and disadvantages for large-scale application. The feasibility of these processes was analyzed based on the energy required to treat per kg of dried biomass, lipid recovery, recovery of solvent, etc., where high-pressure homogenization is found to be the most suitable method if the solid content in the biomass is high. Different pressurized gases can also be used for breaking the algal cell wall to improve the metabolite recovery, but extensive careful studies are needed for its feasibility analysis. The currently used laboratory-scale methods such as microwave irradiation, ultrasound sonication, and ionic liquids are gaining attention for their advantages. In addition to lipid recovery, extraction of protein using these pretreatments was also analyzed from the recent studies and it was found that high-pressure homogenization is the most suited one followed by chemical method. The chemical method needs only a small volume of chemical where a considerable amount (73%) protein was recovered. This method can be studied further for its economic viability. Finally, it was suggested that the lipid and protein recovery from algae can be improved by combining different cell disruption methods.

References

Adam, F., Abert-Vian, M., Peltier, G., & Chemat, F. (2012). "Solvent-free" ultrasound-assisted extraction of lipids from fresh microalgae cells: A green, clean and scalable process. *Bioresource Technology, 114*, 457–465.

Adrio, J. L. (2017). Oleaginous yeasts: Promising platforms for the production of oleochemicals and biofuels. *Biotechnology and Bioengineering, 114*, 1915–1920.

Alam, M. A., Wu, J., Xu, J., & Wang, Z. (2019). Enhanced isolation of lipids from microalgal biomass with high water content for biodiesel production. *Bioresource Technology, 291*, 121834.

Alfenore, S., & Molina-Jouve, C. (2016). Current status and future prospects of conversion of lignocellulosic resources to biofuels using yeasts and bacteria. *Process Biochemistry, 51*, 1747–1756.

Ali, M., & Watson, I. A. (2015). Microwave treatment of wet algal paste for enhanced solvent extraction of lipids for biodiesel production. *Renewable Energy, 76*, 470–477.

Balasubramanian, S., Allen, J. D., Kanitkar, A., & Boldor, D. (2011). Oil extraction from *Scenedesmus obliquus* using a continuous microwave system – Design, optimization, and quality characterization. *Bioresource Technology, 102*, 3396–3403.

Cano-Ruiz, M. E., & Richter, R. L. (1997). Effect of homogenization pressure on the milk fat globule membrane proteins. *Journal of Dairy Science, 80*, 2732–2739.

Cheng, J., Huang, R., Li, T., Zhou, J., & Cen, K. (2015). Physicochemical characterization of wet microalgal cells disrupted with instant catapult steam explosion for lipid extraction. *Bioresource Technology, 191*, 66–72.

Cho, S.-C., Choi, W.-Y., Oh, S.-H., et al. (2012). Enhancement of lipid extraction from marine microalga, *Scenedesmus* associated with high-pressure homogenization process. *Journal of Biomedicine & Biotechnology, 2012*, 1–6.

Choi, S.-A., Lee, J.-S., Oh, Y.-K., Jeong, M.-J., Kim, S. W., & Park, J.-Y. (2014). Lipid extraction from *Chlorella vulgaris* by molten-salt/ionic-liquid mixtures. *Algal Research, 3*, 44–48.

de Moura, R. R., Etges, B. J., dos Santos, E. O., Martins, T. G., Roselet, F., Abreu, P. C., Primel, E. G., & D'Oca, M. G. M. (2018). Microwave-assisted extraction of lipids from wet microalgae paste: A quick and efficient method. *European Journal of Lipid Science and Technology, 120*, 1700419.

Desai, R. K. (2016). *Ionic liquid pre-treatment of microalgae and extraction of biomolecules*. Wageningen: Wageningen University.

Dong, T., Knoshaug, E. P., Pienkos, P. T., & Laurens, L. M. L. (2016). Lipid recovery from wet oleaginous microbial biomass for biofuel production: A critical review. *Applied Energy, 177*, 879–895.

Drira, N., Piras, A., Rosa, A., Porcedda, S., & Dhaouadi, H. (2016). Microalgae from domestic wastewater facility's high rate algal pond: Lipids extraction, characterization and biodiesel production. *Bioresource Technology, 206*, 239–244.

Ellison, C. R., Overa, S., & Boldor, D. (2019). Central composite design parameterization of microalgae/cyanobacteria co-culture pretreatment for enhanced lipid extraction using an external clamp-on ultrasonic transducer. *Ultrasonics Sonochemistry, 51*, 496–503.

Garcia-Gonzalez, L., Geeraerd, A. H., Spilimbergo, S., Elst, K., Van Ginneken, L., Debevere, J., Van Impe, J. F., & Devlieghere, F. (2007). High pressure carbon dioxide inactivation of microorganisms in foods: The past, the present and the future. *International Journal of Food Microbiology, 117*, 1–28.

Garoma, T., & Janda, D. (2016). Investigation of the effects of microalgal cell concentration and electroporation, microwave and ultrasonication on lipid extraction efficiency. *Renewable Energy, 86*, 117–123.

Garoma, T., & Yazdi, R. E. (2019). Investigation of the disruption of algal biomass with chlorine. *BMC Plant Biology, 19*, 18.

Griffiths, M. J., & Harrison, S. T. L. (2009). Lipid productivity as a key characteristic for choosing algal species for biodiesel production. *Journal of Applied Phycology, 21*, 493–507.

Günerken, E., D'Hondt, E., Eppink, M. H. M., Garcia-Gonzalez, L., Elst, K., & Wijffels, R. H. (2015). Cell disruption for microalgae biorefineries. *Biotechnology Advances, 33*, 243–260.

Heo, Y. M., Lee, H., Lee, C., Kang, J., Ahn, J.-W., Lee, Y. M., Kang, K.-Y., Choi, Y.-E., & Kim, J.-J. (2017). An integrative process for obtaining lipids and glucose from *Chlorella vulgaris* biomass with a single treatment of cell disruption. *Algal Research, 27*, 286–294.

Howlader, M. S., DuBien, J., Hassan, E. B., Rai, N., & French, W. T. (2019). Optimization of microbial cell disruption using pressurized CO_2 for improving lipid recovery from wet biomass. *Bioprocess and Biosystems Engineering, 42*, 763–776.

Howlader, M. S., French, W. T., Shields-Menard, S. A., Amirsadeghi, M., Green, M., & Rai, N. (2017a). Microbial cell disruption for improving lipid recovery using pressurized CO_2: Role of CO_2 solubility in cell suspension, sugar broth, and spent media. *Biotechnology Progress, 33*, 737–748.

Howlader, M. S., French, W. T., Toghiani, H., Hartenbower, B., Pearson, L., DuBien, J., & Rai, N. (2017b). Measurement and correlation of solubility of carbon dioxide in triglycerides. *The Journal of Chemical Thermodynamics, 104*, 252–260.

Howlader, M. S., Rai, N., & Todd French, W. (2018a). Improving the lipid recovery from wet oleaginous microorganisms using different pretreatment techniques. *Bioresource Technology, 267*, 743–755.

Howlader, M. S., Venkatesan, S., Goel, H., Huda, M. M., French, W. T., & Rai, N. (2018b). Solubility of CO_2 in triglycerides using Monte Carlo simulations. *Fluid Phase Equilibria, 476*, 39–47.

Kim, Y.-H., Choi, Y.-K., Park, J., Lee, S., Yang, Y.-H., Kim, H. J., Park, T.-J., Hwan Kim, Y., & Lee, S. H. (2012). Ionic liquid-mediated extraction of lipids from algal biomass. *Bioresource Technology, 109*, 312–315.

Lai, Y. S., De Francesco, F., Aguinaga, A., Parameswaran, P., & Rittmann, B. E. (2016). Improving lipid recovery from Scenedesmus wet biomass by surfactant-assisted disruption. *Green Chemistry, 18*, 1319–1326.

Lee, A. K., Lewis, D. M., & Ashman, P. J. (2012). Disruption of microalgal cells for the extraction of lipids for biofuels: Processes and specific energy requirements. *Biomass and Bioenergy, 46*, 89–101.

Lee, S. Y., Show, P. L., Ling, T. C., & Chang, J.-S. (2017). Single-step disruption and protein recovery from *Chlorella vulgaris* using ultrasonication and ionic liquid buffer aqueous solutions as extractive solvents. *Biochemical Engineering Journal, 124*, 26–35.

Liang, K., Zhang, Q., & Cong, W. (2012). Enzyme-assisted aqueous extraction of lipid from microalgae. *Journal of Agricultural and Food Chemistry, 60*, 11771–11776.

Lorente, E., Farriol, X., & Salvadó, J. (2015). Steam explosion as a fractionation step in biofuel production from microalgae. *Fuel Processing Technology, 131*, 93–98.

Lorente, E., Hapońska, M., Clavero, E., Torras, C., & Salvadó, J. (2017). Microalgae fractionation using steam explosion, dynamic and tangential cross-flow membrane filtration. *Bioresource Technology, 237*, 3–10.

Lorente, E., Hapońska, M., Clavero, E., Torras, C., & Salvadó, J. (2018). Steam explosion and vibrating membrane filtration to improve the processing cost of microalgae cell disruption and fractionation. *PRO, 6*, 28.

Lu, W., Alam, M. A., Luo, W., & Asmatulu, E. (2019). Integrating *Spirulina platensis* cultivation and aerobic composting exhaust for carbon mitigation and biomass production. *Bioresource Technology, 271*, 59–65.

Lupatini, A. L., de Oliveira Bispo, L., Colla, L. M., Costa, J. A. V., Canan, C., & Colla, E. (2017). Protein and carbohydrate extraction from *S. platensis* biomass by ultrasound and mechanical agitation. *Food Research International, 99*, 1028–1035.

Martinez-Guerra, E., Howlader, M. S., Shields-Menard, S., French, W. T., & Gude, V. G. (2018). Optimization of wet microalgal FAME production from *Nannochloropsis* sp. under the synergistic microwave and ultrasound effect. *International Journal of Energy Research, 42*, 1934–1949.

Mazanov, S. V., Gabitova, A. R., Usmanov, R. A., Gumerov, F. M., Labidi, S., Ben, A. M., Passarello, J.-P., Kanaev, A., Volle, F., & Le Neindre, B. (2016). Continuous production of biodiesel from rapeseed oil by ultrasonic assist transesterification in supercritical ethanol. *Journal of Supercritical Fluids, 118*, 107–118.

Mukhopadhyay, A. (2015). Tolerance engineering in bacteria for the production of advanced biofuels and chemicals. *Trends in Microbiology, 23*, 498–508.

Orr, V. C. A., Plechkova, N. V., Seddon, K. R., & Rehmann, L. (2016). Disruption and wet extraction of the microalgae *Chlorella vulgaris* using room-temperature ionic liquids. *ACS Sustainable Chemistry & Engineering, 4*, 591–600.

Park, Y.-M., Lee, D.-W., Kim, D.-K., Lee, J.-S., & Lee, K.-Y. (2008). The heterogeneous catalyst system for the continuous conversion of free fatty acids in used vegetable oils for the production of biodiesel. *Catalysis Today, 131*, 238–243.

Patel, A., Arora, N., Mehtani, J., Pruthi, V., & Pruthi, P. A. (2017). Assessment of fuel properties on the basis of fatty acid profiles of oleaginous yeast for potential biodiesel production. *Renewable and Sustainable Energy Reviews, 77*, 604–616.

Phong, W. N., Show, P. L., Le, C. F., Tao, Y., Chang, J.-S., & Ling, T. C. (2018). Improving cell disruption efficiency to facilitate protein release from microalgae using chemical and mechanical integrated method. *Biochemical Engineering Journal, 135*, 83–90.

Portillo, H. A., Howlader, M. S., Campbell, Y. L., French, T., Kim, T., Goddard, J., Hassan, E. B., & Schilling, M. W. (2018). Incorporating fermented by-products of *Lactobacillus diolivorans* in food grade coatings designed for inhibition of *Tyrophagus putrescentiae* on dry-cured hams. *Journal of Stored Products Research, 77*, 77–83.

Ramakrishnan, A. M. (2015). Biofuel: A scope for reducing global warming. *Journal of Petroleum & Environmental Biotechnology, 7*, 1.

Ren, H.-Y., Xiao, R.-N., Kong, F., Zhao, L., Xing, D., Ma, J., Ren, N.-Q., & Liu, B.-F. (2019). Enhanced biomass and lipid accumulation of mixotrophic microalgae by using low-strength ultrasonic stimulation. *Bioresource Technology, 272*, 606–610.

Safi, C., Cabas Rodriguez, L., Mulder, W. J., Engelen-Smit, N., Spekking, W., van den Broek, L. A. M., Olivieri, G., & Sijtsma, L. (2017a). Energy consumption and water-soluble protein release by cell wall disruption of *Nannochloropsis gaditana*. *Bioresource Technology, 239*, 204–210.

Safi, C., Olivieri, G., Campos, R. P., Engelen-Smit, N., Mulder, W. J., van den Broek, L. A. M., & Sijtsma, L. (2017b). Biorefinery of microalgal soluble proteins by sequential processing and membrane filtration. *Bioresource Technology, 225*, 151–158.

Safi, C., Ursu, A. V., Laroche, C., Zebib, B., Merah, O., Pontalier, P.-Y., & Vaca-Garcia, C. (2014). Aqueous extraction of proteins from microalgae: Effect of different cell disruption methods. *Algal Research, 3*, 61–65.

Samarasinghe, N., Fernando, S., Lacey, R., & Faulkner, W. B. (2012). Algal cell rupture using high pressure homogenization as a prelude to oil extraction. *Renewable Energy, 48*, 300–308.

Sathish, A., & Sims, R. C. (2012). Biodiesel from mixed culture algae via a wet lipid extraction procedure. *Bioresource Technology, 118*, 643–647.

Shields-Menard, S. A., Amirsadeghi, M., French, W. T., & Boopathy, R. (2018). A review on microbial lipids as a potential biofuel. *Bioresource Technology, 259*, 451–460.

Singh, J., & Gu, S. (2010). Commercialization potential of microalgae for biofuels production. *Renewable and Sustainable Energy Reviews, 14*, 2596–2610.

To, T. Q., Procter, K., Simmons, B. A., Subashchandrabose, S., & Atkin, R. (2018). Low cost ionic liquid–water mixtures for effective extraction of carbohydrate and lipid from algae. *Faraday Discussions, 206*, 93–112.

Wan, C., Alam, M. A., Zhao, X.-Q., Zhang, X.-Y., Guo, S.-L., Ho, S.-H., & Bai, F.-W. (2015). Current progress and future prospect of microalgal biomass harvest using various flocculation technologies. *Bioresource Technology, 184*, 251–257.

Wang, D., Li, Y., Hu, X., Su, W., Zhong, M., Wang, D., Li, Y., Hu, X., Su, W., & Zhong, M. (2015). Combined enzymatic and mechanical cell disruption and lipid extraction of green alga *Neochloris oleoabundans*. *International Journal of Molecular Sciences, 16*, 7707–7722.

Xu, Z., Wu, J., Zhang, Y., Hu, X., Liao, X., & Wang, Z. (2010). Extraction of anthocyanins from red cabbage using high pressure CO_2. *Bioresource Technology, 101*, 7151–7157.

Yao, S., Mettu, S., Law, S. Q. K., Ashokkumar, M., & Martin, G. J. O. (2018). The effect of high-intensity ultrasound on cell disruption and lipid extraction from high-solids viscous slurries of *Nannochloropsis* sp. biomass. *Algal Research, 35*, 341–348.

Yap, B. H. J., Crawford, S. A., Dumsday, G. J., Scales, P. J., & Martin, G. J. O. (2014). A mechanistic study of algal cell disruption and its effect on lipid recovery by solvent extraction. *Algal Research, 5*, 112–120.

Yap, B. H. J., Dumsday, G. J., Scales, P. J., & Martin, G. J. O. (2015). Energy evaluation of algal cell disruption by high pressure homogenisation. *Bioresource Technology, 184*, 280–285.

The manufacturer's authorised representative in the EU is Springer Nature Customer Service Centre GmbH, Europaplatz 3, 69115 Heidelberg, Germany. If you have any concerns regarding our products, please contact ProductSafety@springernature.com

Printed and bound by CPI Group (UK) Ltd, Croydon, CR0 4YY
30/09/2026
02225247-0017